PHYSICAL Science

Student Edition

purposeful design publications

Colorado Springs, Colorado

Development Team

Vice President for Purposeful Design Publications

Steven Babbitt

Directors for Textbook Development

Don Hulin
Lisa Wood

Editorial Team

Merrilee Berndt
Diane Comstock
Lindsey Duncan
Janice Giles
Julie Holmquist
Macki Jones
Adelle Moxness
Jessica Reid
Ian Work

Design Team

Claire Coleman
Steve Learned
Christian Massey

Cover Design

Mike Riester

Art Illustrations

Aline Heiser

Purposeful Design Publications is grateful to Christian Schools International for the contributions it made to the original content of the Purposeful Design Physical Science course.

PHYSICAL Science

Student Edition

Purposeful Design

Printed in the United States of America
24 23 22 21 20 19 18   1 2 3 4 5 6 7

Physical Science – Student Edition
Purposeful Design Science series
ISBN 978-1-58331-546-0, Catalog #20091

Purposeful Design Publications is the publishing division of the Association of Christian Schools International (ACSI) and is committed to the ministry of Christian school education, to enable Christian educators and schools worldwide to effectively prepare students for life. As the publisher of textbooks, trade books, and other educational resources within ACSI, Purposeful Design Publications strives to produce biblically sound materials that reflect Christian scholarship and stewardship and that address the identified needs of Christian schools around the world.

References to books, computer software, and other ancillary resources in this series are not endorsements by ACSI. These materials were selected to provide teachers with additional resources appropriate to the concepts being taught and to promote student understanding and enjoyment.

Purposeful Design Publications
A Division of ACSI
731 Chapel Hills Dr. • Colorado Springs, CO • 80920
Customer Service Department: 800/367-0798
Website: www.purposefuldesign.com

Table of Contents

Chapter 1: *Composition of Matter*

Chapter 2: *Nature of Matter*

Chapter 3: *Atomic Structure*

Chapter 4: *Chemical Interactions*

Matter

UNIT 1

UNIT 2

Types of Substances

Chapter 1: *Metals*

Chapter 2: *Nonmetals and Metalloids*

Chapter 3: *Organic Compounds*

Chapter 4: *Other Useful Materials*

UNIT 3

Interactions of Matter

Chapter 1: *Solutions*

Chapter 2: *Chemical Reactions*

Chapter 3: *Acids, Bases, and Salts*

UNIT 4

Matter in Motion

Chapter 1: *Forces and Motion*

Chapter 2: *Work and Energy*

Chapter 3: *Types of Machines*

UNIT 5

Energy at Work

Chapter 1: *Thermal Energy*

Chapter 2: *Nuclear Energy*

Chapter 3: *Energy Resources*

UNIT 6

Sound and Light

UNIT 7

Electricity and Magnetism

Chapter 1: *Electricity*

Chapter 2: *Magnetism*

The Scientific Method

Background

The word *science* comes from the Latin word *scire*, which means "to know." A scientist is someone who does science—a person who wants to understand the natural world and all of its complexities.

Even after scientific experiments, many scientific hypotheses are disputed within the scientific community; all are subject to revisions and changes as new data and experimental evidence are collected. Theories and laws, however, are supported by a substantial number of experiments and observations. Ideally, scientists are open to constructive criticism about their hypotheses. When scientists communicate in scientific journals or at conventions, they work toward finding new evidence and creating solutions to resolve conflicting interpretations and viewpoints.

The scientific method is an orderly, systematic approach to solving a problem or answering a question. Although different scientific endeavors require different approaches and steps, the following terminology and steps are often used. You need to become familiar with the vocabulary of science as well as realize that scientists understand that the scientific method is not a rigid set of steps that is always followed. Various orders are possible, and sometimes steps are eliminated or repeated.

Vocabulary

control—the sample in an experiment in which the variables are kept at a base level
hypothesis—a prediction of what you think will happen and which can be tested to see if it is true
inference—an educated guess based on observation
observation—something noticed through the senses
scientific law—a generalization based on observations that describe the ways an object behaves under specific conditions
scientific method—the series of steps that scientists follow when they investigate problems or try to answer questions
theory—an explanation of the scientific laws
variable—a changeable factor that could influence an experiment's outcome

Scientific Method

1. **Identify or define the problem.**

2. **Make a hypothesis**. Not all predictions are hypotheses; they must be measurable. For example, "If the rate of fermentation

is related to temperature, then increasing the temperature will increase gas production" is a hypothesis. On the other hand, "If yeast is heated, then more gas will be produced" is not a hypothesis because it offers no proposition to test. It does not show a relationship or suggest variables. A hypothesis may also be a question ("Does temperature affect the fermentation of yeast?") or a conditional statement ("Temperature may affect the fermentation of yeast").

3. **Experiment, controlling the variables**. Scientists repeat and refine the experiment based on the findings. They will not allow a hypothesis to remain untested; it must be tested and shown valid.

4. **Make observations and record the results**. Most scientists accurately record their observations and measurements in a journal or on a computer.

5. **Make inferences and conclusions**. A conclusion is a statement about what a scientist has learned and whether the scientist's hypothesis was supported. Scientists often learn as much from an incorrect hypothesis as from a correct one. You may feel pressure to have your experiment show that your "guess" was correct, but you need to understand that sometimes scientists arrive at completely unexpected conclusions and that an incorrect hypothesis is never considered a failure. For example, sometimes an unexpected conclusion raises new questions that will lead to another experiment and conclusion. The purpose of science is not to prove what is already known; the purpose is to increase the understanding of God's world!

6. **Apply the findings**. Scientists add their conclusions to their understanding of the natural and technical universe. To do so with wisdom and stewardship benefits all creation.

Measurements

Background

The metric system is a universal system of measurements that serves as a standard for scientific research throughout the world and in many countries for everyday use. It is helpful to become more familiar with the metric system and have practice estimating measurements. Throughout this course, the metric system is used predominantly.

Since 1899, international conferences have been held to standardize the metric system. In 1960, the 11th International Conference of Weights and Measures substantially changed the system, renaming it the International System of Units (abbreviated SI).

Using the SI system of measurement has several advantages. It is based on standards that have been recognized by the international science community for use in the sciences and commerce, making the system universally adopted. All of the unit conversions used in the metric system are based on the number 10, and only decimal multiples of the basic units of length, volume, and mass are employed. The metric system is also used by all of the major countries of the world except the United States.

The instrument frequently used to determine the mass of a small object is the triple beam balance, which is normally capable of measuring masses to the hundredth of a gram. The volume of substances can be determined by using a variety of different methods, depending upon the phase and shape of the substance being measured. Graduated cylinders are used to accurately measure the volume of liquids to the nearest milliliter. Graduated cylinders can also be used to measure granular solids, but this method is inaccurate for substances with large granules. The volume of regularly shaped objects with cubic or rectangular box shapes can be calculated using the formula $V = l \times w \times h$ (Volume equals length times width times height). Water displacement is used to find the volume of irregularly shaped objects that will not dissolve in water. A certain amount of water is poured into a graduated cylinder, and the object is dropped into the cylinder. The volume that the water rises is the volume of the object.

Although students in the United States are more familiar with the Fahrenheit scale, the Celsius (Centigrade) scale is much more common around the world. Scientists use the Celsius scale or the Kelvin scale, which is an absolute scale. The Fahrenheit scale assigns a temperature of 32° for the freezing point of water and 212° for the boiling point of water. The Celsius scale standardized the temperature scale by assigning a value of 0° for the freezing point of water and 100° for the boiling point of water. The size of the divisions used in the Fahrenheit and Celsius scales are not equal. The formula used to convert Celsius to Fahrenheit is $F = 9/5 \times C + 32$. The formula for converting Fahrenheit to Celsius is $C = 5/9 \times (F - 32)$.

One way to record these measurements is on graphs. A graph is a pictorial representation of statistical data or of relationships between variables. While graphs can serve a predictive function because they show general tendencies in the quantitative behavior of data, as approximations they are sometimes inaccurate and misleading. Most graphs include two axes. The horizontal axis represents the independent variables, and the vertical axis represents the dependent variables. In line graphs (the most common type), the horizontal axis often represents time. Bar graphs can be used to depict the relationship between two nontemporal numerical values. Such information can also be expressed in a circular graph (pie graph) that illustrates the part-to-whole relationship. The size of each sector is directly proportional to the percentage of the whole it represents; this graph is often used to emphasize proportions.

Scientific measurements require specific words to be used in descriptions so the language of science is understood worldwide. You need to have a working knowledge of the following terms:

Vocabulary

Celsius scale—a temperature scale in which 0° represents the freezing point of water and 100° represents the boiling point of water

density—the mass per unit of volume of a substance

gram—the standard unit for mass in the metric system

liter—the standard unit for the volume of liquid in the metric system

meter—the standard unit for length in the metric system

metric system—a universal system of measurement based on the number 10 used by scientists around the world

Laboratory Safety

Background

Each lab has specific safety needs, and you must focus on the teacher's instructions. Take only your notebook, student text, and a pen or pencil to the laboratory area. Leave all other items at your desk. Read and review your teacher's procedures to be certain that you understand the instructions.

Safety Rules

Some standard safety rules include the following:

- Wash your hands before and after any hands-on activity or experiment. Wear protective gear appropriate to the activity. Some activities require goggles or gloves. Sanitize safety goggles after each use.
- Report any spills or breakage to your teacher immediately.
- Store equipment and materials in a safe location and make sure that you return them to their designated places after use. Always test and check the materials prior to use.
- Do not eat or handle supplies unless instructed to do so.
- Be aware of allergies that you may have to science activity materials and take any necessary precaution.
- Carry only one jar at a time.
- Never lean on aquarium glass.
- Never prop a meterstick over your shoulder.
- Carry microscopes with two hands.
- Never eat or drink during a science activity.
- Remove loose or bulky clothing to reduce the chance of spreading fire or knocking over equipment.
- Keep equipment (especially breakable or expensive items such as glassware or microscopes) away from the edge of work surfaces.
- Wear safety goggles, especially for activities involving potentially harmful chemicals, glassware, or heat. (When your teacher is wearing goggles, you must wear goggles.)
- If glassware is used and breaks, back away from the accident and do not touch the shattered glass. Designate someone to retrieve a dustpan and broom and alert the teacher.
- Know basic chemical safety. Many chemicals are labeled with three hazards—health, flammability, and reactivity—rated on a scale of 0 to 4. A 0 rating indicates that the substance is not a threat for this hazard, and a 4 means it is a substantial threat.
- Be aware of two common safety symbols—flames (signifying flammability) and a test tube being poured on a hand (signifying a corrosive substance). These symbols are primarily for storage purposes, but they also warn the user of

these two dangers. Different chemicals have different disposal guidelines. Follow your teacher's directions regarding how to properly dispose of the chemicals and materials you use.

Safety Procedures

- Be a good example. Follow the safety rules at all times.
- Inspect the lab area to look for possible hazards before you start an experiment.
- Inspect your equipment.
- Report hazards to your teacher immediately.
- Know the location of your lab's fire bucket, fire extinguisher, and safety shower, if available.

UNIT 1

Matter

Chapter 1: *Composition of Matter*
Chapter 2: *Nature of Matter*
Chapter 3: *Atomic Structure*
Chapter 4: *Chemical Interactions*

Vocabulary

adhesion
amorphous solid
atom
atomic number
atomic weight
chemical bond
chemical change
chemical formula
chemistry
chromatography
cohesion
colloid
combustion
compound
condensation
covalent bond
crystalline solid
density
deposition
diatomic molecule
distillation
electron
element
evaporation
filtration
gas
gravity
group
heterogeneous mixture
homogeneous mixture
ion
ionic bond
isotope
liquid
mass
mass number
matter
metal
mixture
molecule
neutron
nonmetal
nonpolar molecule
nucleus
oxidation
period
periodic table
phase change
physical change
physical science
physics
polar molecule
polyatomic ion
pressure
proton
pure substance
solid
sublimation
surface tension
suspension
thermal expansion
Tyndall effect
vaporization
viscosity
volume
weight

Key Ideas

- Systems, order, and organization
- Evidence, models, and explanation
- Change, constancy, and measurement
- Form and function
- Abilities necessary to do scientific inquiry
- Understandings about scientific inquiry
- Properties and changes of properties in matter
- Motion and forces
- Transfer of energy
- Structure of atoms
- Structure and properties of matter
- Chemical reactions
- Abilities of technological design
- Understandings about science and technology
- Risks and benefits
- Science and technology in society
- Science as a human endeavor
- Nature of science
- History of science

SCRIPTURE

But the pot he was shaping from the clay was marred in his hands; so the potter formed it into another pot, shaping it as seemed best to him.

Jeremiah 18:4

1.1.1 *The Study of Physical Science*

OBJECTIVES

- Describe the two broad fields of physical science.
- Identify the benefits and drawbacks of scientific pursuits.
- Convert units of measure.
- Describe a Christian worldview of science and technology.

VOCABULARY

- **chemistry** the study of how atoms, molecules, and chemicals interact with each other
- **physical science** the study of the nonliving things that make up the world
- **physics** the study of matter, energy, motion, and forces

What would your life have been like as a teenager before the Industrial Revolution? You probably would have lived on a farm, working long hours doing chores before and after school—if you even went to school! Your house would not have had electricity, meaning no TV, cell phones, Internet, blow dryer, or video games. You would have gone to bed not long after sunset because your only light was from a candle or gaslight. You would have walked, ridden a horse, or used a wagon to get from place to place. In many ways, the life of a teenager 200 years ago was more difficult than your life is today. Most of the blessings you now enjoy are the result of the advancements of physical science and technology. The explosion of knowledge in these areas has changed the world.

Physical science is the study of the nonliving things that make up the world. The two major branches of physical science are chemistry and physics. **Chemistry** is the study of how atoms, molecules, and chemicals interact with each other. **Physics** is the study of matter, energy, motion, and forces. By using chemistry and physics, scientists have discovered many of the rules that God designed to govern the world. Understanding these rules has made it possible for people to develop an abundance of new technologies that have greatly changed the world.

Chemists have developed millions of compounds and materials that are used every day. Products such as plastics, pesticides, fuels, fabrics, paper, fertilizers, rubber, alloys, and drugs are but a few of the substances produced through chemistry. If

Homestead during the late Industrial Revolution era.

someone were to come to your house and remove everything that contains man-made chemicals, your house would be quite empty! Similarly, technology applications developed using physics include nuclear power, space exploration, flash drives, video games, television, radio, modern transportation, skyscrapers, and medical equipment such as X-rays and MRI machines.

In addition, all knowledge can be used either wisely or foolishly. The knowledge gained through the study of physical science has produced mixed results. Medicines, machinery, better foods, faster transportation, and more rapid communication methods are among the benefits of advances in physical science. However, through technological advances, unforeseen by-products have also been created, such as air and water pollution, radioactive waste, and other harmful environmental effects. As scientists study the laws of nature and find ways to use them, they should be mindful of the possible adverse effects that can result from their discoveries.

Some people look at pollution and other problems connected to technology and conclude that the drawbacks of science outweigh the benefits technology brings to society. Others think that Christians should not care much about science—that they should concentrate on studying the Bible and not worry about the physical world. But God is the Creator of both Scripture and nature, which enhance each other to give believers a deeper and clearer vision of God's truth. Proverbs 1:1–7 states that God

Two chemists in a lab

CAREER

Physicists

The term *physics* comes from the Greek word *physikos*, meaning "nature." Physicists conduct experiments to understand at the most fundamental level why the universe acts the way it does. Because it is so vast, however, physicists tend to specialize in one aspect of physics. Astrophysicists study what stars and galaxies are made of according to the type of light they shed. Geophysicists have traditionally studied the forces on Earth, such as those that cause earthquakes or the flow of currents. Modern geophysics overlaps astrophysics to include the analysis of outer space and other planets. Both fields depend in large part on indirect evidence because much of what is studied cannot be directly observed.

TRY THIS

Metric Conversions
Using conversion factors, find the equivalent measurements.

1. How many liters are in 2,849 milliliters?
2. How many centimeters are in 49 kilometers?
3. How many milligrams are in 3 kilograms?
4. How many centimeters are in 89 inches?

wants people to pursue knowledge and wisdom. Science is a tool to better know God and creation.

The systematic study of creation through chemistry or physics includes the need for many measurements. For science to progress, scientists all over the world share and compare their research, using a standard system of measurements established for the scientific community. The International System of Units (SI) employs the metric system. By always using the same units, scientists reduce the number of errors and miscommunications. Chemists and physicists know to include and check units when recording and sharing data. For example, if a lab assistant

FYI

Skyscrapers
Genesis 11:3–4 states, "They said to each other, 'Come, let's make bricks and bake them thoroughly.' They used brick instead of stone, and tar for mortar. Then they said, 'Come, let us build ourselves a city, with a tower that reaches to the heavens, so that we may make a name for ourselves; otherwise we will be scattered over the face of the whole earth.' " For thousands of years, humankind has had a desire to reach for the heavens, to create the biggest and the tallest structure possible. The men of Babel wanted to build a structure that could reach into the sky. The Egyptians built enormous pyramids for their pharaohs, overcoming the laws of gravity and physics by building structures that were larger at the bottom than the top.

In the 1800s, New York City was growing at a rapid pace with population and business surpassing the amount of land available to accommodate such growth. What was the obvious solution to the problem? Build up, of course. But building more than five stories high was impractical because the base of buildings had to be very large to support the height. Engineers, physicists, and architects went to work, searching for stronger and lighter building materials and a way to spread the load of the gravitational force without spreading the building. The Bessemer process was developed and iron and steel were mass produced, creating the necessary strong, lightweight materials needed for such lofty structures. William Le Baron Jenney designed the first building to use steel beams. In 1884–1885, the Home Insurance Company building was constructed with 10 stories. Since that time, technology has provided the means to build taller skyscrapers, some reaching nearly 800 m in height.

SCIENCE IN ACTION

Converting Units

1. Start with the original measurement including units, and create a fraction.	450 ft = 450 ft / 1
2. Find the relationship between the original unit and the new unit.	1 m = 3.28 ft
3. Make the relationship into a fraction with the new unit on top and the original unit on the bottom. This is the conversion factor.	1 m / 3.28 ft
4. Multiply the original measurement by the conversion factor.	(450 ft / 1) × (1 m / 3.28 ft)
5. Cancel the original units.	(450 / 1) × (1 m / 3.28)
6. Finish the problem.	450 m / 3.28 = 137.2 m

records the traveling distance of a hovercraft in feet, the lead scientist would instruct her to convert the measurement into meters. First, the assistant would take the initial measurement and make it a fraction: 960 ft/1. Second, the lab assistant would take the relationship of 1 meter = 3.28 feet and create a fraction called *a conversion factor*. Either part of the relationship can be the numerator or the denominator, but the assistant knows to write the fraction with the original unit on the bottom so it will cancel out the unit of the initial measurement. Finally, the assistant would take the measurement of 960 ft/1 and multiply it by the conversion factor 1 m/3.28 ft, producing a corrected measurement of 292.68 m.

In addition, all scientific study uses a systematic method of observation and experimentation to understand creation. Scientists in each field seek to answer different questions about

FYI

Metric Unit Conversion Factors

Length	Volume	Mass
1 meter (m) = 1,000 millimeters (mm)	1 liter (L) = 1,000 milliliters (mL) or 1,000 cubic centimeters (cm^3)	1 gram (g) = 1,000 milligrams (mg)
1 meter (m) = 100 centimeters (cm)		1,000 grams (g) = 1 kilogram (kg)
1,000 meters (m) = 1 kilometer (km)		

FYI

SI Base Units

electric current	ampere	A
length	meter	m
mass	kilogram	kg
thermodynamic temperature	kelvin	K

Relationship Between Metric and English Units

1 inch = 2.54 cm	1 centimeter = 0.393 in.
1 foot = 0.3048 m	1 meter = 3.28 ft
1 yard = 0.9144 m	1 meter = 1.09361 yd
1 mile = 1.609 km	1 kilometer = 0.621371 mi
1 gallon = 3.78 L	1 liter = 0.264172 US gal

the world. As scientists pursue knowledge, they evaluate and correct their previous conclusions, revise old theories, or replace them with new ones.

Scientists who are Christians study chemistry and physics to discover more about how God works in His world and to bring blessing to that world. God gives His people the responsibility of being good stewards of creation. Through the proper study and use of science, Christians can make wise decisions about the use of what God has made.

LESSON REVIEW

1. Name two branches of physical science and explain what they cover.
2. What benefits and drawbacks have come from chemistry and physics?
3. If a building is 685 ft tall, what is its height in meters?
4. A chemist drives 3,412 m to share results with a fellow chemist. What is the distance in feet?
5. How should a Christian view science and technology?

Substances 1.1.2

OBJECTIVES

- Explain the purpose of classifying matter.
- Distinguish between different types of matter.

VOCABULARY

- **atom** the smallest unit into which an element can be divided and still maintain its properties
- **compound** a pure substance consisting of atoms of two or more elements that are chemically combined in a fixed proportion
- **element** a pure substance that cannot be broken down into simpler substances by physical or chemical means
- **matter** anything that has mass and takes up space
- **molecule** a particle consisting of two or more atoms chemically tied together
- **pure substance** a substance of constant composition that possesses definite qualities

Have you ever thought about how many different types of substances make up the world? Start with the foods inside your refrigerator. If you peeked inside, you might see thawing chicken, spinach salad, milk, grapes, mustard, orange juice, leftover burritos, and cake. Then there are the substances that make up your body: blood, bones, various tissues, potassium, iron, hair, fingernails, enamel, DNA, and brain cells—and these are just a few of the substances knit together in your body! Think also about substances you might see if you took a walk—soil, water, pollen, granite, smog, cement, and wood—to name just a few. How would you categorize all of these substances? Would you even know where to begin? Classifying and organizing substances, creatures, forces, planets, and all other things help people talk about and study them in detail.

You may not realize it, but you classify and categorize things in your house every day. Plates and cups are kitchen items, so they are kept in the kitchen. Clean clothes are stored in drawers and closets, and books are set on shelves or a desk. Why do you keep all the cups in the kitchen and the clothes in the dresser? It makes sense because items with similar purposes and characteristics belong together. You classify and categorize constantly. Your math work goes in the math folder and your science work goes in your science folder or maybe it stays in the science classroom. What happens when these things are not categorized? Imagine if every single thing in your home was in one big pile. How would you make sense of it? What would happen if every object in the pile had the same name—*thing*? Every time you needed something, you would look for the thing. Or you would tell your mother that you need more things because you are almost out. In the same way that you have names and classifications for things or objects that you use in your everyday

Composition of Matter

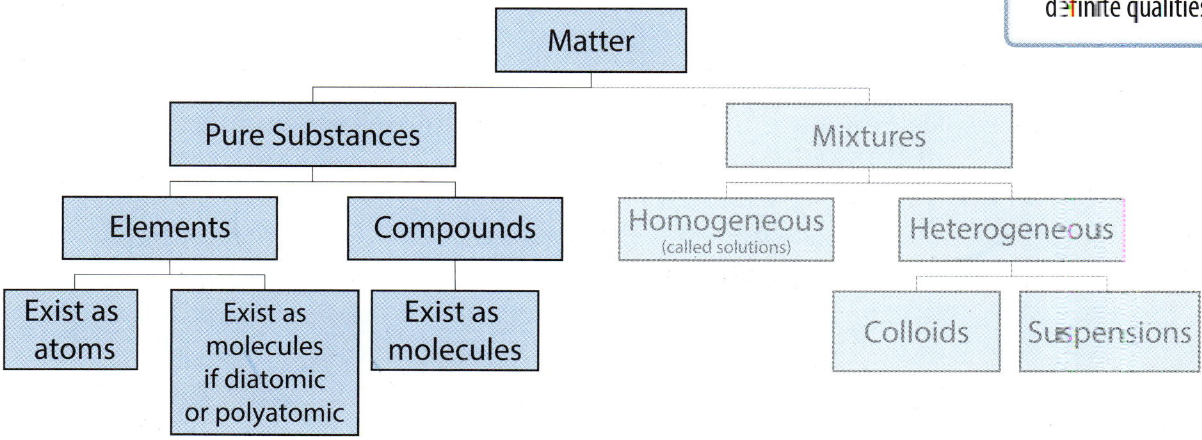

The James Webb Space Telescope is finished with gold on its reflective mirrors. The technicians inspecting the telescope look "glowing" because of the long-exposure photographic technique.

life, scientists use basic categories to classify all of the amazing types of **matter** that are found in the universe. Matter is anything that has mass and takes up space. The words *substance* and *matter* can be used interchangeably.

One type of matter is a **pure substance**, which is a substance of constant composition that possesses definite qualities. Gold, diamonds, water, baking soda, and table salt are all pure substances. The Greek philosopher Aristotle believed that all things in the world were made up of combinations of only four basic substances: earth, air, fire, and water. He theorized that these four basic substances mixed together in different combinations and proportions to form mountains, oceans, sky, living things, and even lightning. For example, he thought that sunlight was a mixture of fire and air, that ice was made of water and earth, and that steam was a combination of fire and water.

Obviously, God used more than four basic substances to create the world, but Aristotle had the right idea. God did use basic substances to build every part of creation. These basic substances are called **elements**. An element is a pure substance that cannot be broken down into simpler substances by physical or chemical means. Today, chemists recognize more than 110 elements, and at least 90 of these elements exist in nature. Gold is an element because gold cannot be broken down into any other substances. Oxygen and hydrogen gas cannot be broken down any further, so these two gases are also elements. Steel is not an element, because steel is a mixture of the elements iron, chromium, and nickel. Water, which consists of the elements hydrogen and oxygen, is not an element either. Scientists have made other elements during nuclear reactions, but most of these elements exist for only a few seconds.

Elements are made of **atoms**. An atom is the smallest unit into which an element can be divided and still maintain its properties. Atoms, then, are the building blocks of all matter.

A particle consisting of two or more atoms bonded together is called a **molecule**. Each molecule of a substance has all of the properties of that substance. For example, one molecule of water (H_2O) has all of the properties of a large body of water. Just as all

TRY THIS

Parts of a Substance

Granulated sugar is made of molecules that contain the elements carbon, hydrogen, and oxygen. Add a small amount of sugar to a pan. Heat it over medium heat on a hot plate, stirring constantly with a wooden spoon. Describe the two distinct substances formed when the sugar is heated. Is sugar a pure substance? Is it a compound or an element? Why?

atoms of a certain element are alike, all molecules of a specific compound are alike. What are compounds?

Steel and water are **compounds**—the second type of pure substance. A compound is a pure substance consisting of atoms of two or more elements that are chemically combined in a fixed proportion. Compounds containing the same kind of atoms in different proportions have very different properties. For example, water is a compound created from two hydrogen atoms and one oxygen atom (H_2O), and it is good to drink. Adding another oxygen atom to the formula creates a new molecule, hydrogen peroxide (H_2O_2), which is a compound that is extremely harmful to drink. The properties of a compound differ from the properties of its individual elements. For example, the compound salt, a stable and tasty food enhancer, can be broken down into the elements sodium and chlorine. Sodium is a silver metal that reacts explosively with water, and chlorine is a poisonous green

FYI

Pure Substances

Elements

Sulfur

Mercury

Silver

Bismuth

Compounds

Water molecule

Salt

Olive oil

Steel

BIBLE CONNECTION

Gold

Gold is a pure substance, an element, that can be shaped into fine jewelry, polished to reflect like a mirror, and hammered into thin sheets to cover statues and other works of art. When God gave Solomon instructions to build the temple, Solomon was to find gold workers to cover many things in gold. "And on the two olive-wood doors he carved cherubim, palm trees and open flowers, and overlaid the cherubim and palm trees with hammered gold" (1 Kings 6:32). Long ago, God instructed Solomon to use the finest materials to build the temple. Today, people still desire gold for its great beauty and will pay more for gold than for many other metals. Part of the cost includes the smelting process, which separates the gold from other elements in the earth. The process purifies the gold. The more pure the gold, the more beautiful it is. The apostle John wrote in the book of Revelation that the main street of heaven is paved with gold, a gold much purer than is seen on Earth. The main street of heaven is paved with gold so clear that you are able to see your reflection. "The twelve gates were twelve pearls, each gate made of a single pearl. The great street of the city was of gold, as pure as transparent glass" (Revelation 21:21). God has given many beautiful things to enjoy on Earth, but the beauty of heaven surpasses all imagination!

gas. The compound salt obviously has very different properties than either sodium or chlorine.

Statue of Aristotle

Look around you at the amazing variety of substances that God placed in the world. These substances are all made of combinations of at least 90 different atoms! Who but God could demonstrate such creativity? And think of the interactions between elements, atoms, molecules, and compounds. What a wonderful example of the complex design of creation!

LESSON REVIEW

1. Why is classification necessary?
2. What is matter?
3. Compare elements and compounds. How are they similar and how are they different?
4. How are an atom and a molecule different? Give an example of each.

God created only two kinds of pure substances: elements and compounds. All other substances are mixtures of these two. A **mixture** is a combination of two or more substances that are not chemically combined. For example, muddy water is a mixture of water and mud. If you pour muddy water into a jar, the water and mud will separate because they are not chemically combined. In contrast, water is not a mixture. Water is made of oxygen and hydrogen, but it is not a mixture because these two substances are chemically combined.

Mixtures can be divided into two subgroups—heterogeneous and homogeneous. A **heterogeneous mixture** is a combination of substances in which the particles are not uniformly distributed; one sample is not like every other. If you consider a square meter of soil and take four different samples from random locations, each sample will contain different quantities of rock, dirt, dead plant material, worms, insects, and other substances. When you look closely at sand, you will see clear quartz, red garnet, black magnetite, and ground-up white sea shells. Soil and sand are both heterogeneous mixtures. Italian salad dressing also classifies as a heterogeneous mixture. Can you think of any others?

A **homogeneous mixture**, on the other hand, is a combination of substances that has the same properties and appearance throughout; one sample is like every other sample. For example, if you dissolve sugar in hot water, the sugar water will look the same throughout. You will not be able to see individual particles of sugar floating around. Gold jewelry is another example of a homogeneous mixture. It is an alloy of metals made from a combination of gold, silver, and copper.

Since the substances in a mixture are not chemically combined, they can be separated from each other. **Distillation** separates

Composition of Matter

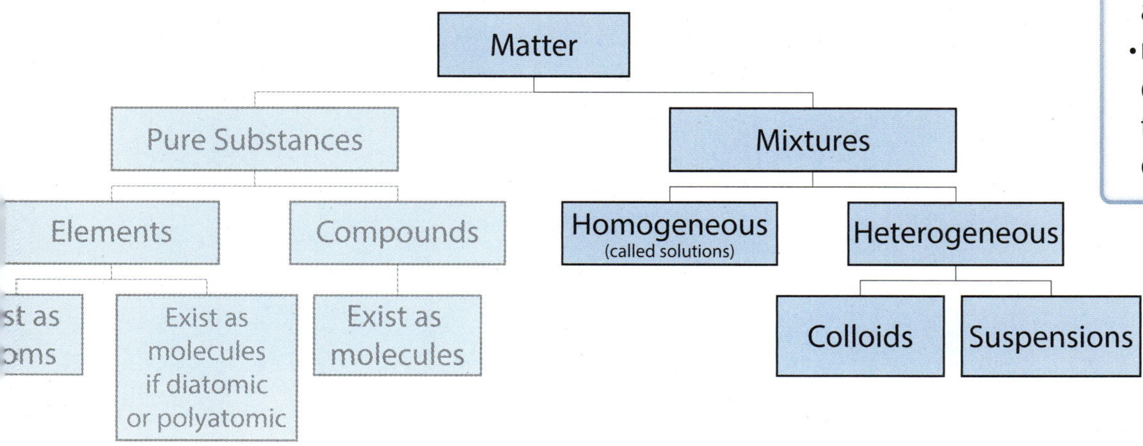

OBJECTIVES

- Identify different types of mixtures.
- Describe methods of separation.
- Separate mixtures using various methods.

VOCABULARY

- **chromatography** a method of separating mixtures according to how the different components are absorbed
- **distillation** a method of separating a liquid mixture by heating it so the liquid evaporates and condenses
- **filtration** the use of paper or other materials to remove solid particles from a liquid mixture
- **heterogeneous mixture** a combination of substances in which the particles are not uniformly distributed
- **homogeneous mixture** a combination of substances in which the properties and appearance are the same throughout
- **mixture** a combination of two or more substances that are not chemically combined

Soil is an example of a heterogeneous mixture.

Tea is an example of a homogeneous mixture.

FYI

Forensic Chromatography

Chromatography separates a mixture into individual components using either gas or liquid. Two different methods are used in forensic science. Thin-layer chromatography can be used to track down bank robbers. Before the bank teller hands the robber the money, she sneaks an ink pack into the bundles of cash. After the robber leaves, the ink pack explodes and releases a bright red liquid onto the money and anything else nearby. The ink cannot be washed out of clothing or upholstery, and it cannot be removed from skin. If a suspect has upholstery or clothing with a red substance on it, thin-layer chromatography can be used to separate the red substance. If it separates into the same components that the ink pack is known to have, this is good evidence that the right suspect has been arrested. The ink used in these packs is unique; no other dye uses the same formula.

High-pressure liquid chromatography (HPLC) separates material used in making bombs. Chromatography experts analyze residue from a bomb to help determine its type and origin. Scientists first examine a sample under a microscope. Sometimes they can see different-sized particles. After this, a sample goes into a solvent to see if any particles will separate into smaller particles with negative or positive charges. Pressure and another liquid are added to move the sample through a tube. The scientists then monitor how long particles stay in the tube. A detector measures the speed, and the rates are measured against other known substances. This evidence is not conclusive but can lead an investigation in a reasonable direction.

the substances in a liquid mixture by heating it so the liquid evaporates and then condenses. For example, distillation can be used to separate salt from seawater. The salt is left behind as the water evaporates. When the vapor condenses, the water is pure. Distillation is also used to produce such products as distilled water, gasoline, alcohol, vinegar, and kerosene.

Filtration is another method of separating liquid mixtures. The process of filtration uses paper or other materials to remove solid particles from a liquid mixture. For example, how do you separate pasta from the water it was boiled in? You pour the pasta and water through a colander or sieve. Filtration works only with heterogeneous mixtures that contain particles of different sizes. The smaller particles slip through the holes in the filter, but the larger particles remain in the filter. Filtration is used in water purification facilities to extract harmful materials from drinking water. Fuel filters in cars remove impurities in gasoline so it cannot damage the cars' engines.

Another widely used method for separating mixtures is **chromatography**. Chromatography is a method of separating mixtures into different components according to how different substances are absorbed. The Greek word *chroma* means "color." This method was first used to separate colored substances found in plants. Scientists use two types of chromatography: paper and gas. Gas chromatography is used to check air quality. Chromatography is used by scientists to detect pesticides in the environment, flammable chemicals in fire investigations, and drugs or alcohol

Filtration

TRY THIS

Chromatography

Using several colors of water-soluble markers, make a single dot with each about 2 cm from the short edge of a piece of filter paper. Add enough water to a small beaker to bring the water level just below 2 cm. Stand the filter paper in the beaker, with the marker dots just above the water line. Let the system sit for about 30 minutes. Check your paper. What happened to the color dots? Describe what you see. Do the activity again with different brands of black water-soluble markers. After 30 minutes describe what you see. How do you think this process could be used in forensic science?

HISTORY

Panning for Gold

The California gold rush began in 1848. James Marshall discovered a nugget of gold as he was building a saw mill for John Sutter. The news of his great find spread rapidly and a mad rush for California gold was on. Even though gold is a pure substance, it is never found completely alone. During the California gold rush, miners would take shovels, pick axes, magnets, and pans on their search for gold. The gold had been found in a river where small gold pieces were part of a heterogeneous mixture that included black silt, sand, and gravel. To separate the gold from the other substances, miners would take their metal pans and wash the mixture. The smooth metal pans were fashioned with slightly angled sides. As the miners swirled the water, the lighter particles would wash over the sides, leaving the heavier particles, including the gold pieces, in the bottom. After the heavy pieces dried, miners could use a magnet to remove some of the iron-rich sand and then pick out the gold from what was left in the pan. Later, this method was improved by employing a cradle. As the cradle was rocked, the gold-infused gravel was shoveled onto a perforated iron plate. Water was poured over it and the finer particles fell through, leaving the gold and other dense particles behind. Through time, more improvements were made to the process but even the newest technologies depended on the basic process of filtration, separating solids from liquids and then separating less dense solids from more dense solids. Individual pans, cradles, sluicing methods, and dredging have all been used to recover minerals like gold, platinum, and tin.

in blood or urine samples. Paper chromatography separates the different pigments used to make ink.

LESSON REVIEW

1. What is a mixture?
2. Differentiate between a homogeneous mixture and a heterogeneous mixture. Give an example of each.
3. How does a mixture differ from a compound?
4. Explain chromatography and its purpose.
5. What is distillation, and why is it an important process?
6. If you are making a strawberry smoothie and you want to remove the seeds before you drink it, what could you use to separate the seeds from the smoothie? What type of separation technique would it be?

Suspensions and Colloids 1.1.4

If you have ever taken liquid medicine for an illness, have you noticed that sometimes the words *shake well* are on the label? Some liquid medications, such as milk of magnesia, have particles that settle to the bottom of the bottle, but others, like cough syrup, do not. Not all mixtures stay mixed. What makes the difference? Heterogeneous mixtures can be divided into two categories: suspensions and colloids.

If the particles of a substance are mixed throughout but are large enough that they will settle, the mixture is called a **suspension**. The particles of a suspension are large enough to block light, and they can be separated by passing them through a filter.

For example, if you shake a bottle of oil-and-vinegar salad dressing, the oil is temporarily suspended in the vinegar. If you do not pour the dressing quickly, you will have to shake it again. Or, if you mix clay with water, only a small portion of the clay dissolves. The clay particles are very large compared to the water molecules. Many small particles of clay are bounced around by water molecules. Because this prevents the clay particles from settling quickly, they remain suspended in the water for a long period of time, but they will eventually sink. Paint is also a suspension, which is why cans of paint must be shaken or stirred to suspend the pigments in the liquid. Other common suspensions are the mixture of flour and water, some liquid antibiotics, and some sunscreens. Your body even contains a very important suspension—red blood cells are suspended in plasma in your bloodstream.

A **colloid**, by contrast, is a heterogeneous mixture in which the particles of a substance are dispersed throughout and are not large enough to settle. The particles in a colloid are smaller than the particles in a suspension and cannot be seen without an electron microscope.

OBJECTIVES

- Compare heterogeneous mixtures.
- Explain the Tyndall effect.

VOCABULARY

- **colloid** a mixture in which the particles of a substance are dispersed throughout and are not large enough to settle
- **suspension** a mixture in which the particles of a substance are mixed throughout but are large enough to settle
- **Tyndall effect** the scattering of light by colloids

Composition of Matter

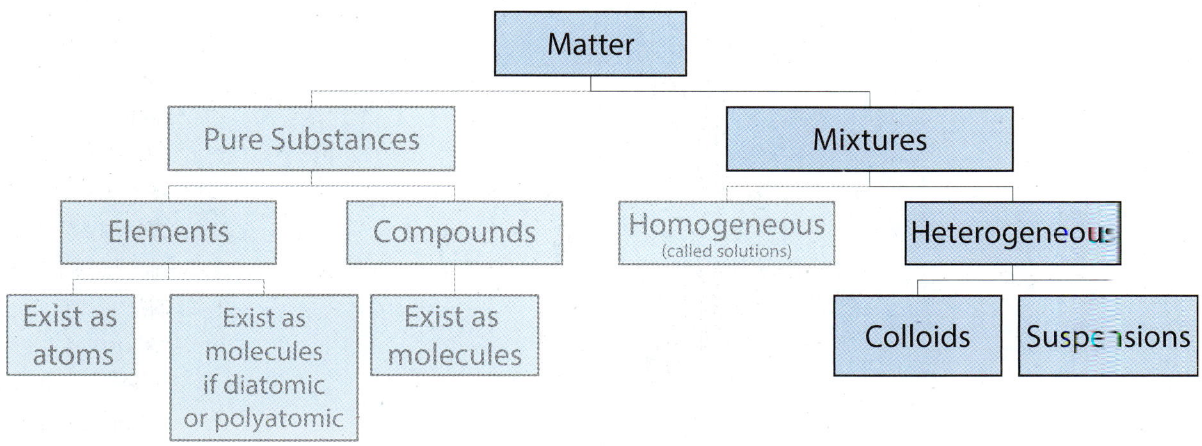

Tyndall effect

If the particles of a colloid are too small to see, what evidence demonstrates their existence? The answer to this question was discovered by John Tyndall, an Irish professor who worked at the Royal Institution, London. In the mid-1800s, he was researching the radiant heat of gases in the atmosphere. During his studies, he examined the diffusion of light and concluded that the sky is blue because the particles in the atmosphere scatter the sun's rays. This scattering of light by tiny particles became known as the **Tyndall effect**. The Tyndall effect provides indirect evidence that the mixture is a colloid. Samples of a homogeneous mixture and a colloid appear the same under a microscope because no separate particles can be discerned. But if a light is directed at both samples, only the colloid scatters the light. Milk is an example of a colloid; its white, cloudy appearance is caused by the Tyndall effect. Fog, also a colloid, consists of water particles mixed with air. A person driving a car in fog has a hard time seeing through fog with bright headlight beams because the fog scatters the light from the headlights, which is why drivers use dim headlights on foggy days. In the same way, a beam of light shining through dusty air shows the dust particles. The particles scatter and reflect light in

TRY THIS

Suspensions

Suspensions are heterogeneous mixtures. In a liquid suspension, the solids can be evenly dispersed when the mixture is shaken, but over time, the solids will settle. Make the following mixtures in Erlenmeyer flasks: oil and water, powdered chalk dust in water, and dried herbs and water. Gently shake each mixture. Describe what you see. Let each mixture settle for about 10 minutes. Describe what you see. Is each mixture a suspension? Why?

many directions. Because of the Tyndall effect, colloids usually look cloudy.

Whereas the particles in a suspension are easily filtered out, the particles in a colloid are too small to be trapped by most filters. Colloids can be separated by using special filters or using a device called *a centrifuge*. A centrifuge spins the mixture at a very high rate, separating the components according to their densities. A centrifuge is often used to separate the components of blood for testing.

LESSON REVIEW

1. How are a suspension and a colloid similar?
2. What is the difference between a suspension and a colloid?
3. What is the Tyndall effect? What does this effect demonstrate about colloids?
4. Why is the Tyndall effect considered indirect and not direct evidence?

TRY THIS

Unmixable Mixture

Some substances resist being mixed together. Completely remove the label from a plastic bottle. Fill the bottle three-fourths full with isopropyl alcohol. Add drops of blue food coloring. Gently shake it. Do you have a mixture or a compound? Fill the rest of the bottle with vegetable oil, leaving a small space at the top. Add glue to the lid and secure it on the bottle. Describe what you see. Shake the bottle vigorously. Describe what you see. Do you have a mixture? Why?

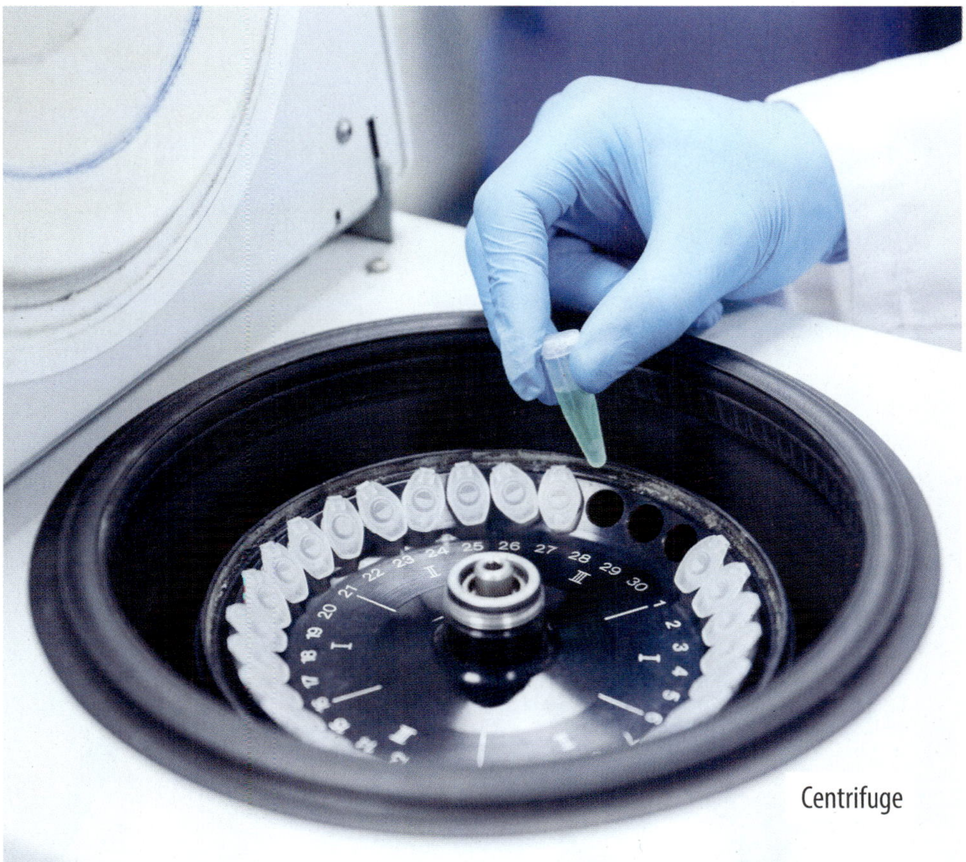

Centrifuge

1.2.1 *Properties of Matter*

OBJECTIVES

- Distinguish between noncharacteristic and characteristic properties.
- Contrast the terms *mass* and *weight*.
- Correlate the terms *mass*, *volume*, and *density*.
- Calculate the density of a substance.

VOCABULARY

- **density** the mass per unit of volume of a substance
- **gravity** an attractive force between two objects with mass
- **mass** the amount of matter in an object
- **volume** the amount of space an object or substance takes up
- **weight** the measure of the amount of gravity exerted on an object

The characteristics of matter are called *properties*. All matter has certain properties, such as mass, weight, volume, and density. Physical properties include the characteristics of a substance that can be observed without changing the basic nature of the substance. For example, you do not have to change what a banana is made of in order to see that it is yellow. Color, odor, shape, mass, volume, and weight comprise a small list of physical properties. Scientists use some physical properties to identify specific substances and to draw conclusions about the nature of these substances. However, some characteristics cannot be used to identify substances. These properties are called *noncharacteristic properties*.

Suppose you want to check whether a substance is sulfur. Sulfur is yellow, but so are pollen, amber, and corn. So color should not be used to identify a substance such as sulfur. Color, then, is a noncharacteristic property. Other noncharacteristic properties include mass, shape, volume, and length. Two substances can have the same color, mass, volume, or length and still be completely different substances.

Mass, the amount of matter in an object, depends on the number of atoms or molecules in the object. The mass of an object

Hot air balloon

remains constant, which means it will not change unless matter is added or taken away from the object. The standard metric units for mass are grams (g) and kilograms (kg).

Some people confuse mass with weight, but mass and weight are different. **Weight** is the measure of the amount of gravity exerted on an object. **Gravity** is a force that pulls two objects together because of their masses. An object would have the same mass on Earth as on the moon, but it would have a lower weight because the moon exerts less gravitational force than the earth does.

Volume is the amount of space an object or substance takes up. The chair you are sitting on, the milk you drink, and the air around you all take up space and have volume. The standard metric units for the volume of a liquid are milliliters (mL) and liters (L); the units for the volume of a solid are cubic centimeters (cm^3) and cubic meters (m^3).

Properties that are unique to a particular substance and can be used to identify that substance are called *characteristic properties*. These properties include density, melting point, and boiling point. For example, sulfur has a density 1.96 times that of water, a melting point of 115.36°C, and a boiling point of 444.75°C. Sulfur forms a toxic gas when burned in air. These specific properties apply only to sulfur, so they can be classified as characteristic properties.

Which is heavier, iron or Styrofoam? If you answered iron, you are correct, sometimes. An iron horseshoe is heavier than a

FYI

Objects with Different Density

Aluminum

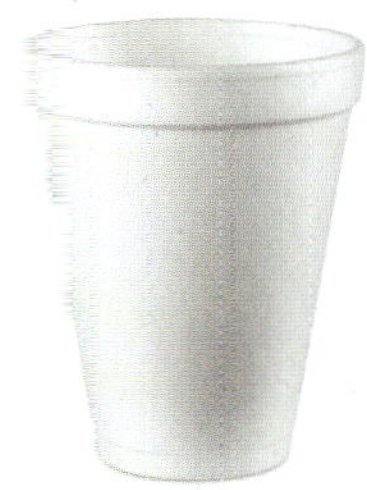

Polystyrene

Gold

BIBLE CONNECTION

Scales

In ancient times, scales were an important tool used to measure the amount of payment for an object. A buyer would place his gold or silver coins on one side of the scale, and the merchant would place his weights on the other. Because gold and silver have specific densities, scales could be used to determine the correct payment for an object. If the merchant was dishonest, he might use weights with greater density so the buyer would have to add more coins to his side of the scale to bring the scales into balance. In God's eyes, dishonest scales are detestable. "The Lord detests dishonest scales, but accurate weights find favor with Him" (Proverbs 11:1).

LAWS & PRINCIPLES

Density

Density is a derived unit and therefore must be calculated using the mass and volume of a substance. The density (*D*) of a substance is equal to its mass (*m*) divided by its volume (*V*).

$$D = m/V$$

TRY THIS

Mass, Volume, Density

Take a sample of clay and shape it into a rectangle. Measure the volume in cm^3 and the mass in grams. Use the formula $D = m/V$ to determine the density of the clay. Remove some clay from the sample and shape the remaining clay into a rectangle. Measure volume and mass and calculate density. Repeat the procedure one more time, removing more clay from the sample and reshaping. Record measurements. Which measurements changed each time? Which measurement remained fairly constant? Why?

Styrofoam ball, but a room full of Styrofoam balls is heavier than an iron nail. What you really mean when you say iron is heavier than Styrofoam is that iron is more dense than Styrofoam. The density of an object is how much matter is packed into the volume of the object. **Density** (*D*) is the mass (*m*) per unit of volume (*V*) of a substance. You can use a simple formula to determine a substance's density: $D = m/V$. The units used in density are grams per centimeters cubed. For example, a sample of water has a mass of 100 g and a volume of 100 cm^3; its density is 1 g/cm^3. At 20°C, water always has a density of 1 g/cm^3 whether it is one drop or a pool full. Every substance has its own specific gravity, which is its density compared with water. Look at the chart below to find the densities of some common substances.

LESSON REVIEW

1. Why are noncharacteristic properties poor indicators of a substance's identity? List three noncharacteristic properties.
2. What is the difference between weight and mass?
3. Define *characteristic property* and list three examples.
4. Consider two substances. The first substance has a volume of 500 mL and the second has a volume of 100 mL. They have the same mass. What can you conclude about the density of each?
5. If a substance has a mass of 200 g and a volume of 200 cm^3, what is it?
6. Find the density of a substance that has a volume of 245 cm^3 and a mass of 175 g. Is it more or less dense than water?

Substance	Density (g/cm^3)
gasoline	0.7
ethanol	0.785
wood	0.85
water (solid)	0.92
water (liquid)	1.0
salt	2.2
aluminum	2.7
steel	7.8
iron	7.9
copper	9.0
silver	10.5
lead	11.3
mercury (liquid)	13.5
gold	19.3

Kinetic Theory 1.2.2

Think about these three things that could be found in your kitchen: ice cubes, water from the faucet, and steam rising from a boiling tea kettle. Ice is cold and solid. Faucet water can be hot or cold and is a transparent liquid. Steam is incredibly hot and takes the form of a gas. Ice, faucet water, and steam are very different from each other, but in one way they are the same; they are all phases, or states, of water. Most matter on Earth can exist in any of the three phases: solids, liquids, and gases. Plasma, a fourth phase, is very rare on Earth but it is the most common phase in the universe. Stars and the aurora borealis both exist as plasma, a hot ionized gas that is influenced by electric and magnetic fields.

Scientists have developed a model called *the kinetic theory* to explain the properties of solids, liquids, and gases. The word kinetic comes from the Greek word *kinetikos*, which means "of motion." Even though the molecules in substances are too small to see, the molecules in all states of matter are moving. Like bouncing balls, these moving particles are constantly bumping into each other. Only at 0 Kelvin does all motion cease. The three assumptions of the kinetic theory are as follows:

- All matter is made up of tiny particles that are in constant, random motion.
- The speed of the particles directly corresponds to their temperature.
- At a given temperature, the total amount of kinetic energy of the colliding particles remains constant.

Imagine a room full of bouncing rubber balls. Some balls would slow down as they hit other balls, but others would speed up

Solar eruption on the sun produces a wave of plasma.

OBJECTIVES

- Explain the kinetic theory.
- Describe thermal expansion.

VOCABULARY

- **thermal expansion** the increase in the volume of a substance resulting from an increase in temperature

FYI

Absolute Zero

Are there upper and lower limits to temperature? Is there a maximum amount of energy that molecules can have? Is there a point at which all motion ceases? Scientists have agreed on a temperature that must exist where all molecular motion ceases. There is absolutely no energy left. This is 0 Kelvin, or absolute zero, which equates to −273.15°C. Absolute zero has not yet been reached, but in 2003, researchers at MIT came close. They attained 0.45 nK, which is 45 hundredths of a billionth of 1 Kelvin.

BIOGRAPHY

Robert Boyle

Robert Boyle was born in Ireland on January 25, 1627. He was the youngest son in a large and wealthy family. At 8 years of age, he began his formal education at Eton College in Berkshire, England. In his 11th year, he was taken out of the classroom to travel across Europe with his older brother, Francis, and tutor, Isaac Marcombes. After the tour, he and his tutor remained in Geneva where he continued his education for several more years.

Boyle returned to England and began writing works that were both theological and scientific in nature. In his book, *Some Motives and Incentives to the Love of God*, he wrote, "And when with excellent Microscopes I discern in otherwise invisible Objects the Inimitable Subtlety of Nature's Curious Workmanship; And when, in a word, by the help of Anatomicall Knives, and the light of Chymicall Furnaces, I study the Book of Nature, and consult the Glosses of Aristotle, Epicurus, Paracelsus, Harvey, Helmont, and other learn'd Expositors of that instructive Volumne; I find my self oftentimes reduc'd to exclaim with the Psalmist, How manifold are thy works, O Lord? In wisdom hast thou made them all." It is clear that Boyle was moved to worship God the creator as he pursued his scientific research. In 1649, he began investigations that focused on the physical nature of air. In 1659, Boyle and Robert Hooke worked together, finished building their famous air pump, and began experimenting on the nature of gases. Boyle concluded that an inverse relationship exists between the volume and pressure of a gas. This conclusion sparked a cascade of research that led to today's kinetic theory.

When Boyle worked, he followed a "mechanical philosophy," which means he used the scientific method of experimentation and observation to study natural forces. He did not want to make generalizations as much as he wanted to discern the details of nature. Boyle had studied the works of Aristotle and was aware of his theories about physical changes in relation to the four elements: earth, wind, fire, and water. Boyle disagreed with Aristotle's theory and sought to demonstrate that change happens in relation to particles and their motion. Boyle used strict experimental procedures and for his adherence to the application and methods of chemistry, he was often referred to as *the father of chemistry*.

Boyle believed that the earth was created by God, and during his lifetime he worked to demonstrate that science and faith went hand in hand. In *The Christian Virtuoso*, he stressed that the pursuit of science would greatly enhance a person's understanding of God. Before he died in 1691, he left his writings to the Royal Society and requested that a series of lectures be created, now known as *the Boyle Lectures*, to defend Christianity and the truth that God created a universe according to definite physical laws.

as they are hit. If the balls never stopped moving, their total energy would not change. As particles in a substance collide, they transfer their energy to each other. Some particles lose energy and others gain energy, but overall there is no loss of energy.

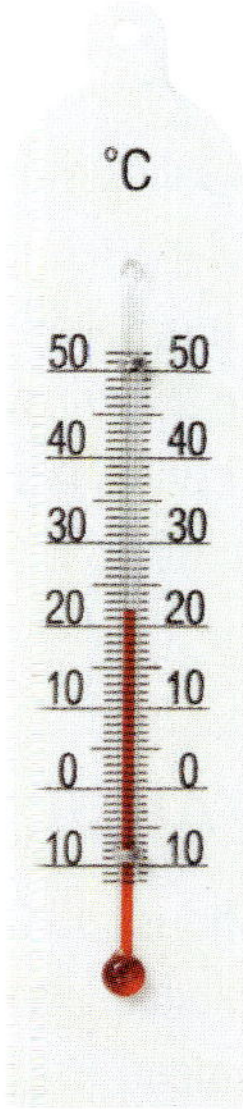

Thermal expansion causes liquid to rise.

Most substances in any phase expand when they are heated because their molecules move farther apart. This increase in the volume of a substance caused by an increase in temperature is called **thermal expansion**. Glass thermometers depend on thermal expansion to indicate changes in temperature. As temperature increases, the liquid in the thermometer expands and moves up the tube. As temperature decreases, the liquid loses energy, decreases in volume, and moves down the tube.

Thermal expansion is the reason that cement sidewalks are made with spaces between the sections and spaces are left between bridge sections and between railroad tracks. The spaces leave room for the concrete or metal to expand in hot weather. Thermal expansion also explains why a tightly closed bottle of soda explodes or bubbles over if it is left in the sun and why automobile tires expand in hot weather.

LESSON REVIEW

1. What are the three basic assumptions of the kinetic theory?
2. Explain why spaces in sidewalks help keep the sidewalk sections from cracking.
3. Why does air pressure change in a tire when the temperature changes?

Spaces separate sections of a sidewalk to accommodate the effects of thermal expansion.

TRY THIS

Atomic Motion

Because atoms and molecules cannot actually be seen, scientists must rely on indirect evidence to support the kinetic theory. If atoms and molecules are in motion, food coloring added to a glass of water should eventually move throughout an entire glass. Test the kinetic theory. Place 1 drop of food coloring in a glass of water and watch what happens. Wait a couple of minutes and describe what you see. Kinetic theory also states that temperature measures the amount of energy in the molecules of a substance. To test this part of the theory, you will need a large ice cube, a beaker of cold water, and a beaker of hot water. Record the temperature of each. With a partner, place 1 drop of food coloring in the hot water. Record the time it takes for the food coloring to completely color the water. Next, release 1 drop of food coloring on the ice cube and record how long it takes for the food coloring to completely color the ice cube. Finally, release 1 drop into the cold water and measure the time it takes to completely color the water. How do the times compare? Why? What type of evidence did you gather?

1.2.3 *Solids*

OBJECTIVES

- Contrast solids with other phases of matter.
- Explain how solids maintain their shape.
- Compare and analyze types of solids.

VOCABULARY

- **amorphous solid** a solid with a random particle arrangement
- **cohesion** the force of attraction between particles of the same kind
- **crystalline solid** a solid whose particles are arranged in a regular, repeating three-dimensional pattern
- **solid** the state in which matter has a definite shape and a definite volume

Look around your classroom and take note of all of the different substances. What do most of them have in common? They are all solid. At room temperature, most common substances exist as solids. **Solid** is the state or phase in which matter has a definite shape and a definite volume. For example, all the desks, chairs, tables, walls, and books in the classroom are solids and are considered to be in a solid state of matter.

Because solids have definite shape and volume, they cannot flow like liquids and gases. All other phases of matter will change shape depending on the shape of the container in which they are placed. For example, if you pour water into a tall, narrow container, it will become tall and narrow. If you pour water into a short, wide container, it will become short and wide. Gases and plasma can even change volume when the temperature or pressure on them alters. By contrast, if you want to change the shape of a solid, you cannot put it into a differently shaped container. Instead, you have to apply force. You can pound clay, roll out dough, crumple a piece of paper, or smash an aluminum can to change the shape of these solids.

FYI

Crystalline Solids

Pyrite

Amethyst

Crystallized sugar: rock candy

Amorphous Solids

Glass

Rubber hose

Plastic tea set

Two factors contribute to solids maintaining their shape and volume: lack of energy and motion, according to the kinetic theory, and **cohesion**, the force of attraction between particles of the same kind. Kinetic theory states that the particles of a solid are moving even though you cannot see the movement. The particles in a solid have the least amount of energy and motion of all phases of matter, so they only vibrate and cannot move past each other. In contrast, liquids, gases, and plasmas have more energy so the particles move more and can overcome cohesion. Because the particles of solids are packed closely together and cannot easily move or flow over one another, solids are generally hard and rigid with definite shapes. You can pick up a block, a pen, or a cup, and they will maintain their shape and volume. That would not happen if you tried to pick up a handful of water.

The small amount of energy available and the close quarters of particles in a solid make it difficult for the particles to overcome cohesion. Because cohesion is a short-range force, it is only effective over microscopic distances. For example, if a solid is broken, gas particles from the air cling to the broken surfaces, which prevents the rejoining of the solid surfaces. The particles of the broken surfaces are not close enough to have sufficient attraction to hold, which is why solids can break. When a vase breaks, the pieces will not go back together unless glue is added.

Scientists separate solids into two categories: crystalline solids and amorphous solids. A **crystalline solid** has particles that are arranged in a regular, repeating three-dimensional pattern. Snowflakes, table salt, sugar, diamonds, and graphite are all

A snowflake is a crystalline solid.

FYI

Lattice Structures of Crystalline Solids

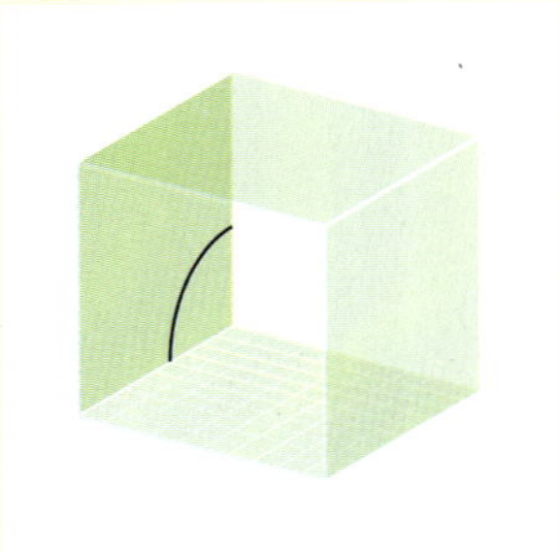

Cubic

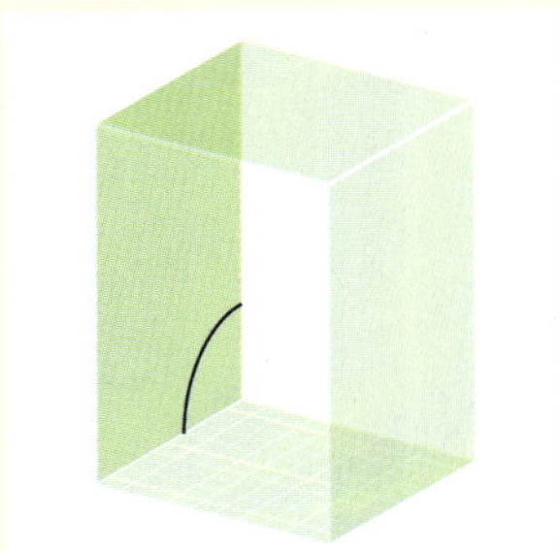

Tetragonal

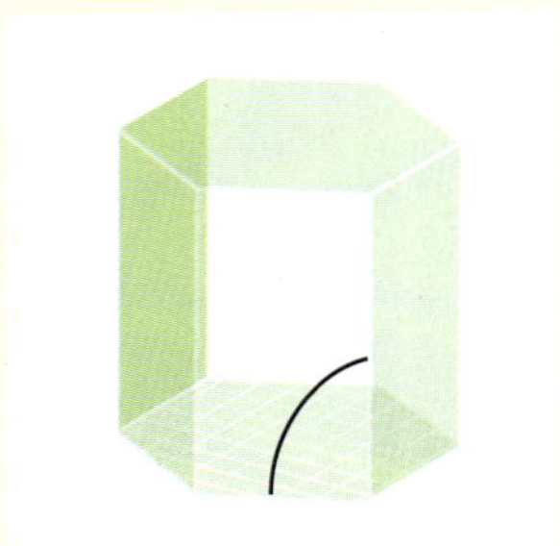

Hexagonal

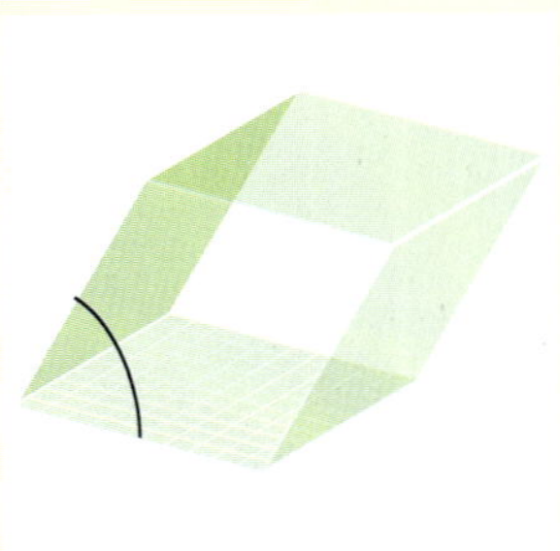

Rhombohedral

TRY THIS

Growing Crystals

Mix 200 mL of sodium silicate with 800 mL of water in a shallow pan. Stir to thoroughly mix the water and the sodium silicate. Allowing enough space between the crystals for growth, add one or more samples of the following compounds: cobalt chloride, copper sulfate, manganese sulfate, lead nitrate, or iron sulfate. Do not disturb the pan once the compounds have been added. When the crystals have stopped growing, siphon off the sodium silicate solution with a large pipette and add water to preserve the crystals. Illustrate what you see. Take a sample of the crystals from the pan and look at the sample with a hand lens or a stereo microscope. Place a metric ruler next to the sample to measure the size of the crystals. Record the size and illustrate what you see. What type of crystalline shape do you think you see?

examples of crystalline solids. The crystalline structure is a characteristic property used by scientists to identify many substances. The patterns, or lattice structures, take many shapes, such as cubic, tetragonal, orthorhombic, monoclinic, hexagonal, rhombohedral, and triclinic. Snowflakes have a hexagonal crystalline structure formed as the result of water vapor changing directly into a solid. Table salt has a cubic structure.

A solid with a random particle arrangement is called an **amorphous solid**. Unlike a crystalline solid, when these solids are studied under the microscope, the particles form random structures without orderly patterns. The lack of organization also affects the melting point of the substance. Amorphous solids will soften as energy is added, but an exact temperature for melting is not well defined. Consider what happens when heat is added to candle wax. The wax will soften and gradually change shape until it finally melts into a liquid, but the exact point at which the solid wax becomes liquid is difficult to measure. Amorphous solids include wax, paraffin, charcoal, glass, and rubber.

Most substances are more dense when they are solid than when they are liquid. If you place a block of solid steel into a vat of melted steel, the solid will sink to the bottom because the particles are closer together. But what happens when solid water is added to liquid water? Add an ice cube to a glass of water and the ice floats because solid water is less dense than liquid water. Why is water different than almost every other substance? Consider the fish that live in lakes and ponds. If frozen water were more dense than liquid water, the ice would build up on the bottom of the lakes and ponds, destroying food sources and shrinking the habitat. The fish and other aquatic life would die. Instead, ice floats to the top of the water, leaving everything underneath intact.

Solids are essential to your everyday life. Because solids maintain shape and volume, you can be confident that when you sit on a chair, it will maintain its shape and hold you up. You know a car will not become the same shape as the road or float away as you are driving down the highway. And the shoes you are wearing will be the same shape and size at the end of the day as they were when you put them on in the morning.

LESSON REVIEW

1. What physical properties make solids different from all other phases?
2. Why do solids maintain their shape and volume?
3. What is the difference between a crystalline solid and an amorphous solid?
4. Except for water, how does the density of a substance in its solid state compare to the density of the substance in its liquid state?
5. Explain how water is different from other substances. Why is this important?

HISTORY

Rubber

A natural product of the rain forest, rubber was first identified by French scientists during a South American expedition in 1735. Thirty-five years later, English chemist Joseph Priestley discovered the substance could rub out pencil marks, which led to its common name. Large chainlike molecules give rubber its great elasticity. However, natural rubber responds to changes in temperature by softening or hardening, a trait that made its use problematic. It also has a strong odor, deteriorates easily, and is sticky. These problems were resolved through the process of vulcanization, which was accidentally discovered by Charles Goodyear in 1839. Goodyear found that heating a combination of rubber and sulfur created a durable elastic product that was ideal for tires. Natural rubber solids are found in latex, a milklike fluid in the bark of *Hevea brasiliensis* trees. Rubber trees are grown commercially on plantations where they are tapped for their latex. Synthetic rubber was developed in response to natural rubber shortages during World War II and constitutes over half of the rubber market today.

1.2.4 *Liquids*

OBJECTIVES

- Identify properties unique to liquids.
- Analyze the benefits of liquids.
- Compare the viscosity of liquids.

VOCABULARY

- **adhesion** the attractive force between unlike substances that holds them together
- **liquid** the state in which matter has a definite volume but not a definite shape
- **surface tension** the greater force acting on particles at the surface of a liquid
- **viscosity** a liquid's resistance to flow

When Jesus talked to the woman at the well, He offered her living water that would quench her soul eternally. She would never thirst again. Anyone reading the account understands her disbelief at the possibility of never drinking water again. Water is as essential to life as breathing. Imagine how different life would be without it. You cook with it, bathe with it, wash dishes, clean clothes, water the lawn, and of course, you drink it several times a day. You need water. In fact, all organisms need water.

By studying water, scientists can observe several properties of all liquids. A **liquid** is the state in which matter has a definite volume but not a definite shape. All liquids are fluid, have space between their particles, are cohesive in some form, and have a specific rate of flow. Liquids flow and can be poured from one container to another. Unlike solids, their shape changes to match the container, but the volume remains the same. Water changes shape as it flows from the faucet into your glass. Syrup spreads to cover the surface of your pancakes. Oil flows through an engine to lubricate all the moving parts.

The fluid motion of liquids is explained by the kinetic theory. Liquids have more energy in their molecules than solids of the same substance, and the added energy creates more motion in the molecules, allowing them to overcome some of the effects of

cohesion. Liquid molecules are able to move and roll over each other, to flow, and to take the shape of the container holding them. If you take a liter of water and a liter of ice cubes and pour them both out, the water will move and flow more than the ice cubes.

Water strider (pond skater)

This greater range of motion implies more space between the molecules. Because the molecules in liquids have space between them and more energy than the molecules in solids, they are capable of diffusion. The molecules of one liquid will move into the molecules of a second liquid. For example, the molecules of food coloring will diffuse throughout the molecules of water.

Cohesion in a solid is so great that it opposes the motion of molecules and prevents movement; the solid maintains its shape. By contrast, water molecules are cohesive and fluid at the same time; the opposing forces work together. Imagine spilling a glass of water on a table. The water pours out of the glass; its shape is no longer the shape of the glass but more like that of the table. As it flows across the table, the water stays together because the molecules are still attracted to each other.

The tendency of liquid molecules to stay together becomes very interesting at the surface of a fluid. For example, if you fill a graduated cylinder with water, the water molecules form a curved surface at the top of the cylinder. A glass of water can

HISTORY

Molasses Tsunami

In January 1919, a tank containing 8.7 million liters of molasses ruptured, allowing its thick, sticky contents to flow through the North End of Boston. This bizarre incident wrecked buildings, overturned vehicles, killed 21 people, and injured 150 others. According to eyewitness accounts, the nearly 12-meter wave of molasses moved at about 56 kph. For nearly 100 years, scientists wondered why the spill was so deadly. In 2016, a team of scientists and engineering students in a fluid dynamics class at Harvard studied the accident. Their findings indicated that the cold weather made the accident more catastrophic than it would have been had it happened when the weather was warmer. The cold air thickened the molasses, which slowed down the flow but also trapped people in the viscous liquid and made rescue very difficult. The Harvard team also determined that the tank ruptured because the type of steel used for the tank walls was too brittle and the tank walls were too thin.

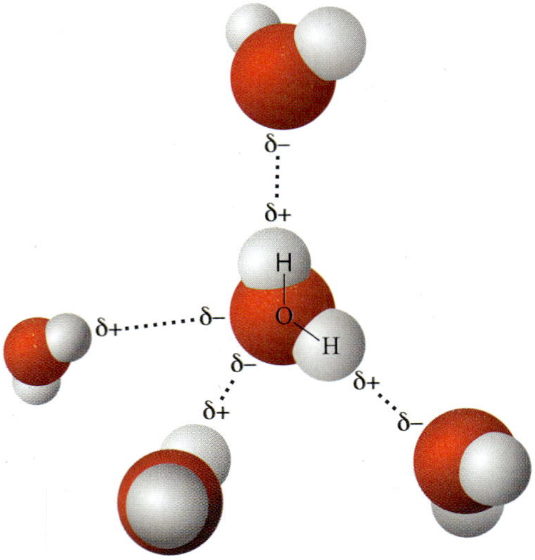

Hydrogen bonding of water molecules

be filled slightly above the rim of the glass and not spill over. The cohesion of the molecules on the surface of liquids is called **surface tension**. Aquatic creatures such as water striders benefit from surface tension; they can skate across the surface of streams and ponds without falling through the surface of the water. The cohesive properties of water also enable it to remain on surfaces of plants, creating small pools of water for insects and animals to drink, and to remain on animals like amphibians, so they can absorb water to maintain proper moisture levels.

What makes water cohesive and why does it create surface tension? A water molecule is composed of two hydrogen atoms and one oxygen atom (H_2O). The hydrogen atoms have a slight positive charge and the oxygen atom has a slight negative charge. The hydrogen atoms of one water molecule are slightly attracted to the oxygen portion of the other molecules. This attraction between molecules is called *hydrogen bonding*, and this force creates cohesion in water and gives it a strong surface tension.

Plant with roots showing

Another important property of liquids is **adhesion**, the force of attraction between unlike substances that holds them together. Some examples of adhesion include water droplets hanging on the tips of pine needles, water soaking into a paper towel, and water adhering to the surface of a car. In contrast, when the car is waxed, the water is repelled and no longer adheres to the car but instead clings more to itself as little beads.

Cohesion, surface tension, and adhesion are not only valuable individually, but together they play an important role in the distribution of water to all parts of a plant. Because water has to move in opposition to gravity to travel from the roots of a plant to all its structures, the fluid motion of water is not enough for transport. The plant depends on the specific properties God placed in all liquids to distribute the water as needed.

Although all liquids are cohesive in some manner and have fluid motion, they do not all flow at the same rate. Think of the many different liquids found in your kitchen, such as water, honey, milk, syrup, and juice. Some of these liquids flow more easily than others. For example, water flows more easily than honey. Honey has a higher resistance to flowing, or **viscosity**, than water. Substances with a higher viscosity flow slower than liquids with a low viscosity. Motor oil has a high viscosity, which enables it to lubricate and protect the moving parts of car engines.

LESSON REVIEW

1. How is a liquid different from a solid?
2. What causes a liquid to be fluid?
3. Explain the difference between cohesion and adhesion.
4. How do the cohesive properties of water benefit life?
5. How does water get from the roots of a plant to the leaves?
6. Describe viscosity using water and honey as examples.

FYI

Variations in Viscosity

Pouring honey

Pouring oil

Pouring shampoo

Pouring water from a faucet

TRY THIS

Surface Tension

Fill a small bowl with water. Place a small sewing needle on a piece of tissue paper. Carefully lower the tissue paper with the needle onto the surface of the water. Wait a couple of minutes and observe what happens. Record your observations. What happened? Why? Add a drop of dishwashing soap to the water, away from the needle. What happened? Why? If your results were not what you expected, explain what variables could have caused incorrect results.

1.2.5 Gases

OBJECTIVES

- Identify properties of gases.
- Apply kinetic theory to states of matter.
- Predict the behavior of gases according to Boyle's law and Charles's law.

VOCABULARY

- **gas** the state in which matter does not have a definite shape or a definite volume
- **pressure** the amount of force exerted on an object per unit of area

The moon is a cold, lifeless place. It is a barren rock compared to Earth, which flourishes with life. The moon cannot sustain life because it lacks water and an atmosphere. The gases that make up the earth's atmosphere are often overlooked because they are invisible, but without gases life would be impossible.

Gas is the state in which matter has no definite shape and no definite volume, unlike solids, which have definite shape and definite volume, and liquids, which have no definite shape but do have definite volume. Kinetic theory states that gas particles have more energy than solids or liquids, and this energy causes greater motion and allows the particles to overcome cohesion. Consider what would happen if you filled three 500 mL plastic storage bags with three different substances: a block, water, and air. If you dropped the block out of its bag, the block would maintain its size and shape because it is a solid. The water in the second bag would be in the shape of the bag. If you poured the 500 mL of water into a container, it would still be 500 mL of water, but now the water would be in the shape of its new container. And, if the air in the third bag came from your lungs, the air would have changed shape from your lungs to the shape of the bag. When the air is released from the sealed bag, would it stay in the shape of the bag? Would it take up the same amount of space? No, it would expand throughout the room and change its shape and volume.

Girl inflating bike tire

All gases share four properties. First, gases have no definite shape or volume. Second, the volume of gases varies with pressure and temperature. To understand this, think of a balloon filled with air. If you press on a balloon, the air inside compresses and takes up less room. If you bring a balloon from a warm room to the cold outdoors, it shrinks—the gases inside take up less room. Third, gases have low density because the particles have plenty of energy to move and are not subject to cohesion. When you walk through air, you simply push the particles out of the way because of air's low density. Just like solids and liquids, density varies from gas to gas. Helium balloons float because helium is less dense than air. Fourth, gases are capable of diffusion. When perfume or air freshener is sprayed into the air, the liquid evaporates and the gases carry the fragrance throughout the room, diffusing with the air.

In 1660, the English chemist Robert Boyle experimented with gas pressures and developed Boyle's law, which states that the volume of a fixed amount of gas at a constant temperature increases as its pressure decreases, and decreases as its pressure increases.

FYI

Pressure in the Deep Sea

On Earth's surface, the atmospheric pressure is something living things are not even aware of. But deep in the dense ocean, the pressure is intense. How can air-breathing surface animals like Weddell seals and sperm whales thrive in depths up to a mile below the ocean surface? Such deep-diving mammals allow their lungs to collapse completely and rely on the oxygen stored in their muscles. Once its lungs collapse, a seal is heavier, so it sinks in the water. It can descend without much effort and retain its stored oxygen for swimming back to the surface. Unlike land mammals, many deep-diving animals lack air-filled sinuses, or they can squeeze tissues to eliminate any pockets of air in their bodies. Part of the difficulty scientists encounter when studying deep-sea creatures is that the animals cannot survive a trip to a surface lab. For example, fish that live near the ocean floor rely on swim bladders with a fixed buoyancy, which means they are specifically designed to thrive in deep water but cannot withstand a move to shallower depths. Even bacteria gathered from the Mariana Trench, more than six miles below the ocean's surface, do not grow in less-pressurized environments. Many of these creatures also have cell membranes that contain more fluid fats than land animals, which allows them to withstand the great pressure of the ocean's depths.

For example, if 50 students were packed into a closet, their volume would be the size of the closet and the pressure would be great. If they were then released into a large room, the pressure would decrease because everyone would spread out, which would increase the volume of the group. Gas molecules behave similarly.

Pressure is defined as the amount of force exerted on an object per unit of area. When a greater pressure is applied to a gas, the molecules of the gas are squeezed closer together, reducing the gas's volume. Boyle discovered that if the pressure on a gas was doubled, the volume of the gas decreased by one half. And the reverse is also true. If the pressure decreased by one half, the volume doubled. Gas and pressure have an inverse relationship. Boyle's law explains why deep-sea fish die when they are brought to the surface. At sea level, the average air pressure is 14.5 pounds per square inch (psi). But for every 10 meters of depth in the water, the pressure increases by 14.5 psi. If a fish lives almost 1,600 meters below sea level, its normal pressure is close to 2,330 psi. When a deep-sea fish is brought to the surface, the pressure decrease is so great that the gases inside the fish expand enough to burst its cells and swim bladder.

TRY THIS

Ballooning

Take three identical balloons and inflate them to be equal in size. Draw a face of the same size on each balloon. Boil a pan of water and place a screen over the pan. Place one balloon on the screen, place the second balloon in the freezer, and leave the third balloon at room temperature. Wait five minutes, and observe the results. How did the temperature affect the volume? Why?

LAWS & PRINCIPLES

Boyle's Law

$$P_1V_1 = P_2V_2$$

Boyle's law states that the starting pressure of a gas (P_1) times the starting volume (V_1) is equal to the final pressure (P_2) times the final volume (V_2). This equation demonstrates an inverse relationship between pressure and volume. Pressure is measured in kilopascals, a metric unit comparable to the customary unit of pounds per square inch. As pressure increases, volume decreases. For example: If $P_1 = 50$ kPa, $V_1 = 2$ L, and $P_2 = 200$ kPa, you can solve for the final volume of the gas (V_2).

$50\text{ kPa} \times 2\text{ L} = 200\text{ kPa} \times V_2$ $\qquad \frac{50\text{ }\cancel{\text{kPa}} \times 2\text{ L}}{200\text{ }\cancel{\text{kPa}}} = \frac{\cancel{200\text{ kPa}} \times V_2}{\cancel{200\text{ kPa}}}$ $\qquad \frac{100\text{ L}}{200} = V_2$ $\qquad 0.5\text{ L} = V_2$

$50\text{ kPa} \times 2\text{ L} = 200\text{ kPa} \times 0.5\text{ L}$. Compare initial pressure (50 kPa) with final pressure (200 kPa). The pressure went up. Compare initial volume (2 L) with final volume (0.5 L). The volume went down.

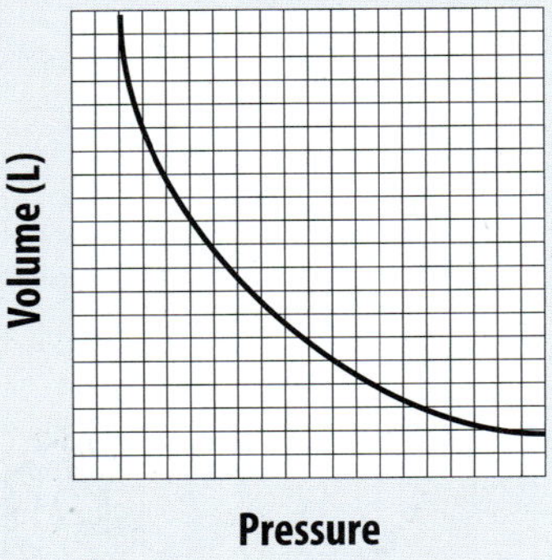

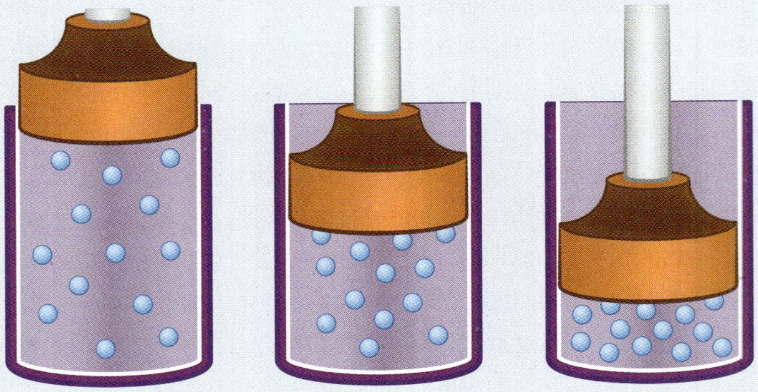

LAWS & PRINCIPLES

Charles's Law

$$\frac{V_1}{T_1} = \frac{V_2}{T_2}$$

Charles's law states that the volume of a gas is directly proportional to its temperature if the pressure and the number of particles are constant. For example, if the starting volume (V_1) is 2 mL, starting temperature (T_1) is 73 K, and the final temperature (T_2) is 292 K, you can solve for the final volume (V_2). The relationship between temperature and volume is direct. As the temperature increases, the volume increases.

$\frac{2\text{ mL}}{73\text{ K}} = \frac{V_2}{292\text{ K}}$ $\qquad 73\text{ K} \times V_2 = 2\text{ mL} \times 292\text{ K}$ $\qquad \frac{73\text{ }\cancel{\text{K}} \times V_2}{73\text{ }\cancel{\text{K}}} = \frac{2\text{ mL} \times 292\text{ }\cancel{\text{K}}}{73\text{ }\cancel{\text{K}}}$ $\qquad V_2 = \frac{584\text{ mL}}{73}$ $\qquad V_2 = 8\text{ mL}$

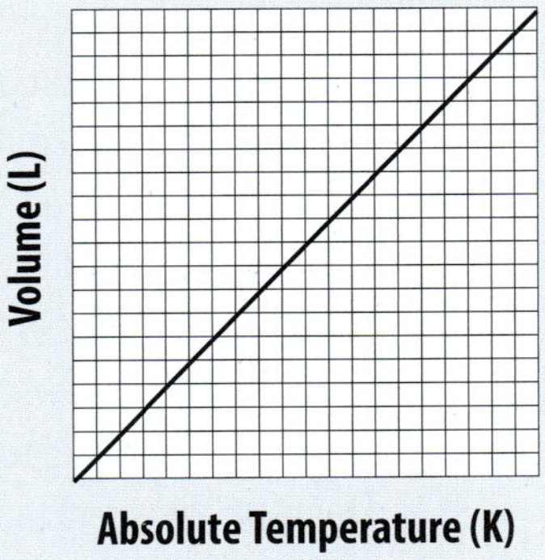

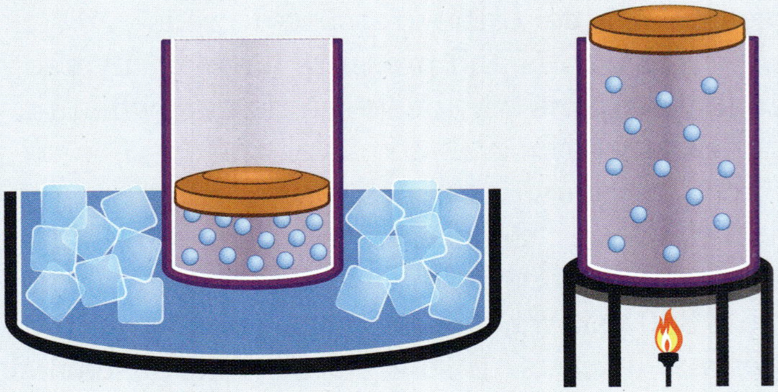

BIOGRAPHY

Jacques Charles

Jacques Charles was a French physicist, mathematician, and inventor who was fascinated by science, particularly the expansion of heated gases. He was born in 1746 and worked as a clerk before pursuing scientific study. Charles developed inventions such as a hydrometer, a device used to measure the characteristics of a liquid. He also improved other scientists' inventions. In 1783, he and brothers Nicolas and Anne-Jean Roberts built one of the first hydrogen balloons. He and Nicolas Roberts were the first people to ever fly in a hydrogen balloon, and when the balloon landed in a field in a small village outside Paris, the residents were frightened by what they thought was a monster descending from the sky. In 1795, Charles was elected to the Academie des Sciences before becoming a physics professor. In 1802, French philosopher Joseph Louis Gay-Lussac introduced Charles's law, which describes how gases expand when they are heated.

About 100 years after Boyle discovered the relationship between the pressure and volume of gases, Jacques Charles discovered another property. When heat is added to a gas, the molecules gain energy and move faster. They move farther apart, which increases the volume of the gas. The reverse is true when the temperature of the gas is lowered. Jacques Charles's discoveries became known as Charles's law, which states that the volume of a gas at a constant pressure increases as its temperature increases, and decreases as its temperature decreases.

Have you ever wondered why the plunger on a turkey thermometer pops up when the turkey is done? This can be explained by the relationship between gas pressure and temperature. When the temperature inside the turkey climbs, the particles of the air trapped under the plunger move faster and farther apart, increasing the volume of air. Finally, the pressure of the increasing volume of air makes the plunger pop up. Charles's law also explains why a balloon inflated indoors will shrink if you take it outside on a cold day, or why a slightly under-inflated raft will swell up when you place it on hot sand.

LESSON REVIEW

1. What are the four properties of gases?
2. In the context of kinetic theory, compare solids, liquids, and gases.
3. What is Boyle's law?
4. Describe Charles's law.
5. What would happen to a balloon that you inflated to capacity in a cold room and brought into a warm room? Why?

1.2.6 *Phase Changes*

OBJECTIVES

- Infer the behavior of molecules during a phase change.
- Explain what is necessary for a phase change to occur.
- Identify characteristic properties associated with phase changes.
- Define the law of conservation of mass.

VOCABULARY

- **condensation** the changing of a substance from a gas to a liquid
- **deposition** the changing of a substance from a gas directly to a solid
- **evaporation** the slow vaporization at the surface of a liquid
- **phase change** the changing of a substance from one phase to another
- **physical change** a change in a substance's physical properties that does not change the type of matter in the substance
- **sublimation** the changing of a substance from a solid directly to a gas
- **vaporization** the changing of a substance from a liquid to a gas

Every day you see changes in matter. You leave ice cubes on the counter, and they melt. You step on a pencil, and it breaks. You break spaghetti so you can cook it. These are examples of physical changes. A **physical change** is a change in a substance's physical properties that does not change the chemical composition of the substance. Tearing, crushing, and breaking objects cause physical changes. Only physical properties are altered during physical changes; no new substances are produced. For example, water is made of two atoms of hydrogen connected to one atom of oxygen; freezing water into ice does not change its chemical makeup. Each frozen water molecule is still composed of two hydrogen atoms and one oxygen atom.

An important type of physical change is a **phase change**, which is the changing of a substance from one phase, or state of matter, to another. Water freezing into ice cubes, ice cream melting, steam rising from boiling soup, water vapor condensing on a soda can, and snow turning into slush are all examples of phase changes.

How do substances change from one phase to another? They change through the addition or subtraction of energy. Change in pressure is also a factor but will not be addressed in this lesson. A solid ice cube melts when it is taken out of the freezer and placed in a cup on the counter. The heat from the room is added to the ice cube. A pan of water turns to steam when it is placed on the stove and heated up. Give the substance enough energy and the molecules gain sufficient kinetic energy to break free of one another and make a phase change. Or, remove energy from

Ice melting

a substance, and the molecules can no longer move as much as they did. They become subject to cohesion.

Molecular Differences of States of Matter

Melting occurs when a substance changes from a solid to a liquid. The temperature at which this happens for a particular substance is called *the melting point*. The melting point of water, for example, is 0°C. As the particles of a substance gain more energy, they vibrate more rapidly and break the forces keeping them in a rigid position. With their increase in energy, the particles of melted substances can roll over one another, allowing them to flow. The amount of heat required to melt a certain amount of a particular substance is called *the heat of fusion*. Substances have unique melting points that are used for identification.

Freezing is the opposite of melting and occurs when a substance changes from a liquid to a solid. In freezing, substances lose energy. Their particles become fixed in place and can no longer roll over each other. The molecules in a glass of water have a certain amount of energy. When the water is placed in the freezer, the energy is lost and the molecules move slower and slower until they no longer have enough energy or motion to overtake cohesion. The water becomes frozen. The temperature at which a substance freezes is called *the freezing point*. The freezing point and the melting point for a particular substance are usually the same. The freezing point for water is 0°C, and the freezing point for mercury is –39°C. The melting point is a characteristic property that is used to identify substances.

Ice forming

TRY THIS

Breathing Water

Did you know that when you exhale you breathe out water vapor? Exhale. Do you see any water vapor? Why? Hold a small mirror in front of your mouth and exhale hard. What do you see on the mirror? Why?

Another phase change is **vaporization**, which occurs when energy is added to a liquid and it changes to a gas. When water is boiled, the molecules are given so much energy that they completely break free of their hydrogen bonds and turn to steam. Fast vaporization throughout a liquid is called *boiling*. The boiling point, a characteristic property, is the temperature at which the rising vapor bubbles in a liquid burst at the surface, and it is the highest temperature at which that substance can exist as a liquid.

It takes a lot of energy to boil a liquid because the molecules have to be separated from other molecules to make a gas. The amount of heat required to vaporize a certain amount of a substance is called *the heat of vaporization*. The heat of vaporization for a substance is much higher than the heat of fusion. Liquid molecules need more energy to break free and move about as individual gas molecules. Adding more heat to a liquid will increase the total amount of energy available to its molecules, allowing more liquid molecules to overcome the attractive forces holding them together and break free as gas molecules.

Vaporization can also happen slowly, at the surface of a liquid. This is called **evaporation**. Evaporation can occur at room temperature. An aquarium losing water over a period of several days is an example of evaporation.

Condensation is the opposite of vaporization and happens when a gas changes to a liquid. Have you ever noticed dew forming on grass or on the car early in the morning? At night, the air

Water boiling

FYI

Substance	Melting Point/Freezing Point (°C)	Boiling Point (°C)
hydrogen	–259	–253
oxygen	–218	–183
mercury	–78	–33
water	0	100
salt	801	357
copper	1,085	2,562
iron	1,538	2,861

becomes cooler and water vapor in the air condenses on surfaces such as plants, vehicles, and spider webs, forming dew. When gas molecules are cooled, they lose energy and stick together to form a liquid.

Gases phase change into plasma through ionization when charged particles travel through gas. Fluorescent and neon bulbs are filled with gas that is ionized to produce light. The reverse process, changing from plasma to gas, is called *deionization* or *recombination.*

Water condensing

Types of Phase Transition

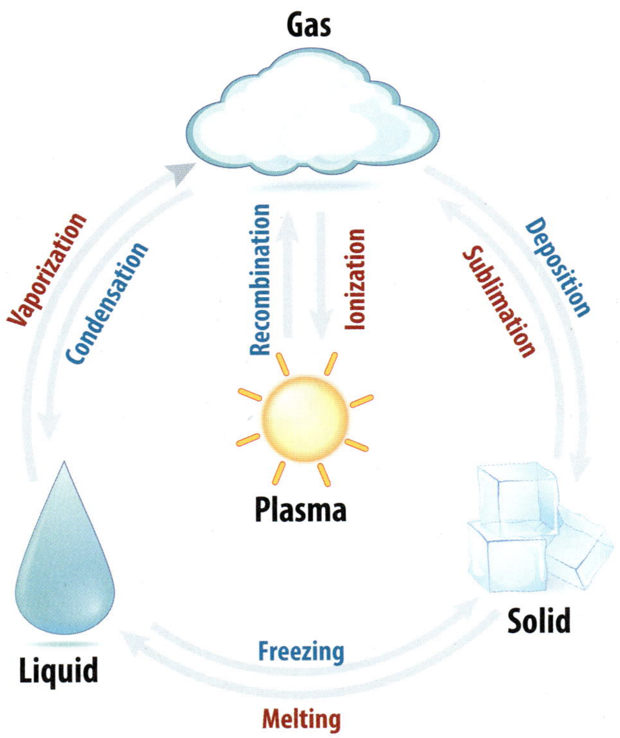

The final two phase changes happen when a substance skips a phase. For example, **sublimation** happens when a substance changes directly from a solid to a gas. When dry ice, or frozen carbon dioxide, is exposed to warm air, it changes to a vapor without becoming a liquid first. That is why it is often used on-stage for special effects such as fog. It is extremely cold at a temperature of –78.5°C. When it makes contact with the air, two things happen. The solid carbon dioxide sublimates to vapor and the cold vapor causes the water vapor in the air to condense. The result is fog that you see. Sublimation is the reason that ice cubes sometimes shrink in the freezer. **Deposition** is the reverse process by which water vapor freezes into snow.

Despite a phase change, or any other physical change, the amount of matter before and after never changes. When a substance undergoes a physical change, atoms are neither created nor destroyed. If an ice cube melts into a puddle, both phases have the same number of molecules. The total mass of the substance before the change always equals the total mass of the substance after the change. This law is called *the law of conservation of mass*. How might the world be different if mass could be created or destroyed?

LESSON REVIEW

1. Describe what happens to molecules when they freeze.
2. What is needed for water vapor to condense into a liquid?
3. What happens to the molecules of a substance when it melts?
4. If you have two substances that appear identical, how can melting them help you determine what they are?
5. If a volume of water freezes and is less dense than when it was water, were any molecules destroyed in the process? Why?

Ripping paper or freezing water changes the physical properties of the paper and water, but they do not change the substances themselves. The paper is still paper, and the water is still water. These are physical changes.

On the other hand, a **chemical change** is a change that occurs when a substance changes into an entirely different substance with unique properties. A chemical change always results in a change in physical properties, such as color or smell, as well as a change in chemical properties. A chemical property is a characteristic defined by the way a substance reacts with other materials. Some chemical properties are flammability (the ability to burn) and the ability to combine with oxygen to form rust.

One common chemical change is **combustion**, the burning of substances. The ashes of a substance after it burns have different chemical properties than the substance had before it burned. Another common chemical change is **oxidation**. A substance oxidizes when it combines with oxygen to form a new substance. Iron combines with oxygen to form rust. Methane combines with oxygen in the air to make carbon dioxide and water vapor. Other chemical changes include silver reacting with sulfur to produce silver sulfide, the black material known as *tarnish*; an antacid, calcium hydroxide, reacts with stomach acid, or hydrochloric acid, to make calcium chloride and water. Each of these changes results in a substance unique to the original components.

OBJECTIVES

- Contrast chemical changes with physical changes.
- Describe types of chemical changes.
- Identify the evidence of chemical changes.

VOCABULARY

- **chemical change** a change that occurs when a substance changes into an entirely different substance with unique properties
- **combustion** the burning of substances
- **oxidation** a chemical change in which a substance combines with oxygen

FYI

Fall Leaf Color

Have you ever wondered what makes leaves turn color in fall? Actually, the brilliant reds and yellows of leaves in autumn are the result of chemical changes. Chlorophyll is the green pigment in leaves; it is very unstable, so plants have to constantly synthesize it. Chlorophyll synthesis requires lots of sunlight and warm temperatures, which is why leaves are so green in summer. The other pigments in plant leaves, carotene and anthocyanins, behave differently than chlorophyll. As temperatures become cooler and daylight decreases in fall, leaves reduce chlorophyll production. Carotene, which is stable, shows up as a yellow color when chlorophyll pigments fade. Anthocyanins form as low temperatures trigger a rise in the sugar content of cell sap, and these pigments result in red leaves.

TRY THIS

Chemical Changes

Unlike physical changes, chemical changes produce new substances. Combine some of the following substances and record what changes you observe. Mix iodine with liquid starch. Mix iodine with isopropyl alcohol. Burn a piece of toast. Add vinegar to milk. Heat sugar in a pan on a hot plate. Keep an iron nail on a damp cloth for several days. What changes did you see that suggest a new substance was produced?

God designed all of the elements to interact with each other in various, predictable ways to form different kinds of matter. Without seeing the molecules regroup, how do you know a chemical change has taken place? In addition to a new substance forming, chemical changes create gases, new colors, odors, and temperature changes. For example, leaves change color in autumn, fruit ripens, milk spoils, wood burns, and metal rusts. Each of these events demonstrates a chemical change.

Since God created a world of order, certain elements always interact with each other in a certain way. Many elements in creation interact to maintain the life that God created. Photosynthesis requires carbon dioxide and water to produce energy and oxygen. Cellular respiration uses oxygen for a combustion reaction to produce energy, water, and carbon dioxide. Because of the predictable ways that elements interact, scientists study chemical changes and use their knowledge to produce many different substances, such as medicine, plastic, nylon, perfume, air freshener, and soap. Look around you. Nearly everything at school and at your home has been made or treated with chemicals made by chemical reactions.

LESSON REVIEW

1. What is the difference between a physical change and a chemical change?
2. What is combustion?
3. How does rust form?
4. List things that happen when a chemical change takes place.
5. What everyday substances are made as a result of chemical changes?

TRY THIS

Colorful Compounds
Put on safety goggles. Fill two beakers with 200 mL of copper(II) nitrate. Add several drops of ammonia water to one of the beakers of copper(II) nitrate solution. Compare the colors of the two copper nitrate solutions. Fill two other beakers with 200 mL of cobalt(II) nitrate solution. Add several drops of ammonia water to one of the beakers of cobalt nitrate solution. Compare the colors of the two cobalt nitrate solutions. What changes did you see? Why did the solutions change color when ammonia water was added?

CAREER

Flavorist
When most people think about careers in chemistry, they might not think about a job in the food industry. However, the field of food science offers chemists the chance to improve food for consumers. For example, flavorists are chemists who use their lab skills to influence how various foods taste. Flavorists work for food production companies and for flavor producers, which create flavors for use in food. Most flavorists have a strong background in organic chemistry, and many pursue several years of additional training with a master flavorist.

1.3.1 *Early Models of the Atom*

OBJECTIVES

- Analyze methods used to determine the structure of an atom.
- Compare the five main atomic theories.
- Describe the current electron configuration around the nucleus.

Suppose you took a piece of copper and cut it in half. Then, you proceeded to divide one section in half multiple times. Imagine being able to segment a piece of copper so small that it could not be divided further and still be copper. That tiny piece would be one atom of copper. An atom is the smallest unit into which an element can be divided and still maintain its properties. This definition of an atom tends to produce many questions. Is there anything smaller than an atom? How do different atoms interact? One common question that many people ask is "What makes up an atom?"

For thousands of years, scientists have been attempting to discover the structure of an atom. This type of research has posed interesting challenges because atoms are extremely small. They cannot be seen with the unaided eye. To visualize an atom's size, imagine a hydrogen atom magnified to the same size as an apple; the apple would then need to be the same size as planet Earth.

To be able to replicate such a small part of God's creation, scientists have developed models to represent what they think an atom looks like. Models depict objects, ideas, systems, or processes that describe a phenomenon that is difficult to see or experience directly. Scientists also use models to better visualize very large objects, such as the solar system. Models are flexible tools that can be changed, updated, or discarded. Over the years, more information has been gathered through advancing technologies, so the atomic models have changed to reflect current findings.

The first concept of the atom was developed more than 2,400 years ago by the Greek philosopher Democritus, who proposed that all matter was composed of invisible particles. He named those particles *atomos*, which means "indivisible" or "not able to be cut." Democritus believed that atoms were hard particles of different shapes and sizes and that every substance was made up of its own type of atom. Other Greek scholars during that time, such as Aristotle, disagreed with Democritus. They believed there was no limit to the number of times matter could be divided. Aristotle's opinion became the prevailing thought at the time. As a result, the idea of the atom was laid aside for more than 2,000 years.

If an atom was the size of an apple, an apple would be the size of Earth.

During the 1800s, John Dalton, an English chemist and meteorologist, became interested in Democritus's concept of the atom. Dalton's observations of the behavior of air led to his experimentation with various gases. Some of these experiments revealed that gases act like individual particles that combine with each other in fixed-mass proportions. Dalton concluded that these particles were the atoms proposed by Democritus. He thought that atoms resembled solid spheres. Dalton's work eventually led to the acceptance of the idea of atoms. In 1803, Dalton summarized the results of his experiments and observations in the first atomic theory. The purpose of an atomic theory is to explain the design and behavior of atoms. Dalton's atomic theory can be summarized in the following four key ideas:

Dalton's solid sphere atomic model

1. All matter is made of atoms—indivisible and indestructible building blocks.
2. All atoms of a given element have the same mass and properties. Atoms of different elements have distinct masses and properties.
3. Compounds are combinations of two or more different types of atoms in fixed proportions.
4. A chemical reaction is a rearrangement of atoms.

Cathode Ray Tube

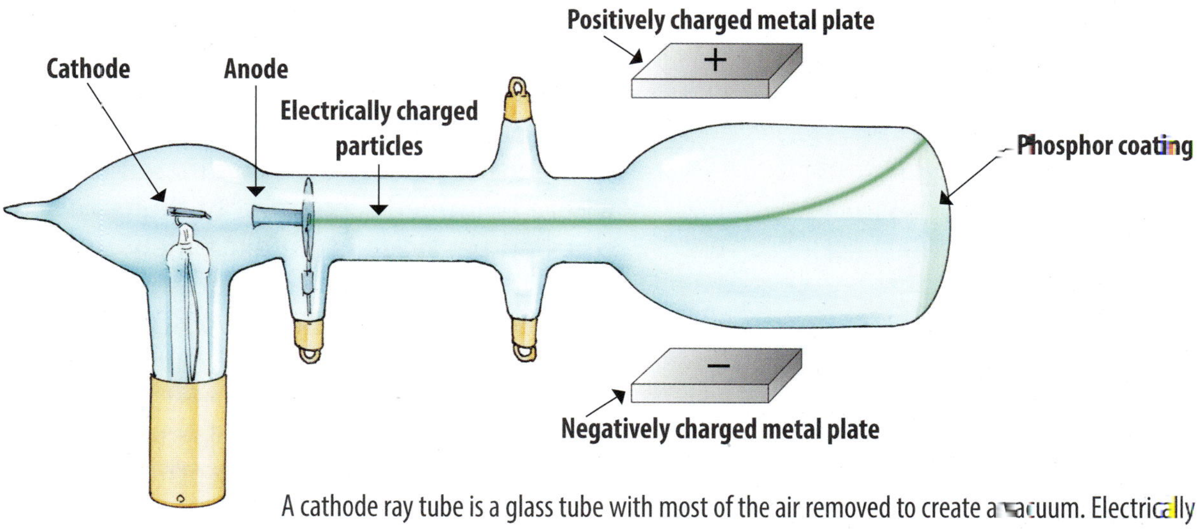

A cathode ray tube is a glass tube with most of the air removed to create a vacuum. Electrically charged particles pass through two conductors in the tube and form a beam that strikes the opposite end of the tube. The phosphor coating on that end of the tube glows when the particles collide with the surface.

The raisins represent electrons in Thomson's plum pudding model.

In 1897, British chemist J. J. Thomson made a discovery that would change some aspects of Dalton's atomic theory. Thomson made his discovery while experimenting with cathode ray tubes. Thomson decided to place a positively charged metal plate and a negatively charged metal plate on either side of a cathode tube to determine if a beam of electrically charged particles had a positive or negative charge. When the beam passed through the metal plates, the particles were deflected away from the negatively charged plate toward the positively charged plate. Thomson concluded that the particles in the beam had a negative charge because like charges repel and opposite charges attract. He was also able to determine that the particles had a much smaller mass than any atom previously measured. These two discoveries debunked Dalton's idea that atoms are indivisible. Thomson explained that atoms are made up of smaller particles. He named the negatively charged particles *electrons*. Thomson knew that the negative charge of the electron must be balanced with a positive charge because atoms are neutral, meaning they have no charge. Thomson created a model of the atom that consisted of a sphere of positive charges with negative particles evenly distributed. This model is often referred to as *the plum pudding model* because the electrons are suspended throughout the atom like raisins in a plum pudding.

Ernest Rutherford

Although the work of J. J. Thomson was very beneficial, there was still much to learn as scientists continued to gather more information regarding the structure of the atom. In 1909, British physicist Ernest Rutherford made a discovery that added to Thomson's understanding of the atom. Rutherford conducted an experiment where he aimed a beam of high-speed, positively charged particles at an extremely thin sheet of gold foil that was 0.00004 cm thick. He assumed that most of the particles would travel straight through the gold foil. Many particles did travel straight through. However, to Rutherford's surprise, some of the particles were deflected and a few bounced straight back. Because a majority of the particles passed straight through the foil, Rutherford concluded that most of the atom was empty space. Rutherford explained that the few particles that bounced back were deflected by a tiny, hard mass in the center of each atom. He had discovered the atomic nucleus, which he also suggested contained most of the atom's mass. Rutherford developed the nuclear model of the atom with negatively charged electrons surrounding a positively charged nucleus.

Niels Bohr

Rutherford's nuclear model of the atom raised some questions in the science community, such as the exact location of electrons, and why negatively charged electrons do not collapse into the positively charged nucleus. Danish scientist Niels Bohr sought to answer these questions. He theorized that electrons travel around the nucleus in specific orbits, or shells. According to Bohr's theory, electrons cannot reside between shells. Bohr explained that when an atom absorbs energy, electrons jump from a lower shell to a higher shell. The result is a less-stable atom. Electrons quickly fall back down to a lower shell to regain stability. When this happens, energy is often released in the form of light. In 1913, Bohr's shell model looked very similar to a solar system model with the planets orbiting the sun in distinct patterns.

More scientific evidence revealed that Bohr's model did not accurately reflect the true structure of the atom. In 1926, Austrian physicist Erwin Schrödinger developed a new model of the atom called *the quantum mechanical model*, also known

FYI

Nanotechnology

Nanotechnology is a fast-growing field of study. Nano scientists seek to operate and manufacture materials and devices that are the size of atoms or small groups of atoms. Materials at the nanoscale are unique because they often have distinct physical and chemical properties. For example, gold that is normally nonreactive is extremely reactive at the nanoscale. Scientists from many fields—physics, biology, chemistry, materials science, and engineering—want to utilize these unique features to create materials that require less energy to manufacture, produce less waste, and ultimately improve health, safety, conservation efforts, and quality of life.

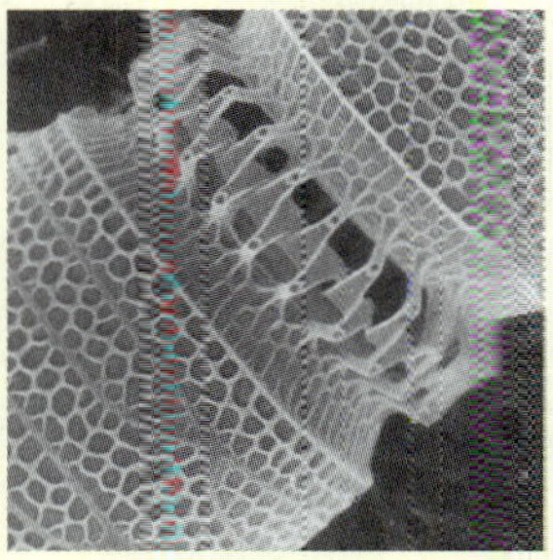

Nanotube

Many of the materials that scientists have manufactured on the nanoscale were inspired by God's handiwork. They demonstrate His extreme efficiency, precision, and power. Some examples include enzymes and catalysts, DNA, and linear and rotary molecular motors that carry out muscle contractions. Scientists have created "nanowhiskers" that are about 10 nm long and are hooklike. They prevent stains on natural and synthetic fibers. Some nanocrystals create invisible sunscreens that block ultraviolet light. Others are incorporated into bandages to prevent infection and kill harmful bacteria. The medical field has particularly benefited from the field of nanotechnology. Some options include drugs that can be delivered to a specific location and released at predetermined times for optimal treatment. Other scientists are working on retinal implants and neural probes that can be implanted into brain tissue to activate and control motor functions.

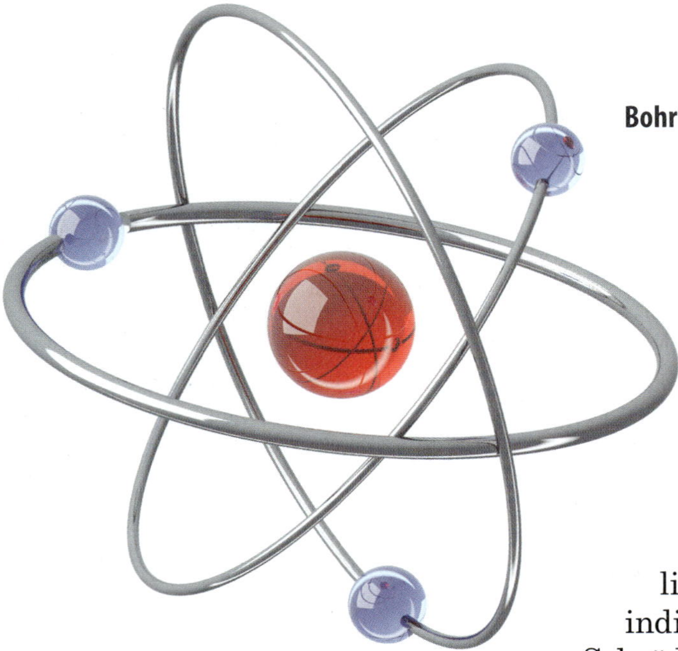

Bohr's quantized shell model

as *the electron cloud model*. In this model, electrons exist in electron cloud regions around the nucleus rather than finite orbits. These electron cloud regions help scientists predict where a particular electron is most likely to be found.

Consider a population density map. Each dot on the map represents 1–100 people. The locations that appear brighter have more dots and represent where most people live. Dense regions in the electron cloud model indicate where electrons spend most of their time.

Schrödinger's model is the one used today. However, as more information is gathered and more experiments are conducted, new models will be constructed in the future.

CHALLENGE

Micro Art

God's handiwork is often appreciated at the macroscopic level. Vast oceans, tall mountain peaks, and giant elephants all point to a Creator. But, take a closer look, and you will find even more astounding discoveries. Science and technology have paved the way for individuals to explore this microscopic world. Some people have used their talents to create microscopic artistic expressions. For example, some have recreated the Last Supper scene in the eye of a needle and inscribed The Lord's Prayer on the head of a pin. Using a microscope and a pin or other shaping device, create your initials out of a thin layer of silt grains on a slide with the dimensions of 3 mm x 3 mm or less.

LESSON REVIEW

1. What did the cathode ray experiment indicate about atoms?
2. Why did a majority of the positively charged particles pass straight through the gold foil in Rutherford's experiment?
3. Why do electrons quickly fall to lower shells?
4. Compare the five main atomic theories.
5. Describe the electron configuration around the nucleus in the current atomic theory.

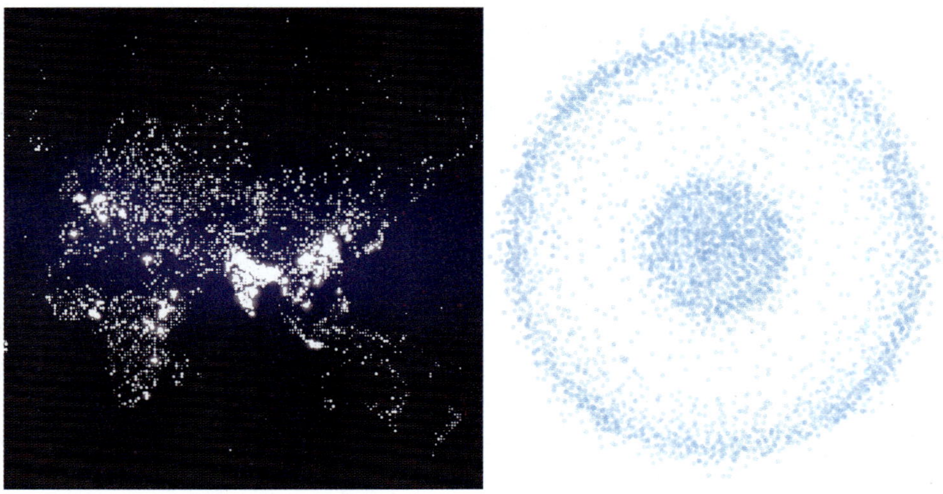

Similar to a population density map, dense regions in the electron cloud model represent areas where electrons frequent.

Structure of the Atom 1.3.2

During the early 1900s, scientists knew that an atom had a comparatively massive nucleus that contained protons and was surrounded by electrons. However, they also knew that the number of protons in a particular nucleus could not account for the observed mass. They concluded there must be something else in an atom. In 1932, Englishman James Chadwick bombarded beryllium nuclei with alpha particles—positively charged particles that resemble helium nuclei—in an experiment. The beryllium atoms emitted a type of radiation that had not been previously observed. Chadwick deduced that these subatomic alpha particles had about the same mass as a proton, but had no electric charge. He also concluded that the particles were a main component of the atom's nucleus and called them **neutrons**. The neutrons formed a strong attractive force with the protons that was powerful enough to overcome the repulsive forces between the protons and keep the nucleus of an atom together.

With Chadwick's discovery, a picture of an atom with all its major parts was formed. At the center of the atom is the **nucleus**, which contains protons and neutrons and is surrounded by an electron cloud. A **proton** is a positively charged subatomic particle in the nucleus of every atom. The charge of a single proton is +1. Each kind of atom has a specific number of protons. You can identify an element if you know the number of protons contained in the nucleus. For example, an atom with one proton in its nucleus is always a hydrogen atom. A shorthand method for writing hydrogen is $_1H$. Helium atoms always contain two protons in the nucleus, so a shorthand method for writing helium is $_2He$. The number of protons in one atom of an element is called the **atomic number**.

All subatomic particles contain very little matter. The amount of matter in an atom that can be measured is contributed almost entirely by the nucleus, which is about 99.9% of an atom's mass. Electrons are extremely small, so their contribution to the total

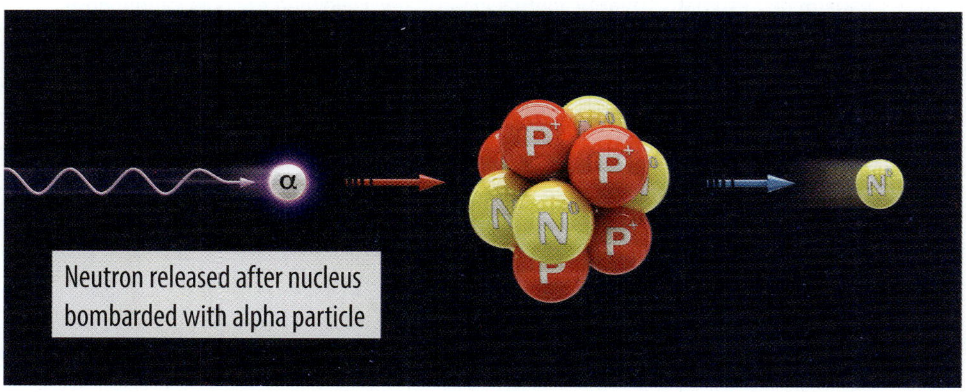

Neutron released after nucleus bombarded with alpha particle

OBJECTIVES

- Determine the number of particles in a neutral atom using the atomic number and mass number.
- Identify particles in isotopes and calculate the atomic weight.

VOCABULARY

- **atomic number** the number of protons in one atom of an element
- **atomic weight** the weighted average mass of the isotopes of an element
- **electron** a negatively charged subatomic particle outside the nucleus of an atom
- **isotope** an atom of an element that has the same atomic number but a different number of neutrons and a different mass number
- **mass number** the sum of the number of protons and neutrons in an atom's nucleus
- **neutron** a subatomic particle that has no charge in the nucleus of an atom
- **nucleus** the center of an atom that contains protons and neutrons and is surrounded by an electron cloud
- **proton** a positively charged subatomic particle in the nucleus of every atom

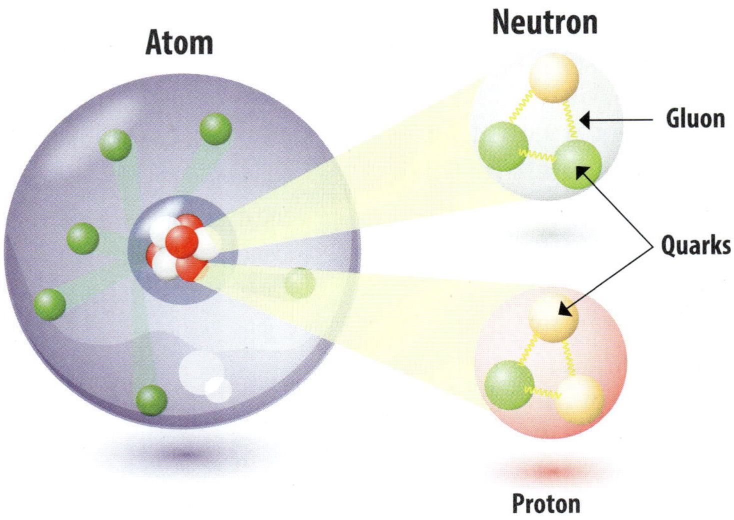

mass of an atom is negligible. Scientists developed the atomic mass unit (amu) to more easily calculate the mass of the particles in atoms. If the mass of a proton was measured using the metric system, it would be 1.6726×10^{-24} g. This quantity is equivalent to 1 amu. Since the mass of a neutron is almost identical to the mass of a proton, the mass of a neutron is also 1 amu. The sum of the number of protons and neutrons in an atom's nucleus is called the **mass number**. Nitrogen has 7 protons and 7 neutrons, so the mass number of nitrogen is 14.

Even though all atoms of one kind of element have the same number of protons, they may have a different number of neutrons. For example, naturally occurring titanium atoms have 22 protons and 26 neutrons. The mass number of titanium is 48. This form of titanium can be abbreviated as follows: Ti-48 or ^{48}Ti. However, there are at least 11 other known varieties of

FYI

Quarks

Protons have a positive charge and neutrons have no charge. How do they stay together in the nucleus of an atom? These two subatomic particles are made up of even smaller particles called *quarks* that are held together by the strong force that is transferred in the "glue" of atoms called *gluons*. There are six different varieties, or flavors, of quarks—up, down, charm, strange, top, and bottom. These flavors differ in charge and mass. Many scientists think that protons and neutrons each contain three quarks. As science progresses, even smaller elementary particles may be discovered or a completely new theory may be proposed.

BIOGRAPHY

Satyendra Nath Bose

Indian physicist, Satyendra Nath Bose, was born in Calcutta on January 1, 1894. He received a Master of Science degree in mixed mathematics from Calcutta University in 1915 and later became a research scientist at the University of Calcutta where he studied the theory of relativity. Bose wrote many scientific articles, including a paper regarding his derivation of Planck's quantum radiation law without using classical physics. This paper caught the attention of Albert Einstein who recognized the importance of such work and submitted Bose's paper to a prestigious scientific journal in Germany. Following the publication, Bose worked for two years with Einstein, Marie Curie, and other scientists in crystallography and X-ray laboratories. Bose and Einstein's collaboration led to the prediction of one of the two basic classes of subatomic particles, known as the Bose-Einstein Condensate. This class of particles consists of bosons—subatomic particles with integral spins. Unfortunately, Satyendra Bose did not receive the credit that he was due for such a remarkable discovery that forever changed people's understanding of the fundamental structure of matter. Many scientists have been awarded Nobel prizes for their research as it relates to the Bose-Einstein Condensate. Bose, the man whom the particle is named after, never received a Nobel Prize for his work on particle statistics.

FYI

Subatomic Particle	Charge	Atomic Mass (g)	Atomic Mass (amu)
proton	+1	1.6726×10^{-24}	1.007
neutron	0	1.6749×10^{-24}	1.009
electron	–1	9.1093×10^{-28}	0.0005

titanium. All forms of titanium have 22 protons, but each variety has a different number of neutrons and a different mass number. For example, Ti-46 has 24 neutrons and Ti-51 has 29 neutrons. An atom of an element that has the same atomic number but a different number of neutrons and a different mass number is called an **isotope**. To understand isotopes, picture a Red Delicious apple. Most Red Delicious apples look much the same in shape and color. But if you cut a few of these apples into pieces, you see differences inside. The apples may have six seeds, four seeds, or even two. Even though they have different numbers of seeds, they are all still Red Delicious apples. Similarly, even though isotopes of an element have different numbers of neutrons, they are all the same element.

Most elements have one or more isotopes. Some isotopes are more prevalent than others. The listed weight of an atom is typically not a whole number. It is often in decimal form. The weighted average mass of the isotopes of an element is called **atomic weight**. For example, the weight of chlorine is 35.453 amu. Chlorine exists in two isotopic forms: Cl-35 accounts for 75.78% of the atoms, and Cl-37 accounts for 24.22% of the atoms.

Isotopes of Hydrogen

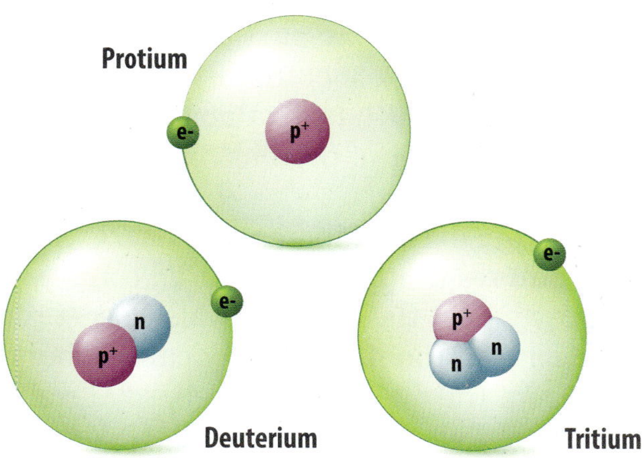

SCIENCE IN ACTION

Calculating Atomic Weight

Most atoms have one or more isotopes. Some of these isotopes have been discovered in the laboratory even though they may only exist for a short period of time. One isotope of seaborgium has a half-life of only 800 milliseconds. Scientists measure the atomic mass of each isotope and the percent abundance of each isotope to determine the atomic weight that is listed on the periodic table. The atomic weight is the weighted average mass of all of the isotopes. Lithium has two isotopes. What is the atomic weight of lithium?

Step 1: Determine the percent abundance for each isotope.

Li-6: 7.59%

Li-7: 92.41%

Step 2: Multiply the atomic mass of each isotope by the percent abundance. Remember to convert each percent to a decimal when completing this mathematical operation.

6 g × 0.0759 = 0.4554 g

7 g × 0.9241 = 6.4687 g

Step 3: Add each mass to determine the atomic weight.

0.4554 g + 6.4587 g = 6.9141 g

Step 4: Check solution to make sure that the answer makes sense. The atomic weight should be close in value to the atomic mass of the most abundant isotope.

The calculated atomic weight of 6.9141 g is close to 7 g, which is the atomic mass of the most prevalent isotope.

The atomic weight of chlorine is calculated as follows: (35 amu × 0.7578) + (37 amu × 0.2422) = 35.453 amu. Notice that the atomic weight of chlorine is closer in value to the more common isotope.

An electron is another particle of the atom. Although electrons do not identify the atom or contribute to the atomic mass, they play an important role when different atoms interact. An **electron** is a negatively charged subatomic particle outside the nucleus of an atom. The charge of a single electron is equal but opposite to that of a proton, which means it has a value of –1. A neutrally charged atom has an equal number of protons and electrons.

Electrons in an atom exist in increasing energy levels or shells. The electrons that are the most stable and contain the least amount of energy are in the energy levels closest to the nucleus. Electrons that contain the most energy can be found in energy levels farthest from the nucleus. Each energy level can hold a maximum number of electrons. The first energy level can hold

TRY THIS

Make an Atomic Model

Design and build a creative three-dimensional atomic model of an element with an atomic number between 3 and 20. Each subatomic particle needs to be made of a different material, and there should be distinct energy levels.

2 electrons, the second energy level can hold 8 electrons, the third energy level can hold 18 electrons, and the fourth energy level can hold 32 electrons. Chemical properties of an atom depend largely on the outermost electrons, which are called *valence electrons*.

The size of an atom is also determined by its number of electrons and energy levels. Atoms are so small that it may seem impossible to imagine their size. The period at the end of this sentence could hold over 1,000 atoms. Atoms range in size from 1×10^{-8} cm to 5×10^{-8} cm. These numbers can be somewhat cumbersome when performing mathematical operations. To make the numbers simpler, scientists utilize the ångström unit. One ångström is equal to 10^{-10} m, or one ten-billionth of a meter. Most atoms have a radius that measures between one and two ångströms. To provide a comparison, imagine the nucleus of an atom being magnified to the size of a grape. The electrons that surround the grape would reach a distance the size of a football stadium.

If the nucleus of an atom was magnified to the size of the period shown above, the closest electron would extend to the distance of approximately 19 *Physical Science* textbooks set side-by-side (about 4.07 meters away).

LESSON REVIEW

1. Describe the structure of a neutral atom and the properties of each component.
2. How do the number of protons and electrons compare in an electrically neutral atom?
3. A sample of iodine has an atomic number of 53, a mass number of 127, and 54 electrons. How many protons and neutrons does iodine contain? Is this iodine neutral? Why?
4. Boron-10 accounts for 19.9% of all boron atoms. Boron-11 accounts for 80.1% of all boron atoms. Calculate the atomic weight of boron.

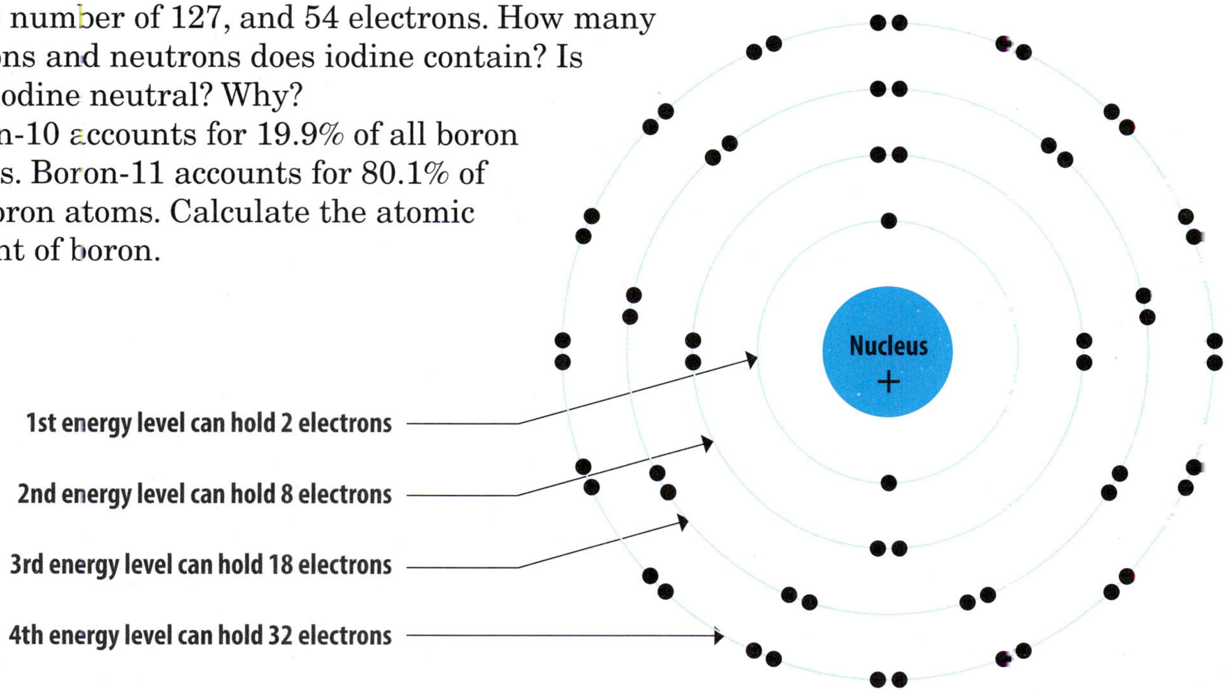

Stable atoms have full valence shells.

1.3.3 *Periodic Table of the Elements*

OBJECTIVES

- Compare Mendeleev's arrangement of the periodic table to the modern periodic table.
- Interpret information provided on the periodic table.
- Relate the number of valence electrons to the properties of elements in a group.

VOCABULARY

- **group** a vertical column of elements in the periodic table
- **metal** an element that is usually hard, shiny, malleable, ductile, and a good conductor of heat and electricity
- **nonmetal** an element that is usually soft, not malleable or ductile, and a poor conductor of heat and electricity
- **period** a horizontal row of elements in the periodic table
- **periodic table** an organized arrangement of elements in order of increasing atomic number with repeating physical and chemical property patterns

More than 200 years ago, chemists and physicists explored to find order among the observed elements in nature. They noticed that certain elements shared similar properties. For example, some elements readily react with other elements and others lack reactivity. These similarities prompted various scientists to create an organized list of elements. One of the first lists was created by Johann Döbereiner in 1817. He arranged elements in groups of three that he called *triads*. In his list, Döbereiner grouped together lithium, sodium, and potassium. He also grouped chlorine, bromine, and iodine together. These triads were grouped by appearance and reaction. He then found that the elements were related by atomic mass. Peter Kremers later created a list that compared triads of elements that were arranged perpendicularly. These early lists have been revised and improved as science has advanced and new elements have been introduced.

In 1869, Russian chemist Dmitri Mendeleev recognized that similarities among elements occurred in repeating patterns, which he called *periodic characteristics*. Mendeleev built on the work of other scientists and created the first periodic table that included all 63 elements that were known at the time. He arranged the elements that shared similar physical and

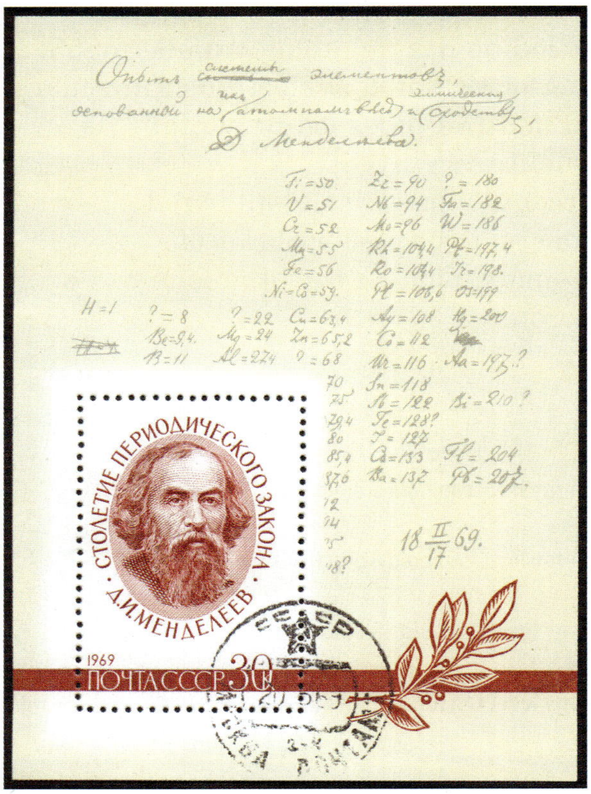

Mendeleev's early periodic table

chemical properties according to their atomic masses rather than their atomic numbers, because scientists had not discovered the subatomic structure. Interestingly, Mendeleev left open spaces for undiscovered elements and he predicted their atomic weights. Changes were made to Mendeleev's periodic table at the beginning of the 20th century by Henry Moseley. The modern **periodic table** is an organized arrangement of elements in order of increasing atomic number with repeating physical and chemical property patterns. This version of the periodic table is currently used today.

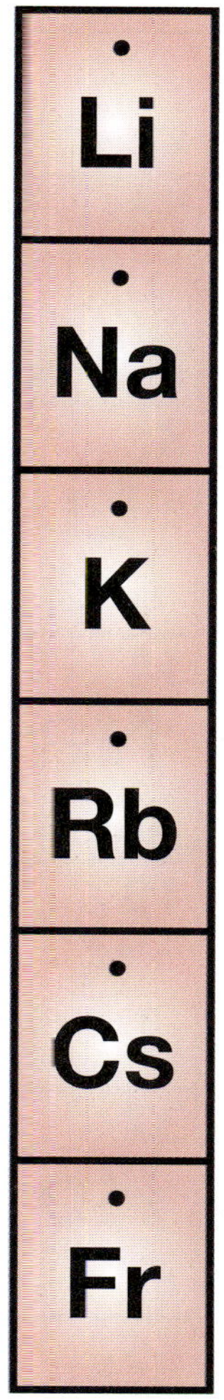

Elements in a group have the same number of valence electrons and similar chemical properties.

The periodic table developed by Mendeleev and Moseley contains a wealth of information about elements. The information contained in each element square includes the atomic number, element symbol, element name, and mass number. The boiling point, freezing point, and electron configuration can be determined by element placement in the table. The table is divided into columns and rows. Each vertical column of elements in the periodic table is called a **group**, or family. There are eight main groups on the left and right side of the periodic table. Sandwiched between groups II and III is a block of elements known as *the transition metals*. Just as human family members share similar traits and characteristics, elements of the same group on the periodic table share similar properties. This happens because elements in a group all have the same number of valence electrons—the electrons in the outermost energy level. The group number is equal to the number of valence electrons. For example, hydrogen, lithium, sodium, and the other elements of group I all combine with other elements in similar ways because they each have one valence electron. Most of these elements are soft, silvery metals that are highly reactive elements.

Sodium metal

Reaction of sodium and water

PERIODIC TABLE OF ELEMENTS

Group →

Period ↓

Key:

- 1 — atomic number
- H — symbol
- 1.008 — atomic mass
- hydrogen — name

* indicates a synthetic element

Alkali Metal · Alkaline Earth · Transition Metal · Basic Metal · Metalloid · Nonmetal · Halogen · Noble Gas · Lanthanide · Actinide

Period	IA 1	IIA 2	IIIB 3	IVB 4	VB 5	VIB 6	VIIB 7	VIIIB 8	9
1	1 H 1.008 hydrogen								
2	3 Li 6.941 lithium	4 Be 9.012 beryllium							
3	11 Na 22.99 sodium	12 Mg 24.305 magnesium							
4	19 K 39.098 potassium	20 Ca 40.078 calcium	21 Sc 44.956 scandium	22 Ti 47.867 titanium	23 V 50.942 vanadium	24 Cr 51.996 chromium	25 Mn 54.938 manganese	26 Fe 55.845 iron	27 Co 58.933 cobalt
5	37 Rb 85.468 rubidium	38 Sr 87.62 strontium	39 Y 88.906 yttrium	40 Zr 91.224 zirconium	41 Nb 92.906 niobium	42 Mo 95.96 molybdenum	43* Tc (98) technetium	44 Ru 101.07 ruthenium	45 Rh 102.906 rhodium
6	55 Cs 132.905 cesium	56 Ba 137.327 barium		72 Hf 178.49 hafnium	73 Ta 180.948 tantalum	74 W 183.84 tungsten	75 Re 186.207 rhenium	76 Os 190.23 osmium	77 Ir 192.217 iridium
7	87 Fr (223) francium	88 Ra (226) radium		104* Rf (265) rutherfordium	105* Db (268) dubnium	106* Sg (271) seaborgium	107* Bh (270) bohrium	108* Hs (277) hassium	109* Mt (276) meitnerium

Lanthanides	57 La 138.906 lanthanum	58 Ce 140.116 cerium	59 Pr 140.908 praseodymium	60 Nd 144.24 neodymium	61* Pm (145) promethium	62 Sm 150.36 samarium
Actinides	89 Ac (227) actinium	90 Th 232.038 thorium	91 Pa 231.036 protactinium	92 U 238.029 uranium	93* Np (237) neptunium	94* Pu (244) plutonium

Group →

Period ↓

Li	H	Br
solid	gas	liquid
at 25°C		

10	IB 11	IIB 12	IIIA 13	IVA 14	VA 15	VIA 16	VIIA 17	VIIIA 18
								2 He 4.003 helium
			5 B 10.811 boron	6 C 12.011 carbon	7 N 14.007 nitrogen	8 O 15.999 oxygen	9 F 18.998 fluorine	10 Ne 20.18 neon
			13 Al 26.982 aluminum	14 Si 28.086 silicon	15 P 30.974 phosphorus	16 S 32.065 sulfur	17 Cl 35.453 chlorine	18 Ar 39.948 argon
28 Ni 58.693 nickel	29 Cu 63.546 copper	30 Zn 65.38 zinc	31 Ga 69.723 gallium	32 Ge 72.64 germanium	33 As 74.922 arsenic	34 Se 78.96 selenium	35 Br 79.904 bromine	36 Kr 83.798 krypton
46 Pd 106.42 palladium	47 Ag 107.868 silver	48 Cd 112.411 cadmium	49 In 114.818 indium	50 Sn 118.710 tin	51 Sb 121.76 antimony	52 Te 127.6 tellurium	53 I 126.904 iodine	54 Xe 131.293 xenon
78 Pt 195.078 platinum	79 Au 196.967 gold	80 Hg 200.59 mercury	81 Tl 204.383 thallium	82 Pb 207.2 lead	83 Bi 208.980 bismuth	84 Po (209) polonium	85 At (210) astatine	86 Rn (222) radon
110* Ds (281) darmstadtium	111* Rg (280) roentgenium	112* Cn (285) copernicum	113* Nh (284) nihonium	114* Fl (289) flerovium	115* Mc (288) moscovium	116* Lv (293) livermorium	117* Ts (294) tennessine	118* Og (294) oganesson

63 Eu 151.964 europium	64 Gd 157.25 gadolinium	65 Tb 158.925 terbium	66 Dy 162.5 dysprosium	67 Ho 164.93 holmium	68 Er 167.259 erbium	69 Tm 168.934 thulium	70 Yb 173.04 ytterbium	71 Lu 174.967 lutetium
95* Am (243) americum	96* Cm (247) curium	97* Bk (247) berkelium	98* Cf (251) californium	99* Es (252) einsteinium	100* Fm (257) fermium	101* Md (258) medelevium	102* No (259) nobellium	103* Lr (262) lawrencium

TRY THIS

Scavenger Hunt

Explore the periodic table using the information you have learned in this lesson. Reference the periodic table in this book to answer each question.

Group VIII elements do not easily react with other elements because they have full valence shells and are very stable. Most unstable atoms that do not have a full valence shell will combine or react in order to attain a stable state. The number of electrons in a valence shell can be easily represented by an electron dot diagram. These diagrams include an element symbol surrounded by dots that represent the number of valence electrons. A single electron dot is placed on each of the four sides of the element symbol before they are paired. For example, the carbon symbol is surrounded by four single electron dots. The oxygen symbol is surrounded by two pairs of electron dots and two single electron dots. The dot diagrams help scientists determine the chemical properties of elements and their reactivity.

A horizontal row of elements in the periodic table is a **period**. Even though the elements in the same period are very different from each other, they are arranged in the same row because they have the same number of energy levels. One energy level is added to the elements' shells in each successive period's row. Period 1 elements have one energy level. Elements in period 2 gain an additional energy level for a total of two levels that each contain electrons. The first element in a period is always a very reactive solid. For example, in period 4, potassium (K), a very reactive solid, is the first element. These elements combine easily with other elements. The last element in a period exists in a gaseous

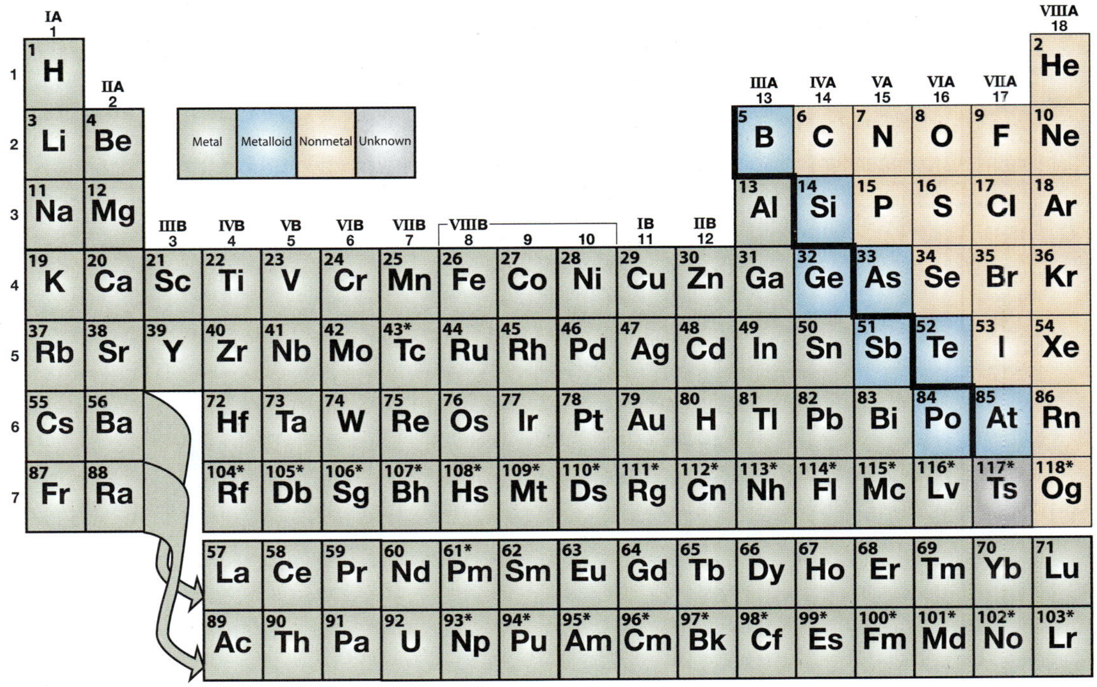

state and will not combine with other elements. Chemists call these elements *the noble gases*. Most noble gases are found in the atmosphere. Helium, neon, and argon are noble gases.

The periodic table can be further divided into three basic regions—metals, nonmetals, and metalloids. Metals are located on the left portion of the periodic table. These elements are the largest group on the chart. **Metals** are elements that are usually hard, shiny, malleable, ductile, and good conductors of heat and electricity. Malleability is a substance's ability to be flattened into thin sheets by hammering or by the application of intense pressure. Nickel and aluminum are examples of malleable metals. Ductility is a substance's ability to be drawn or pulled into the form of a wire. Copper is an example of a ductile metal. **Nonmetals** are on the right portion of the periodic table. They are elements that are usually soft and not malleable or ductile, and they are poor conductors of heat and electricity. Oxygen, bromine, and fluorine are examples of nonmetals. A zigzag line separates the metals from the nonmetals. Metalloids exist along this line. They are elements that have properties of both metals and nonmetals. Germanium, antimony, and tellurium are examples of metalloids.

The patterns Mendeleev saw are evident in the periodic table. Larger atoms tend to be farther to the left in a period and farther down a group column because they have an additional energy level. Metals on the left portion of the table tend to lose electrons to form positively charged atoms called *cations*, and nonmetals on the right portion of the table tend to gain electrons to form negatively charged atoms called *anions*. Chemists use the periodic table to determine, at a glance, many different characteristics of elements. As you continue your study of chemistry, you will find that the periodic table of the elements is a helpful tool for understanding how elements react with each other.

LESSON REVIEW

1. How did Mendeleev arrange his periodic table? How did Moseley arrange the modern periodic table?
2. Of the elements boron, oxygen, indium, lithium, argon, potassium, aluminum, and nickel, which three have the most in common? Why?
3. How does the number of valence electrons determine the reactivity of an element?
4. How many electrons are required to provide selenium with a full valence shell?
5. What two elements exist as liquids at room temperature?

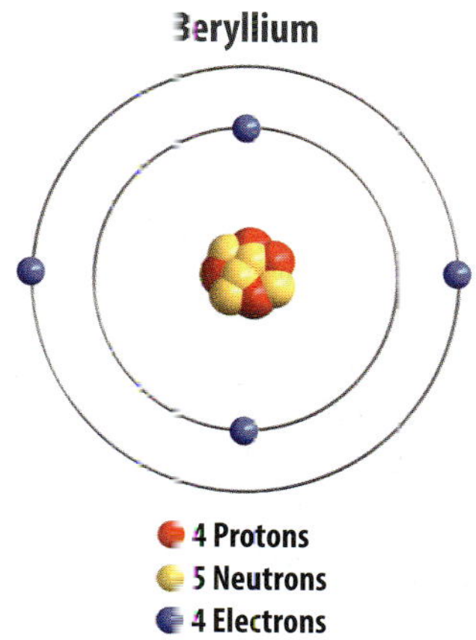

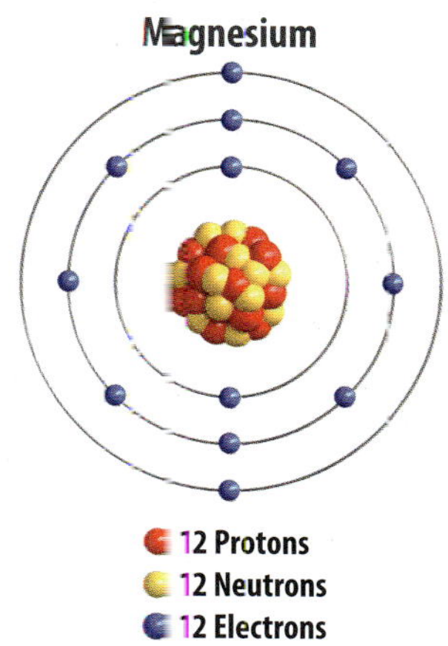

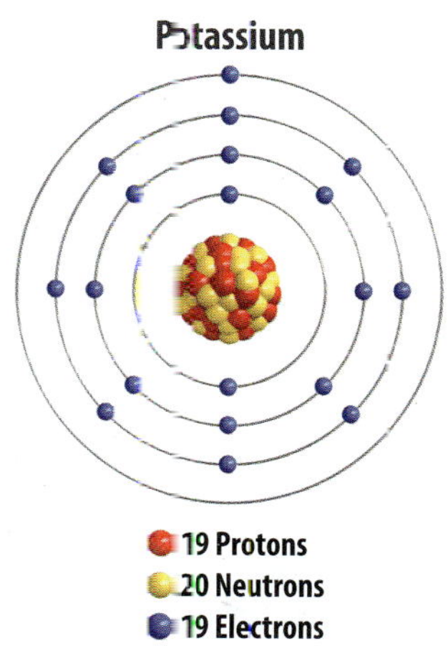

The number of energy levels increases for elements located in higher periods located in the successive lower rows of the periodic table

1.4.1 *Chemical Formulas*

OBJECTIVES

- Determine the identity and quantity of elements in a chemical formula.
- Distinguish between molecular, structural, and empirical formulas.
- Calculate the mass of simple and complex molecules.

VOCABULARY

- **chemical formula** an expression that identifies the types and numbers of elements present in a compound

At least 90 different types of atoms, one for each element, make up the vast combinations of matter in the universe. This can be a difficult concept to comprehend. However, God designed atoms in such a way that the electric forces that keep subatomic particles together also allow for a variety of combinations between different atoms. When these atoms combine, the result is usually a stable structure in either a molecule or a compound.

An example of a stable structure that forms when carbon dioxide combines with water is glucose. The process that produces glucose is known as *photosynthesis*. Glucose is comprised of six carbon atoms, twelve hydrogen atoms, and six oxygen atoms. That is a lot of information. Scientists developed a convenient way to depict the structure of glucose and other molecules to describe their composition. A **chemical formula** is an expression that identifies the types and numbers of elements present in a compound.

There are three basic types of chemical formulas—molecular, structural, and empirical. Molecular formulas indicate the identity and quantity of all elements present in a molecule. The identity of an element is represented by a symbol that can be located on the periodic table. Symbols are one- or two-letter abbreviations that are derived from the element name. For example, the symbol for fluorine is F and the symbol for chlorine is Cl. The first letter is always uppercase, and the second letter is always lowercase. Subscripts relate to the number of atoms in a molecule. The molecular formula of glucose is $C_6H_{12}O_6$. The subscript 6 next to the carbon symbol indicates there are six carbon atoms in one glucose molecule. If an element symbol in a formula does not have a subscript, it is assumed there is only one atom of that particular element contained in the molecule. For example, the molecular formula for methane is CH_4. The

Types of Carbon Dioxide Formulas

Molecular

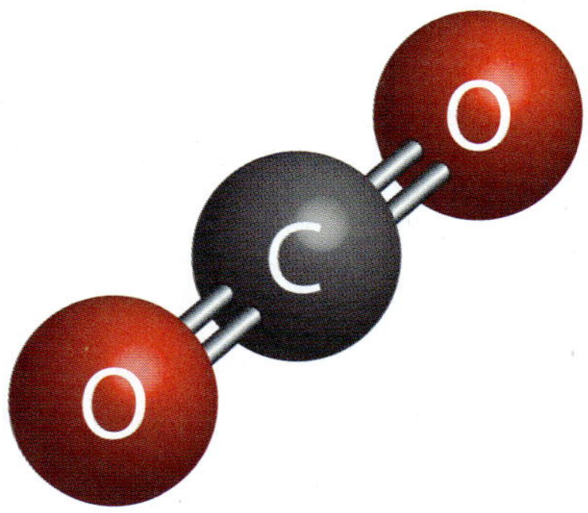

Structural

Space-filling

absence of a subscript for carbon indicates there is only one atom of carbon in one methane molecule.

Subscripts are important because they help distinguish one molecule from another. Alkanes are a group of organic compounds that contain only hydrogen and carbon atoms. The unique physical and chemical properties of each alkane are determined solely by the number of hydrogen atoms bound to the number of carbon atoms. If one carbon is bound to four hydrogen atoms, the product is methane. Ethane consists of two carbon atoms bound to six hydrogen atoms. The chemical formula for propane is C_3H_8. Notice that with the addition of one carbon atom and two hydrogen atoms, a completely new substance is formed.

Alkanes

Name	Structural Formula	Molecular Formula	Empirical Formula
Methane	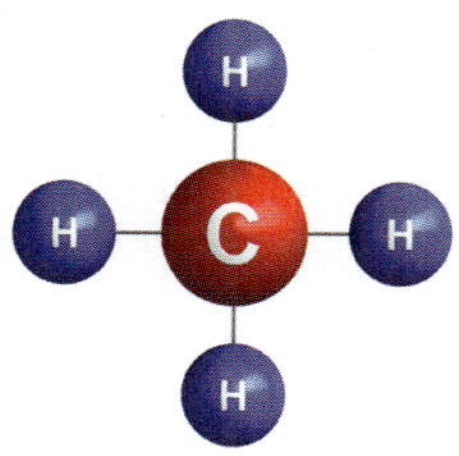	CH_4	CH_4
Ethane	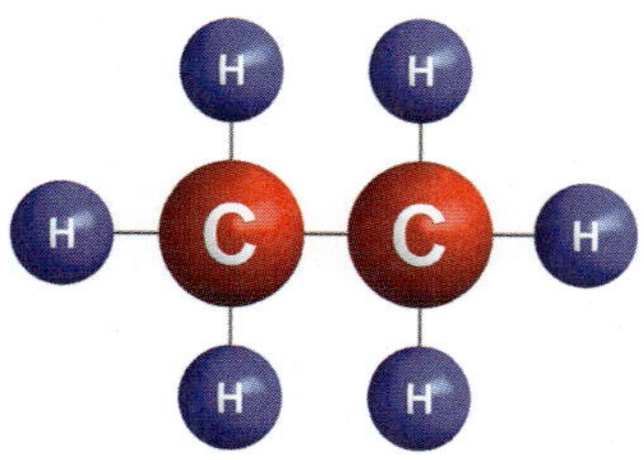	C_2H_6	CH_3
Propane	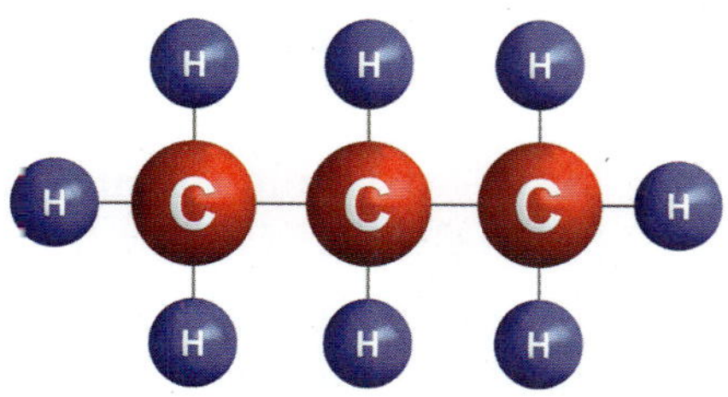	C_3H_8	C_3H_8
Butane	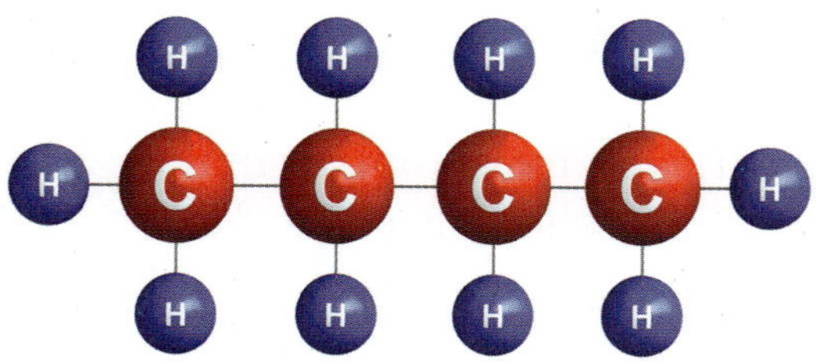	C_4H_{10}	C_2H_5

People benefit from the use of aluminum sulfate in water treatment plants.

Many compounds have more complex molecular formulas. The formula for aluminum sulfate is $Al_2(SO_4)_3$. This compound is used in water treatment plants to clarify drinking water. The portion of the formula inside the parentheses is treated as a single unit. This means that the subscript 3 affects both the quantity of sulfur atoms and oxygen atoms.

$$Al_2(SO_4)_3$$

$$Al = 2 \text{ atoms}$$
$$S = (1 \times 3) = 3 \text{ atoms}$$
$$O = (4 \times 3) = 12 \text{ atoms}$$

Aluminum sulfate contains two aluminum atoms, three sulfur atoms, and twelve oxygen atoms.

Structural formulas provide a visual picture of a molecule. They show how atoms join together. This type of formula helps scientists predict how molecules will interact. They are especially beneficial when identifying molecules that have identical molecular formulas. For example, the molecular formulas of glucose and fructose are the same. However, the way the atoms join together is very different. This explains why fructose releases energy slower than glucose does and is much sweeter than glucose.

Glucose

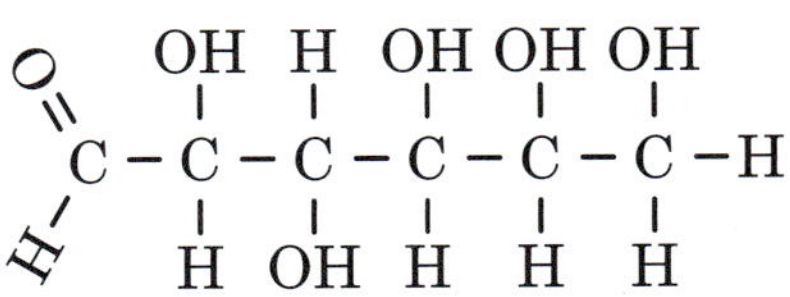

Fructose

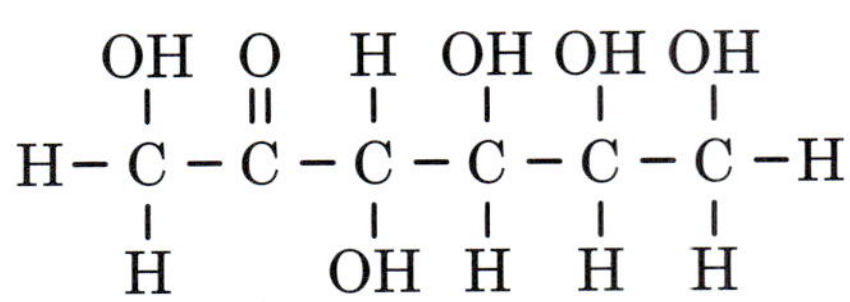

Empirical formulas show the simplest whole number ratio of atoms. The molecular formula for ethane is C_2H_6. The greatest common factor for this molecule is two. If both subscripts are divided by two, the simplest whole number ratio is 1:3. This means that for every one carbon atom there are three hydrogen atoms. The empirical formula for ethane is CH_3.

Chemical formulas, specifically molecular formulas, help scientists determine the masses of molecules. Molecular mass is calculated by finding the sum of the atomic masses. Water, H_2O, contains two hydrogen atoms that each have a mass of 1.008 g and one oxygen atom that has a mass of 15.999 g. The molecular mass of water is (2 × 1.008 g) + 15.999 g for a total of 18.015 g.

LESSON REVIEW

1. How many total atoms are in $(NH_4)_2Cr_2O_7$?
2. Rewrite the formula AG2C2O4 with the correct format and provide the names of the component elements.
3. What are the molecular and empirical formulas for the following structural formula?

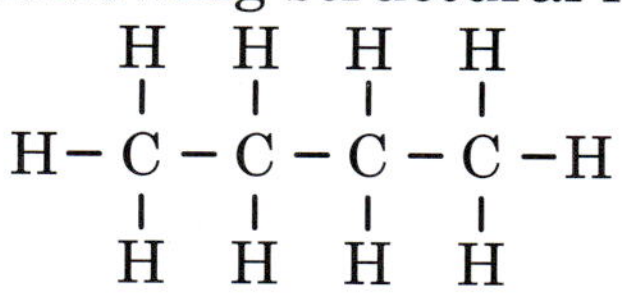

4. Explain which type of formula is the least informative.
5. Which molecule has the greater molecular mass, NaOH or $CaCl_2$?

With the addition of three oxygen atoms, sodium hypochlorite (NaClO)—a standard bleach cleaner—becomes sodium perchlorate ($NaClO_4$)—a DNA extractor in the laboratory.

1.4.2 *Chemical Bonds*

OBJECTIVES

- Explain why atoms bond.
- Predict how atoms achieve a full valence shell.
- Describe the inert nature of noble gases.

VOCABULARY

- **chemical bond** a strong force of attraction that holds the atoms of a compound together

All of the different types of matter are a result of atoms combining with one another in a variety of ways. You may wonder what causes atoms to combine to form compounds. Atoms have a tendency to look for ways to reach a stable state, which requires that they have full valence shells. Few atoms have full valence shells. In an effort to reduce the amount of energy required to exist in an unstable state, uncombined atoms bond with other atoms to achieve full valence shells. A **chemical bond** is a strong force of attraction that holds the atoms of a compound together. The same electric force that keeps the subatomic particles inside an atom together also keeps multiple atoms together. Negatively charged electrons from one atom are attracted to the positively charged nucleus of another.

Electrons in the lower energy levels are closer to the nucleus. The strong attraction between these electrons and the nucleus prevents them from bonding with other atoms. Excited electrons in the outermost energy level of an atom are free to interact with the nuclei of other atoms. These electrons determine how an atom bonds with other atoms.

Atoms can gain, lose, or share valence electrons. In some instances, it may require less energy for an atom to release the outer electrons and drop down to the next lower energy level in order to have a full valence shell. For example, potassium has one valence electron in the fourth energy level. A potassium atom (K) has two options: it can gain seven electrons or lose one electron. Much less energy is required to lose just the one electron. When potassium bonds with another atom, the electron in the fourth energy level is released. The potassium atom now has a full third energy level and an overall positive charge because it has 19 protons and 18 electrons. As a general rule, if an atom has three or fewer valence electrons, it will lose electrons during a chemical

TRY THIS

Bonding Shape

Using a hand lens, observe the structure of individual granules of table salt, Epsom salt, and sucrose. Sketch your observations. What do you think accounts for the different crystal structures? Could you identify a compound using only the crystal shape?

Atoms bond to achieve a stable state.

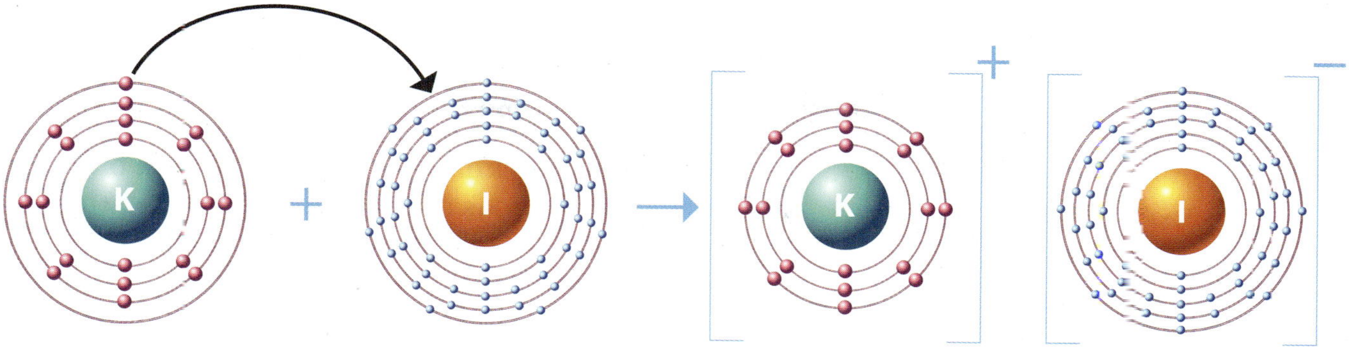

A potassium metal loses one valence electron to an iodine nonmetal.

reaction. Most metals have few valence electrons and will lose those electrons when they bind with nonmetals. If an atom has five or more valence electrons, it will most likely gain electrons. Most nonmetals fit into this category. They will gain or share electrons to complete their valence shells.

There is an order to the organization of atoms. Most atoms have a tendency to want eight electrons in their outermost energy levels. This tendency is called *the octet rule*. For example, fluorine has seven valence electrons. It will want an additional valence electron because it needs only one more electron to fill up its outer energy level. Oxygen, on the other hand, has six valence electrons. It will want two valence electrons.

There are exceptions to the octet rule. Remember that the first energy level contains a maximum of two electrons. An individual hydrogen atom is partially stable with one valence electron in the first energy level. Hydrogen is more stable when it is bound to another atom because it completes a full valence shell of two electrons. Lithium, beryllium, and boron behave in a similar manner when they bind to nonmetal atoms. It is energetically more favorable to have these atoms lose their valence electrons in the second energy level and drop down to the first energy level that holds two electrons.

God designed a group of elements that are uniquely stable because their valence electron shells are full. These elements are known as *the noble gases*. Because their valence shells are full, these elements will not combine with other elements under ordinary conditions. They are relatively unreactive or inert. The noble gases include helium, neon, argon, krypton, xenon, radon, and organesson. Since noble gases do not normally combine with other elements, they form molecules made of individual atoms.

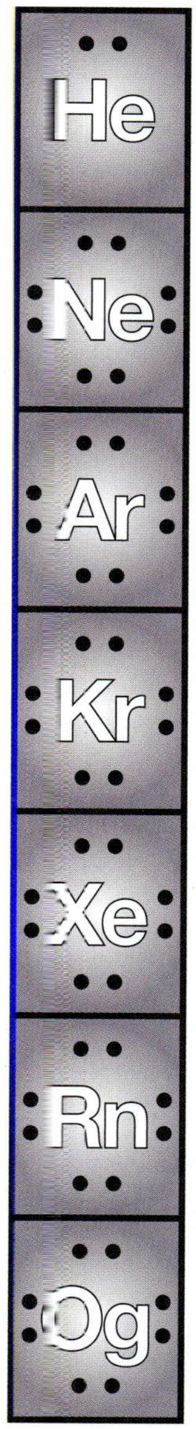

Noble gases have full valence shells.

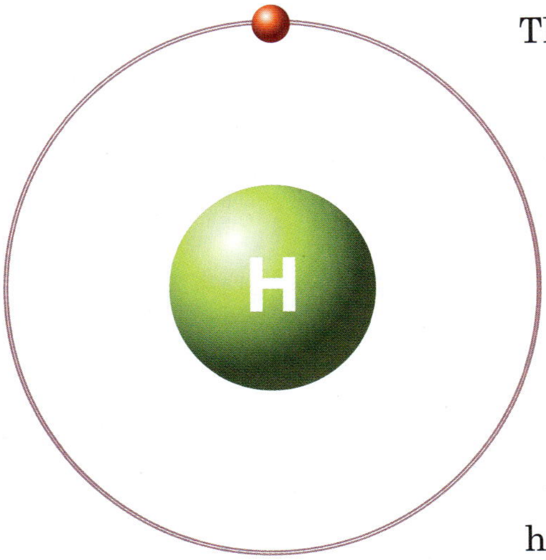

Hydrogen has only one energy level.

The two main types of bond formation are covalent bonding and ionic bonding. Covalent bonding involves a sharing of electrons between two or more nonmetals. Ionic bonding involves a transfer of electrons from a metal to a nonmetal. There is a third type of bonding that occurs when metals interact with other metals. This is called *metallic bonding*. The types of chemical bonds that hold compounds together help determine their appearance, chemical behavior, and chemical properties. Graphite and diamonds are both made of carbon, but they have very different features because of variations in their chemical bonds. Diamonds are the hardest naturally occurring substance known because the atoms are arranged in an interlocking crystal structure. Graphite atoms are arranged in a sheetlike pattern with layers that can be easily sheared off if enough pressure is applied. This is why pencil markings remain on the paper after you write a word. Sand and glass are both made of silicon and oxygen, but they too are dramatically different because of their chemical bonds. By studying how chemical bonds form, scientists have been able to create a multitude of products people use every day.

FYI

Titanium Welding

Many believe that titanium is too difficult to weld. This is not the case if a few key steps are followed. If welded properly, titanium is shiny, reflective, and visually appealing. Because titanium is a very reactive element, it cannot be welded in a normal atmosphere. When titanium is heated in the presence of oxygen, it produces an oxide that makes this metal brittle and easily breakable. Instead, welders pump argon gas over the surface of the titanium metal as it is being heated. The unreactive noble gas produces an atmosphere around the metal that prevents oxidation.

LESSON REVIEW

1. What type of electrons participate in chemical bonding?
2. How many electrons in a phosphorus atom can form a bond?
3. Why are atoms prone to bond?
4. Does calcium most likely lose, gain, or share electrons when bonded to a nonmetal? Why?
5. Why do noble gases not react with other atoms?
6. What type of bonding involves the sharing of electrons?

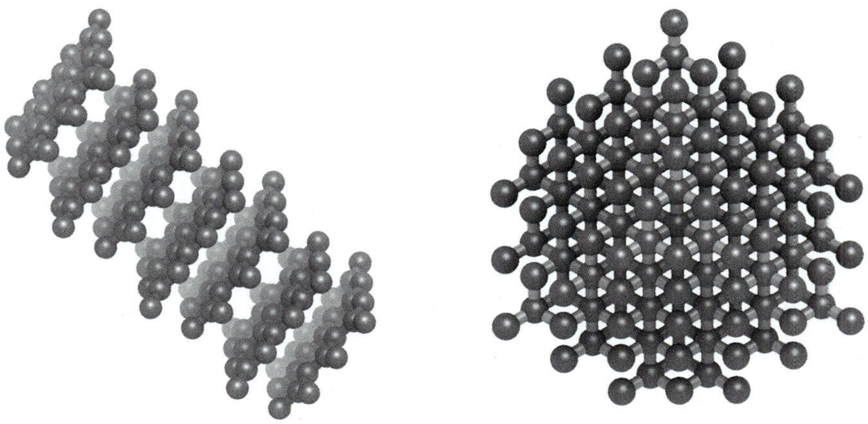

Graphite atoms (left) are arranged in sheetlike layers. Diamonds (right) have a strong, interlocking crystal structure.

OBJECTIVES

- Describe covalent bonds.
- Interpret the tendency of different atoms to form polar molecules.
- Compare polar covalent and nonpolar covalent molecules.
- Determine formulas and write names for covalent molecules.

VOCABULARY

- **covalent bond** the type of bond in which valence electrons are shared between atoms
- **diatomic molecule** a molecule made of two atoms of the same element
- **nonpolar molecule** a molecule with an equal charge distribution
- **polar molecule** a molecule with a balanced charge that is not equally distributed

Many of the substances that are observed in the natural world are the result of the union of nonmetals. Carbon dioxide and oxygen are exchanged during the process of respiration. Water flows through a cycle that nourishes and sustains life. These substances are created when electrons are shared rather than lost or gained. The type of bond in which valence electrons are shared between atoms is a **covalent bond**. This bond occurs if the atoms in a molecule have a lower total energy compared to the energy of the same atoms before they formed the molecule.

A molecule that is made of two atoms of the same element is a **diatomic molecule**. The electrons in diatomic molecules are equally shared. Two of the most important gases in the atmosphere—nitrogen and oxygen—form diatomic molecules. There are five other diatomic molecules that share their electrons equally: hydrogen, fluorine, chlorine, bromine, and iodine.

Each covalent bond shares two electrons and is represented by a single line in a structural formula. A bond between two atoms that shares one pair of electrons is called *a single covalent bond*. Each atom contributes one valence electron to ensure that both atoms have a full valence shell. Two hydrogen atoms can form a single bond. The structural formula for diatomic hydrogen can be represented as H–H. Each hydrogen atom is stable with a full valence shell of two electrons. A methane molecule has four single covalent bonds between one carbon atom and four hydrogen atoms. The central carbon atom has four valence

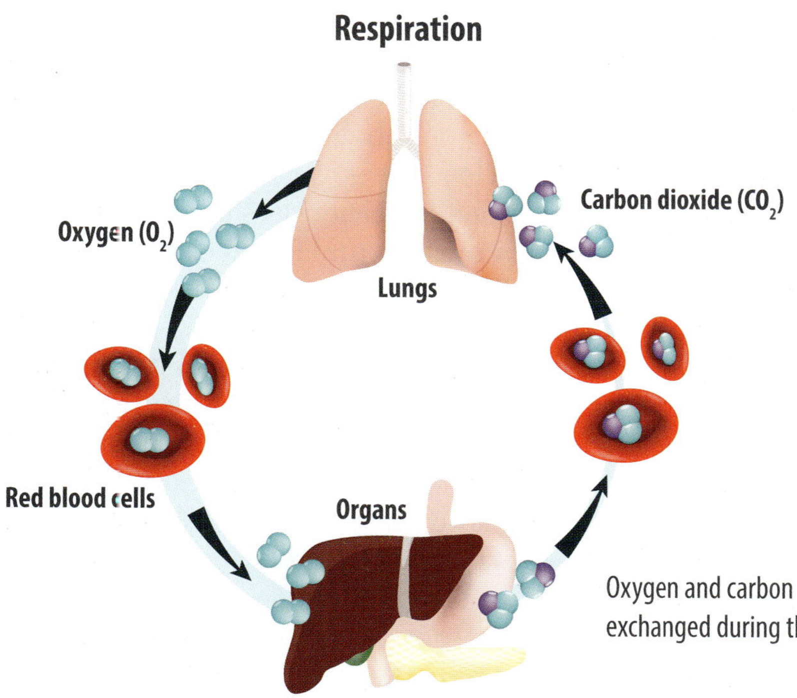

Oxygen and carbon dioxide molecules are exchanged during the process of respiration.

electrons. Each hydrogen has one valence electron that bonds to one of the outer electrons surrounding carbon. When the atoms combine, each hydrogen atom shares two electrons and the carbon atom shares eight electrons.

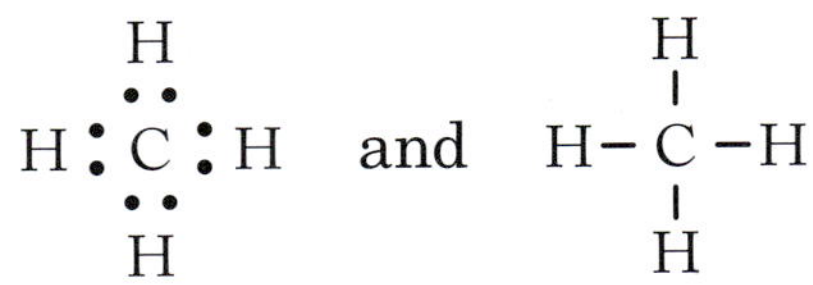

Single, double, and triple bonds connect these diatomic molecules.

Molecules also develop from multiple bonds. Two atoms can share two pairs of electrons for a total of four electrons. This type of bond is called *a double covalent bond* and is represented by two lines in a structural formula. Two oxygen atoms form a double covalent bond in the oxygen molecule. Each oxygen atom has six valence electrons. Two of the electrons are unpaired and available to bond. The structural formula of an oxygen molecule shows two oxygen atoms connected with a double line and each atom also has two lone pairs of electrons.

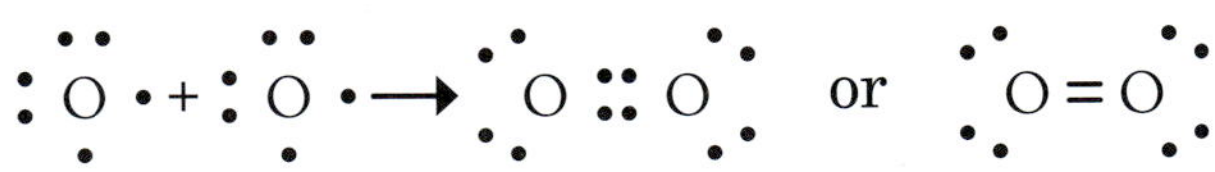

Diatomic oxygen shares four electrons.

A bond between two atoms that share three pairs of electrons for a total of six electrons is called *a triple covalent bond*. An example of a molecule that forms a triple covalent bond is diatomic nitrogen.

Not all molecules share their electrons equally. Some atoms have a stronger attraction to electrons that causes these atoms to pull electrons closer to their nuclei. The strength of the attraction is dependent on a number of factors. Scientists observe that the attractive forces between a nucleus and an electron decrease with increasing distance. The more space that is put between these two subatomic particles, the less likely a strong attraction will form. Consider a magnet. You would not feel the magnetic force if you held two magnets far apart. Two other factors that affect the attraction between a nucleus and an electron are the nuclear charge and the total number of electrons. An atom with a large nuclear charge tends to attract more electrons. However, the atomic size increases with an elevated atomic number. More protons means more electrons that reside in several shell layers. Numerous inner electrons tend to block the nucleus from the valence electrons. This reduces the attraction as well. The

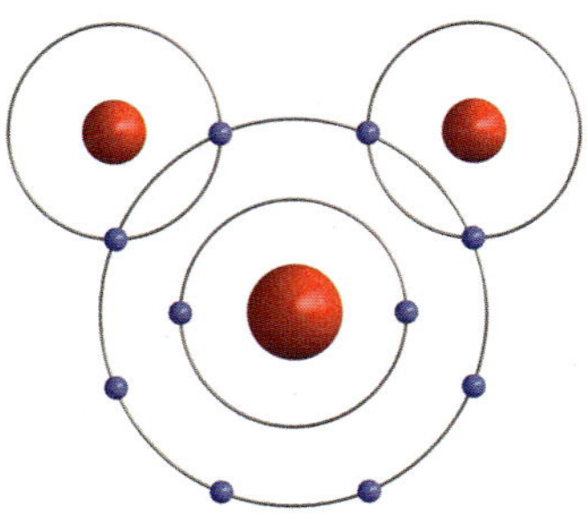

Electrons are shared in covalent bonds.

periodic trend for electron attraction increases as you go up a period and to the right of a group. Fluorine has the greatest electron attraction.

Some atoms share electrons equally. Others do not. Think of it as a game of tug-of-war. Both teams are holding the rope, but the stronger team pulls the rope closer.

When a molecule contains some atoms that have a stronger attraction to electrons than others, the result is an unequal charge distribution. The atom with a stronger attraction gains a slightly negative charge, and the atom with a weaker attraction gains a slightly positive charge. For example, hydrogen and bromine form a single bond to create hydrobromic acid. The two shared electrons are more closely attracted to the bromine atom where they spend most of their time because of bromine's great positive nuclear charge. Hydrogen gains a partial positive charge and bromine gains a partial negative charge.

A water molecule behaves in much the same way. Similar to bromine, the oxygen atom attracts the shared electrons more than the hydrogen atoms. A partial negative charge forms around the oxygen atom and a partial positive charge forms around the two hydrogen atoms. Molecules with a balanced charge that is not equally distributed are **polar molecules**. Water is a polar molecule because it is not symmetric. This feature allows water to dissolve other polar compounds like salt and vitamins—a crucial factor for many cellular processes.

Nonpolar molecules are molecules with an equal charge distribution. Diatomic molecules are nonpolar because they have the exact same attraction to electrons. Molecules that are symmetrical are also nonpolar. For example, a carbon dioxide molecule contains one central carbon atom surrounded by two oxygen atoms. The electrons are equally attracted to both of the oxygen atoms.

When people discuss compounds, they often refer to a common name. For example, you would not typically say that a glass of dihydrogen monoxide was very refreshing when you are talking about a glass of water. However, common names do not provide descriptive information about the compound. The common name ammonia does not indicate that this compound contains three hydrogen atoms bonded to one nitrogen atom. Scientists decided to create a systematic naming system that everyone

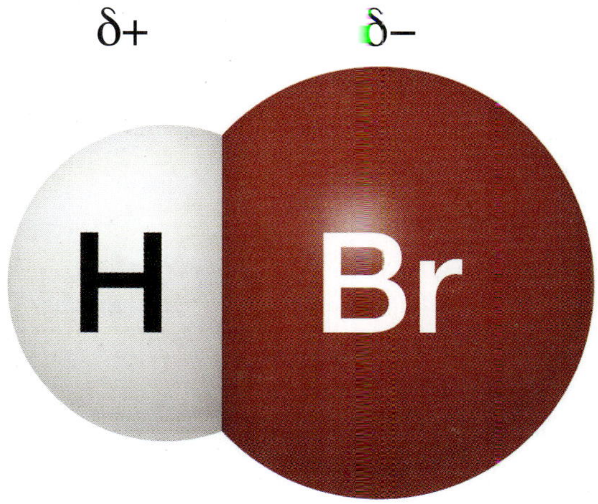

Hydrogen bromide is a polar molecule.

can use. This system is especially helpful because some pairs of nonmetals can combine in a wide variety of ways to form multiple compounds. For example, one carbon atom can combine with one atom of oxygen or two atoms of oxygen.

Common Name	Chemical Name	Formula
baking soda	sodium bicarbonate	$NaHCO_3$
liquid bleach	sodium hypochlorite	$NaClO$
Borax	sodium tetraborate decahydrate	$Na_2B_4O_7 \cdot 10H_2O$
Epsom salt	magnesium sulfate heptahydrate	$MgSO_4 \cdot 7H_2O$
gypsum	calcium sulfate dihydrate	$CaSO_4 \cdot 2H_2O$
laughing gas	dinitrogen oxide	N_2O
limestone	calcium carbonate	$CaCO_3$
milk of magnesia	magnesium hydroxide	$Mg(OH)_2$

The naming system employs Greek prefixes to specify how many of each type of atom belong to a binary covalent compound. The first element is given a prefix if there is more than one atom. *Mono–* is never included with the first element. The second element is written with a prefix and the suffix changes to *–ide*. For example, the chemical names of PH_3, N_2O_3, CO, PCl_5, and SF_6 are phosphorus trihydride, dinitrogen trioxide, carbon monoxide, phosphorus pentachloride, and sulfur hexafluoride.

LESSON REVIEW

1. Describe the type of bond that two or more nonmetals create when they join together.
2. Will fluorine or selenium have a greater tendency to form a polar molecule when bound to other atoms? Why?
3. Describe the appearance of polar and nonpolar molecules.
4. What is the name of P_2H_4 and SeS_6?
5. What is the formula for dinitrogen dihydride and arsenic trioxide?

Greek Prefixes					
Number	Prefix	Number	Prefix	Number	Prefix
1	mono-	5	penta-	8	octa-
2	di-	6	hexa-	9	nona-
3	tri-	7	hepta-	10	deca-
4	tetra-				

FYI

CarbFix, CO_2

Excess carbon dioxide emissions from power plants have been a major concern for the global community. Researchers from Iceland, the United States, and France may have discovered a solution to this concern. They have been testing new ways to trap this greenhouse gas in a permanent and efficient way. Their method involves injecting CO_2 into volcanic rock along with heavy carbon and excess water. When these substances reach underground basalt layers, they react with metals to form new carbonate minerals such as calcite. This process is called *carbonation*. After injecting CO_2, the researchers gathered samples from nearby wells to monitor the geochemistry, pH levels, and other characteristics. They assumed the carbonation process would take at least a decade to complete. However, they were surprised to discover that after only a year and a half, 95% of the injected CO_2 had converted into calcite. Their research is very important because it provides one way that CO_2 can be permanently stored underground without any concern of accidental leakage into the atmosphere.

Ionic Bonds 1.4.4

Calcium is an essential nutrient that ensures the health of bones and teeth. This element also regulates heart contractions, helps with the transmission of nerve impulses, and assists with muscle contractions. In order for the body to receive these benefits, calcium must be ingested in a stable form. Unbound calcium is a soft metal that burns in air and reacts quickly with warm water. This type of calcium is not safe to ingest. When calcium exchanges electrons to bond with phosphorus and oxygen, the result is a stable form of calcium phosphate that is found in milk. Once milk has entered your system, certain reactions take place so your body can access the usable form of calcium. The products of these reactions are a phosphate group that contains more electrons and an overall negative charge, and calcium that contains fewer electrons and an overall positive charge. An atom or group of atoms that has a charge is an **ion**.

Atoms will gain or lose electrons to form ions because these atoms continually seek a stable state with full valence shells. When an atom loses electrons, there are fewer electrons than protons in the nucleus. This creates a positively charged ion,

OBJECTIVES

- Describe ionic bonds.
- Analyze oxidation numbers to predict if an atom will lose or gain electrons.
- Determine formulas and write names for ionic compounds and hydrates.

VOCABULARY

- **ion** an atom or group of atoms that has a charge
- **ionic bond** the force of attraction between oppositely charged ions
- **polyatomic ion** a group of atoms that are covalently bound and have a charge

Calcium-rich foods, such as beans, milk, cheese, and almonds, help strengthen bones.

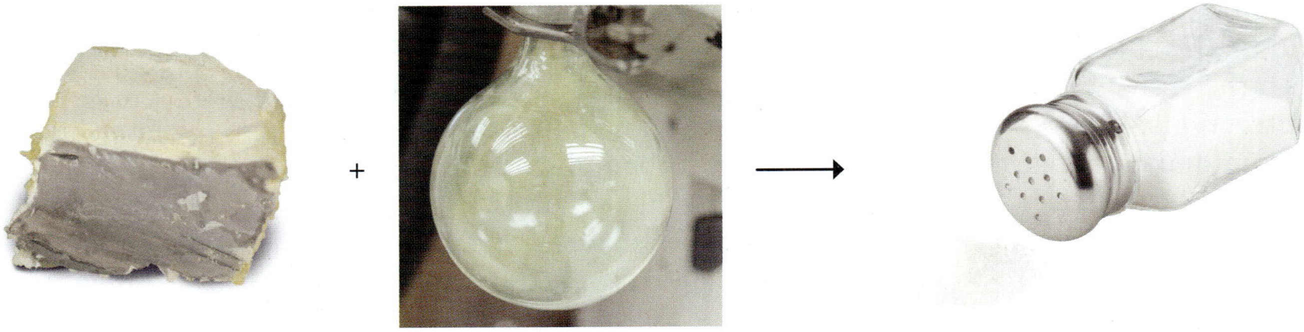

Sodium Metal **Chlorine Gas** **Table Salt**

called *a cation*. Atoms of most metals have few electrons in their outer energy level. When metals bond with nonmetals, the metal atoms tend to lose these valence electrons. Some atoms gain electrons, creating a charge imbalance because there are more electrons than protons in their nuclei. This creates negatively charged ions, called *anions*. Atoms of nonmetals tend to gain electrons from other atoms, forming negative ions.

The ion charge is indicated with a superscript. For example, a neutral calcium atom has two valence electrons. When calcium reacts with a nonmetal, it loses two electrons. The calcium ion has 20 protons and 18 electrons. Two additional protons give the calcium ion an overall charge of +2. The shortcut way to write a calcium cation is Ca^{2+}. A neutral nitrogen atom has five valence electrons. When nitrogen reacts with a metal, it will gain three electrons. The nitrogen ion has 7 protons and 10 electrons. Three additional electrons give the nitrogen ion an overall charge of –3. The shortcut way to write a nitrogen anion is N^{3-}.

The opposite charges of cations and anions are what attract these ions to each other and keep the compounds together. They form ionic bonds. An **ionic bond** is the force of attraction between oppositely charged ions. When an atom gains electrons, one or more atoms must lose an equal number of electrons to maintain the neutral charge of the compound. For example, a sodium atom loses one valence electron to become a positive sodium ion (Na^+). A chlorine atom gains an electron to become a negative chloride ion (Cl^-). The positively charged sodium ion and the negatively charged chloride ion attract each other to form an ionic bond of sodium chloride, NaCl. The positive charge of the sodium ion is balanced by the negative charge of the chloride ion, making sodium chloride neutral.

Ionic bonds can also be formed by an atom gaining one or more electrons from more than one atom. For example, calcium fluoride (CaF_2) is created when two fluorine atoms gain one electron each from one calcium atom. The +2 charge on the calcium ion balances the –2 charge on the fluoride ions. When ionic compounds are formed, the electrons surrounding each atom are transferred to other atoms. The total number of electrons of the combined atoms does not change.

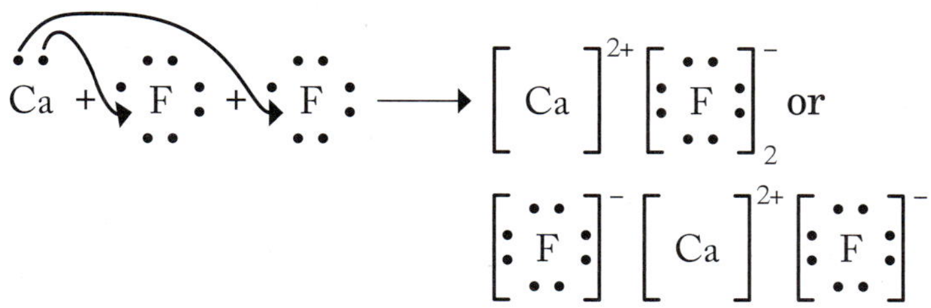

Ionic bonds create ionic compounds that are made of strong crystal lattice structures. At room temperature, ionic compounds are solid. In the solid state, they do not conduct electricity. However, when ionic compounds are dissolved in water, the free-floating ions conduct electricity very well. The ionic compounds in your blood carry electrical impulses from the brain to your nerves and muscles.

Scientists have created a naming system for ionic compounds in accordance with the International Union of Pure and Applied Chemistry (IUPAC). The simplest compounds to name are binary compounds that consist of two elements. These two elements are typically a metal and a nonmetal. To name any compound,

General Group Oxidation Numbers	
Group Number	**Oxidation Number**
IA	1+
IIA	2+
IIIA	3+
IVA	4+
VA	3–
VIA	2–
VIIA	1–
VIIIA	0

Common Elements with Multiple Oxidation Numbers	
Element Name	**Oxidation Numbers**
Chromium, Cr	2+, 3+
Copper, Cu	1+, 2+
Iron, Fe	2+, 3+
Lead, Pb	2+, 4+
Tin, Sn	2+, 4+

SCIENCE IN ACTION

Writing Chemical Formulas for Binary Ionic Compounds
What is the formula for beryllium phosphide?

Step 1: Identify cation, anion, and oxidation numbers.

Cation: Be^{2+}
Anion: P^{3-}

Step 2: Transfer the charge superscripts to the subscript of the other atoms without the sign.

$Be^{2+} \; P^{3-} \longrightarrow Be_3P_2$

Step 3: Reduce the subscripts to the smallest whole number ratio.

Be_3P_2
(Not necessary for this compound. It is already the smallest whole number ratio of 3:2.)

Step 4: Check to make sure the compound is neutral.

Be: quantity of 3 atoms × +2 charge = +6
P: quantity of 2 atoms × –3 charge = –6
0 charge

the oxidation number must first be determined. This number indicates how many electrons an atom will lose or gain when it combines with another atom to form a stable bond. The oxidation number for an ion is equal to its charge. For example, calcium loses two electrons to form an ion with a +2 charge. Calcium's oxidation number is +2. Oxygen gains two electrons to form an ion with a –2 charge. Oxygen's oxidation number is –2.

Some elements have multiple oxidation numbers. For example, lead can have an oxidation number of +2 or +4 depending on how it combines. When naming these elements with multiple oxidation numbers, a roman numeral is included within parentheses next to the metal's name. Lead with a +2 oxidation number is written as lead(II) along with the anion in the compound.

It is important for the positive charge from the cation to equal the negative charge from the anion to ensure a neutral compound. For example, when lithium and iodine combine, it is a 1:1 ratio. The +1 charge from the lithium ion balances the –1 charge from the iodide ion. Lithium iodide is a neutral compound. It becomes a little more complex when the charges do not automatically balance each other in a 1:1 ratio. For example, the magnesium ion has a charge of +2 and the bromide ion has a charge of –1. In order for the +2 charge of the magnesium ion to be balanced, there must be two bromide ions that combine to form the neutral compound of magnesium bromide, $MgBr_2$. What about boron and sulfur? The boron ion has a +3 charge, and the sulfide ion has a –2 charge. To determine how these two elements combine, you need to figure out the least common multiple. Two boron ions with a +3 charge create a total +6 charge. Three sulfide ions with a –2 charge create a total –6 charge. The formula for boron sulfide is B_2S_3.

The identity of the elements, oxidation numbers, and least common multiples will help determine the chemical formula of an ionic compound. Formulas begin with the symbol of the element that has the positive oxidation number or charge. This includes hydrogen and all metals. The next step includes determining the oxidation number of the cation. Next, write the symbol of the element that has the negative oxidation number, or charge, and determine the oxidation number of the anion. Nonmetals, except hydrogen, have negative oxidation numbers. The charge superscript of each atom becomes the subscript of the other atom without the sign. For example, when potassium and oxygen combine, the +1 charge from the potassium ion becomes the subscript for oxygen. Also, the –2 charge from the oxide ion becomes the subscript for potassium. With the signs removed, the formula for potassium oxide is K_2O. A subscript of one does not

SCIENCE IN ACTION

Writing Binary Ionic Compound Names

What is the name of PbO?

Step 1: Write the name of the cation.
Pb is the cation. Its name is lead.

Step 2: Determine if the cation can have more than one oxidation number. If so, write the oxidation number in the form of a roman numeral.
Lead has two oxidation numbers: 2+ or 4+.
Oxygen has only one oxidation number of 2–.
The compound is neutral, so the 2– oxidation number of oxygen balances the 2+ oxidation number of lead.
lead(II)

Step 3: Write the prefix of the anion followed by the suffix *–ide*.
Oxygen is the anion.
The prefix of oxygen is *ox–*. Adding *–ide* makes the anion name oxide.

Step 4: Write the full name.
lead(II) oxide

need to be included next to the oxygen. The last step in writing a formula is reducing the subscripts to the smallest whole number ratio.

If a chemical formula is provided, you can easily write the name of a binary ionic compound. The first step is to write the name of the cation. Determine if the cation has more than one oxidation number. If it does, remember that the compound is neutral and the anion has only one possible oxidation number. Write the oxidation number of the cation in the form of a roman numeral inside of parentheses next to the name. Next, write the prefix of the anion and add the suffix *–ide*.

Common Polyatomic Ions		
Ion Name	**Ion Formula**	**Ion Charge**
Ammonium ion	NH_4^+	1+
Acetate ion	$C_2H_3O_2^-$	1–
Carbonate ion	CO_3^{2-}	2–
Chlorate ion	ClO_3^-	1–
Chlorite ion	ClO_2^-	1–
Cyanide ion	CN^-	1–
Hydroxide ion	OH^-	1–
Hypochlorite ion	ClO^-	1–
Nitrate ion	NO_3^-	1–
Nitrite ion	NO_2^-	1–
Perchlorate ion	ClO_4^-	1–
Permanganate ion	MnO_4^-	1–
Phosphate ion	PO_4^{3-}	3–
Sulfate ion	SO_4^{2-}	2–
Sulfite ion	SO_3^{2-}	2–

Some compounds contain more than two elements. For example, potassium nitrate has a formula of KNO_3. Potassium is the positive ion in this compound, and nitrate is the negative ion. A nitrate ion consists of one nitrogen atom bound to three oxygen atoms with an overall –1 charge. This group of atoms acts as a single unit. A group of atoms that are covalently bound and have a charge is called a **polyatomic ion**. It is important that the subscripts in the polyatomic ion never change. Otherwise, the polyatomic ion would be a completely different substance.

SCIENCE IN ACTION

Writing Chemical Formulas for Compounds with Polyatomic Ions
What is the formula for ammonium sulfate?

Step 1: Identify the element or polyatomic ion that has a positive charge.
The formula for ammonium is NH_4^+ with a +1 charge.

Step 2: Identify the element or polyatomic ion that has a negative charge.
The formula for sulfate is SO_4^{2-} with a –2 charge.

Step 3: Balance the oxidation numbers to create a neutral compound.
Two NH_4^+ ions are required to balance the –2 charge of SO_4^{2-}.
If there is more than one polyatomic ion, place parentheses around the ion and add the subscript to the outside of the parentheses.

$$NH_4^{1+} \quad SO_4^{2-} \longrightarrow (NH_4)_2SO_4$$

There are two ammonium ions. Add parentheses around NH_4 and a subscript of 2 to the outside of the parentheses.
Solution: $(NH_4)_2SO_4$

Step 4: Check to make sure the compound is neutral.

NH_4: quantity of 2 ions × +1 charge =	+2
SO_4: quantity of 1 ion × –2 charge =	–2
	0 charge

Writing the names of compounds that contain polyatomic ions is very similar to naming binary ionic compounds. First, write the name of the cation. There is one polyatomic cation called *ammonium*, NH_4^+. Next, write the name of the anion. For example, the cation in Na_3PO_4 is sodium. The name of the anion, PO_4^{3-}, is phosphate. Sodium phosphate is the complete name.

Determining chemical formulas for compounds that contain polyatomic ions is also very similar to writing formulas for binary ionic compounds. First, write the symbol of the element or polyatomic ion that has the positive charge and the oxidation number. Next, write the symbol of the element or polyatomic ion that has the negative charge and the oxidation number. Balance the oxidation numbers to create a neutral compound. If there is more than one polyatomic ion, place parentheses around the ion and add the appropriate subscript outside of the parentheses.

Hydrates are another group of ionic compounds. These compounds contain water molecules in their structure. Cobalt(II) chloride hexahydrate is an example of a hydrate. The formula for

FYI

Opal Hydrates

Minerals have very distinct crystalline structures. Opals do not have a true crystalline structure, so they are included in the mineraloid category. Silicon dioxide (SiO_2) is the primary compound that makes up the opal chemistry. Surrounding the silicon dioxide molecules are water molecules. Water concentration levels vary between 5% and 10%, sometimes greater. The amount of water contained in the opal structure determines the color of the opal. When the light strikes varying amounts of water, a great diversity of brilliant colors is produced. This water not only produces colorful jewelry, it can also help geologists determine the temperature of the host rock at the time the opal formed.

this compound is $CoCl_2 \cdot 6H_2O$. It contains six water molecules for every unit of cobalt(II) chloride. When water molecules are attached to the ionic structure of cobalt(II) chloride, the crystals appear pink. As the compound is heated, the water evaporates and an anhydrous compound is produced. The anhydrous crystals appear blue. Many anhydrous compounds used in the laboratory have special instructions to be kept tightly sealed because these compounds can even extract water from the air. The formulas for hydrate compounds include the ionic portion followed by a "•" and the quantity of H_2O molecules. For example, the formula for magnesium sulfate heptahydrate is $MgSO_4 \cdot 7H_2O$. Notice the name follows the rules for ionic compounds with the addition of a Greek prefix followed by the word *hydrate*. The name for $BaCl_2 \cdot 2H_2O$ is barium chloride dihydrate.

LESSON REVIEW

1. What types of elements combine to form ionic bonds, and how do the electrons interact?
2. Why are ionic compounds neutral?
3. What is the oxidation number of fluorine? Will this atom most likely lose or gain electrons when it bonds?
4. Are the following compounds ionic or covalent: calcium chloride, phosphorus tribromide, and potassium nitride?
5. What is the name of CuO and AlF_3?
6. What is the formula for iron(III) oxide and ammonium sulfate?

UNIT
2

Types of Substances

Chapter 1: *Metals*
Chapter 2: *Nonmetals and Metalloids*
Chapter 3: *Organic Compounds*
Chapter 4: *Other Useful Materials*

Vocabulary

alkali metal
alkaline-earth metal
alkane
alkene
alkyne
alloy
amalgam
biogas
biomass
ceramic
composite
corrosion
ductility
energy farming
glass
halogen
hydrocarbon
isomer
malleability
metallic bond
metalloid
noble gas
ore
organic compound
polymer
reduction
transition metal

Key Ideas

- Systems, order, and organization
- Change, constancy, and measurement
- Form and function
- Abilities necessary to do scientific inquiry
- Understandings about scientific inquiry
- Properties and changes of properties in matter
- Structure of atoms
- Structure and properties of matter
- Chemical reactions
- Abilities of technological design
- Understandings about science and technology
- Personal health
- Populations, resources, and environments
- Natural and human-induced hazards
- Risks and benefits
- Science and technology in society
- Science as a human endeavor
- Nature of science
- History of science

SCRIPTURE

Do not store up for yourselves treasures on Earth ... but store up for yourselves treasures in heaven ... for where your treasure is, there your heart will be also.

Matthew 6:19a, 20a, 21

2.1.1 *Properties of Metals*

OBJECTIVES

- Describe metallic bonding.
- Relate the structure of metals to their physical properties.

VOCABULARY

- **corrosion** the gradual wearing away of a metal as a result of a chemical reaction in which the metal is changed into a metallic compound
- **ductility** a substance's ability to be drawn or pulled into the form of a wire
- **malleability** a substance's ability to be flattened into thin sheets by hammering or by the application of intense pressure
- **metallic bond** the force of attraction between the metal atoms and the surrounding sea of electrons in the metal

In ancient times, caravan traders carried brightly colored powder pigments across deserts. The rarest of these precious materials were transported across entire continents to the palaces of kings. These pigments were produced by metals extracted from Earth's crust. They included zinc white, cadmium yellow, cobalt blue, iron oxide red, and chromium oxide green. Traders sold these exotic colored pigments to wealthy rulers who valued the pigments as much as they valued semiprecious gems.

The discovery of different metals occurred over a long period of time. One of the first metals to attract attention was gold, more than 6,000 years ago. Some of gold's attributes that make it most appealing are its brilliant luster and resistance to corrosion or tarnish. Also, gold is easy to mold and shape. Different historical civilizations that especially appreciated gold include the Egyptians, Assyrians, and Minoans of Crete. Copper and silver were discovered around this same time.

Before 3,000 BC, a new material created excitement. Metal workers discovered they could combine two metals to make an exceptionally strong material. Bronze is a combination of the

Egyptians were one of the first civilizations to appreciate the rarity of gold.

metals copper and tin, and it gradually replaced stone as the material used in making weapons and tools. As a result, technology exploded, civilizations flourished, and networks of trade routes extended across three continents. This time in history is known as *the Bronze Age*.

Without a protective coating, steel is subject to corrosion when it reacts with water vapor.

Today, people depend on metals and metal alloys for a multitude of manufacturing needs. A metal alloy is a substance made when two or more metals are combined or when a metal and a nonmetal are united. Copper and aluminum wires carry electricity. Buildings and bridges are built with iron and steel. Steel is an alloy of iron and carbon. Aluminum alloys are used to make car parts, airplanes, and space vehicles. The properties of metals allow for a wide variety of uses.

Of the more than 100 elements, 91 are metals and metalloids. These elements can be found on the left side of the zigzag line on the periodic table. Hydrogen is an exception. A metal is an element that is usually hard and shiny. It is a good conductor of heat and electricity, has a high melting point, and tends to lose electrons. Most metals are malleable and ductile. Mercury is the only metal that exists in the liquid state at room temperature. All other metals exist in the solid state.

Malleable metals can be pounded or rolled like stiff pizza dough.

Malleability is a substance's ability to be flattened into thin sheets by hammering or by the application of intense pressure. Aluminum is a malleable metal that is commonly rolled into foil. Different metals can be hammered into the shapes of pots, pans, and utensils. **Ductility** is a substance's ability to be drawn or pulled into the form of a wire. Copper and gold are highly ductile metals.

Copper is used in building wiring because it is a great conductor and relatively inexpensive.

The physical properties of metals relate to the structure of their atoms. Most metals have between one and three electrons in their outer energy level. When a metal bonds with a nonmetal, an ionic bond is formed where the outer electrons are transferred from the metal to the nonmetal. The result is a compound that is chemically stable.

Because valence electrons are easily taken away from metals by nonmetals, metals react easily with water and oxygen. These chemical reactions often result in the corrosion of a metal. **Corrosion** is the gradual wearing away of a metal as a result of

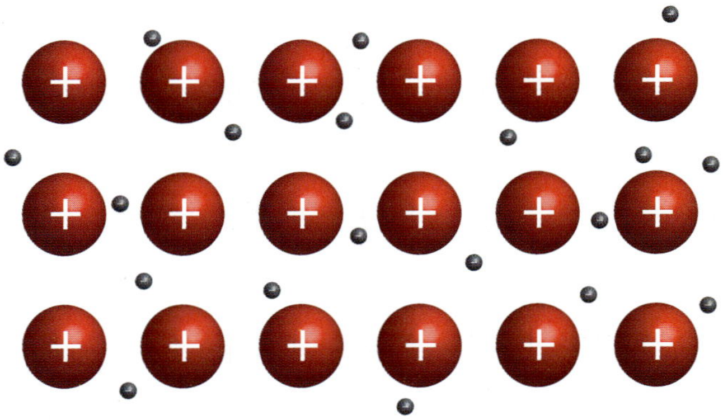

Metal nuclei are surrounded by a sea of electrons.

a chemical reaction in which the metal is changed into a metallic compound. Rusting is a common example of corrosion. The rust that results when the element iron combines with oxygen in the air is called *iron(III) oxide*.

Covalent bonding exists mostly between nonmetals, but metal bonding is usually either ionic (between metals and nonmetals) or metallic (between metals). When metal atoms come in close proximity to one another, their valence electrons create a sea of negative charge around each positively charged nucleus. The electrons are shared among all of the metal atoms. There is not a gain or loss of electrons. The total number of electrons remains the same, and the solids are neutral. The force of attraction between the metal atoms and the surrounding sea of electrons in the metal is called a **metallic bond**. A greater number of valence electrons means that the metals will form stronger

FYI

Computer Chips

Scientists estimate that more than 50 of the 90 naturally occurring elements are used in modern computers. Some of the most common include silicon, germanium, gold, copper, silver, aluminum, platinum, and nickel. Most materials used in computers are semiconductors, which can function in computer components at room temperature. Semiconductors have both the properties of a metal, which conducts electricity, and of an insulator, which does not. Silicon is a semiconductor. It is abundant, inexpensive, pure, easy to work with, and it forms stronger bonds than materials like germanium. Although germanium was initially considered a better material for computer use than silicon, it is expensive and its weaker bonds can fail. Another common computer material is copper. Like silicon, copper is plentiful, cheap, and easily crafted. Silver is the only metal that is more conductive than copper, but it is expensive, so its use in computer manufacture is limited.

bonds. Few valence electrons form weak bonds. For example, the elements that belong to group IA on the periodic table only have one valence electron. The weak bonds that are formed reveal why these elements are soft and have relatively low melting points.

The ease at which electrons move in metallic bonds explains different physical properties. The flow of charged particles creates electric currents. Metals already have a flow of negatively charged electrons, so they are able to conduct electricity and heat. Silver is the best conductor of electricity. The structure of metals also explains why they are malleable and ductile. When metals are forced into new shapes, the electrons and nuclei rearrange. Because metallic solids have no individual bonds, their electrons can move into new positions without breaking the metal objects.

LESSON REVIEW

1. Contrast ionic compounds and metal solids.
2. What type of bond is formed when oxygen reacts with a metal? Explain the process that can occur when this type of bond is formed.
3. Why do some metals have higher melting points than others?
4. Explain why many metals do not easily crack when they receive an impact.

TRY THIS

Bobby Pin Heat Treatment

Control: Count the number of times you are able to bend a bobby pin back and forth before it breaks. *Anneal*: Hold one end of a bobby pin with the tongs and place the other end in the flame until the metal glows red. Allow the pin to cool to room temperature on a piece of wire gauze. Count the number of times you are able to bend the heat-treated bobby pin back and forth before it breaks. *Harden*: Hold one end of an untreated bobby pin with the tongs and place the other end in the flame until the metal glows red. Immediately submerge the hot pin in a beaker of cold water. Dry off the pin and count the number of times you are able to bend the pin back and forth before breaking. *Temper*: Hold one end of an untreated bobby pin with the tongs and place the other end in the flame until the metal glows red. Submerge the hot pin in a beaker of cold water. Dry the pin and gently reheat it until a blue oxide coating appears. Allow the pin to cool to room temperature on a piece of wire gauze. Count the number of times you are able to bend the pin back and forth before breaking. Which bobby pin was the easiest to break? What does heat do to the metallic bonds?

FYI

Rubbery Metal

Researchers at Virginia Polytechnic Institute and State University have developed a metal composite with the strength of conventional metals and the vibration-dampening attributes of rubber. Why? Vibrations in metal objects can cause the metal to crack or fail, which could be catastrophic in planes or automobiles. On a smaller scale, if you have ever used a power tool or a metal tennis racket, you have experienced the jarring action of vibrations. Virginia Tech's new composite dampens such vibrations just like rubber does. The composite was created by embedding ceramic particles in a metal matrix; the ceramic pieces diffuse the vibrations and the metal provides strength and stability.

2.1.2 *Ores*

OBJECTIVES

- Discuss why metals are found in elemental form or combine with nonmetals to form ores.
- Summarize the process of transforming ores into metals.

VOCABULARY

- **ore** a naturally occurring rock from which a useful metal or mineral is recovered
- **reduction** the process of converting an ore into a free, uncombined metal

When you look at a shiny new penny, a toaster, or a glittering gold bracelet, do you ever wonder where these metals were found? Did people find them in the earth all shiny and new? If not, what did the metals look like?

Among the first metals discovered and used were gold and silver. These two metals are rare, so why did people find these metals so easily? It is true that gold and silver are not the most abundant elements on Earth, but these metals are less likely to be chemically combined with nonmetals. This characteristic made them easier to find and mine.

Most metals found in Earth's crust are combined with other substances, because metals easily lose their valence electrons and combine with oxygen or other nonmetals to form metal compounds. Metals must be separated from ores before they can be made useful. An **ore** is a naturally occurring rock from which a useful metal or mineral is recovered.

Eventually people discovered ores containing tin, and they mixed copper and tin to make bronze. Then people also began melting down iron ores to make weapons, tools, and many other useful products. As people continued to learn more about chemistry and fire, a wide variety of other ores was mined and transformed into useful products.

Most Abundant Elements in Earth's Crust

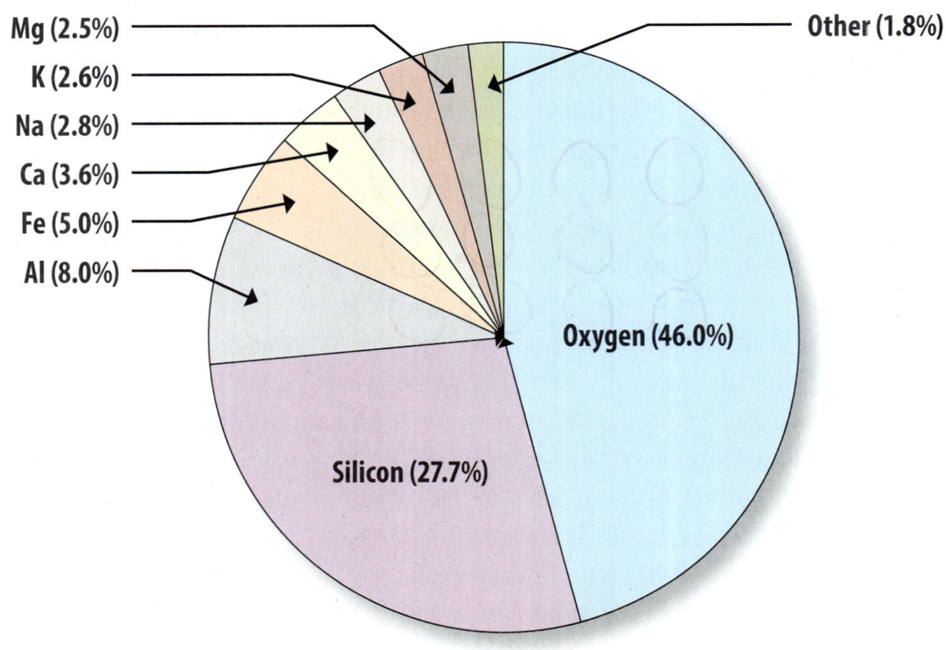

Many ores contain oxygen, silicon, or sulfur. Nearly 50% of Earth's crust is made of oxygen, and 28% is made of silicon. This makes oxygen and silicon readily available to combine with metals. Four common groups of ores are sulfides, sulfates, silicates, and oxides. Sodium sulfide and copper(II) sulfate are examples of sulfide and sulfate compounds that each contain sulfur. Silicates are compounds formed of silicon, oxygen, and metals such as aluminum, sodium, iron, calcium, magnesium, and potassium, and they are the most abundant of minerals. Silicates include quartz, emerald, asbestos, and talc. Glass is a mixture of silicates. Oxides are compounds containing oxygen. Iron(III) oxide and magnesium oxide are two examples of oxides.

Granite veins at the rock surface experence weathering.

Ores are often found in ore-filled fissures called *veins*. The veins vary in thickness from one centimeter to more than 100 meters. Some veins are deep within the ground. Others lie close to the earth's surface. Veins of ore at Earth's surface are exposed to weathering. They are often redeposited in gravel or sand. Metals are recovered using various methods depending on the type of ore. Possible methods involve separating with magnets, electricity, or by crushing and separating.

The process of removing useful minerals from the ground is called *mining*. Mining minerals from the ground has a great impact on the land because valuable minerals must be separated

Copper mine n South Africa

TRY THIS

Making Copper

Pour 100 mL of copper(II) chloride solution in a beaker. Attach one wire with an alligator clip to one of the graphite rods and attach the second wire with an alligator clip to the other graphite rod. Insert the two graphite rods in each of the U-tube openings. Place the U-tube, graphite rods, and wires in the copper(II) chloride solution. Ask your teacher to inspect your set-up before you proceed. Connect the wires to the terminals of a 6-volt battery and observe the copper precipitate forming at the bottom of the beaker.

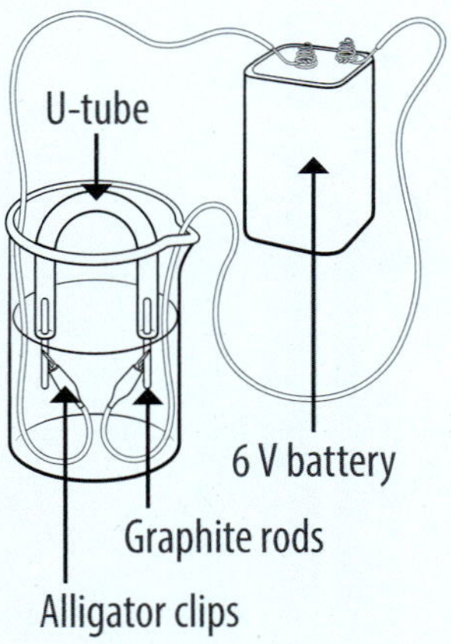

Steel metallurgical plant in Poland

from unusable ones. This type of mining operation leaves large amounts of waste. Dumping the rubble back into the hole is not an easy solution. Before they are mined, rocks and minerals are tightly compressed. However, once they are broken and processed, spaces are created between fragments, increasing the volume. Wet residues are kept in ponds. The ponds where the wet wastes are kept may break, releasing toxic chemicals into rivers, lakes, or oceans. Dry waste rock is stored in piles. If the discarded mineral fragments are exposed to water, they release acids and toxic metal compounds into the ground. Dry rubble can blow around as dust. Large, unstable piles of rubble may also create landslides.

Some minerals, such as gold and silver, must be processed with toxic chemicals after they are removed from the ground. The chemicals used for processing, such as potassium cyanide, can pollute nearby soil and streams. Today, the disposal of wastes is better regulated to ensure that damage is reduced.

The process of converting an ore into a free, uncombined metal is called **reduction**. A reactive metal is more likely to lose its valence electrons. It is more difficult to separate a reactive metal from the ore. Because gold, silver, platinum, copper, and lead are less reactive than other metals, they are easily separated from ores by roasting—heating the ores in air. Iron, chromium, zinc, and manganese are more reactive metals. These ores must be heated with carbon (coke). Extremely reactive metals such as aluminum, magnesium, sodium, calcium, potassium, and lithium are difficult to separate from their ores. Electrical currents passed through certain molten salts are generally used to separate these metals from their ores.

All metal ores are nonrenewable resources—natural resources that cannot be replaced or that can be replaced only over a period of thousands of years. Technology is helping people make better use of these dwindling resources. For example, copper ores once contained between 35% and 88% copper. These copper-rich ores are no longer plentiful, but technology makes it possible for miners to extract this metal from ores that contain less than 1% copper. It is important to remember that these God-given gifts will not be available forever, so people must not waste them.

LESSON REVIEW

1. Why is gold typically found in an elemental form in nature?
2. What are the two most prevalent elements in Earth's crust?
3. Describe three types of reduction processes.
4. Why are ores considered nonrenewable resources?

Molten iron is poured into a mold after being separated from the ore.

2.1.3 *Alkali Metals*

OBJECTIVES

- Infer the reactivity of alkali metals according to their electron configuration.
- Deduce the benefits of the alkali metal properties.

VOCABULARY

- **alkali metal** the elements in group IA on the periodic table, except for hydrogen

About 77% of the elements on the periodic table are metals. They are arranged in groups according to shared properties. These properties reflect the way that God designed each atom. Metal atoms with a common design show similar features.

The elements in group IA on the periodic table, except for hydrogen, are called **alkali metals**. Lithium, sodium, potassium, rubidium, cesium, and francium belong to the alkali metal family. The total number of electrons in each alkali metal is different, but they each have only one valence electron. When alkali metals bond with nonmetals, they easily lose this electron to form a positively charged ion that has a +1 charge.

The effortless donation of an electron makes alkali metals the most reactive metals. For this reason, alkali metals are never found in nature as free metals. They are always combined with other elements as ores or in the form of salts. Alkali metals react rapidly and violently with water to produce hydrogen gas and metal hydroxides. They also react with oxygen to form oxides. To prevent any unexpected reactions, chemists typically store alkali metals in unreactive substances such as oil. Imagine what would happen if you picked up a pebble of a free alkali metal and tossed it into a stream! Alkali metals even react violently to water vapor in the air. The fact that alkali metals are only found as ores shows God's wisdom.

All alkali metals have one valence electron.

Alkali metals have relatively low melting points and high degrees of luster. They are very good conductors of electricity and are extremely malleable and ductile. However, alkali metals are also very soft. Some can be easily cut with a knife. One way chemists identify alkali metals in the laboratory is by using flame tests. When alkali salts are exposed to an ignition source, the electrons become excited and energy is emitted in the form of distinct, identifying colors. Lithium—the lightest metallic element—produces a red color when exposed to a flame. Sodium burns bright yellow, potassium produces a blue flame, cesium burns deep blue-violet, and rubidium produces a red-violet flame. This method of identifying elements is not always reliable because some elements share similar flame colors. Scientists conduct other analysis tests to determine the true identity of any element.

Alkali metals may be dangerous as free elements, but as compounds they are a great blessing. For example, alkali metals combine with elements from group VIIA to form salts such as sodium chloride, which is table salt. Potassium is used in fertilizers, detergents, and explosives. Some atomic clocks contain

Potassium Sodium Pump

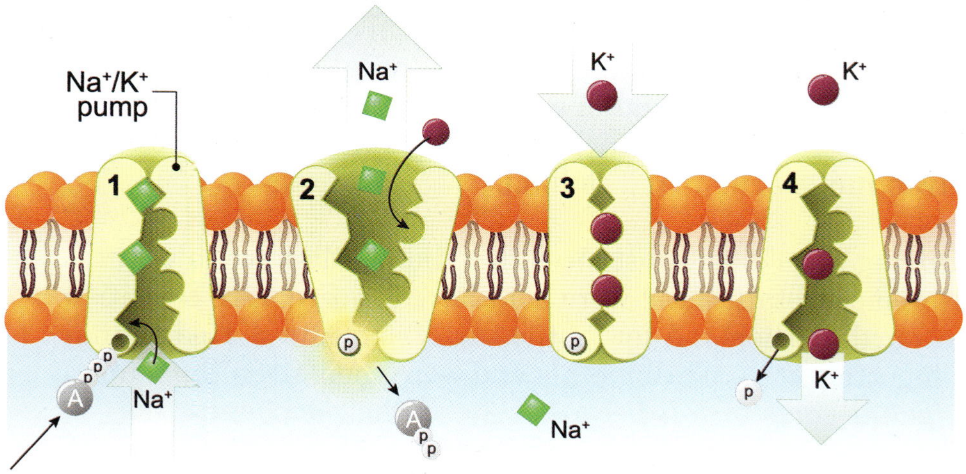

Sodium and potassium ions help maintain healthy cells.

One way that scientists help to identify alkali metals is by conducting a flame test.

rubidium compounds. Both rubidium and cesium are used to make photocells. Sodium and potassium ions are two alkali metals crucial for good health. They help with the functioning of the nervous system, and both elements control the amount of water retained by cells. Lithium is used in medicines in the form of lithium carbonate, which reduces tremors and helps treat some psychological disorders. Francium is the rarest alkali metal. It is radioactive because as its nucleus deteriorates, particles and energy are given off. The small amount of francium that does exist on Earth is usually found in uranium minerals.

LESSON REVIEW

1. Explain why alkali metals are extremely reactive.
2. Why do alkali salts produce a colored flame when ignited?
3. How can alkali metals be useful?
4. Why is very little known about francium?

Many foods, such as bananas, are rich in potassium.

2.1.4 Alkaline-Earth Metals

OBJECTIVES

- Relate the number of valence electrons to alkaline-earth metal properties.
- Recognize the uses of alkaline-earth metals.

VOCABULARY

- **alkaline-earth metal** the elements in group IIA on the periodic table

As you gazed into the night sky and admired a dazzling fireworks show, did you ever wonder what produced the vibrant colors? The ancient Chinese discovered that a particular group of elements and their compounds give off extraordinary colors when they combust. These elements are called **alkaline-earth metals**. They are found in group IIA on the periodic table. Alkaline-earth metals include beryllium, magnesium, calcium, strontium, barium, and radium.

During the 1700s, substances that did not dissolve in water, were nonmetallic, and were not altered by fire were classified as earth elements. A few of the earth elements shared similar properties with alkali metals and were then given the name *alkaline-earth metals* because some of the compounds that they produced were basic. *Alkaline* is another term for "basic." Like most metals, alkaline-earth metals are malleable and ductile, and they are good conductors of electricity. They have a gray-white luster but tarnish quickly when they are exposed to air. Alkaline-earth metals are also very reactive, but not as reactive as alkali metals. For this reason, alkaline-earth metals are not found in their elemental form in nature because they readily combine with nonmetals. They also have higher melting points and boiling points than the related alkali metals.

The two valence electrons of alkaline-earth metals contribute to their reactivity. These electrons are readily donated to nonmetals

Strontium: red, *calcium*: orange, *magnesium or barium*: white, *barium*: green, *strontium and copper mixture*: purple

Colorful fireworks shows are the result of the two valence electrons in all alkaline-earth metals becoming excited.

when they react. When an alkaline-earth metal loses electrons, it has more protons than electrons and becomes positively charged. The nonmetal that gains the electrons becomes negatively charged because it has more electrons than protons. An ionic salt forms because the oppositely charged ions are attracted to each other. For example, when calcium reacts with bromine, the salt calcium bromide is formed. Calcium bromide acts as the clear brine fluid that is found in oil and gas. Another type of reaction occurs between water and alkaline-earth metals. The product is a hydroxide and hydrogen gas. For example, when magnesium reacts with water vapor, magnesium hydroxide and hydrogen gas are produced. You may be familiar with magnesium hydroxide. It is the active ingredient in antacids that helps neutralize stomach acid and relieve heartburn.

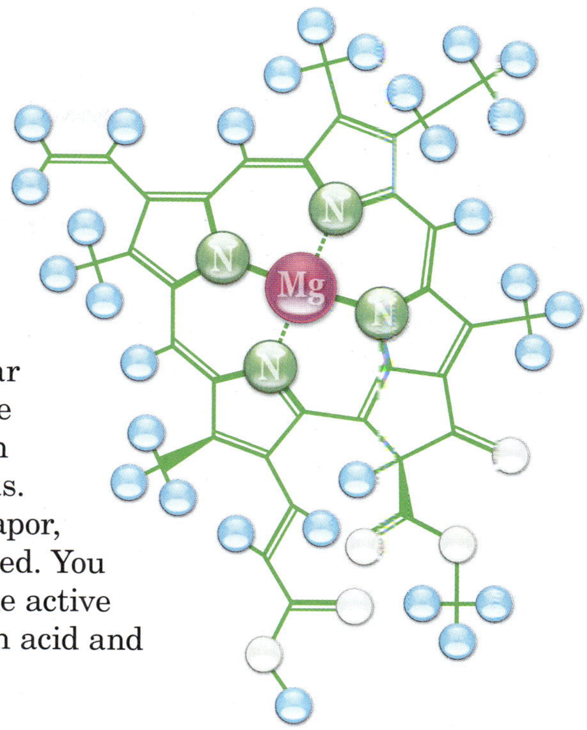

Chlorophyll is a magnesium compound.

Alkaline-earth metals have many beneficial qualities. Beryllium has a very low density. It is lightweight, very strong, and stable over wide temperature ranges. These attributes make beryllium an ideal material for space vehicles and telescope reaction plates. Beryllium is also used to manufacture radio parts. Magnesium and calcium are two of the six most common elements on Earth. Magnesium, the most widely-used alkaline-earth metal, is combined with aluminum to make lightweight, strong alloys. These magnesium compounds are used in flares, space vehicles, and medicines. Talc and asbestos contain magnesium. When magnesium is burned, it gives off an intense white light that is used in flashes, flares, and fireworks. Deposits of magnesium compounds are common both on land and in the sea. Magnesite—magnesium carbonate—reserves are prevalent across parts of Russia and China. Calcium is used to make plaster, mortar, cement, and softened water. This metal is essential for the growth of healthy teeth and bones in many animals. Calcium carbonate is abundant in limestone, chalk, and the shells of snails, clams, and oysters. In fact, the chalk cliffs of England are made of the microscopic calcium carbonate shells of marine animals.

Magnesium hydroxide helps alleviate heartburn.

Barium is used in paints and glass making. This element is also helpful in the medical field because it absorbs X-rays well. When an individual ingests

TRY THIS

Egg Dyeing

What is the chemistry behind egg dyeing? The right acidic environment is important to ensure that the dye will adhere to the surface of an egg. Add 200 mL of distilled water to each of three beakers. Prepare two vinegar solutions with different amounts of vinegar. For example, add ½ tsp of vinegar to beaker 1 and 1 tbsp of vinegar to beaker 2. The beaker without the vinegar will be your control. Add an equal amount of dye to each beaker. Place an egg in each beaker and wait for at least five minutes. Remove the eggs, and determine which solution produced the most vibrant egg color. Place a few marble chips in the best solution and wait for at least five minutes. Remove the marble chips and note any color change. Compare your results.

barium liquid, it coats the lining of the digestive system and helps reveal some differences that may not have been seen on standard X-ray images. Radium is rare and radioactive. It is often associated with uranium. During the early 1900s, radium was frequently used to treat cancer. Modern medicine has since replaced radium treatment with less expensive alternatives.

Pyrotechnicians use magnesium in fireworks. They also work with the compound strontium sulfide that produces a bright red color when ignited. Strontium-90 is a radioactive element produced in nuclear explosions. In the early 1960s, scientists discovered that strontium builds up deposits inside bones and teeth. This buildup of strontium can cause serious health issues. Concerns about strontium-90 finding its way into mothers' milk contributed to the Nuclear Test Ban Treaty of 1963, which required that nuclear testing be performed underground.

LESSON REVIEW

1. How does the number of valence electrons affect alkaline-earth metal properties?
2. Why is beryllium an ideal material for space vehicles?
3. What are two of the most common elements on Earth? What is the most widely used alkaline-earth metal?
4. What are some uses of two alkaline-earth metals?

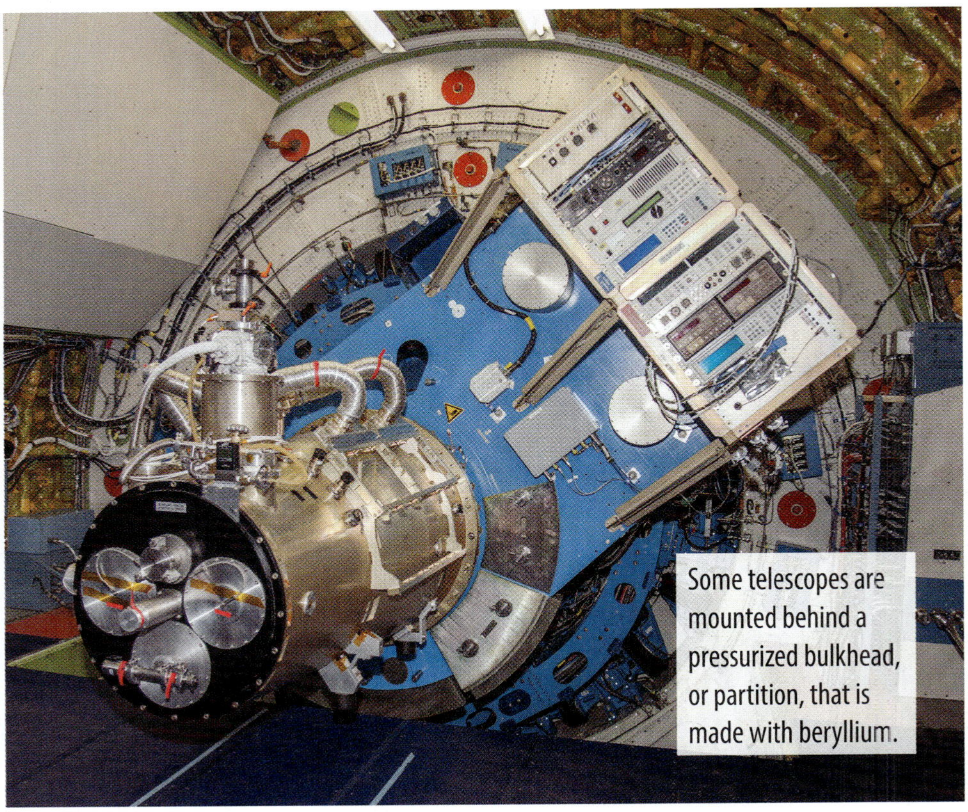

Some telescopes are mounted behind a pressurized bulkhead, or partition, that is made with beryllium.

The metals that you are probably most familiar with belong to the middle section of the periodic table known as the **transition metals**. They are between groups IIA and IIIA. Transition metal groups are identified with a roman numeral and the letter B. For example, titanium is in group IVB. The main transition elements can be found in period (row) 4 through 7 and include elements with atomic numbers between 21 and 30, 39 and 48, 72 and 80, and 104 and 112. There are two rows of elements that have been isolated from the rest of the periodic table. They are called *the inner transition elements*. The first row is called *the lanthanide series* and it includes elements with atomic numbers 57–71. These elements share a similar chemistry with lanthanum. The second row is called *the actinide series* and includes elements with atomic numbers 89–103. These elements are radioactive and unstable, and they share a similar chemistry with actinium. The two rows of inner transition metals were isolated from the rest of the periodic table to make the table more compact.

OBJECTIVES

- Discuss physical and chemical properties of transition metals.
- Summarize the industrial and biological uses of transition metals.

VOCABULARY

- **transition metal** the elements in groups II B through IIB on the periodic table

Transition metals contain valence electrons that exist in two shells instead of one. This makes transition metals highly versatile with a broad range of physical and chemical properties. In general, most metals are hard, strong, malleable, ductile,

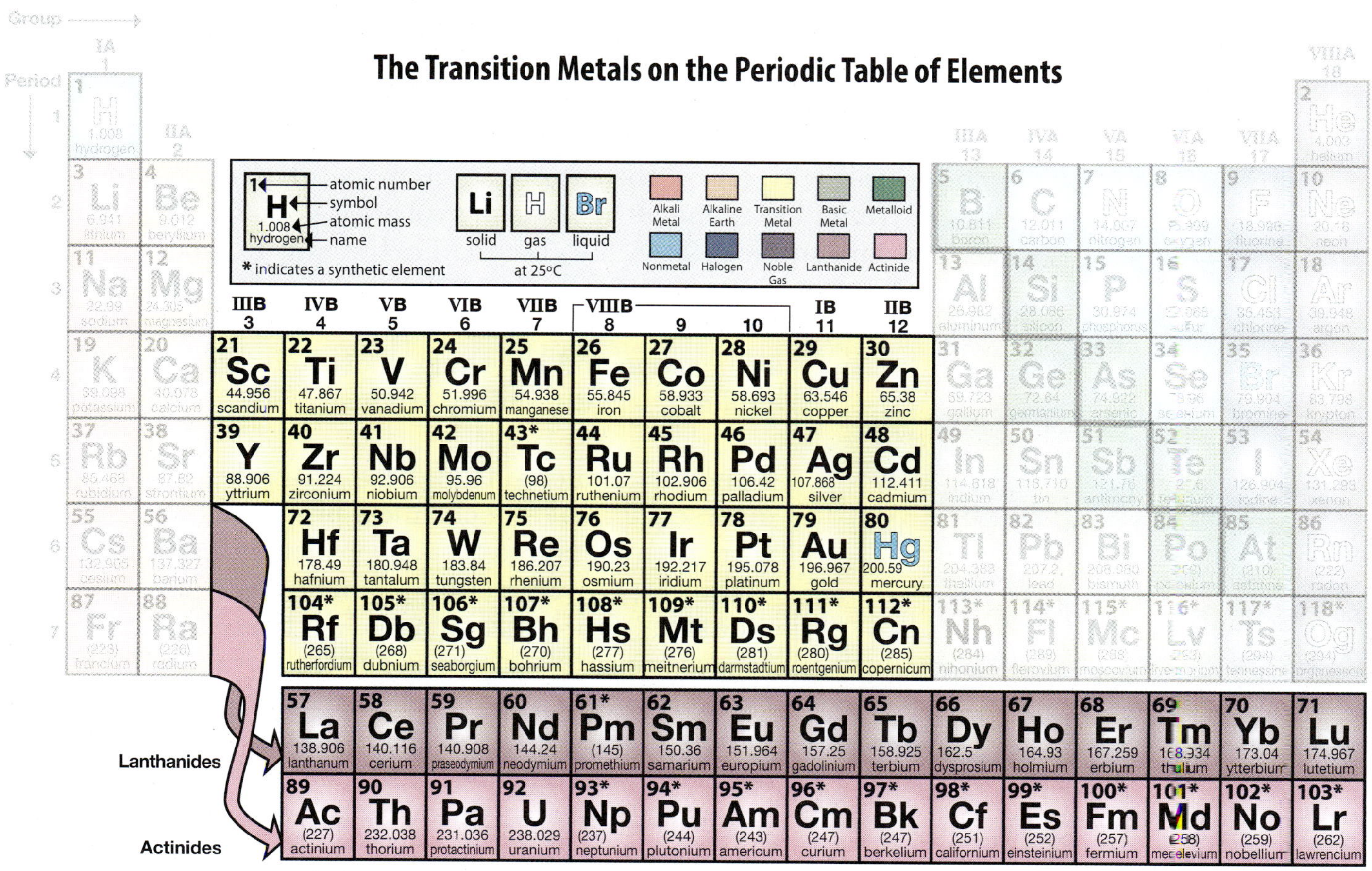

The Transition Metals on the Periodic Table of Elements

The Eiffel Tower is made of wrought iron.

and lustrous. They have high melting and boiling points and are good conductors of heat and electricity. Silver is the best conductor of electricity. Many transition metals are paramagnetic, which means they are temporarily attracted to an externally applied magnetic field. Aluminum is an example of a metal that is paramagnetic. Other transition metals, such as iron and nickel, are ferromagnetic, which means they are strongly attracted to an external magnetic field and can retain the property of magnetism after the external field has been removed.

Iron is the most abundant transition metal in Earth's crust. It was one of the first metals to be discovered along with silver, gold, and copper. A majority of the other transition metals were discovered in the late 1700s. Discovered in 1925, Rhenium was the most recent transition metal to be discovered in nature. Unlike alkali metals or alkaline-earth metals, transition metals often occur uncombined in nature. They exist in their elemental form. When they do combine to form compounds, they are easily distinguished by color.

The closest chemical similarities occur between the first three elements in each group column. However, scandium, yttrium, lanthanum, zinc, cadmium, and mercury do not share many similarities with any of the other transition metals. The fourth row of transition elements consists of synthetic elements from atomic number 93 to 116. Technetium (atomic number 43) and promethium (atomic number 61) are also synthetic. These elements have never been isolated in nature. Many are radioactive and have very short half-lives, which makes them difficult to study.

Nickel-plated guitar strings are brighter and have a little more output than stainless steel strings. They are corrosion resistant and age nicely with normal use.

Scientists and engineers have utilized the diverse physical and chemical properties of transition metals to create versatile materials and incorporate them into many scientific and industrial processes. Iron, nickel, titanium, and copper are commonly used as structural and electrical materials. Nickel is added to different metals to increase strength. It has also been applied as a protective coating on different substances because nickel resists corrosion. Copper is a popular material that is used in electrical wiring because it is an excellent conductor and is relatively inexpensive.

Zinc and cadmium belong to group IIB on the periodic table. Similar to nickel, these two elements are used to coat other metal surfaces. This process is called *plating* or *galvanizing*. A common metal that requires this type of coating is iron. Over a short period of time, iron begins to rust. When zinc is applied to iron surfaces, zinc reacts with water vapor to form an oxide before iron can oxidize. The zinc oxide coating prevents any moisture from coming in contact with the underlying metal layers. Iron is protected, and the structure remains durable for a long time.

Transition metals can be used for their beneficial physical properties. The corrosion-resistant properties of titanium and titanium alloys make these metals ideal for use in or on the human body. They do not tarnish in the presence of body fluids. Many prosthetic apparatuses, pacemaker cases, hip and knee replacements, and heart-valve devices are made with titanium. Some transition metals also act as important catalysts. A catalyst is a substance that speeds up a reaction. Many of the

TRY THIS

Hand Warmers

Place 50 g of iron powder in a small plastic bag. Add 2 g of sodium chloride to the bag. Seal the bag and mix the contents. Add 10 mL of water and thoroughly knead the mixture. What do you notice about the temperature of the bag? Why do you think there is a temperature change?

The Friendship of Nations Arch in Ukraine is covered with titanium that is three times stronger than steel, more resistant to weathering, and lighter weight.

BIBLE CONNECTION

Treat creation responsibly

God created a world rich with uncombined and combined elements. He gave people the choice to use these elements in a manner that would be pleasing to Him. Worldly riches such as those made from gold and silver do not last forever, but a relationship with God is eternal.

metal catalysts are used in the production of petroleum and polymers. With the presence of metal catalysts, industrial processes that were normally very difficult to perform can now be conducted under relatively low temperatures and pressures. One example of a transition metal catalyst is iron, which helps with the production of ammonia by increasing the interactions between nitrogen and hydrogen.

Mercury is a unique transition element. It is the only metal to exist as a liquid at room temperature. Mercury has been used in thermostats, thermometers, and batteries. This metal is not an essential element in the body. It is poisonous and can accumulate in cells through inhalation, ingestion, or skin absorption. One of the most common ways that people ingest mercury is by eating fish or shellfish that lived in mercury-contaminated water. Medical treatment is available if someone has mercury poisoning.

Transition metals are not only important for industrial applications. They also have many important biological roles. Hemoglobin proteins that aid red blood cells in the transport of oxygen from the lungs to other parts of the body contain iron. When a person's iron levels are too low, the person may become anemic and require medical attention. Copper is also prevalent in the body. Human serum contains a molecule called *ceruloplasmin* that is produced by the liver to transport copper in blood plasma. Another molecule that contains copper is tyrosinase. It is an enzyme that controls the production of melanin—a pigment that

FYI

Aluminum Bike Frames

Historically, most bicycle frames have been made of steel, which is very strong and durable, but heavy. In the 1970s, some bike manufacturers began using aluminum, which is much lighter weight than steel. However, because aluminum is about one-third as dense as steel, the early aluminum frames were overly flexible and fragile. Improvements over the next two decades increased aluminum frames' durability but made the frames stiff and uncomfortable. Recent advances in manufacturing are allowing bicycle designers to produce improved aluminum frames with the strength and durability of steel and the lightweight qualities of carbon fiber at a much lower price. Aluminum is easy to work with and inexpensive, unlike other bike materials like titanium. A new process called *hydroforming* is being used to press aluminum into complex bicycle designs that provide a smooth ride. Because of its lightness, aluminum is popular for racing bikes and for mountain bikes.

contributes to hair color, skin color, and eye color. Nickel works with liver enzymes to convert unsaturated fats to saturated fats. Without nickel, the liver would not function properly and organisms could not easily digest their food.

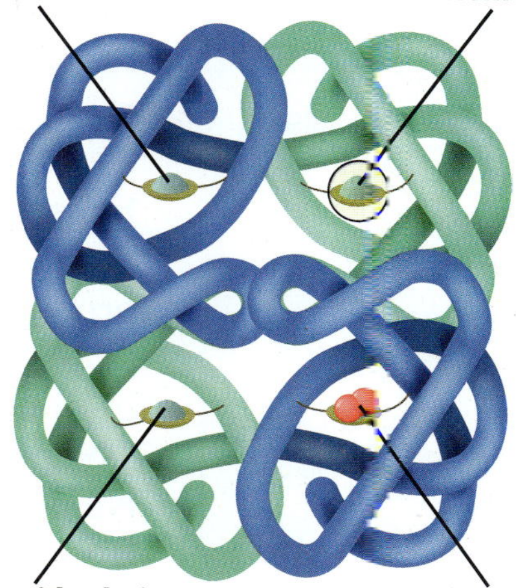

Hemoglobin proteins contain iron molecules.

The least studied transition metals are the actinides because they are so unstable. Two actinides that have been extracted from the earth's crust are uranium and thorium. Uranium contains a tremendous amount of potential energy. Some scientists have chosen to use this energy in nuclear reactors and to produce weapons. High-quality camera lenses used to be made with thorium because of the way this element reduced light distortion. However, concerns were raised regarding the type of impact a radioactive substance would have on the health of the community and the environment.

LESSON REVIEW

1. What is the difference between paramagnetism and ferromagnetism?
2. Why is it beneficial to apply a zinc coating to metals such as iron?
3. Why is titanium an ideal metal to use in the human body? What are some examples of titanium materials that are used in the body?
4. Describe two examples of transition metals that serve industrial and biological uses.

HISTORY

History: Coins

The coins you have in your pocket are so familiar you probably do not think about them, but before the 5th or 6th century BC, such forms of payment did not exist. Historians do not agree on when the first coins were minted or on what civilization first used them for payment. The earliest known coins were made of electrum, which is an alloy of gold and silver. Many early coins were minted by individual merchants for use in trade or as a pledge. Other coins were created by local rulers. The use of coins seems to have spread from the Mediterranean region outward through trade routes, which extended to Egypt, Persia, India, Europe, and Asia. Some Greek coins were even bilingual for use in particular regions of the Greek Empire. The first standardized gold coins were minted around 215 BC in China during the Qin dynasty. Later, paper money was invented by the Chinese as well.

2.2.1 Nonmetals

OBJECTIVES

- Compare the physical and chemical properties of nonmetals with metals.
- Categorize elements according to their properties.
- Describe the unique characteristics, places of origin, and uses of nonmetal and metalloid elements.

VOCABULARY

- **metalloid** an element that has similar physical and chemical properties to both metals and nonmetals

Of the 118 known elements, about 15% are nonmetals. Even though there are only a few nonmetals, they comprise the majority of elements that are essential for the health and function of your body. Nonmetals also represent a large portion of the elements found in Earth's crust.

Except for hydrogen, nonmetals are located on the right side of the periodic table. Nonmetals have physical and chemical properties that are in direct contrast to metals. They are brittle, so they are not malleable and ductile. Nonmetals cannot be hammered into sheets, drawn into wire, or otherwise formed. They do not have a shiny luster, so they do not reflect light well. Nonmetals are poor conductors of electricity and heat because their electrons are strongly attracted to their nuclei.

A majority of metals exists as solids at room temperature, but nonmetals are usually gases or solids at room temperature. The nonmetals nitrogen, oxygen, fluorine, and chlorine are gases under ordinary conditions. Bromine is a dark red liquid. The nonmetals carbon, phosphorus, sulfur, selenium, and iodine are all solids at room temperature. The gases in group VIIIA on the periodic table are also all gases at 25°C. They are the only elements that have full valence shells. These gases are sometimes referred to as *inert gases* because they do not readily react with other elements.

Nonmetal atoms can form ionic or covalent bonds. Because nonmetals have four or more valence electrons, they tend to gain valence electrons when they react with metals. Nonmetals that gain an electron become negative ions and form ionic bonds. They may also share electrons with other nonmetallic atoms. This forms covalent bonds.

The first nonmetal on the periodic table is carbon. This element belongs to group IVA. Carbon may not be the most plentiful element in Earth's crust, but it is widely distributed in the form of coal and petroleum products. It is also found in all animal and plant tissue. It has been discovered that carbon forms more compounds than all the other elements combined. One use for carbon is radiocarbon dating. Carbon 14 is utilized in this dating process to help scientists determine the age of rock strata and deceased organisms.

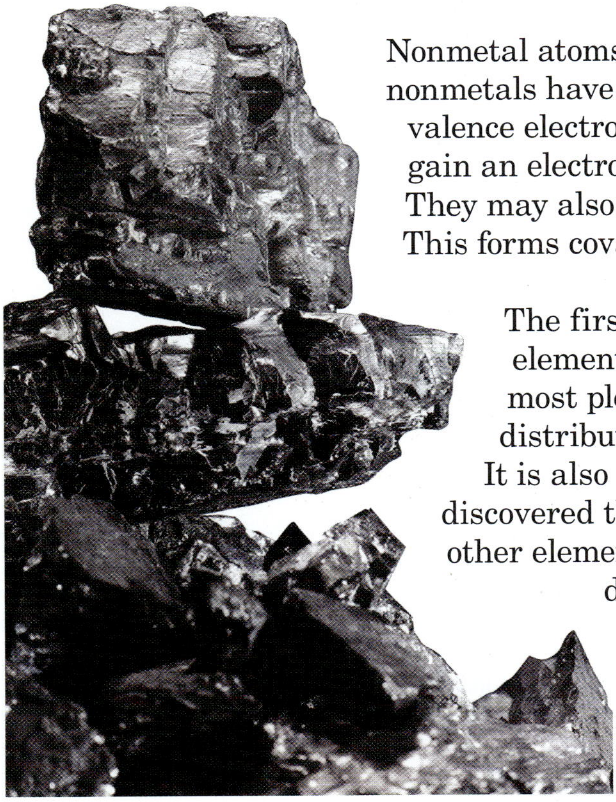

Nonmetals are brittle and lack luster.

FYI

CO_2 Detection

As you are preparing to watch a movie, the idea of popcorn springs into your mind. You set the microwave for 10 minutes and decide there is plenty of time to take out the trash. All of a sudden, that distinct blaring sound penetrates the entire home. The smoke detector is going off because your popcorn burnt. But, how exactly does the smoke detector recognize that there is smoke? There are two types of smoke detectors—ionization and photoelectric. Ionization detectors have specialized chambers that contain a battery that sends a voltage to one positive plate and one negative plate. There is also a very small amount of a radioactive material, Americium-241, inside the chamber that emits alpha particles at a constant rate. These particles resemble a helium atom that contains two protons and two neutrons. The alpha particles ionize oxygen and nitrogen atoms from the air that travel through the chamber. This reaction produces free electrons that travel toward the positively charged plate and induce a constant current. Whenever carbon dioxide enters the chamber, it causes this current to become unbalanced. A change in current triggers the alarm to sound. Ionization detectors can also produce false alarms any time dust or other types of gases enter the chamber and disrupt the current.

Nitrogen is a nonmetal in group VA on the periodic table. It is an odorless, tasteless, colorless gas that is the most plentiful element in Earth's atmosphere. About 80% of the atmosphere is nitrogen. This element is essential to all living things. It can be found in its elemental form in meteorites, the sun, some stars, and in the gases near mines, volcanoes, and some mineral springs. One of nitrogen's uses includes preventing spoilage of foods that may be exposed to mold or insects or may be prone to oxidation. Liquid nitrogen has also been used to freeze dry items and is used in refrigeration systems.

Oxygen is also a tasteless, odorless, colorless gas. It is located in group VIA on the periodic table. Plants produce oxygen that animals and other organisms require to live. This element can form compounds with virtually all other elements. It is a main ingredient in water and is the most plentiful element in Earth's crust. Sulfur also belongs to group VIA on the periodic table. Pure sulfur appears as a pale yellow, brittle solid that is insoluble in water. It is odorless and tasteless, and it is a poor electrical conductor. Sulfur can react with any metal except platinum

Nonmetals exist in all three states. Liquid nitrogen readily converts to a gas without being boiled

BIBLE CONNECTION

Brimstone

The people who lived during the time of Moses were very familiar with a number of nonmetals. In particular, there are references to a substance they called *brimstone*. This substance is another name for *sulfur*. Elemental sulfur is a brittle, yellow solid. The people knew that when sulfur burned, it produced a noxious gas that was harmful to inhale. When the Bible mentions brimstone, it usually associates brimstone with a form of punishment for disobedience to God. Deuteronomy 29:22–23 explains that when the people would not obey the covenant of the Lord, they would be subjected to a burning wasteland of salt and sulfur where no vegetation would grow. God used such an analogy to ensure the people would know the gravity of their actions.

and gold. It also covalently bonds with many other nonmetals. Large quantities of sulfur are mined every year, usually to produce sulfuric acid. Sulfur can be found near volcanoes or in underground deposits where salt domes are located in limestone rock. Two predominant underground deposits are located in Louisiana and off the Gulf of Mexico. Sulfur is used to make detergents, petroleum products, pigments, and storage batteries.

Selenium is a nonmetal located beneath sulfur on the periodic table. Selenium is extremely rare and is hardly ever found in its elemental form. When scientists discover selenium in nature, it is usually bound to copper, lead, silver, or mercury. Many of these selenium compounds are toxic to people and animals. Scientists have discovered uses for selenium that include electronic equipment, glass making, and the production of pigments.

To the left of the nonmetals on the periodic table are the **metalloids**. They make up the remaining 7% of the elements and have similar physical and chemical properties to both metals and nonmetals. Like the other elements, these are a very important part of creation, and their observable properties are related to the way God designed their structures. The metalloid elements include boron, silicon, germanium, arsenic, antimony, tellurium, and astatine. They are found along the zigzag line of the periodic table. Most metalloids conduct heat and electricity, but not as well as metals. Metalloids are ductile and malleable. They are poor conductors of electricity at low temperatures but good conductors of electricity at higher temperatures. Metalloids typically form covalent bonds in compounds. They can also form ionic bonds.

Boron is a metalloid in group IIIA. It is a hard, brittle solid that is never found in its elemental form in nature. Boron always forms compounds with other elements. The compound boric oxide (B_2O_3) is an important compound that can be used to make heat-resistant glass. Boric acid can also be used as an antiseptic and eye wash. Another compound containing boron is borax. Industries often place this compound in cleaners and water softeners.

The metalloid arsenic can be found in many minerals. It is poisonous when consumed from groundwater. Elemental arsenic is relatively inert. However, when arsenic reacts with hydrogen, a very poisonous gas called *arsine* is produced. Normally, arsenic is found combined in nature. When combined with sulfur, arsenic appears as yellow arsenic or metallic arsenic. Metallic arsenic is used to harden alloys, and arsenic compounds are often used as herbicides and pesticides.

Lenses that contain germanium enhance and amplify infrared light so night vision is possible

Silicon is the most common metalloid, and scientists often consider it to be the most important one. Although silicon does not occur free in nature, silicon compounds make up about 95% of Earth's crust. It exists in sand particles, rocks, and soil. Silicon has four valence electrons and a crystalline structure similar to carbon atoms in diamonds. Chemists purify silicon to make iron, steel, copper, and bronze alloys. Engineers have harnessed the semiconductor nature of silicon to manufacture computer chips. Semiconductors have an electrical conductivity between an insulator and a metal conductor. A single computer chip can contain millions of transistors that process billions of bits of information each second. Silicon is necessary to computer function.

LESSON REVIEW

1. What type of bond would lithium and chlorine form? What type of bond would nitrogen and oxygen form? In the ionic compound, which atom would form a negatively charged ion?
2. How are solid metals different from solid nonmetals?
3. Would the following description pertain to a metal, nonmetal, or metalloid: this element is malleable and ductile but only conducts electricity at high temperatures?
4. Describe two nonmetals that are essential to all living things.
5. Name two uses of the metalloids boron and arsenic.

HISTORY

Egyptian Antimony

Antimony is a brittle metalloid that has a silver-blue luster. There are very small amounts of elemental antimony found in nature. It mostly occurs as the mineral stibnite, which is a combination of antimony and sulfur. Ancient civilizations used stibnite in a variety of ways. Chaldean artifacts that contain antimony were found to be dated around 4,000 BC. In 2 Kings, there is reference to Jezebel putting on eye makeup. Historians believe the makeup was black stibnite. During the 1st century AD, philosophers and naturalists mention antimony being used in medicine. Today, antimony is primarily used in alloys to create materials such as cable coverings, tanks, pipes, diodes, infrared detectors, and flame retardants in paint.

TRY THIS

Growing a Silicate Mine

Set a pan in an area that will ensure the contents will not be disturbed over the next few days. Pour 800 mL of water in the pan. Thoroughly mix 200 mL of sodium silicate with the water. Gently add about 5 g of aluminum chloride hexahydrate, cobalt(II) chloride, copper(II) chloride, and iron(III) chloride to the solution with a spatula, leaving space between each salt so the variety of crystals can grow. Observe the crystals' formations. After a few days, siphon off the sodium silicate solution with a large pipette and add water to preserve the crystals.

2.2.2 *Hydrogen*

> **OBJECTIVES**
> - Theorize why hydrogen is considered a nonmetal.
> - Establish the importance of hydrogen.
> - Distinguish among the three hydrogen isotopes.

What is colorless, odorless, and tasteless, and is the most abundant nonmetal in the universe? It is hydrogen. This element is one of the major components of water that is found in rivers, lakes, oceans, and the atmosphere. Hydrogen is also found in all animal and plant tissue just as carbon is. Hydrogen gas is sparsely scattered in the vast amount of space between stars, and a majority of the stars consist of hydrogen. Reactions involving hydrogen provide the energy, or fuel, for the sun, and much of that energy is released in the form of heat and light.

A hydrogen atom consists of one proton and one electron. It is in group IA of the periodic table, above the alkali metals. Hydrogen can act like the members in the alkali family because it can donate its one electron during chemical reactions. However, hydrogen can also gain an electron to have a full valence shell, similar to the halogen elements in group VIIA. Under normal temperature and pressure conditions, hydrogen exists as a diatomic molecule, H_2, in the gaseous state. Hydrogen is the lightest gas. With such a low density, substances are able to diffuse through hydrogen molecules very quickly. This property also makes hydrogen the greatest heat conductor. It also has extremely low melting and boiling points because the attractive forces between molecules are weak.

Hydrogen atoms and molecules are very reactive. The molecules will separate into two hydrogen atoms any time they come in contact with an extremely hot substance or in the presence of an electric discharge. Hydrogen is versatile because it can combine

Hydrogen naturally forms diatomic molecules and reacts with most elements and compounds.

or react with most elements and compounds. The temperature at which the reaction occurs is important. Reactions that occur at room temperature tend to be extremely slow. The reaction rate increases when the temperature is raised. Individual hydrogen atoms react with other elements to create a hydride. For example, one calcium atom reacts with two hydrogen atoms to form calcium hydride, CaH_2, which is a useful drying agent. Elemental hydrogen can also react with metallic oxides to isolate the elemental metal. When pure hydrogen is burned in air, it reacts with the oxygen to form water, which also releases heat.

Ammonia is injected into soil to act as a fertilizer.

Hydrogen has an incredible number of industrial uses, and millions of pounds of hydrogen are consumed daily around the world. The main process that involves hydrogen is the manufacture of ammonia, NH_3. It is produced by the Haber-Bosch process where a catalyst promotes the reaction between nitrogen and hydrogen at elevated temperature and pressure conditions. Consumers use ammonia in many household cleaning products and fertilizers. Another use of hydrogen is the hydrogenation of

HISTORY

Hindenburg

After 80 years of speculation and investigation, researchers at the Southwest Research Institute in San Antonio, Texas, have determined the cause of the fire on the airship *Hindenburg*. Their findings indicate that static electricity sparked the fire that killed 35 of the 97 people onboard the zeppelin on May 6, 1937. The *Hindenburg* flew through a thunderstorm, which caused a buildup of static electricity. When the members of the ground crew grasped the mooring ropes, it grounded the airship and caused a spark. The spark would have been inconsequential except there was a small leak of hydrogen gas, which is flammable. As the gas ignited, the rear section of the zeppelin quickly caught fire. The intense fire spread quickly and in less than a minute, the *Hindenburg* was consumed. The team subjected small replicas of the airship to different conditions, examined the original footage of the disaster, and studied eyewitness accounts to determine the cause of the fire. The team's findings confirm the original investigations' conclusions from 1937.

Deuterium and tritium react inside hydrogen bombs during the fusion process.

organic compounds. Many unsaturated vegetable and animal oils are bound to hydrogen atoms to create products such as margarine and peanut butter. Hydrogen is the preferred propellant for rocket fuel along with oxygen or fluorine. Other hydrogen products include glass, vitamins, cosmetics, semiconductor circuits, soaps, and lubricants.

Peanut butter s a type of hydrogenated food.

There are three hydrogen isotopes. The most prevalent isotope is protium that has a mass of one because it has one proton and no neutrons. This isotope accounts for 99.98% of the naturally occurring element.

Deuterium is the second hydrogen isotope that has a mass of two. This isotope has one proton and one neutron. It accounts for nearly 0.02% of all hydrogen. If there are high concentrations of deuterium in water habitats, animals and plants do not thrive, and eventually die. Scientists have found a use for deuterium as a moderator in nuclear reactors to help slow down the neutron particles without significantly capturing them.

The third hydrogen isotope is tritium, which has a mass of three because it has one proton and two neutrons. This isotope is radioactive and rarely occurs in nature. The small amount that does reach Earth in the form of rain is the result of nuclear reactions caused by cosmic rays in the upper atmosphere. Scientists measure the amount of tritium in water supplies to examine how water is circulating in the oceans, lakes, rivers, and the atmosphere. Apart from atmospheric reactions, tritium can be artificially made in nuclear reactors. Both deuterium and tritium are involved in the fusion process that occurs inside hydrogen bombs. This process is still very volatile, and scientists have not yet determined how to harness this power. When scientists develop a way to control the fusion process, it is possible that an unlimited supply of clean energy would be available.

LESSON REVIEW

1. What does a low density indicate about hydrogen molecules?
2. Is hydrogen a metal or a nonmetal? How could hydrogen be considered both a metal and a nonmetal? Explain.
3. Would the smell of perfume permeate quicker through a room filled with hydrogen gas or nitrogen gas? Why?
4. When do hydrogen molecules separate into individual atoms?
5. Why is hydrogen so important?
6. What hydrogen isotope is found in animal and plant tissue? Name two uses of the other two hydrogen isotopes.

2.2.3 *Halogens*

> **OBJECTIVES**
> - Infer how the electron configuration determines the reactivity of halogens.
> - Outline the different uses of halogens.

> **VOCABULARY**
> - **halogen** an element that belongs to group VIIA on the periodic table

Toothpaste, colorful cosmetics, and industrial bleach are all products of chemical compounds that contain a special group of elements called **halogens**. These nonmetal elements belong to group VIIA on the periodic table. Halogens include fluorine, chlorine, bromine, iodine, and astatine. The elements in the halogen family are reactive nonmetals. They share similar properties because they each have a total of seven valence electrons.

Halogens are highly reactive because they need only one valence electron to achieve a stable state. When halogen atoms react with metals, the halogen gains an electron and becomes an anion with a –1 charge. This electron configuration is similar to the noble gas elements found in group VIIIA on the periodic table. An example of a halogen and metal reaction is chlorine reacting with sodium to form sodium chloride. The product of the halogen and metal reaction is a salt. This is how halogen elements received their family name. The Greek prefix *hal–* means "salt" and the suffix *–gen* means "to produce."

Halogens, in blue above, are highly reactive nonmetals.

Group VIIA elements can also form covalent bonds with other nonmetals. These elements will never be found free in nature because of their extreme reactivity. At room temperature and normal atmospheric pressure, a halogen atom will form a diatomic molecule with another identical halogen atom. When a

> **TRY THIS**
>
> **Toothpaste Comparison**
>
> Label partitioned petri dishes with the names of the four different brands of toothpastes being tested. Also, write the word *before* on one-half of each dish and the word *after* on the other half of each dish. Fill the petri dishes with agar and place them in the refrigerator so the agar solidifies. Remove the petri dishes from the refrigerator and allow them to reach room temperature. Rinse your mouth with water for about five seconds. Moisten a sterilized swab with distilled water and rub the swab over the surface of your teeth. Roll the swab on the half of the agar labeled *before* in the first petri dish. Brush your teeth with the first brand of toothpaste to be tested for a determined period of time. Moisten a second sterilized swab with distilled water and rub the swab over the surface of your brushed teeth. Roll the swab on the half of the agar labeled *after* in the first petri dish. Cover and place the dish in a cool area, away from direct sunlight. Repeat these steps over the next three days with the other brands of toothpaste. Make sure you brush your teeth for the same amount of time during each trial. After five days have elapsed, measure the bacterial growth with a ruler. Create a graph that compares before and after bacterial growth for each brand. Do you think the amount of fluoride in each toothpaste determines its effectiveness?

halogen bonds with a different atom, compounds called *halides* are produced. All halogens can combine with hydrogen to form hydrogen halides. These compounds are colorless gases. The hydrogen atoms in hydrogen halides may also form hydrogen bonds with surrounding compounds. Hydrogen bonds are weaker than the intermolecular covalent bonds. Like most nonmetals, halogens have relatively low melting and boiling points. The melting points and boiling points increase as you move down the group. Astatine has the highest melting point.

Fluorite is the mineral form of calcium fluoride.

Halogens are an important part of creation, and God designed them to be a very important part of your health as well. Fluorine is the most abundant halogen in Earth's crust. This element normally exists as a gas at room temperature. Compounds that contain fluorine are often added to toothpastes and water supplies to produce a hard, acid-resistant layer on the surface of your teeth. This tough layer helps to protect teeth from developing cavities. The halogen iodine is a solid at room temperature that easily sublimes. It is found in some antiseptics and it is part of the growth-regulating hormone thyroxine—a hormone your body needs. A lack of iodine may cause the thyroid gland to enlarge—a condition called *goiter*. This condition is usually reversible and can be prevented with a sufficient supply of iodine. Diets that include fish and iodized salt are good sources of iodine.

Chlorine compounds kill harmful bacteria.

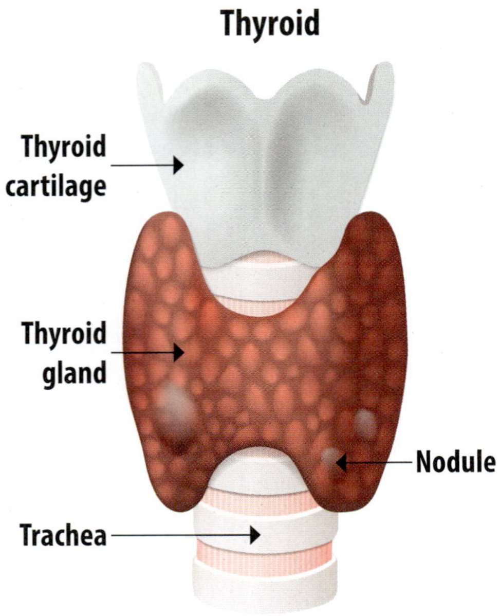

Iodine helps the thyroid gland produce an important growth-regulating hormone.

The halogen chlorine rarely occurs alone in nature, except in small amounts in volcanic gases. Chlorine is a greenish-yellow, pungent gas at room temperature. It is extremely harmful to inhale. Chlorine compounds, on the other hand, are abundant on Earth. These compounds that combine chlorine with alkali or alkaline-earth metals form salts such as potassium chloride, magnesium chloride, and calcium chloride. These compounds occur naturally in seawater and in salt deposits formed by the evaporation of ancient seas. Chlorine is used for bleaching pulp wood for paper, bleaching textiles, making plastics, manufacturing insecticides, and producing dry-cleaning agents. The chlorine ion OCl^- is commonly added to water in swimming pools and drinking water to kill bacteria. It is also used to improve drinking water by changing chemicals that cause bad odors and tastes.

Bromine is the only nonmetal that exists as a liquid at room temperature. This halogen has a strong odor. In fact, its name is derived from the Greek word *bromos*, which means "stench." Bromine is found in seawater and salt deposits. Israel produces large quantities of bromine from deposits in the Dead Sea. Bromine is toxic and reacts easily with other elements. It was used to make a gasoline additive and is an ingredient in pesticides. Bromine is vital for fireproofing plastics and textiles. Silver bromide, a photosensitive compound, is used to produce photographic film.

LESSON REVIEW

1. Halogens try to attain a similar electron configuration to what periodic table family?
2. How does the electron configuration affect the physical and chemical properties of halogens?
3. At room temperature and normal atmospheric pressure, halogens will typically form what type of compound?
4. Why is iodine important for your health? What can you do to maintain a good level of iodine?
5. Describe the two natures of chlorine.

In 1894, English physicist Lord Rayleigh isolated a gas that he presumed was nitrogen. However, he soon discovered that this new gas was denser than nitrogen. Lord Rayleigh collaborated with Scottish chemist Sir William Ramsay to discover argon. Although argon makes up about 1% of the atmosphere, earlier scientists never noticed it. Argon most likely escaped notice because it does not react with other substances. It is an unreactive element because it has eight valence electrons. Argon already has a stable electron configuration and is not prone to gain or lose any electrons under ordinary conditions. Because this gas would not react during experiments, Ramsay named it *argon* after the Greek word for "lazy."

The periodic table was designed before argon was discovered, so Ramsay proposed that another group be added to the table to accommodate the discoveries of argon and helium. Group VIIIA was added and given the name **noble gases**. Elements that are found in the noble gas family include helium, neon, argon, krypton, xenon, radon, and organesson. All of these elements have eight valence electrons and will not react with other substances under ordinary conditions. This was an important discovery that helped Niels Bohr determine that electrons exist in distinct shells around an atom's nucleus. It also helped explain how other elements react in order to achieve an electron configuration similar to these inert noble gases. These other atoms will either gain or lose electrons in an attempt to

OBJECTIVES

- Identify the physical and chemical properties of noble gases.
- Discuss the practical uses of noble gases.

VOCABULARY

- **noble gas** a chemically stable element that belongs to group VIIIA on the periodic table and does not readily react with other elements under ordinary conditions.

Scuba divers need a mixture of helium and oxygen while underwater for long periods of time.

Argon welding machine

have eight electrons in the outermost shell. In 1898, Ramsay and his assistant, William Travers, discovered neon, krypton, and xenon by preparing liquid air and fractionally distilling it until each gas boiled off at a separate boiling point. Radon was discovered in 1900 by German scientist Friedrich Dorn.

All of the noble gases are colorless, odorless, and tasteless nonmetals. They are nonflammable because they do not react with oxygen. All occur naturally in small quantities as single atoms in the atmosphere. They have very low boiling points and melting points. Helium is the lightest element except hydrogen. Because it is less dense than air, it is used in balloons and blimps to make them float. Helium is much safer than hydrogen because it does not burn. It also has the property of low solubility in fluids. This makes helium ideal for use by scuba divers. Helium is mixed with oxygen in air tanks to prevent nitrogen from dissolving in the divers' blood and forming bubbles. This mixture helps prevent a potentially fatal condition called *the bends.*

Helium was first detected in the sun before it was discovered on Earth. It is the most plentiful element in the universe except for hydrogen. Helium is formed during nuclear fusion reactions on the sun and stars. It is also formed naturally on Earth from the decay of radioactive substances such as uranium. Many

VIIIA 18			
2	He	4.003	helium
10	Ne	20.18	neon
18	Ar	39.948	argon
36	Kr	83.798	krypton
54	Xe	131.293	xenon
86	Rn	(222)	radon
118*	Og	(294)	organesson

Noble gases are relatively inert because they have a full valence shell.

Astronomy gives a glimpse into the wonder of God's amazing handiwork beyond Earth.

commercial industries recover helium from natural gas wells.

Chemists have utilized argon to increase the life-span of incandescent light bulbs and to create a whiter light in light bulbs by permitting the filament to burn at higher temperatures. In industry, argon often provides a nonflammable atmosphere for welding and refining metals. If certain metals, such as aluminum, were refined in an oxygen environment, the metal would react with the oxygen.

Neon and other gases burn to produce multicolored lights.

A common use for the noble gas neon is neon lights. The light of neon bulbs is produced by passing electrons through tubes of gas. True neon lights produce a bright red-orange color. If other gases are added or neon gas is replaced, different colors are produced. For example, xenon produces a blue color. That is how "neon" signs are made in a wide variety of colors.

The last three main noble gases are krypton, xenon, and radon. At first, noble gases were called *inert gases* because they were thought to be unable to react with any other substance. However, scientists have been able to force the creation of xenon, radon, and krypton compounds in the laboratory. Krypton has no major industrial uses. Xenon is used in electronic flashes for cameras. It is also a good anesthetic, but it is too expensive for general use. Radon is a radioactive gas that is usually isolated as a product of the decomposition of radium compounds. Long-term exposure to radon can cause lung cancer, but radon has also been used in the past to treat cancer. The risk factors associated with this method of cancer treatment is a concern. Other methods are preferred. Organesson is a synthetic element that was first created in 2002. Very little is known about the physical and chemical properties of this element because few atoms have been made.

LESSON REVIEW

1. List some physical properties that all noble gases share.
2. Why are noble gases so unreactive?
3. How is helium formed?
4. What are some practical uses for noble gases?

TRY THIS

Hydrogen Versus Helium
Ignite a wood splint with the Bunsen burner. With a pair of tongs, hold the glowing splint near the opening of a helium balloon and slowly release the helium gas. Do not touch the splint to the balloon. What do you observe? How does this compare to hydrogen gas?

2.3.1 *Simple Organic Compounds*

OBJECTIVES

- Distinguish between organic and inorganic compounds.
- Describe the molecular structures of saturated and unsaturated hydrocarbons.
- Define and provide examples of functional groups and the resulting compounds.

VOCABULARY

- **alkane** a saturated hydrocarbon that has single bonds between the carbon atoms
- **alkene** an unsaturated hydrocarbon with at least one pair of carbons joined by a double covalent bond
- **alkyne** an unsaturated hydrocarbon with at least one pair of carbons joined by a triple covalent bond
- **hydrocarbon** an organic compound containing only hydrogen and carbon atoms
- **isomer** a compound that has the same chemical formula as another compound but has a different molecular structure
- **organic compound** a compound that contains carbon

Throughout history, scientists have been trying to create an organized system they could use to categorize different substances. Two of the categories they defined in the 1780s were organic compounds and inorganic compounds. Initially, organic compounds included anything that could be derived from living things. Inorganic compounds included everything else. These simple definitions were revised by Antoine Lavoisier, who in 1784 discovered that organic compounds were made primarily of carbon, hydrogen, and oxygen.

During this time, many people believed in a mysterious philosophy called *vitalism*. According to this philosophy, a "vital force" was necessary for the creation of an organic compound. Those who believed in vitalism thought that organic compounds could never be synthesized in the laboratory. In 1828, German chemist Friedrich Wohler dispelled the belief in vitalism by making the organic compound urea in his laboratory. Between 1828 and 1850, many organic substances were created from inorganic compounds.

Today, nearly five million synthetic organic compounds are known, and 100,000 new compounds are created each year. Some of these compounds include plastics, gasoline, rubber, and soap. These compounds all have one chemical structure

Organic Compounds

People

Diamonds

Plastics

Rubber

Soap

Freon

similarity. They each contain carbon atoms. An **organic compound** is a compound that contains carbon. These compounds often contain hydrogen and oxygen, and they can also contain nitrogen, sulfur, phosphorus, and other elements. However, not all compounds that contain carbon atoms are organic. A carbon dioxide molecule has one carbon atom, but it is not considered an organic compound. Also, compounds that have a cyanide group (CN^-) or a carbonate group (CO_3^{2-}) are not organic compounds. They are inorganic compounds.

Organic compounds are nonpolar and cannot dissolve in water.

Organic compounds share several other characteristics. They usually form nonpolar, covalent bonds without charges on either end of their molecules, unlike many inorganic compounds that form ionic bonds. Because most organic compounds are nonpolar, they do not generally dissolve in water. Like substances dissolve like substances. Remember, water is a polar molecule. Many inorganic substances dissolve in water because they are polar molecules. Organic compounds dissolve in other nonpolar organic liquids such as alcohol, carbon tetrachloride, and cooking oil. They exist as gases, liquids, or solids with low melting points.

Inorganic Compounds

Carbon dioxide

Water

Calcite

Magnesite

Baking soda

Marble

Many have strong odors and low boiling points. Most cannot conduct electricity.

How can carbon atoms create so many different compounds? Look at how God created a carbon atom. Carbon atoms have four unbound valence electrons. Each carbon atom can form four covalent bonds with other carbon atoms or other elements. The covalent bonds between the carbon atoms can be single, double, or triple bonds. In a single covalent bond, two atoms share one pair of electrons. Two atoms share two pairs of electrons in a double covalent bond. In a triple covalent bond, two atoms share three pairs of electrons.

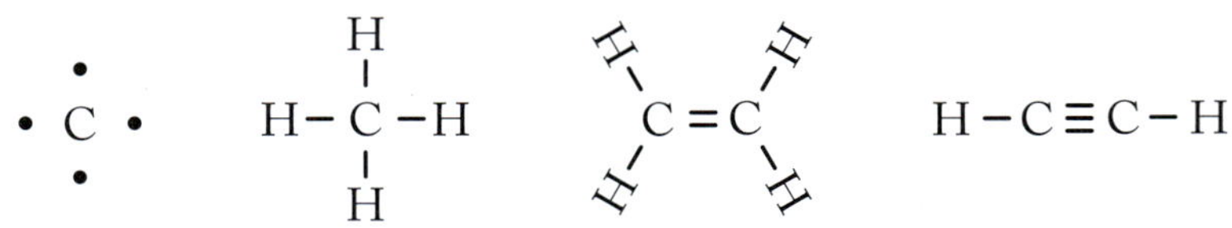

Single Bond **Double Bond** **Triple Bond**

Carbon bonds do not always form in a straight line. Chains of carbon atoms can form molecules with a straight-chain structure, a branched-chain structure, or a ring structure. The carbon atoms in the middle of a straight-chain structure are bonded to no more than two hydrogen atoms. Terminal carbon atoms, the ones at the ends, can bond to three hydrogen atoms. At least one carbon atom in a branched-chain structure may be bonded with three hydrogen atoms, which makes the structure resemble a tree. There may be as few as three carbons in the cyclic structure of a ring with at least one hydrogen bonded to each.

HISTORY

DEET

Historically, people have used such substances as smoke, tar, and mud to repel mosquitoes. However, none of these methods was really helpful. The first effective ingredient used in mosquito repellents was citronella oil, which is extracted from the citronella plant. Citronella's repellent quality was discovered by accident in 1901 when it was used as a hairdressing fragrance. However, citronella is flammable and evaporates very quickly, so researchers for the US military began formulating synthetic mosquito repellents. In 1955, scientists created DEET, which became the most commonly used active ingredient in mosquito repellents. Unlike other repellents, DEET works for many species of mosquito. Sprays containing concentrations of 20%–50% DEET are considered safe for use on skin. DEET and other repellents work by blocking a mosquito's homing receptors on its antennae. As a result, the mosquito cannot sense the presence of a person wearing a repellent.

TRY THIS

One Bad Apple

Before beginning the investigation to see if ripe fruit will spoil others, describe the appearance of the banana and apples. Put one unripe apple into a plastic bag as a control. Seal the bag and place it away from sunlight. Put the second unripe apple with the ripe banana into the other plastic bag, seal it, and put it close to the first bag. After three to five days, conduct a test for the presence of starch. Fill a shallow, nonmetallic pan with a small amount of iodine-potassium iodide solution. Slice each apple in half and place the cut-side portion down in the solution. Wait one minute. Remove the apples from the iodide solution and rinse with water. Compare the sliced apples. What do you notice?

FYI

Jewel Beetles

For up to five years, larval jewel beetles spend the majority of their time munching on tough acacia wood. Their jaws have to be made of a strong material because this type of vegetation is extremely hard to chew. Most arthropods' jaws incorporate metals and other minerals that provide the necessary toughness and strength. Larval jewel beetles' jaws do not contain any metals. Instead, their jaws gain strength from a carbon-based, organic material called *chitin*. The chitin fibers are covered with protein molecules that form a hard exterior. Scientists have compared this material to different metals, and have determined that the hardness of chitin is equal to the hardness of stainless steel.

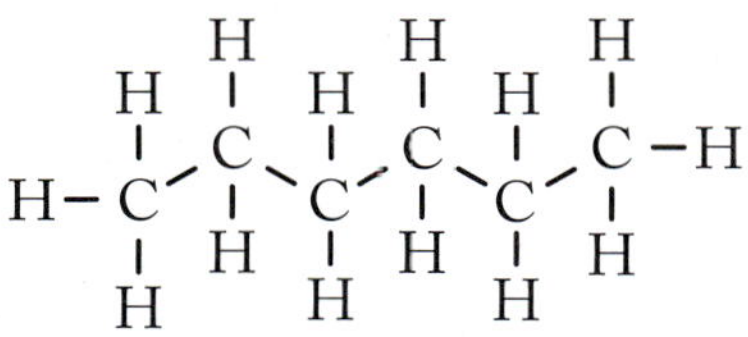

Straight Chain

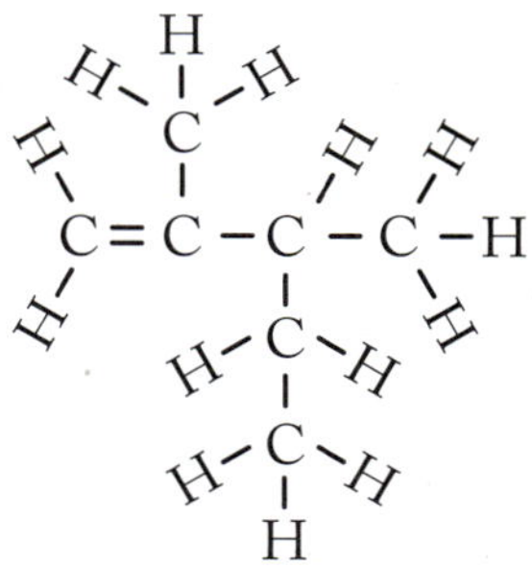

Branched Chain

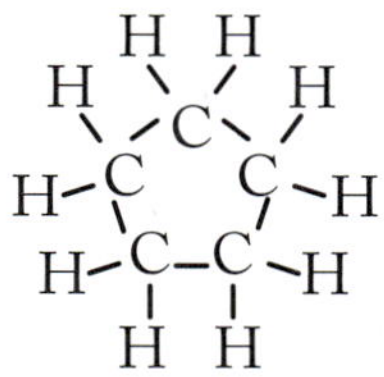

Ring Structure

The simplest forms of organic compounds are called **hydrocarbons**—organic compounds containing only hydrogen and carbon atoms. There are two types of hydrocarbons: saturated and unsaturated. Think of a saturated hydrocarbon like a sponge. When you submerge a sponge in a bucket of water, it absorbs as much water as it can possibly hold. The sponge is completely saturated with water. Saturated hydrocarbons are hydrocarbons that have single bonds between the carbon atoms. Each carbon atom is completely surrounded (saturated) with hydrogen atoms. As the number of carbon atoms increases, the boiling point also increases.

Saturated hydrocarbons are also called **alkanes**. All alkane compound names end in *–ane*. The carbon chains of alkanes can be either straight, branched, or ring-shaped. Methane is the simplest alkane, having one carbon atom surrounded by four hydrogen atoms. It is one of the main components in natural gas.

Unsaturated hydrocarbons have double or triple covalent bonds between two or more carbon atoms. Two types of unsaturated hydrocarbons are alkenes and alkynes. **Alkenes** are unsaturated hydrocarbons with at least one pair of carbons joined by a double covalent bond. The simplest alkene is ethene, also known as *ethylene*, which has a formula of C_2H_4. Ethylene is a natural plant hormone that speeds up the ripening process of different fruits, including bananas. All alkene compound names end in *–ene*. **Alkynes** are unsaturated hydrocarbons with at least one pair of carbons joined by a triple covalent bond. The simplest alkyne is ethyne, also called *acetylene*, which has a formula of C_2H_2. Welders use ethyne as a fuel in specific metal-welding processes. All alkyne compound names end in *–yne*.

Hydrocarbons		
Saturated	**Unsaturated**	
Alkane	**Alkene**	**Alkyne**
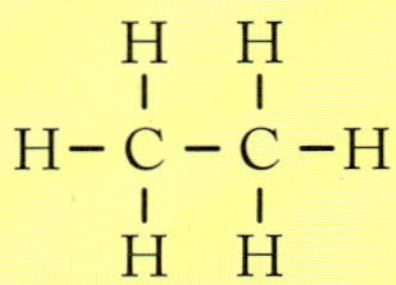	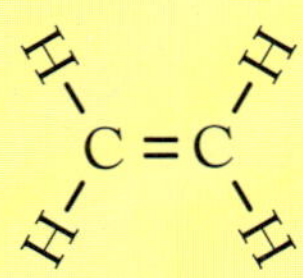	$H-C\equiv C-H$

As the number of carbon atoms in a carbon chain increases, the number of possible structures increases. Compounds that have the same chemical formula but are designed with a different molecular structure are called **isomers**. For example, pentane has a formula of C_5H_{12} and forms a straight-chain structure. Isopentane forms when one of the carbon atoms branches off of one of the central carbon atoms. Isopentane has the same chemical formula as pentane but a different molecular structure. It also has different physical properties. Every isomer has different properties and is considered a different compound even though it has the same chemical formula. This is why structure is so important, because the structure of a molecule determines its function.

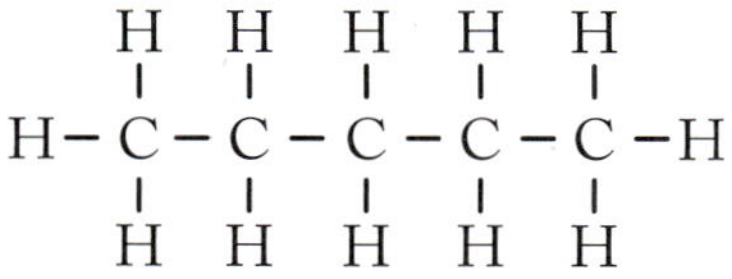

Pentane, C_5H_{12}

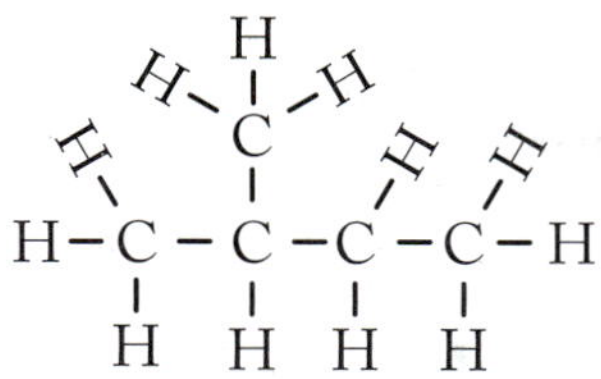

Isopentane, C_5H_{12}

In organic compounds, carbon atoms also bond with many other elements such as oxygen, sulfur, nitrogen, and the halogens. These variations allow for the possibility of even more organic substances. When chemists replace a hydrogen atom with a different element or group of elements, the result is a substituted hydrocarbon. Substituted hydrocarbons have different physical and chemical properties than the original hydrocarbon. Chemists use the process of removing one or more hydrogen atoms and replacing them with different elements for the purpose of manufacturing many useful products.

Substituted hydrocarbons containing halogens are called *halocarbons*. Some halocarbon examples include chloromethane, CH_3Cl, which is used as a refrigerant; tetrachloroethene, C_2Cl_4, which is used in dry cleaning; and dichlorodifluoromethane, also called *freon*, CCl_2F_2, which was used in refrigerators and air conditioners. Freon is no longer used in new refrigerators because many scientists believe it is a leading contributor to ozone depletion.

Tetrachloroethene is a synthetic chemical that is widely used in dry cleaning processes.

Groups of elements that replace hydrogen atoms in hydrocarbons are called *functional groups*. Two functional group examples include hydroxyl groups and carboxyl groups. When a hydroxyl group, –OH, replaces one or more hydrogen atoms, the hydrocarbon becomes an alcohol. The simplest alcohol is methanol, which has the chemical formula CH_3OH. Chemists use methanol as a solvent and they make plastics and synthetic fibers with it. Methanol is also added to the fuel of stunt cars and race cars because it is less flammable. It gives the cars better throttle response, valve-guide-wear improvements, and longevity to the cars' oil. Ethanol, C_2H_5OH, is another simple alcohol produced naturally by yeast or bacteria feeding

Methanol is converted into olefins that are used to make plastics.

on the sugars in grains in a process called *fermentation*. Ethanol is used as a solvent for organic compounds, used in medicines, and mixed with gasoline.

Ants and stinging nettle plants naturally produce formic acid.

A carboxyl group consists of a carbon atom that forms a double bond with one oxygen atom and a single bond with a hydroxyl group, –COOH. Organic acids form when a carboxyl group replaces one or more hydrogen atoms on a hydrocarbon. Methanoic acid, HCOOH, is the most common organic acid. It is commonly called *formic acid*. Formic acid gives some plant nettles and ant bites their sting. Another organic acid, acetic acid, CH_3COOH, is derived from ethane. Vinegar contains acetic acid.

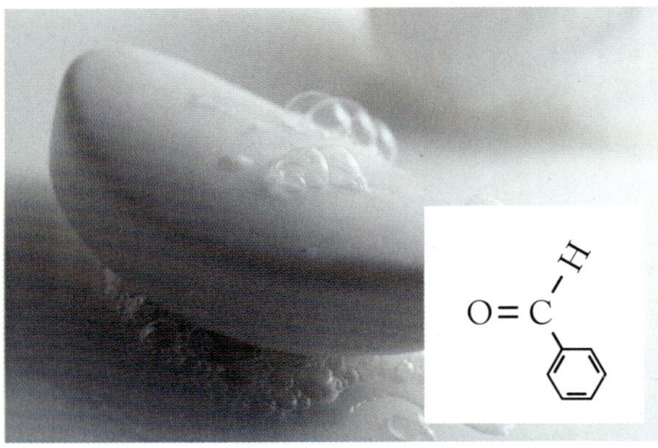

Benzaldehyde produces an almond fragrance, and chemists use it to add scents to food, pharmaceuticals, and soap.

Cyclic hydrocarbons are organic compounds that form a ring structure. Cyclic hydrocarbons can contain single, double, or triple bonds. They usually contain five or six carbon atoms. The most common cyclic hydrocarbon is benzene. It is made of six carbon atoms and six hydrogen atoms with alternating single and double bonds among the carbon atoms. This structure is especially stable because the valence electrons are equally shared among all the carbon atoms. For this reason, the benzene structure is often depicted as a hexagon with a circle in the center.

Compounds that contain benzene rings are called *aromatic compounds*. They were given this name because many of the earlier known benzene compounds, such as the oils from fruit and spices, produced a pleasant smell. However, there are other benzene compounds that have a foul smell such as naphthalene, which is found in moth balls. Chemists have found ways to use the foul smelling property of different aromatic compounds, such as benzyl mercaptan, to help save lives. Natural gas is colorless, odorless, and harmful. People cannot detect the presence of natural gas. Chemists have added benzyl mercaptan, which is a compound found in skunk spray, to natural gas sources to help people detect leaks.

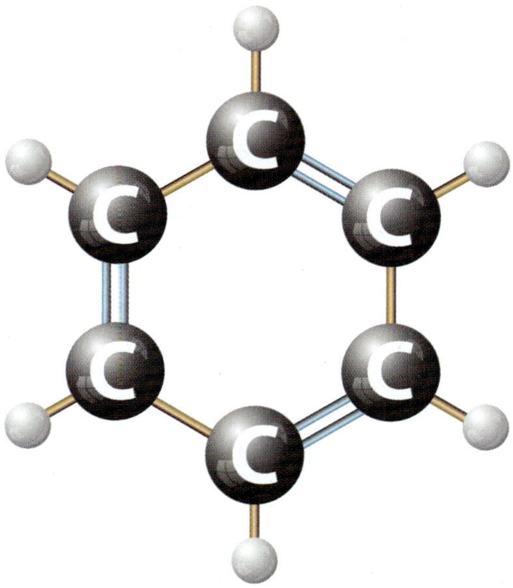

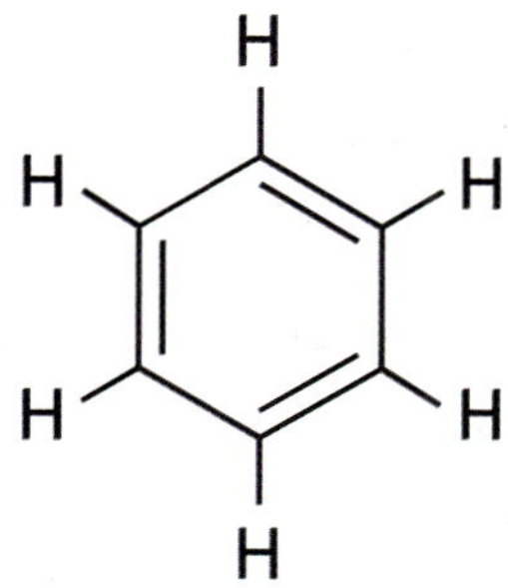

Benzene

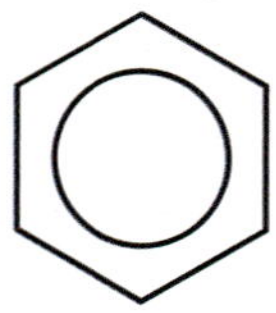

All of the carbon atoms in a benzene molecule share valence electrons.

LESSON REVIEW

1. What was the original definition of an organic compound? How did that definition change to distinguish organic compounds from inorganic compounds?
2. Why are alkanes better at dissolving cooking-oil spills than water?
3. How do the three main groups of hydrocarbons differ? How are they the same?
4. How can two molecules have the same chemical formulas but different structural formulas?
5. Why would a scientist want to replace a hydrogen atom with a functional group? Name an example of a functional group and the resulting compound.
6. What do all aromatic compounds have in common?

2.3.2 Fuels from Organic Compounds

OBJECTIVES

- Identify the source and summarize the refinement process of hydrocarbons.
- Evaluate alternative energy sources.

VOCABULARY

- **biogas** a gas fuel derived from the decay of organic matter
- **biomass** an energy source that is obtained from organic matter, such as crops or agricultural waste products, used for the production of fuel
- **energy farming** the growing of organic substances to produce fuel

God provided Earth with the mechanism for the creation of fossil fuels. Two primary fossil fuels that people have learned to rely on over a long period of time are petroleum and natural gas. When organisms decay, their remains eventually congregate to form thick deposit layers under the surface of the earth and ocean floors. The tremendous pressure from above and the heat from the center of the earth cause the deposits to be converted into oil-rich shale and natural gas. The process that creates fossil fuels takes many, many years. Young-earth supporters believe that the planet is 6,000–12,000 years old and geologic features, such as rock layers and fossils, were caused by major events like the Flood. Others believe that the earth is very old and the processes that created fossils are the result of slow changes and natural progression beginning 4.6 billion years ago. Both theories support that fossils collect, over time and under pressure with intense heat, and then create fossil fuels.

Petroleum and natural gas pools are often found near each other. People have collected these two resources because they are rich in methane and other usable hydrocarbons. In order for these two fossil fuels to be usable, raw petroleum must be separated into its components, or fractions. The process called *fractional distillation* occurs inside large towers that are divided into multiple levels. The petroleum is boiled, and as the vapor rises, fractions begin to condense at different layers depending on their boiling points. Some of the fractions that are collected include gasoline, fuel oil, kerosene, and diesel oil.

Old-earth scientists state that petroleum forms when decayed organisms experience intense heat from the inside of the earth and are placed under tremendous pressure from overlying sediment layers over long periods of time.

Earth's supply of petroleum and natural gas, along with coal, which is a third fossil fuel, may one day be depleted because it takes a tremendous amount of time for these substances to be naturally replenished. They are considered nonrenewable resources. This means that people need to look for ways to reduce the dependence on these particular organic compounds. The burning of fossil fuels also contributes to pollution. Scientists are searching for alternative energy resources to replace the use of fossil fuels. Organic chemistry holds several promising possibilities.

Processing plants separate raw petroleum in large towers where fractions are extracted.

One exciting field of research in organic chemistry is the study of biomass. **Biomass** is an energy source that is obtained from organic matter, such as crops or agricultural waste products, to be used for the production of fuel. The process of photosynthesis initiates the formation of biomass. Plants facilitate a reaction between carbon dioxide and water to produce carbohydrates. This reaction requires solar energy, and the converted energy from the sun becomes stored energy within the chemical bonds of the carbohydrates. This energy forms the building blocks of biomass. Plants, animal wastes, and any other organic substance can be burned as biomass.

Fractional Distillation

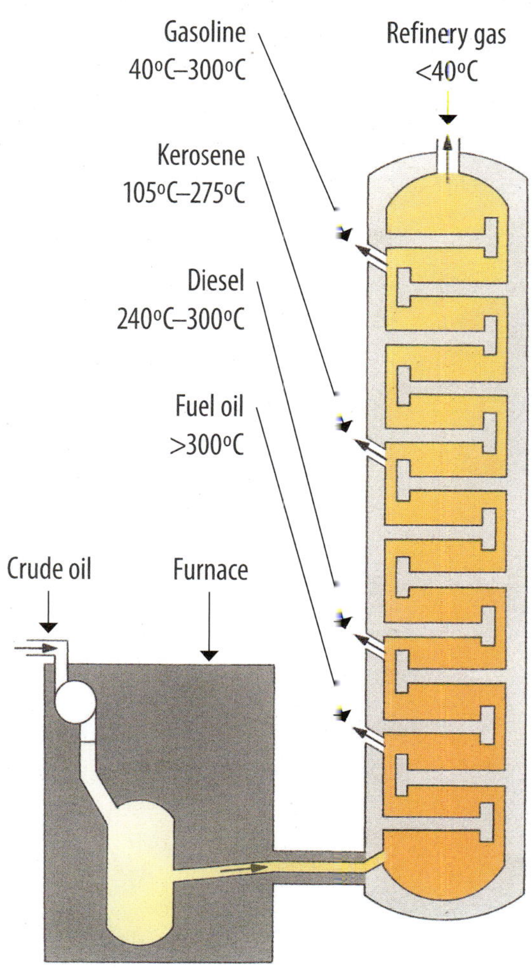

Biomass has met the world's energy needs for thousands of years. Throughout history and in many regions around the world, people have burned wood as their primary biomass source for cooking food and heating their homes. Today, the majority of wood-fueled homes exist in developing countries. Organic chemists are working to identify new biomass sources that burn more efficiently, cause less pollution, and can be grown on low-quality land. Some of the fuel sources being researched include corn husks, sugar-cane fibers, sunflowers, and seaweed.

Through fermentation, the sugar and starch in organic matter converts to ethanol, which can be used as a fuel or mixed with gasoline to make a blended fuel. This type of fuel burns cleaner than gasoline alone. Many gas stations have signs that indicate the gasoline may contain up

The local pig farm supplies this biogas plant with natural starting materials.

Sugarcane crops have been planted in Brazilian fields that were once covered by dense rain forest.

to 10% ethanol. In South America, much of the fuel is ethanol made from sugarcane. **Biogas** is a gas fuel derived from the decay of organic matter. Bacteria acts on the organic matter to produce methane gas, which can be used to produce methanol. Methanol can also be mixed with gasoline to reduce pollution; however, fuel economy decreases with its addition.

Researchers believe that biomass fuels can soon be cost-competitive. There are now biomass plantations—areas of land devoted to growing crops to be converted into fuels. The growing of organic substances to produce fuel is called **energy farming**. Energy farming could help developing countries better manage their land that has already succumbed to deforestation. For example, land in Brazil that was once covered in rain forests could be transformed into biomass plantations.

Biomass can also be used to produce electricity. People can burn wood chips, some crops, and other agricultural wastes to generate steam for electricity-generating turbines. In regions of the world where biomass is relatively inexpensive to produce, biomass can become the preferred fuel source over fossil fuels. A biomass power plant built in Värnamo, Sweden, has proven that biomass can be very efficient and can provide an extremely clean source of energy.

Biomass is considered an efficient fuel alternative. However, there are a few disadvantages to its use. The burning of wastes saves on

TRY THIS

Organic Fermentation

Fill three 500 mL beakers with warm water, one beaker with boiling water, and one beaker with cold water. Label each beaker. Add one packet of yeast and two teaspoons of flour to each beaker. Place one teaspoon of sugar in one warm water mixture, the boiling water mixture, and the cold water mixture. Stir well. Add one teaspoon of salt and one teaspoon of vinegar to the remaining two warm water mixtures. Stir well. Record your observations every 5 minutes for a total of 15 minutes. What is the product of the fermentation experiment? Which environment would yield the most fuel?

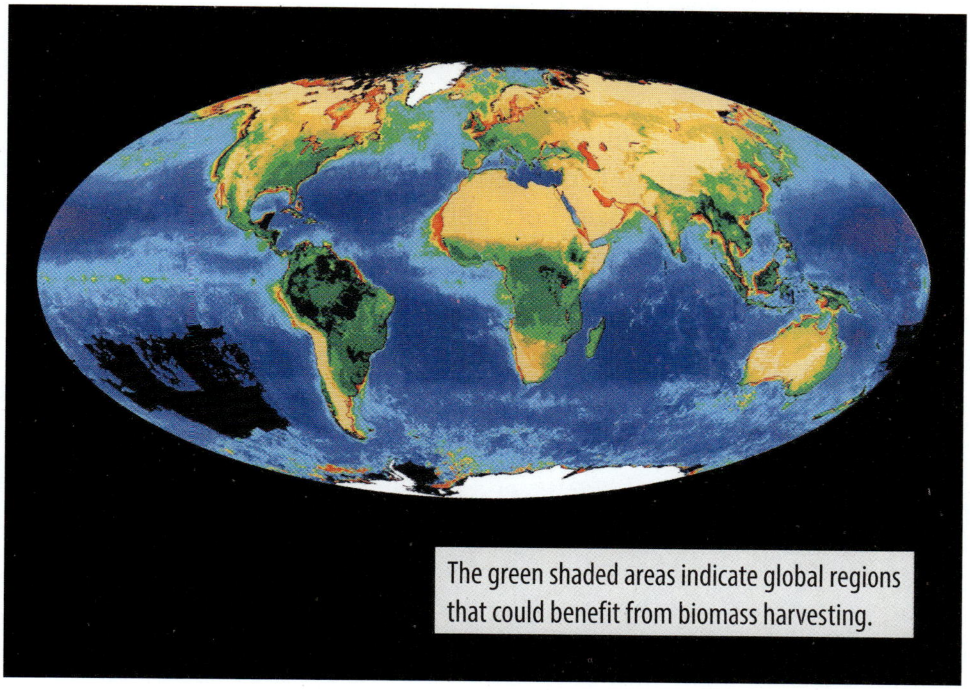

The green shaded areas indicate global regions that could benefit from biomass harvesting.

carbon dioxide emissions, but it increases methane gas, which is harmful to the Earth's atmosphere. Another disadvantage is the land cleared for short-term plantations results in deforestation, robbing birds and other land animals of their homes. At the same time, land allocated to biomass could be used for food crops. The amount of water it takes to sustain the plantations could exceed what is available as photosynthesis requires a great deal of water to be efficient. These plantations are not productive in all climates. Although many countries are working toward a more sustainable, affordable, clean fuel alternative, much research is needed before this can happen.

LESSON REVIEW

1. What is the source of petroleum? What is the source of biomass?
2. What is the fractional distillation process and what are some of its products?
3. Why should people consider alternative energy sources?
4. Why is biomass a good alternative energy source? What is a major concern about creating biomass as a source of energy?

BIOGRAPHY

Albert Cornelius Antoine

Born in New York City to immigrant parents in 1925, Albert Antoine was always encouraged to get an education. His parents, who grew up in the British West Indies, moved to the United States to provide a better life for their children. Antoine lived far away from both the high school and City College of New York—the distance required a daily train trip for him to attend. After being drafted into the service in World War II, his test scores led the Army to send him to Ohio State University to study civil engineering. After the war, he earned his PhD and began teaching college in Atlanta. He was drawn to research, however, and left teaching to work as a research chemist for NASA, where he worked 42 years. Most of Antoine's research focused on improving fuels for aircraft, including fuel efficiency, alternative fuel sources, batteries, and fuel cells.

2.4.1 *Alloys*

OBJECTIVES

- Compare the properties of alloys and the elemental components.
- Give examples of different alloys and their uses.

VOCABULARY

- **alloy** a solid metallic solution or compound made of a metal and one or more elements
- **amalgam** an alloy that contains mercury

Thousands of years ago, people made dwelling places, tools, clothing, and weapons out of stone, wood, leather, bones, clay, ivory, and copper. These materials were not very durable and frequently had to be replaced. Archaeologists believe that a new material was accidently produced more than 5,000 years ago by a group of people that lived between the Tigris and Euphrates Rivers in Western Asia. It is believed that the rocks they used to enclose their fire pits were rich in copper and tin. When these rocks were heated to high temperatures, the metals melted, mixed, and produced a new substance called *bronze*. Stronger tools and weapons were made out of bronze for about 2,000 years during the Bronze Age, but then bronze was replaced by iron as the preferred metal.

Bronze is one of many substances that belongs to the group known as *alloys*. An **alloy** is a solid metallic solution or compound made of a metal and one or more elements. The additional elements that combine with the metal may be another metal or a nonmetal such as carbon. A majority of the materials used today are made of alloys. They are unique because they combine the sets of properties from each element to create a material with a completely new set of properties. The combination of elements often produces a substance that is stronger, harder, lighter, or more durable than individual elements. Metallurgists, scientists who specialize in metals, create alloys because they can make materials with a specific set of properties.

The Sumerian civilization accidentally discovered the first alloy—bronze.

It is important that scientists have a good understanding of each element's physical properties so they know what type and percentage of elements to combine. For example, steel is an alloy of iron and carbon. Elemental iron is a soft metal that is usually not the preferred material for building structures. However, iron can be made substantially stronger with the addition of the nonmetal carbon. Only a small percentage of carbon atoms needs to be added to iron. If too much carbon is added, the product becomes brittle. Steel is a very strong, inexpensive material used to support buildings, bridges, and cargo ships.

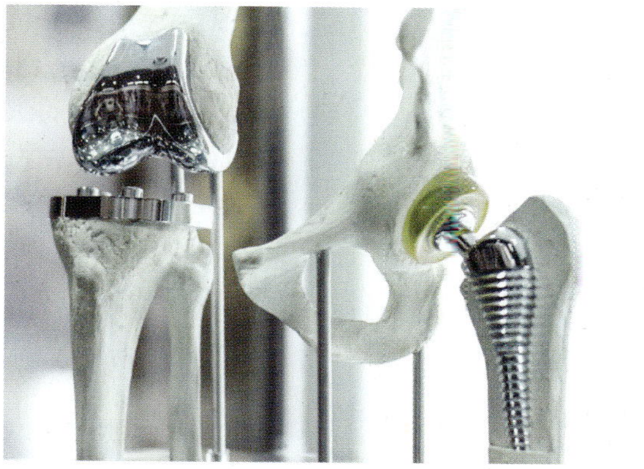
The human body accepts stainless steel replacements of joints because of stainless steel's unreactive property.

There are different types of steel alloys, depending on the amount of carbon added and the way in which the steel is made. One important steel alloy is stainless steel, which is a combination of iron, carbon, chromium, and nickel. Stainless steel is used in cutlery and kitchenware. It is also a preferred alloy for surgical instruments and medical implants because it does not react with elements and compounds in the body. The combination of each element's properties makes an unreactive substance that resists corrosion.

Metallurgists typically make alloys by melting each component, removing any impurities, mixing the molten elements, and allowing the mixture to cool and harden. Many processes have been used to make alloys. One of these processes includes the Bessemer process that was developed around 1856. This process revolutionized the steel industry, because large amounts of steel could be mass produced. Sir Henry Bessemer discovered

TRY THIS

Sandy Clay

Prepare three round, 1 cm thick flat discs of modeling clay that each measures about 15 cm in diameter. Sprinkle ¼ tsp of sand on the first disc, ½ tsp of sand on the second disc, and 1 tsp of sand on the third disc. Fold the discs in half to seal the sand and knead the clay until it is warm and the sand is thoroughly incorporated. Roll each sample to form a cylinder. Each sample should form the same size cylinder. Grasp both ends of each cylinder and slowly pull the clay apart. Describe the appearance of the clay before, during, and after it is separated. Did each clay sample separate in the same way? Compare the fracture surfaces of each sample. What properties of the clay did the sand affect? How is this activity related to alloys?

100% pure gold
(24-karat)

92% gold and 8% copper
(22-karat)

83% gold and 17% copper
(20-karat)

that impurities could be removed by injecting molten iron with air blasts. The oxygen in the air reacted with excess carbon to produce oxides that could be removed. Scientists have continued to find new ways to improve alloy-making processes. The Bessemer process was replaced by the open-hearth process, which has been replaced by the basic oxygen process. Each advancement has made the process more efficient with less impurity contamination.

Gold jewelers make items of gold by mixing metals. Typically, gold jewelry is not 100% pure gold. It is usually a mixture of gold, copper, and sometimes silver. Elemental gold is extremely soft and malleable on its own. The addition of copper makes gold jewelry more durable and less expensive. Jewelers measure the gold content in jewelry in units called *karats*. Each karat represents one twenty-fourth of the mixture. Pure gold is 24 karats. Gold jewelry that is 18 karats is 18 parts gold and 6 parts copper. Although 24-karat gold jewelry is more valuable than 18-karat gold, the 18-karat gold jewelry is harder and will not scratch as easily because it has a higher percentage of hard metals.

Common Alloys

Alloy	Component Elements	Uses
Brass	Copper, zinc	Decoration, locks, gears, doorknobs, plumbing and electrical equipment, musical instruments
Bronze	Copper, tin, aluminum	Coins, springs, turbines, blades
Cast iron	Carbon, iron	Pots, pans, utensils
Nichrome	Chromium, iron, nickel	Resistance wire, dental fillings
Rose gold	Gold, copper	Jewelry
Solder	Lead, tin, antimony	Electronics, plumbing, sheet metal
Stainless steel	Carbon, chromium, nickel	Cutlery, sinks, washing machine drums, window fittings, ship containers, nuts, bolts
Surgical stainless steel	Carbon, chromium, molybdenum, nickel	Surgical instruments, surgical implants, MRI scanners
Wood's metal	Bismuth, lead, tin, cadmium	Fire sprinklers, solder and casting, antique repair

FYI

Stronger Than Titanium

For many years, metallurgists have been attempting to make a steel alloy that is as strong as titanium without the brittle property. A group of scientists at Pohang University of Science and Technology in South Korea have made great strides toward this material becoming a commercially available reality. If this group of scientists succeeds, a flexible, strong, lightweight, inexpensive steel will be available to construct automobiles, shipping containers, and any other structure that requires strength and durability. Extremely strong and lightweight steel is produced with the addition of the metal aluminum. However, this addition also makes the product very brittle. Instead of bending under intense pressure, the steel alloy breaks. The South Korean scientists discovered that the issue is created by veins of boron crystals. They determined that if they can control where the crystals form, they can make steel that does not easily break. For these scientists, the product is not the ultimate goal. They are more interested in the manufacturing process that could dramatically affect the production of many other alloys.

Some alloys are made to resist corrosion, conduct electricity, and be lightweight. One of the most important metals used both in elemental form and in alloys is aluminum. Aluminum and magnesium combine to form the most lightweight alloys. The low densities of aluminum alloys make them ideal for the construction of aircraft, space vehicles, and automobiles. They provide a means to conserve fuel. Aluminum alloys are also strong and good conductors of heat and electricity, and they resist corrosion. They are used to make pots, pans, and electric wiring.

Amalgams are alloys that contain mercury. Mercury can form alloys with nearly every metal except iron. One of the most common mercury alloy uses was for tooth fillings. This material was used to prevent future tooth deterioration. Dentists rarely use this amalgam any more because of health concerns.

Lightweight aluminum is used in the construction of aircraft.

LESSON REVIEW

1. How are alloys made? Describe an example of a process used to make alloys.
2. Are alloys considered an improvement of the elements they contain? Explain.
3. What properties of stainless steel make it a good choice for surgical instruments?
4. How much pure gold is in a piece of 10-karat gold jewelry?
5. What makes amalgams unique?

2.4.2 Ceramics and Glass

OBJECTIVES

- Relate the properties of ceramics to their uses.
- Compare the different varieties of glass.

VOCABULARY

- **ceramic** a nonmetallic, inorganic substance that contains high proportions of silicon and oxygen
- **glass** a type of ceramic that is composed mainly of silicates that are bound to alkali metals under high temperature conditions

Have you ever made anything out of clay in art class? It is hard to imagine that thousands of years ago clay was the primary material that people used for building and making household items. The oldest known clay artifact dates back to more than 5,000 years ago. In fact, the walls that Joshua and his army called down around the city of Jericho were made from hardened mud and straw bricks. Biblical historians believe the wall fell around 1,400 BC.

Clay is one of the most plentiful materials on Earth. It consists primarily of silicon, oxygen, and aluminum. Clay also includes small amounts of magnesium, sodium, potassium, and water. For thousands of years, people have mixed clay with water, molded the mixture, and then heated it to form ceramic products such as pottery or bricks.

A **ceramic** is a nonmetallic, inorganic substance that contains high proportions of silicon and oxygen. Ceramic materials also include clay and feldspar—a mineral that contains calcium, sodium, or potassium. When ceramic materials are gradually heated between 1,000°C and 1,400°C, they become hard and

The walls surrounding Jericho were made from a ceramic material.

rigid. Clay must be heated slowly because if rapid heating occurs, the water inside the raw materials evaporates with such intense pressure that the material explodes. Clay particles expand during the heating process, called *firing*, and the spaces between the particles shrink. The result is a highly dense, strong material that can withstand extremely high temperatures. However, the very dense structure also makes the ceramic brittle. Ceramics may break easily if struck with a hard object or if the temperature changes too quickly.

The ceramic firing process takes place inside of a very hot kiln.

Ceramics have many properties that make them useful. They do not react easily with substances such as oxygen, water, or acids that cause some materials to rust or burn. Ceramic materials are also durable and resistant to wear. Some inert ceramic materials are used to make medical devices, such as those used in hip and knee replacements, because they do not react with body fluids. Ceramics retain heat well, which makes them ideal for kitchenware. They also have low thermal and electrical conductivity. Today, sewer pipes, decorative blocks, pottery, thermal insulators in furnaces, and various electrical products are made from ceramics.

Modern ceramics include a mixture of standard raw materials such as clay and shale, along with metals, carbon, nitrogen, and sulfur. The addition of metallic elements makes the advanced ceramics harder and stronger than traditional ceramics. They are made to withstand higher temperatures better than traditional ceramics. Modern ceramics better resist wearing and corrosion. Some uses of modern ceramics include drilling and cutting tools.

A unique type of ceramic is glass. Think of all the ways that you depend on glass. Food jars, building and automobile windows, drink bottles, mirrors, tables, appliances, cell phones, aircraft structures, optical glass, and fiber optic cables are all made out of glass. **Glass** is composed mainly of silicates that are bonded to alkali metals under high temperature conditions. It is an amorphous solid, which means that it has a random structure. The crystals do not have

Some fiber optics contain ceramics because they have low thermal and electrical conductivity.

TRY THIS

Ceramic Strength Test
Shape three pieces of modeling clay into rectangles that each measure 20 cm x 5 cm x 1.5 cm. Place one piece of clay in the freezer and allow it to cool for at least one hour. Set another piece of clay under a heat source for at least one hour. Hold both ends of the room-temperature piece of clay with the 5 cm flat side facing up. Have your partner place increasing amounts of weight on the clay until it begins to dip or break. Record the maximum amount of weight that the clay can hold. Repeat this procedure with the cold and hot clay. Summarize your results. Obtain two more pieces of clay. Choose two materials from rice, beans, coins, or marshmallows that will act as a reinforcement, thoroughly mix each material into a piece of clay, and shape the pieces of clay into rectangles with equal dimensions to the first set. Repeat the strength test with the two reinforced pieces of clay. Record the amount of weight that each piece of clay can hold, and summarize your results.

CAREER

Chemist

The field of chemistry is very broad and involves many different industries. Chemists conduct research and experiments with a variety of chemical substances. They may seek to gain a better understanding of the properties of the chemical, improve the quality of existing chemical products, or create new products. Sometimes, chemists try to uncover the identity and properties of chemicals in an unknown compound. Chemists usually work in a laboratory or another type of controlled environment where they utilize special equipment to conduct their experiments. These laboratories may be associated with a government agency or a private industry. Chemists often work in teams and with advanced computer programs to develop new technologies. Colleges and universities offer great opportunities for chemists to conduct their research and educate students. A chemist generally prefers to choose a specific field of study. The four main fields include biochemistry, nuclear chemistry, neurochemistry, and theoretical chemistry. Biochemists study chemical compounds specifically related to living organisms. Some areas of interest include the development of pharmaceutical drugs, genetic engineering, agricultural products, and DNA therapies. Nuclear chemists research and conduct experiments that relate to radioactivity and nuclear matter. They may study the effects of radiation on living organisms, or they may develop new technologies to produce efficient radioactive energy. Neurochemists study chemical compounds and processes that are involved with the nervous system. Theoretical chemists seek to study and develop scientific ideas and theories that may then be used in industry. Some of their work has helped advance medical treatments and has developed new alternative fuel sources.

enough time to form a pattern because they are cooled rapidly. Because of this, glass is brittle and transparent.

Various types of glass have been developed from different materials, but the three main materials within glass are sand, limestone, and sodium carbonate. The most common glass, soda-lime, is about 70% silicon dioxide (SiO_2), 9% calcium oxide (CaO), and 15% sodium oxide (Na_2O). Glass is made by heating these three oxides, which look like white powder. Many other oxides can be added to ordinary glass to give it special properties. Adding boron oxide (B_2O_3) prevents the glass from expanding when heated, so this type of glass is used in laboratory glassware. The addition of lead oxide allows glass to reflect light better, which makes it a good choice for cut glass and lenses. Silica glass is made of 100% silicon dioxide. Normally, silicon dioxide crystallizes as the mineral quartz, but it forms a transparent glass if molten silica is cooled rapidly. Silica glass is difficult to work with and melts at much higher temperatures than ordinary glass, but it can be cooled or heated rapidly without breaking. For this reason, silica glass is also used for laboratory glassware.

Colored glass can be created by adding different elements and compounds. Iron gives glass a green color, cobalt(II) oxide (CoO) turns glass blue, and chromium(III) oxide (Cr_2O_3) gives glass a deep green color. Adding tin(IV) oxide (SnO_2) creates opaque white glass. Silver chloride ($AgCl$) creates photochromic glass, which is used in sunglasses that darken when exposed to sunlight and become clear when in the presence of indoor light.

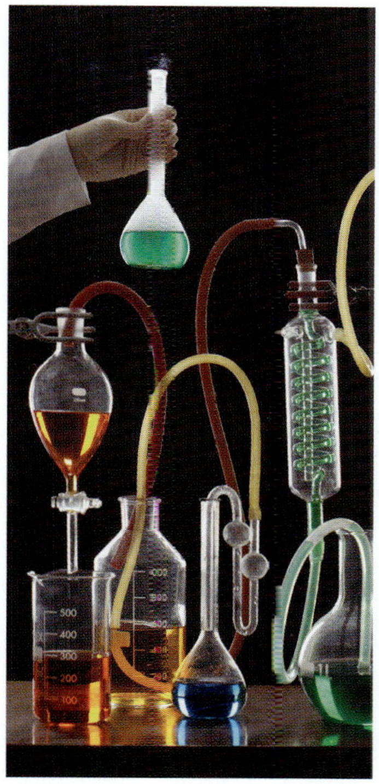

Laboratory glassware is stronger than ordinary glass because it is fortified with boron oxide.

LESSON REVIEW

1. How are ceramics made?
2. What are several desirable ceramic features? Name one example of a material that utilizes one of these features.
3. Why is glass considered an amorphous solid?
4. Describe two types of glass that are used to make laboratory glassware.
5. What is the compound that makes photochromic glass? Describe this type of glass.

Colored glass is made with the addition of metal elements and compounds.

2.4.3 Recycling

OBJECTIVES

- Express the three major reasons for recycling.
- Describe the recycling methods of glass and aluminum.

Have you ever considered how many things are thrown away each day? Think about all of the materials that your household throws away in a year. Many of these products are used only once before being hauled off to a landfill. When billions of people around the world throw away a few things every day, it adds up to a lot of trash.

One way to reduce waste is by recycling, which is the collecting and processing of used materials to make new products. There are three major reasons for recycling materials. First, recycling materials conserves fossil fuels. Fossil fuels are a nonrenewable resource, and they are being used at an extremely rapid rate. Examples of fossil fuels include petroleum, natural gas, and coal. When manufacturers produce glass and aluminum, fossil fuels supply the necessary energy. Many plastics are also derived from fossil fuels. Recycled plastics conserve the nonrenewable resources they are made from. Second, recycling materials cuts down on air pollution and the production of waste chemicals. To make materials such as finished metals, glass, and paper, fossil fuels must be burned. Burning fossil fuels produces air pollution and releases waste chemicals. Third, finding places to dispose of waste products is becoming a major issue, especially for urban areas. Landfills cannot be the only solution because they are filling up quickly.

Waste Product Disposal

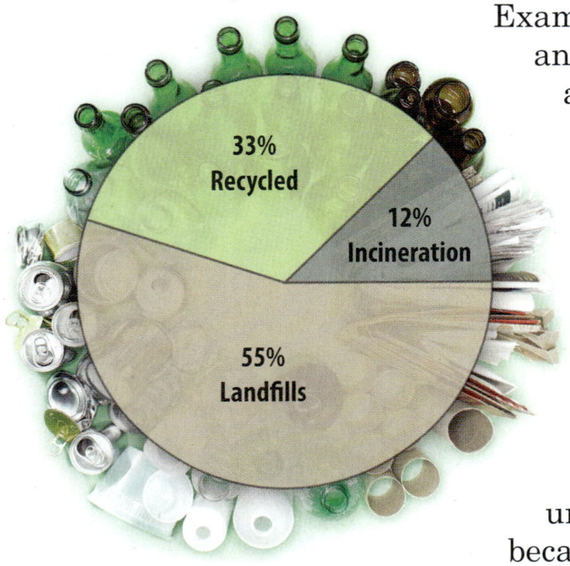

Glass, plastics, paper, and metals all require different recycling methods. Recycling glass is a relatively easy procedure compared to that for recycling plastics or metals. Discarded glass is first broken up into small pieces called *cullet*. At the glass factory,

Fifty-five percent of America's trash is buried in landfills.

Finding new uses for old materials is another way to practice recycling.

cullet is passed through powerful magnets that remove any metal pieces. A vacuum removes plastic coatings and paper labels. The cullet is then added to a mixture of silicon dioxide, calcium oxide, and sodium oxide. These are the same materials that are used to manufacture glass. The cullet lowers the melting temperature of the mixture, which is beneficial to the environment because less heat is needed and fewer fossil fuels are burned. In closed-loop recycling, the original object is reshaped into a new object.

Have you ever wondered where aluminum soda cans go after you toss them into the recycling bin or take them back to the grocery store? The cans are first washed and crushed. The

Glass bottles are broken down into small pieces before they can be recycled to form more glass bottles, concrete pavement, or flooring material.

TRY THIS

Making Paper

Tear the matte waste paper into small pieces. Place the small pieces in a food processor with warm water, and blend the mixture until a fine pulp forms. Attach the mesh or screen to the picture frame by applying duct tape or heavy-duty staples to the outer edges. Pour the paper pulp into a large tray and insert the picture frame into the mixture. Gently agitate the frame and remove from the tray. Allow excess water to drain from the frame. Press the cloth or sponge into the surface of the pulp in the frame to press out a majority of the water that remains. Let the pulp dry overnight in the frame, or flip the frame onto a towel and allow the pulp paper to dry on the towel. How might this recycled paper be used? Why does the process begin with small pieces of paper?

RECYCLING FACTS

It takes about 5% of the energy to recycle an aluminum can as it does to make the original aluminum can. The 95% saved energy can power a television for two hours.

It takes about 35 recycled, 2-liter plastic bottles to make enough polyester fiber to fill one sleeping bag.

Recycling one ton of newsprint saves 17 mature trees.

One mature tree filters 27 kg of pollutants from the air each year.

The average American household throws away enough recyclable material that could reduce CO_2 emissions by 380 kg each year.

It takes more than 4,000 years for a glass bottle to degrade in a landfill.

Recycling one glass bottle can power a 14-watt LED light bulb for 100 hours.

Aluminum scrap ready for recycling

aluminum-can scrap is rolled out into thin sheets and remelted with pure metal to strengthen the scrap. Then, it is used to make cans or other aluminum products. Because recycled cans contain small amounts of silicon from glass, iron, and copper, scrap aluminum cannot be used for making items such as aircraft parts.

Engineers are designing products that reduce the amount of waste and pollution throughout the entire industrial process. They refer to such products as *designed for disassembly* and *designed for recycling*. These scientists are using creation as their model. In natural ecosystems, all waste products are used by other organisms. In this way, raw materials are recycled through food chains and conserved. Industrial ecology tries to mimic nature. It focuses on structuring industries so they can reuse and recycle waste products from other industries.

The town of Kalundborg, Denmark, is an example of an industrial ecosystem. The town's oil refinery uses waste heat from the power plant. The refinery sells sulfur that has been removed from crude oil to a chemical company. It also sells sulfur as calcium sulfate to a company that uses this product to make wallboard to replace the gypsum wallboard normally used. The refinery also uses excess steam from its power plant to heat water for the aquaculture farming and to warm greenhouses and homes. The goal of industrial ecologies is to reduce pollution, energy use, and waste. A perfect system would mean that every waste from one industry would become the raw material of another industry.

LESSON REVIEW

1. Summarize the three main reasons to recycle.
2. Describe the recycling process of glass and aluminum.
3. How do companies that belong to industrial ecosystems reduce their environmental impact?
4. What are the three foci of industrial ecosystems?

2.4.4 *New Materials*

OBJECTIVES

- Explain the composition of polymers.
- Examine the properties and uses of polymers.

VOCABULARY

- **composite** a material made of two or more substances that are combined to produce materials with useful properties
- **polymer** a very large compound that is made of small, repeating units called monomers

Scientists and engineers have spent many years developing products and processes that have made a great impact on society. The Bronze Age ushered in the development of alloys. During the Iron Age, people changed the way they farmed crops and created weapons. But probably one of the most significant technological advancements occurred during the 20th century, also known as *the Plastic Age*. The invention of polymers revolutionized how scientists understood materials. Prior to the 1900s, many scientists believed molecules with a molecular weight greater than 1,000 AMUs could not exist. Hermann Staudinger, a German chemist, disproved this theory when he determined rubber and cellulose were made of 10,000 or more atoms. With this new understanding, scientists and engineers began creating new clothing materials, construction compounds, toys, and packaging materials. Many of the materials you come in contact with on a daily basis are made of a type of polymer.

Polymers are very large compounds that are made of small, repeating units called *monomers*. A typical polymer contains 10,000 monomers, but polymers can reach sizes of up to 200,000 monomers. The structural model of a polymer often includes the letter *n*, which means there are numerous repeating monomer units. Long polymer chains may contain identical units, but most are made of two or more different types of monomers. Many polymer chains contain hydrocarbons that are made of hydrogen and carbon atoms. These types of polymers are created from oil or natural gas sources. Other types of chains contain oxygen, nitrogen, or halogen atoms.

Not all polymers are synthetic or man-made. Many natural polymers exist outside of the laboratory. They compose the structural parts of animal and plant cells. Proteins are an example of a natural polymer that is made of amino acids. Cellulose is a polysaccharide—a molecule made of many sugar molecules. Nucleic acids that contain your genetic information

Polymer Milestones

1891
Rayon

1907
Bakelite

1912
Cellophane

1930
Neoprene

1933
Polyethylene

1938
Nylon and Teflon

Monomer

$CH_2{=}CH_2$

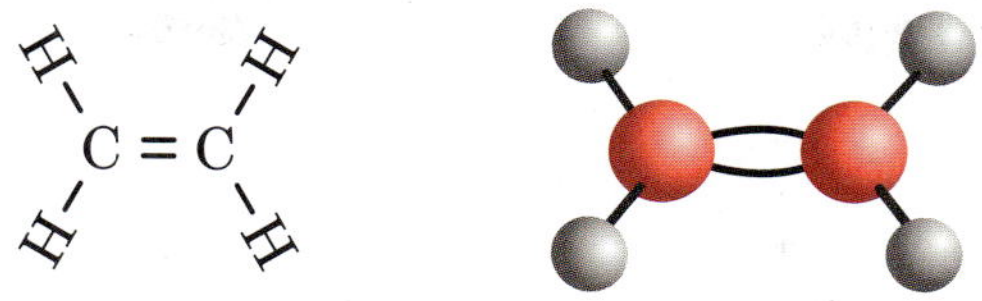

Polymer

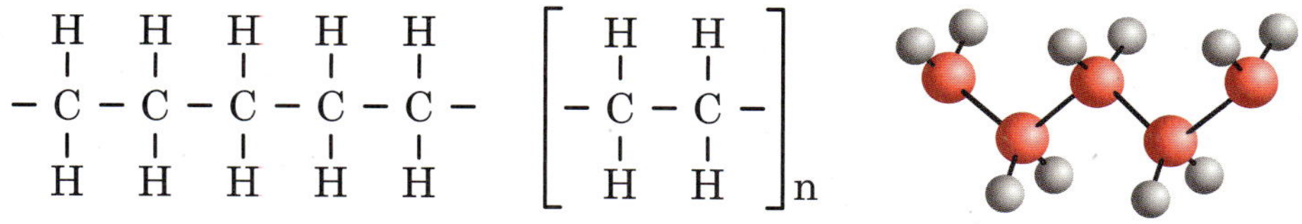

are natural polymers as well. Diamonds, quartz, and feldspar are mineral polymers.

One way that scientists can create very long polymer chains in the laboratory is by adding one monomer at a time to the end of existing chains. These very long chains tend to bend and coil to create macromolecules that look like a plate of spaghetti. Polyethylene is a synthetic polymer that contains about 10,000 monomers. It is translucent and softens when heated. This polymer has been used in packaging materials, coatings, and bottles and containers. Polypropylene is harder than polyethylene because it contains between 50,000 and 200,000 monomers. This material is used in the textile industry. Polystyrene is glassy and transparent, and it can be colored any shade. Plastic toys contain polystyrene. Some examples of synthetic polymer products include concrete, plastics, glass, and rubber.

In general, polymers are lightweight, strong, flexible, durable, and relatively inexpensive. They are also very versatile because scientists can change the order and arrangement of monomers in practically unlimited ways, which affects the properties of the polymer. A particular group of synthetic polymers, aramids,

1950
Polyester

1954
Polypropylene

1965
Kevlar

1970
Thermoplastic polyester

1980
High-density polyethylene

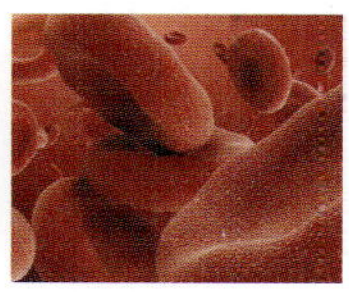

1994
Artificial red blood cells

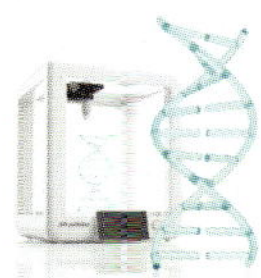

2005
3D biomedical printing

Common Polymers

Name	Monomer	Properties	Uses
Polyethylene	$-(CH_2-CH_2)_n-$	Soft waxy solid	Film wrap, plastic bags
Polypropylene	$-[CH_2-CH(CH_3)]_n-$	Soft elastic solid or hard strong solid	Carpet, upholstery
Poly (vinyl chloride) (PVC)	$-(CH_2-CHCl)_n-$	Strong rigid solid	Pipes, siding, flooring
Polystyrene	$-[CH_2-CH(C_6H_5)]_n-$	Hard rigid, clear solid	Toys, cabinets, foam packaging
Polytetrafluoroethylene (Teflon)	$-(CF_2-CF_2)_n-$	Resistant smooth solid	Nonstick surfaces, electrical insulation
Poly (methyl methacrylate) (Plexiglas)	$-[CH_2-C(CH_3)CO_2CH_3]_n-$	Hard transparent solid	Lighting covers, signs, skylights
cis-Polyisoprene (natural rubber)	$-[CH_2-CH=C(CH_3)-CH_2]_n-$	Soft sticky solid	Tires, motor mounts
Polychloroprene (Neoprene)	$-[CH_2-CH=CCl-CH_2]_n-$	Tough rubbery solid	Synthetic rubber

have been created to be lightweight and five times stronger than steel. These properties make aramids ideal for the production of fireproof clothing worn by race car drivers and firefighters, bulletproof vests, and puncture-resistant clothing.

Many materials have been replaced with polymer compounds because of their versatility and desirable properties. For example, some ivory piano keys have been replaced with plastic. Many natural fibers, such as cotton, wool, linen, and silk, have been replaced by synthetic fibers. Polyester fibers, such as fleece and

Aramid polymers are used in the clothing worn by motorcyclists to protect them from abrasive surfaces.

BIOGRAPHY

Percy Julian

Percy Julian was born in 1899 in Montgomery, Alabama. Julian's city offered no public education for African Americans beyond eighth grade at that time, so he entered DePauw University in Indiana after skipping high school. He graduated with honors. After earning his PhD in chemistry, Julian continued his chemistry research. He investigated plant products, especially traditional medicinal plants. In 1935, he used one of these plants, the African calabar bean, to develop a synthetic chemical called *physostigmine*, or *eserine*. This chemical treated glaucoma, a disease that causes blindness. Julian later found that hormones from soybean products could be useful in treating certain cancers. He used another soy product to develop a foam that could put out oil and gas fires. This was used by the Navy during World War II.

In 1948, the Mayo Clinic discovered cortisone, which relieves rheumatoid arthritis. Cortisone was extracted from the adrenal glands of oxen and was very expensive. The following year, Julian and his team had successfully created a cortisone substitute that was relatively inexpensive. He made it possible for thousands of people to benefit from the use of cortisone. Julian developed many medical products that became widely available and inexpensive. His contributions promoted extensive research, which led to the development of medicines that suppress the immune system. These medicines make organ transplants possible.

Gore-tex, are warmer and provide better protection against water than many traditional clothing materials.

Many synthetic materials are a combination of several different compounds. These substances are called **composites**. They are materials made of two or more substances that are combined to produce materials with useful properties. For example, manufacturers weave small threads of glass into plastic to create fiberglass. The glass reinforces the plastic to create a strong material that is also lightweight. A combination of the two materials reduces the brittleness of plastic so the substance does not easily fracture or break under a great force. Fiberglass composites are used in the manufacturing of automobile and boat exteriors.

Today, composite materials are very popular. They are used to build and repair bridges, replace timber structures in harbors, and create strong, flexible fishing poles. These materials can sustain extreme temperatures. Other composite materials include cement, plywood, and fleece. Researchers are continually

TRY THIS

Play Putty

Place 4 tsp of glue in a foam cup. Add a few drops of food coloring, and thoroughly mix with a wooden splint. Using a graduated cylinder, measure 10 mL of liquid starch. Pour the liquid starch into the foam cup. Stir the mixture until the putty is dry enough to handle. Place the mixture on a paper towel, and knead it for about five minutes. Manipulate the putty. Pull it apart slowly then quickly. Describe its properties. Store the putty in a sealable plastic bag.

TRY THIS

Making Nylon

Measure 50 mL of hexamethylenediamine/sodium hydroxide solution in one graduated cylinder. Measure 25 mL of adipoyl/chloride solution in a separate graduated cylinder. Pour the contents of the first cylinder into a 250 mL beaker. Add a few drops of food coloring. Carefully pour the adipoyl chloride/hexane solution into the beaker. Try not to mix the two solutions. The adipoyl chloride/hexane solution should float on the surface. Observe the polymer film that forms where the two solutions meet. This film is nylon. With a pair of tongs, carefully grasp the film toward the center of the beaker. Wrap the nylon polymer around a wooden rod, and slowly roll the rod to draw the nylon from the solution. Rinse the nylon with water. What physical properties does the nylon have?

creating and studying new polymer materials. New plastics have already been developed to create flexible computer screens and artificial muscles. This is an exciting field that is constantly growing.

LESSON REVIEW

1. Why is the discovery of polymers important?
2. Describe polymer chains.
3. Provide two examples of polymers found in nature.
4. How do polyethylene and polypropylene compare?
5. Describe the properties of aramids and three uses.
6. Explain how scientists and engineers continue to develop new composites. Use an example of a composite material in your answer.

Kayaks are made of composite materials because they create a lightweight, flexible structure.

UNIT 3

Interactions of Matter

Chapter 1: *Solutions*
Chapter 2: *Chemical Reactions*
Chapter 3: *Acids, Bases, and Salts*

Vocabulary

acid
activation energy
aqueous solution
base
boiling point elevation
catalyst
chemical equation
chemical reaction
coefficient
combustion reaction
concentration
decomposition reaction
desalination
endergonic reaction
endothermic reaction
end point
equivalence point
exergonic reaction
exothermic reaction
freezing point depression
gaseous solution
indicator
inhibitor
liquid solution
neutralization reaction
nonpolar substance
pH scale
polar substance
product
reactant
reaction rate
salinity
salt
solubility
solute
solution
solvent
strong acid
strong base
synthesis reaction
titration
water-soluble solid
weak acid
weak base

Key Ideas

- Systems, order, and organization
- Evidence, models, and explanation
- Change, constancy, and measurement
- Abilities necessary to do scientific inquiry
- Properties and changes of properties in matter
- Motion and forces
- Transfer of energy
- Structure and properties of matter
- Chemical reactions
- Conservation of energy and increase in disorder
- Personal health
- Populations, resources, and environments
- Natural and human-induced hazards
- Risks and benefits
- Science and technology in society
- Nature of science
- History of science

SCRIPTURE

Let your conversation be always full of grace seasoned with salt, so that you may know how to answer everyone.

Colossians 4:6

3.1.1 Dissolving

OBJECTIVES

- Define the parts of a solution.
- Explain the factors that affect dissolving rates.
- Identify the significance of polar and nonpolar characteristics when one substance dissolves another substance.

VOCABULARY

- **nonpolar substance** a substance that has molecules whose electric charges are equally distributed
- **polar substance** a substance that has molecules whose electric charges are unequally distributed
- **solute** the substance in a solution that is dissolved
- **solution** a homogeneous mixture of two or more substances
- **solvent** the substance in a solution that is present in the largest amount and acts as the dissolving agent

It is a hot summer day and you are thirsty. You go to the kitchen to make some lemonade from a powdered mix. After adding just the right amount of powder into the water, you stir until all the powder has dissolved. Then you pour a glass of lemonade. Later, you can pour another glass of lemonade without having to stir the lemonade again. Why has the lemonade powder not fallen to the bottom of the container?

The lemonade does not need to be stirred because it is now a solution. A **solution** is a homogeneous mixture of two or more substances. When two substances are mixed together to form a solution, one is dissolved into the other. The substance that is present in the larger amount and acts as the dissolving agent is called the **solvent**. The substance being dissolved is called the **solute**. The solvent dissolves the solute. When you add lemonade powder to water, the water is the solvent and the lemonade powder is the solute.

The solute particles of a solution are too small to be seen. They are the size of ions or molecules. The sugar particles dissolved in the lemonade and will stay dissolved in the lemonade indefinitely. They will not need to be stirred again. If a liquid solution is poured through a paper filter, both the solvent and solute will pass through the filter without leaving any residue.

Not all solutions form at equal rates. Three things determine how fast a solid solute will dissolve into a solvent to form a solution.

Temperature. Solute crystals dissolve more quickly in a warm solvent than in a cold solvent. The added energy of the heat makes the molecules move faster, creating more frequent collisions between solute and solvent molecules.

TRY THIS

Saltwater from Dirt

Using scissors, poke a small hole in the bottom of two foam cups. Place a coffee filter inside one of the cups and place that cup inside the other. Add about 50–80 g of soil and about a tablespoon of salt to the coffee filter. Gently mix. Create three small "feet" from clay. Center the cup on the feet and place the footed cup on a square of black paper. Pour enough water into the soil mixture to saturate the soil. Let the system sit for a couple of days. What do you see on the black paper? How did they get there? Identify the solvent and the solute. Why did the soil not pass through with the water?

Surface area. The greater the total surface area, the faster the solute will dissolve. Smaller crystals of solute will dissolve more rapidly than larger crystals because there is more surface area. Have you ever added a sugar cube instead of sugar granules to a hot drink? If so, you know it takes much longer for the cube to dissolve. The sugar cube has less total surface area from which the hot drink can pull away particles.

Movement. When the solvent is shaken or stirred, solute particles are moved around, bringing fresh solvent particles into contact with the solute particles and making the solute dissolve faster.

Think about salad dressing made from oil and vinegar. You have to shake and quickly pour the dressing on the salad before the oil and vinegar separate. Oil and vinegar will not form a solution because solvents and solutes must conform to the *like dissolves like* rule. There are two basic types of substances: polar and nonpolar. A **polar substance** has molecules whose electric charges are unequally distributed. This means that opposite sides of the molecules have opposite charges. Water is an example of a polar substance. Many acids are also polar. Polar solvents dissolve polar solutes, so the solvent water can dissolve salt, another polar substance. The slight charges on a water molecule attract the charged particles of the salt.

A **nonpolar substance** is a substance that has molecules whose electric charges are equally distributed. Oils are nonpolar

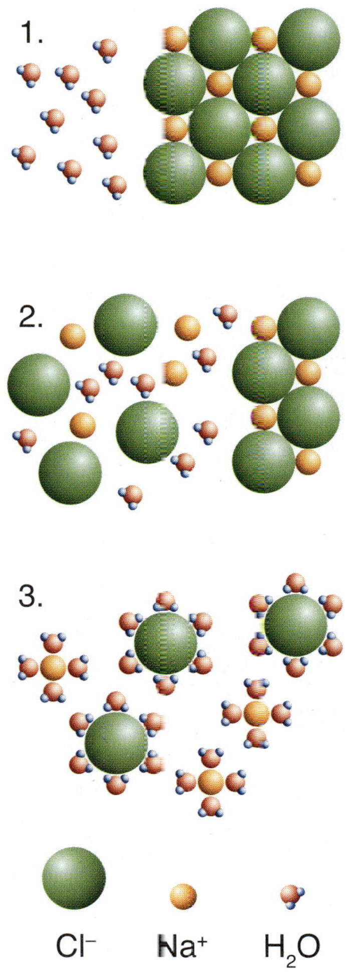

When salt is dissolved in water, the charged particles of the water molecules surround the oppositely charged ions of Na^+ and Cl^-.

Scientist stirring a solution

Duck covered in oil

FYI

Dialysis

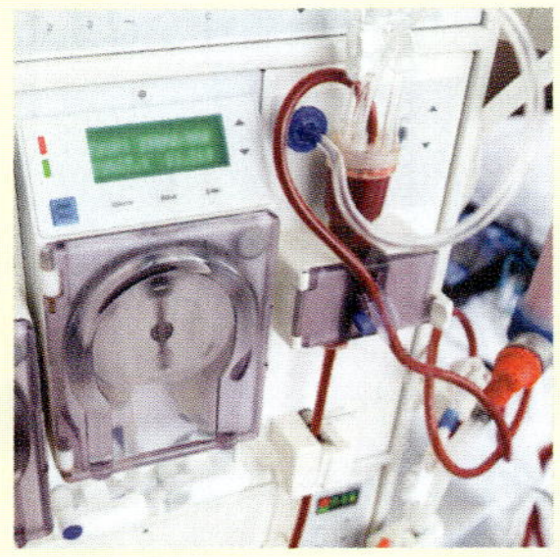

In the body, the kidneys filter wastes and extra fluids from the blood, which maintains the proper balance of water and minerals in the body. When a person's kidneys fail, dialysis is used to replace this filtration function. In Greek, *dialysis* means "dissolution," and it works by using the natural osmosis of solutes and fluids across a semipermeable membrane. In the process, blood flows past one side of the membrane, and dialysis fluid flows past the other. Tiny dissolved waste particles and extra fluids in the blood diffuse through the membrane and are filtered from the body with the dialysis fluid. There are two types of dialysis. In hemodialysis, a person's blood is circulated about three times a week through a filter outside the body. In peritoneal dialysis, the smooth abdominal wall inside the body is used to filter the blood through diffusion. This exchange process is repeated four or five times a day. The procedures return the blood to a healthy balance.

solvents and will dissolve other nonpolar substances such as gasoline and grease. Ethanol (C_2H_5OH) is a solvent with both polar and nonpolar properties, and it will dissolve both polar and nonpolar substances.

Water, a polar substance, rolls off a duck's back because its feathers are coated in oil, a nonpolar substance. Oil and water do not mix. What happens when a duck lands in a body of water containing an oil spill? The oil from the spill dissolves in the oil on the duck's back. Now the duck is heavily laden with oil and cannot fly or swim. A solution of warm water and dishwashing soap dissolves the oil and cleans the bird.

LESSON REVIEW

1. What is a solution?
2. Give an example of a solution, and identify the solvent and the solute.
3. Describe three ways to make sugar cubes dissolve faster in water. Explain why these methods were chosen.
4. What is the difference between polar and nonpolar substances? Which will form solutions?

CHALLENGE

Solve the Water Crisis

Design an experiment to remove harmful contaminants from water in a developing country. Consider the following possible contaminants: chlorine, oil from an oil spill, leaking fuel tanks, lead, arsenic, barium, chromium, and mercury. Explain why it is important to remove these contaminants.

What type of substance comes to mind when you think of solutions? Most people think of a liquid, but solutions can be liquids, solids, or gases. A solution can be any combination of these states of matter.

Two or more gases mix together to form a **gaseous solution**. Even though mixtures of gases are not generally referred to as solutions, they have all the characteristics of solutions. Because the oxygen, nitrogen, argon, carbon dioxide, and other gases in the atmosphere are well mixed by the wind, the air you breathe is a solution. Any sample taken from the atmosphere will have all the gases, and they will not be distinguishable.

A **liquid solution** consists of a solid, gas, or other liquid that is dissolved into a liquid solvent. Water is a common liquid solvent. A solution in which water is the solvent is called an **aqueous solution**. In chemistry, the symbol (aq) is used to indicate an aqueous solution. The solute in aqueous solutions can be any of the three phases. When you add sugar to tea, the sugar (a solid) is the solute. When you add lemon juice to water, the lemon juice (a liquid) is the solute. One of the solutes in carbonated drinks is carbon dioxide, a gas. Oxygen gas also dissolves in water. Most marine and freshwater organisms depend on small amounts of oxygen dissolved in oceans and lakes for survival.

A solid that dissolves in water to form an aqueous solution is called a **water-soluble solid**. In order for a substance to dissolve in water, it must be ionic or polar, like salt and sugar. These two substances, however, dissolve in slightly different ways. Salt is composed of ions and when it dissolves, the polar water molecules are attracted to the ions. The more negative oxygen atoms attract

Examples of Solutions

Solute	Solvent	Solution
Oxygen (gas)	Nitrogen (gas)	Air
Carbon dioxide (gas)	Water (liquid)	Carbonated soda
Hydrogen (gas)	Platinum (solid)	Hydrogen platinum plates (used in industry)
Water (liquid)	Air (gas)	Atmospheric water vapor
Ethylene glycol (liquid)	Water (liquid)	Antifreeze
Mercury (liquid)	Silver (solid)	Some dental fillings
Salt (solid)	Water (liquid)	Ocean water
Copper (solid)	Gold (solid)	14-karat gold

OBJECTIVES

- Compare types of solutions.
- Explain how soap dissolves in both water and oil.
- Distinguish between water-soluble and fat-soluble vitamins.

VOCABULARY

- **aqueous solution** a solution in which water is the solvent
- **gaseous solution** a solution in which two or more gases are mixed together evenly
- **liquid solution** a solution in which a solid, gas or other liquid is dissolved into a liquid solvent
- **water-soluble solid** a solid that can be dissolved in water to form an aqueous solution

Sea life depends on the oxygen that dissolves in the water.

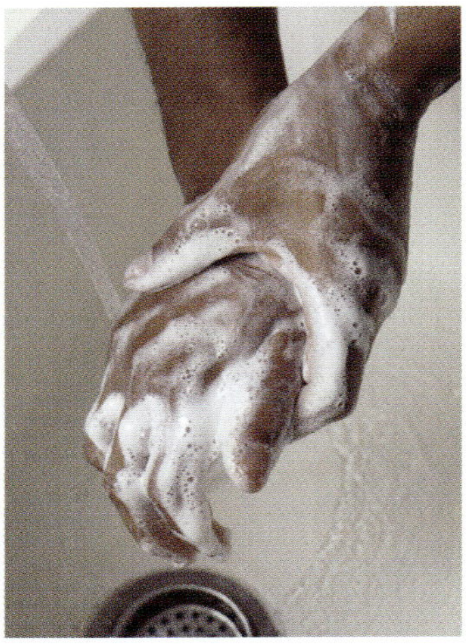

Hand washing to remove dirt and oil

FYI

Ionization

When neutral molecules gain or lose electrons they become ionized. For example, when hydrogen chloride gas dissolves in water, ions are formed by the reaction of the solvent and solute particles. Dissociation and dispersion are physical changes. Ionization is a chemical change.

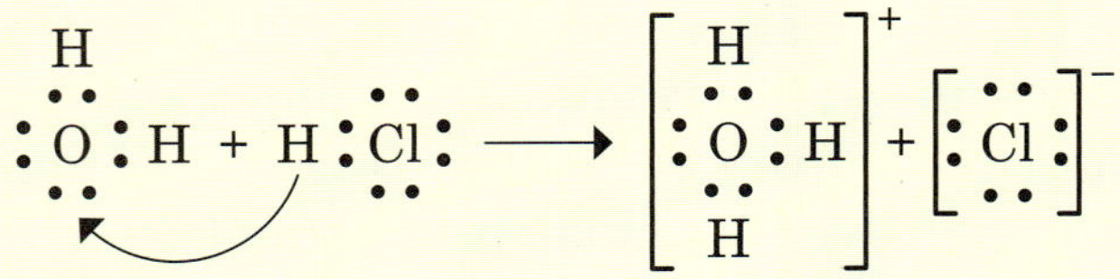

the sodium ions (Na^+), and the more positive hydrogen atoms attract the chloride ions (Cl^-). This process is called *dissociation.*

When sugar dissolves, the process is slightly different. The sugar breaks into small pieces and scatters throughout the whole solution through a process called *dispersion.* The water molecules attract the sugar molecules, separating them from each other. As more sugar separates, more molecules are exposed to the water until all the sugar dissolves.

Water-soluble substances are also present in soap solutions, such as those used to wash hands, dishes, and clothes. Many soaps and detergents are designed to dissolve in water and dissolve oil. How is this possible? Water is polar and oil is nonpolar. Soaps are made from molecules, such as sodium stearate, which have both polar and nonpolar ends. The polar end dissolves the soap and the nonpolar end traps the oil, which washes away.

Sodium Palmitate

$$CH_3(CH_2)_{13}CH_2 - \overset{\overset{\displaystyle O}{\|}}{C} - ONa$$

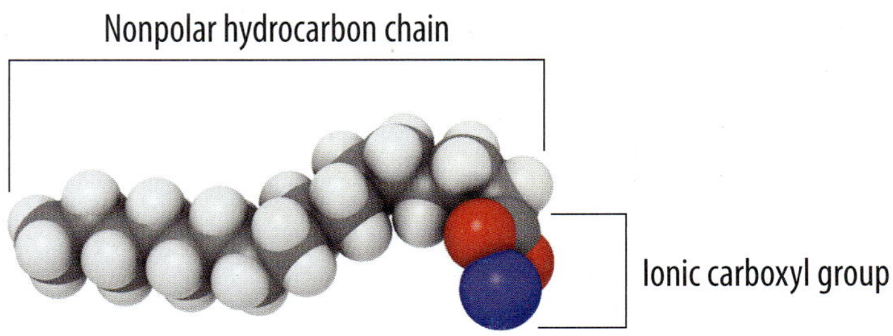

Sodium palmitate and sodium stearate are useful molecules in soap. The long hydrocarbon chain is nonpolar and dissolves oil. The short end is ionic and dissolves easily in water.

HISTORY

Scurvy

Caused by a lack of vitamin C, scurvy was first identified by both the Greeks and Egyptians around 3700 BC. However, it was not until 1753 that a British naval doctor, James Lind, proved that the disease could be prevented by eating citrus fruits such as lemons and oranges. In the years that followed, British ships began carrying citrus to keep crew members healthy, which is why British sailors were called *limeys*. The human body cannot produce vitamin C, which is necessary to maintain healthy bones, teeth, skin, tissues, and blood cells. Vitamin C is essential for the absorption of iron and the production of the protein collagen, a primary component of connective tissues like tendons and ligaments. To maintain a healthy concentration of the nutrient, teenagers need about 75 mg of vitamin C daily, an amount that can be found in a large orange or a cup of strawberries.

Most foods contain vitamins that are necessary for the proper function of various systems in the human body. Some vitamins are water-soluble and some are fat-soluble. The body uses the water-soluble vitamins B and C as needed. Any excess dissolves in the water of the cells and is excreted out of the body. Because vitamins B and C are not stored, they need to be replenished frequently. In contrast, the fat-soluble vitamins A, D, E, and K are also used as needed, but any excess amounts dissolve in fat cells, stored for later use. They are not flushed out. Having a reserve of vitamins can be beneficial, but it can also be harmful. These vitamins are toxic at high levels. Eating a balanced diet provides the proper amount of vitamins, but taking too many fat-soluble supplements can be a serious problem. It is important to read labels and consume only the recommended amounts.

Not only do some solids dissolve in certain liquid solvents, but a solid can also act as a solvent. It might seem odd to think of one solid dissolving another. Two or more solids melt together to form

Water-Soluble Vitamins

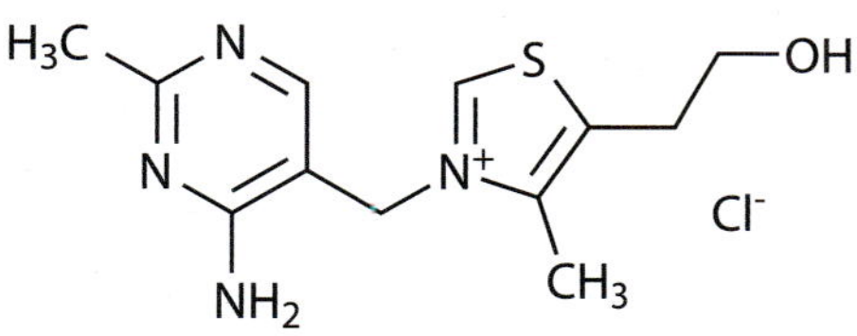

Thiamine chloride (B1)

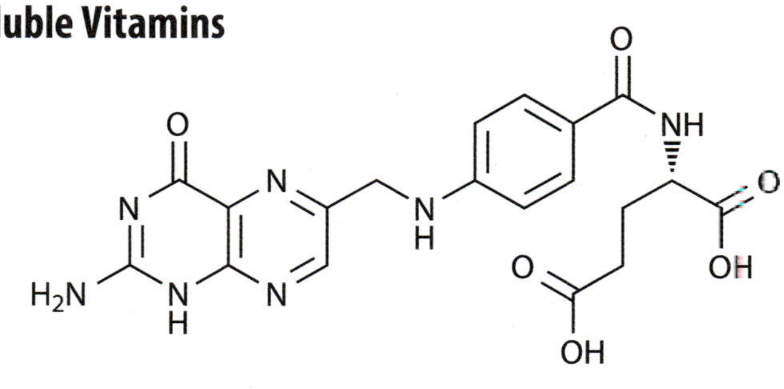

Folic acid (M)

TRY THIS

Vitamin C

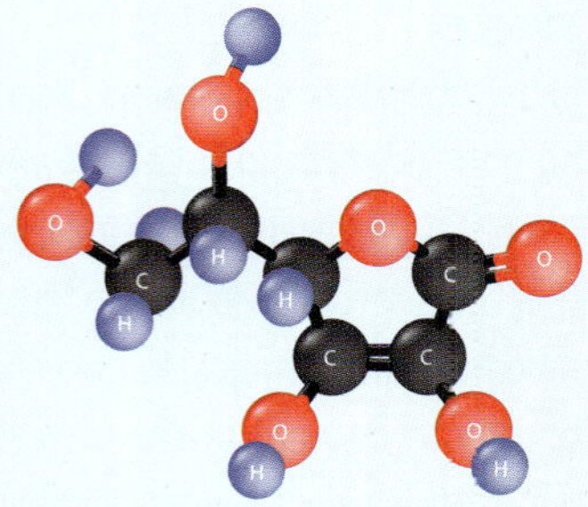

Vitamin C is a water-soluble substance necessary for proper function of the immune system. Choose several drinks and foods to test for the presence of vitamin C. Be sure to use one item that you know has vitamin C. This is your control. If any of the substances are solids, grind them with a mortar and pestle. Pour a sample of each food into its own beaker. For the ground food, place it in a beaker and add distilled water to make an aqueous solution. Place 15 drops of indophenol solution into each test tube. Take a clean pipette and start dropping the first selection of food into a test tube. Count the number of drops it takes to turn the indophenol clear. Record your data. Repeat the procedure with the other food and drink items, using a clean pipette and a separate test tube for each sample. The fewer drops needed to change the indicator to a clear color, the more vitamin C the food item has. Which items have vitamin C? Which item has the most?

alloys, which are solid solutions made of two or more metals or of a metal and a nonmetal. Brass, stainless steel, and gold used in jewelry are all alloys. Copper dissolves into gold to create gold jewelry. Copper combines with zinc to make brass, and iron and chromium form stainless steel. Metallurgists use alloys because the alloys have properties that each of their original elements do not have. They can be stronger and more resistant to corrosion, have higher melting points, and conduct heat or electricity better than the original metals. For example, even though solid gold is worth more than a copper-gold alloy, jewelry made of copper and gold is stronger and harder than solid gold.

The alloy NiP (nickel and phosphorus) comprises one part of the disk in a computer's hard drive.

Even solids and gases form solutions. Small atoms such as hydrogen, lithium, or boron can mingle with metal atoms. Hydrogen, for example, dissolves in platinum.

LESSON REVIEW

1. What is an aqueous solution? Give an example of an aqueous solution that is a crucial part of creation.
2. Describe how soap works with water to clean oil.
3. How are vitamins B and C different from vitamins A, D, E, and K?
4. What is an alloy?
5. Why are alloys useful?

Oceans and Saltwater 3.1.3

It is hot outside, but you have things to do. You play basketball with friends and then go on a bike ride. When you get home, you mow the lawn. But now you are hot and also very thirsty. Would drinking a nice cold glass of saltwater be satisfying?

Drinking saltwater is not the thing to do! Although you think of freshwater, not saltwater, as being essential to life, salty ocean water sustains a wide variety of ocean life. In fact, most organisms on Earth would die without saltwater.

One taste of ocean water reveals that it is not pure water. However, it is not just water and salt either. This aqueous solution contains pure water, gases, and solids. Ocean water is 95% pure water, with the other 5% consisting of dissolved gases and nearly 80 different minerals. Each cubic meter of ocean water holds 1.5 kg of dissolved substances. Even gold and platinum compounds are dissolved in ocean water!

The three primary gases dissolved in ocean water are carbon dioxide (CO_2), oxygen (O_2), and nitrogen (N_2). Carbon dioxide dissolves most easily in ocean water and comprises the greatest portion of solutes. Marine organisms convert carbon dioxide into organic carbon compounds, such as carbohydrates, which they need to live. Oxygen is essential for fish, ocean invertebrates, and other marine organisms. Plankton, the small organisms that serve as food for fish and the foundation of marine food chains, depend on nitrogen gas for photosynthesis.

OBJECTIVES

- Identify solutes in the ocean.
- Explain what factors affect the salinity of the oceans.
- Describe methods of desalination.

VOCABULARY

- **desalination** the process of removing salt from ocean water to obtain freshwater for drinking, irrigation, or industrial use
- **salinity** the amount of salt in water

Coastline of a saltwater ocean

Body of Water	Average Salinity
Open ocean	3.3%
Red Sea, between Africa and Saudi Arabia	4.2%
Gilbert Bay in Great Salt Lake, Utah	Fluctuates between 6% and 27%
Gunnison Bay in Great Salt Lake, Utah	About 30%
Dead Sea, between Israel and Jordan	36%

The most obvious feature of ocean water is its saltiness. Where does the saltiness come from? Carbon dioxide in the atmosphere makes rain and snow slightly acidic, which then allows the precipitation to readily dissolve many of the minerals in rocks. Wind, water runoff, and other erosion processes eventually carry these dissolved minerals to rivers, and ultimately to the seas. In fact, each year Earth's rivers transport 400 billion kilograms of minerals to the oceans, and most of these are salts. No wonder ocean water is salty! More than 99% of the ocean's salt content is made up of only six elements. Aside from the elements hydrogen and oxygen in water, the two most common substances found in the ocean are compounds of chlorine and sodium. Chlorine, in the form of a chloride ion, accounts for 1.94% of the ocean, and sodium accounts for 1.05%. These two elements, which combine to form common table salt, account for more than 85% of the ocean's salt. Other trace elements found in seawater include vanadium, zinc, copper, and mercury.

The amount of salt in ocean water is called **salinity**, and it is generally expressed in parts per thousand (ppt). Salinity can

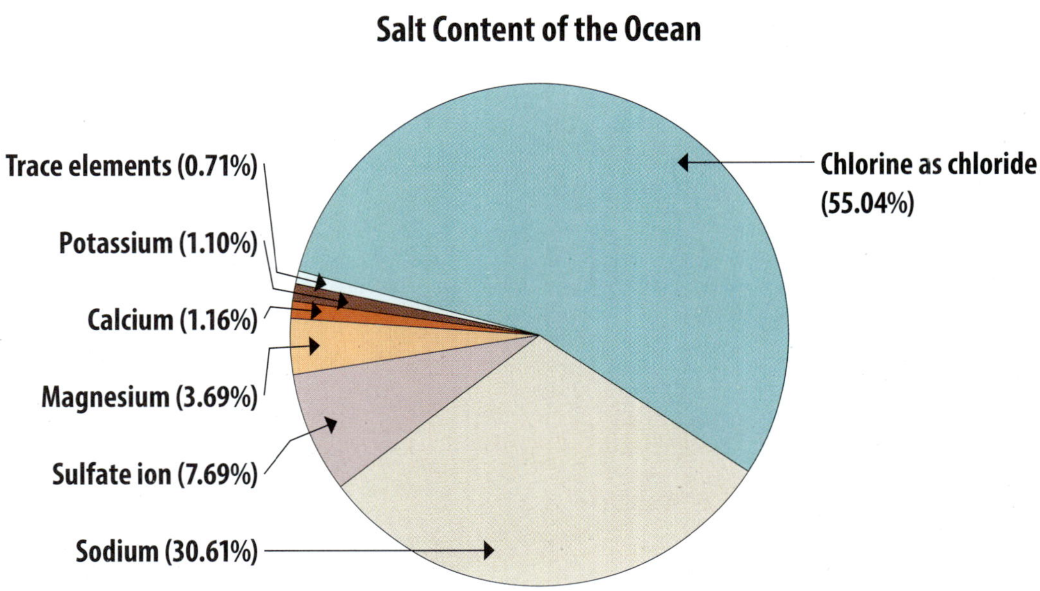

Salty shore of the Dead Sea

vary depending on the location and season. Enclosed and mostly enclosed seas such as the Dead Sea and the Caribbean Sea have a higher salt content ranging from 37 to 42 ppt. Coastal areas near large rivers such as the Amazon, Mississippi, or Yangtze have a low salt content of less than 33 ppt. Why would ocean water near rivers have less salt? The freshwater pouring in dilutes the ocean water.

FYI

Freshwater Versus Saltwater

Why do freshwater fish live only in freshwater and saltwater fish live only in saltwater? The cells of a freshwater fish are designed to tolerate the solute levels in freshwater. If it were put in saltwater, it would find itself in a hypertonic solution. The solute concentration of saltwater is too high, and the freshwater fish would become dehydrated as its liquid contents were drawn out of its cells. The reverse would be true for a saltwater fish placed in freshwater. Its cells are accustomed to a hypertonic solution. When it enters the hypotonic freshwater, the high concentration of solutes in its cells would draw the water in, causing them to swell to the point of killing the fish.

God designed some fish to survive in both environments. Salmon are anadromous, which means they are born in freshwater and live the rest of their lives in saltwater. North American and European eels are catadromous, which means they are born in saltwater and spend their lives in freshwater.

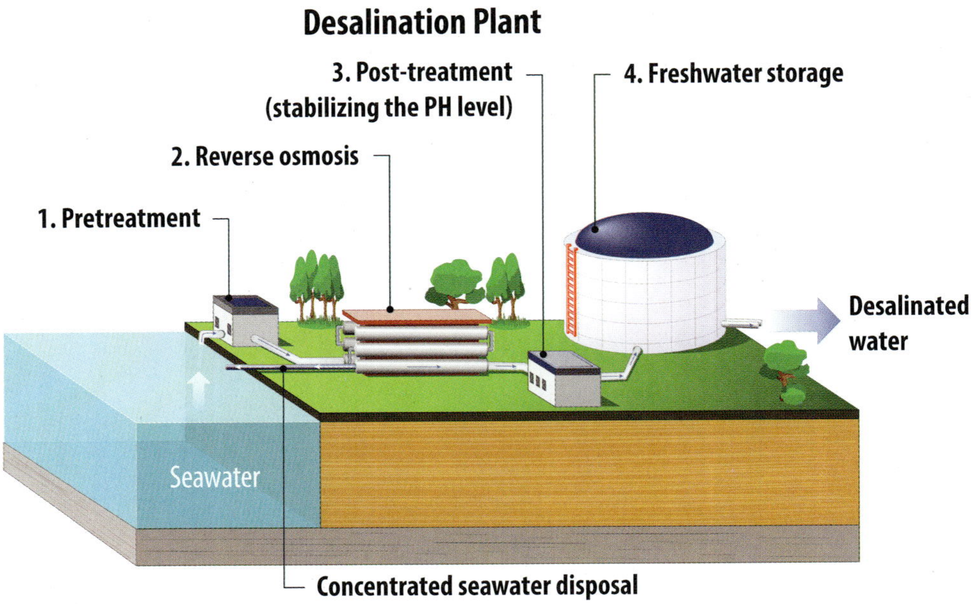

The salinity throughout a body of water varies. Differences in salinity, along with differences in water temperature, create ocean currents. Colder or saltier water is more dense and sinks, forcing warmer or less salty water to the surface. These temperature and salinity differences form convection currents throughout the world's oceans. The continuous churning of the water facilitates the distribution of nutrients to sea life.

Salt from ocean water is an important product for many countries. Sea salt is cheaply extracted in countries with hot, dry climates, like Spain and Israel. First, shallow pans are flooded with ocean water. Then the water is allowed to stand until the water evaporates, leaving salt deposits.

Harvesting salt in India

Reverse Osmosis

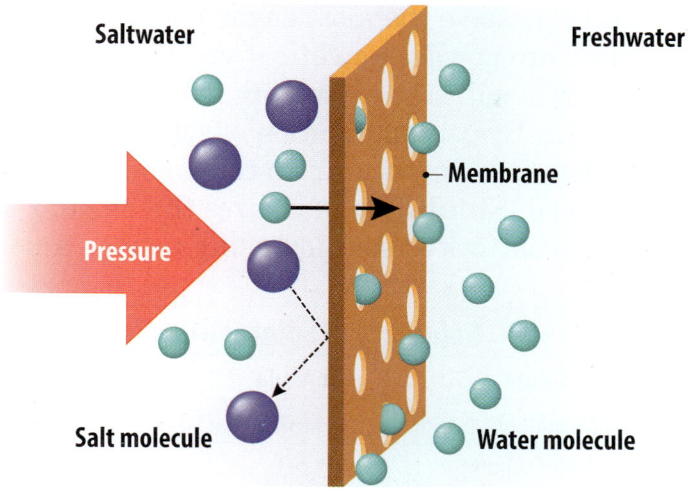

Desalination is the process of removing salt from ocean water to obtain freshwater for drinking, irrigation, or industrial use. Countries that lack large freshwater supplies, like Saudi Arabia, Israel, and Caribbean nations, use desalination to obtain more drinking water. The oldest and most widely used method of desalination is distillation. In distillation, ocean water is heated to the point of vaporization. Coils filled with cold water condense this water vapor into pure water, leaving behind all the solutes found in ocean water. Distillation is an effective method for removing salt, but it is expensive. Other methods include reverse osmosis and electrodialysis. Reverse osmosis involves forcing saltwater against a membrane. The freshwater passes through the membrane and the salts remain behind. In electrodialysis, electrical potential drives the positive and negative ions of the dissolved salts through separate filters, after which the freshwater is left behind.

Reverse osmosis desalination plant in Barcelona, Spain

LESSON REVIEW

1. Identify the solvent, two solid solutes, and three gaseous solutes in ocean water.
2. What percentage of ocean water is pure water? What makes up the rest of ocean water?
3. How does salt get into ocean water?
4. How do salinity levels change?
5. Describe three desalination processes.

3.1.4 *Solubility and Concentration*

OBJECTIVES

- Calculate the concentration of a solution.
- Classify solutions according to their concentrations.
- Compare the solubility of gases to liquids and solids.

VOCABULARY

- **concentration** a measure of the amount of solute dissolved in a certain amount of solvent
- **solubility** a measure of the amount of solute needed to make a saturated solution using a given amount of solvent

If you add a spoonful of sugar to a cup of hot coffee, all the sugar will dissolve. If you add another spoonful, that sugar will dissolve too. To dissolve the most sugar possible, you should crush the sugar into powder and stir it vigorously. At some point, however, the coffee will stop dissolving more sugar. It will have reached its limit. Every combination of solvent and solute, at a specific temperature, has a limit to the amount of solute that can be dissolved. **Solubility** measures the amount of solute needed to make a saturated solution using a given amount of solvent.

Solubility of Solutes at Different Temperatures

Solute	Formula	0°C	60°C	100°C
potassium nitrate	KNO_3	13.9	106	245
cerium sulfate	$Ce_2(SO_4)_3$	20.8	3.87	n/a
table salt	NaCl	35.7	37.1	39.2
potassium bromide	KBr	53.6	86	104
sodium chlorate	$NaClO_3$	79.6	135	204
table sugar	$C_{12}H_{22}O_{11}$	179	287	487

The solvent and solute of a solution can have a variety of different proportions. For example, a lot of sugar (solute) added to iced tea (solvent) makes it very sweet and just a little sugar added makes it less sweet. **Concentration** measures the amount of solute dissolved in a certain amount of solvent. The simplest statement of the concentrations in a solution is in terms of percentages by weight or volume. For instance, in a 300 mL tea and sugar solution in which 3 cm^3 of the solution is sugar, the concentration of sugar equals 1%. A 300 mL tea and sugar

Temperature Effects on Solubility

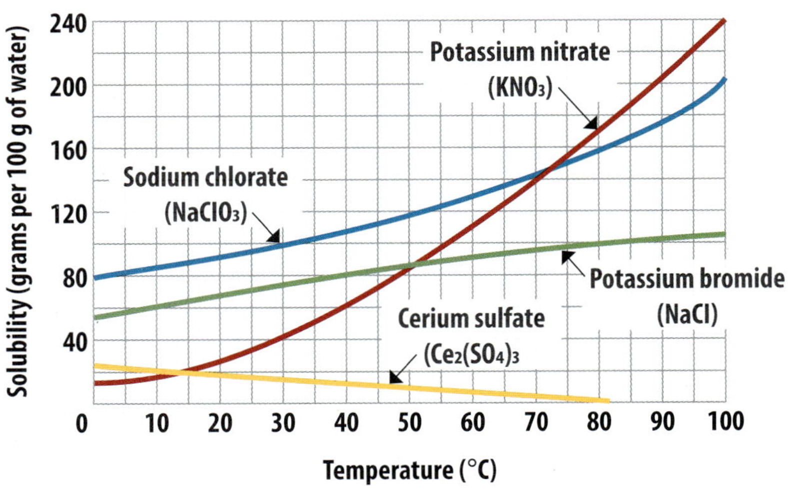

SCIENCE IN ACTION

Concentration

To determine the concentration of a solution, divide the volume of the solute by the volume of the solution. Then, multiply by 100%. For example, given 500 mL of soda that contains 10 mL of corn syrup, what is the concentration of corn syrup? 10 mL/500 mL × 100% = 2%. The concentration of corn syrup is 2%. What is the concentration if the solute is increased to 150 mL? 150 mL/500 mL × 100% = 30%. The first solution is a dilute solution and both solutions are unsaturated.

solution in which 30 cm^3 of that solution is sugar results in a 10% sugar concentration. This means 10% of the solution is sugar and 90% of the solution is tea.

Very low concentrations may be expressed in parts per million (ppm), a unit typically used to measure air pollution levels. In a million molecules of air in which three of those molecules are carbon monoxide, the concentration of carbon monoxide equals 3 ppm. The same concentration written as a percent is 0.0003%, a number so small that it is difficult to comprehend. In these instances, using ppm makes more sense.

One method of classification sorts solutions according to the relative concentration of a solute. A solution containing a small amount of solute is called *a dilute solution*. A solution containing a large amount of solute is called *a concentrated solution*. Frozen fruit juice is a concentrated solution comprised of very little water and a large portion of solutes. Ready-made fruit juice is usually labeled *from concentrate* to indicate that water has been added to the concentrated solution.

A solution containing the maximum amount of a solute that a solvent can hold at a given temperature and pressure is called *a saturated solution*. An unsaturated solution has less solute than a saturated solution. The saturation level of a solution usually rises with added heat. When the heated saturated solution cools, the extra solute typically separates as crystals. If crystals do not form, a *supersaturated solution* is formed, which contains more dissolved solute than a normal saturated solution under the same conditions. If a

Rock candy forms from a supersaturated solution of sugar and water.

crystal of solute is added to a supersaturated solution, crystals will form in the solution.

How easily a solute dissolves into a solvent depends on the concentrations of each substance, the temperature, and pressure. If a few grams of sodium chloride, or table salt, are added to 500 mL of water, it dissolves easily. If 400 g of salt are added to 500 mL of water, the dissolving process is more difficult. The temperature of the solvent also affects solubility. At a temperature of 0°C, a maximum of 35.7 g of sodium chloride dissolves in 100 mL of water. But when the temperature increases to 100°C, the salt's solubility also increases to 39.2 g. For this reason, more sugar dissolves in hot tea than in iced tea. Usually, raising the temperature of solids and liquids increases solubility.

By contrast, raising the temperature of gases lowers the solubility. Soda is a carbonated drink that releases carbon dioxide. Heat makes gas molecules move faster, enabling them to escape more easily from a liquid solvent. That is why warm soda goes "flat" faster than cold soda.

Pressure does not always affect the solubility of a liquid or solid solute. But it does affect the solubility of gases because gas molecules have an indefinite volume and are not connected. Increasing the pressure on a gas forces more gas molecules into

BIOGRAPHY

William Henry

English chemist William Henry (1775–1836) was born and raised in England. After studying at Edinburgh University in Scotland, he became a physician and returned to Manchester, England, to practice medicine. Because of health problems resulting from an earlier accident, Henry was forced to retire after five years.

A friend of Henry's, the famous chemist John Dalton, encouraged him to turn his attention to chemistry. Henry researched gases, and he enjoyed discussing his results with Dalton. Henry calculated the molecular formula for ammonia by exploding it with oxygen. He also found ways to analyze gas mixtures and to purify coal gas. His most important work focused on the solubility of gases under different temperatures and pressures. His claim in 1803 became known as *Henry's law*.

Henry wrote a very successful textbook about chemistry, *Elements of Experimental Chemistry*. The textbook was updated in 11 editions over a period of 30 years. He was honored as a fellow of the Royal Society and was awarded the Royal Society's Copley Medal in 1808 for his work.

HISTORY

Soda

Joseph Priestly of England spent much of his life as a scientist researching and investigating electricity and gases. The late 18th century was plagued by a multitude of diseases spreading rampantly through growing urban populations. Priestly invented an apparatus that produced carbonated water, with hopes that the water would provide help to those plagued by scurvy and other diseases. The carbonated water proved ineffective, but the introduction of it eventually was flavored, bottled, and mass produced as a worldwide pleasure.

the liquid. Henry's law, named for the English chemist William Henry, summarizes the effect of pressure on gases. It states that the amount of gas that will dissolve in a liquid is directly proportional to the pressure applied to that gas. In other words, when more pressure is applied to a gas, more of the gas will dissolve. When soda is made, the syrup solution is poured into cans or bottles. Then, carbon dioxide is forced into the syrup to make the bubbly soda. The bottles or cans are sealed tightly to maintain the pressure. When you open a can of soda, you release the pressure inside the can, causing the gas to escape. That is another reason soda gets "flat" after the container has been opened—the pressure decreases, allowing the carbon dioxide to break free from the solution.

Henry's law explains a sometimes fatal condition for scuba divers called *decompression sickness*, or *the bends*. The pressure exerted on the liquid gas solution in a diver's blood increases with depth. At sea level, the diver experiences 1 atm of pressure, which is equivalent to 14.7 psi (pounds per square inch). The standard for atmospheric pressure is 1 atm. But just 10 m under water, the pressure doubles to 29.4 psi. At 1,000 m, the pressure reaches 1,484.7 psi. When a diver breathes from an air tank underwater, pressure forces more nitrogen into the diver's bloodstream. Because of the increased pressure, this nitrogen dissolves easily in the diver's blood. The problem occurs when the diver ascends to the surface. If the diver rises too quickly, the excess nitrogen in the blood escapes and bubbles up. These gas bubbles block small veins and arteries and collect in the tissues, cutting off the diver's oxygen supply. This leads to nausea, dizziness, pain in the joints and abdomen (causing the person to bend forward in

TRY THIS

Fizz Factor

Open a cold can of soda and a warm can of soda. Which fizzes more? Why?

LAWS & PRINCIPLES

Henry's Law

The amount of a gas that will dissolve in a liquid is directly proportional to the pressure applied to that gas. The more pressure applied, the more gas that dissolves into the solvent.

the "bends"), paralysis, and disorientation. If the condition goes untreated, the diver may be permanently injured or even die.

A diver can avoid the bends by rising slowly to the surface to allow excess nitrogen time to dissipate. Because the chance of getting the bends is greater the deeper a person dives, recreational scuba diving is usually limited to a depth of 36 m.

LESSON REVIEW

1. What is the concentration of salt in a 550 mL solution containing 11 cm^3 of salt?
2. Describe four different types of solutions as classified by their concentrations.
3. What is the effect of temperature on the solubility of a gas in a liquid?
4. How does temperature affect the solubility of a liquid solute?
5. What is Henry's law?
6. What is decompression sickness and how does Henry's law explain it?

Scuba divers

Boiling Point and Freezing Point 3.1.5

Adding antifreeze to the water in a car's radiator keeps it from freezing in temperatures below 0°C. Sprinkling rock salt on snowy roads and sidewalks in the winter prevents ice from forming. Adding salt or sugar to water raises its boiling point. Why do these things happen?

Adding a solute to a solvent changes both the freezing point and boiling point of the solvent. Lowering the freezing point of a solvent by adding a solute is called **freezing point depression**. The solute particles interfere with the phase change of the solvent molecules. This means that the solution stays in the liquid phase at a lower temperature than the solvent would alone. For example, although water freezes at 0°C, a solution of ethylene glycol (antifreeze) and water remains a liquid at 0°C. The solution will freeze only when the temperature is lowered by several degrees below 0°C; how much lower depends on the amount of solute added.

In order for a substance to boil, its vapor pressure must equal the opposing external pressure. Solvent molecules in a solution need to move faster to escape into the air as gas molecules than they do on their own. This means more energy, or a higher temperature, is needed to evaporate them. Raising the boiling point of a solvent by adding a solute is called **boiling point elevation**. For example, when the solute salt is added to the solvent water, the salt particles slow down the water molecules and keep them from vaporizing, or changing from liquid to gas, as rapidly. Saltwater boils at a higher temperature, so it takes longer to start boiling than pure water does.

OBJECTIVES

- Describe how adding a solute decreases the freezing point of a solvent.
- Explain how adding a solute elevates the boiling point of a solvent.
- Infer how change in altitude affects boiling point.

VOCABULARY

- **boiling point elevation** the raising of a solvent's boiling point by the addition of a solute
- **freezing point depression** the lowering of a solvent's freezing point by the addition of a solute

Adding antifreeze to a car's coolant system depresses the freezing point so the coolant stays liquid in freezing temperatures.

TRY THIS

Salty Ice

Put a wet string on top of an ice cube. Sprinkle salt on top of the string. Let the ice cube sit for one minute, and then carefully lift up the string. What happens? Why?

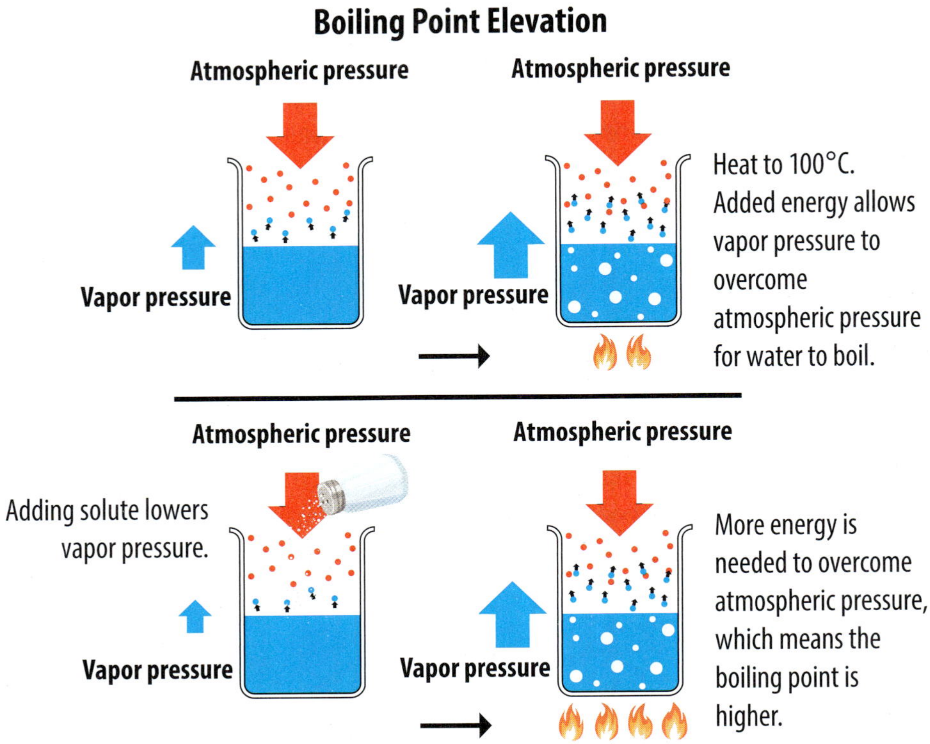

Another factor that can change boiling point is atmospheric pressure. At sea level, when water reaches a temperature of 100°C, the vapor pressure of the water equals the opposing air pressure, and the water boils. But what happens to the atmospheric pressure at high altitudes? There are fewer molecules of air, so there is less air pressure. This results in an equalization of vapor and air pressure at lower temperatures. It takes less energy to overcome a lower air pressure. At an elevation of 1,830 m, water boils at or near 95°C, which

means the water cannot reach 100°C. Foods cook at a lower temperature, resulting in longer cooking times.

LESSON REVIEW

1. How does adding a solute to a solvent affect the freezing point?
2. How does adding a solute to a solvent affect the boiling point?
3. Why do some packaged foods have different cooking instructions for higher elevations?

FYI

Antifreeze for Fish

For most people, winter brings extremely cold temperatures and the need for extra clothing and heat to stay warm. Many animals hibernate and sleep through the winter and do not wake up until spring when temperatures begin to warm. A few places on the planet experience winterlike conditions all year long. The waters around Antarctica are extremely cold and hostile to most aquatic life. However, there is a group of fish called *notothenioids*. Equipped with an antifreeze glycoprotein, these fish thrive in waters that have an average temperature of –1.5°C. Research shows that the special proteins bind to the ice crystals and prevent the fish from freezing. Many of the ice crystals reside in the spleen, which leads scientists to believe a process exists to eliminate the ice from the bloodstream. Glycoprotein, a natural antifreeze, This natural antifreeze is used to increase the shelf life of frozen foods and preserve tissues for transplant.

3.2.1 *Reactants and Products*

OBJECTIVES

- Identify the parts of a chemical reaction.
- Specify the five indicators of a chemical reaction.

VOCABULARY

- **chemical reaction** the process in which one or more substances are transformed into different substances
- **product** a new substance that is formed following a chemical reaction
- **reactant** a substance that is present before a chemical reaction that then undergoes changes during a chemical reaction

Chemical reactions are happening all around you. The trees outside may not look like they are doing very much, but they are actually busy converting carbon dioxide and water into food through the process called *photosynthesis*. Similarly, your cells are converting oxygen into a usable form of energy that allows your muscles to contract and your brain to function. The formation of rust, the exploding volcano model in a science fair, the fermentation of cheese and yogurt, and the burning of wood in a fireplace are all examples of chemical reactions.

A **chemical reaction** is the process in which one or more substances are transformed into different substances. But not all chemical interactions are reactions. When liquid water freezes and becomes ice, the water molecules are not transformed into a different substance. They are still water molecules. This is an example of a physical change. A chemical change must occur for the interaction to be considered a reaction. During a chemical reaction, atoms and molecules collide. The force of the collisions causes the molecular bonds to break. New bonds are formed and new substances are created.

Chemical reactions consist of two main parts: reactants and products. **Reactants** are substances that are present before a chemical reaction and undergo changes during a chemical reaction. **Products** are the new substances that are formed following a chemical reaction. For example, rust (iron oxide) is the product of the reaction between iron and oxygen, which are

The fermentation of cheese is an example of a chemical reaction.

the reactants. During this reaction, the bonds that hold the oxygen atoms together are broken. New bonds form between the oxygen atoms and the iron atoms.

Reactants → Product

Iron + Oxygen → Iron oxide

If reactions were written in word form only, they would be very cumbersome to read. For example, the photosynthesis reaction could be written as follows: six units of carbon dioxide react with six units of dihydrogen monoxide to produce one unit of glucose and six units of diatomic oxygen. Scientists created chemical equations to simplify reaction descriptions. Chemical equations include the formula and quantity of each reactant and product. The chemical equation of the photosynthesis reaction is written as follows:

$$6CO_2 + 6H_2O \rightarrow C_6H_{12}O_6 + 6O_2$$

Unlike the chemical interaction of liquid water becoming ice, the products of a chemical reaction have completely different properties compared to the reactants. For example, chlorine is a poisonous green gas, and sodium is a metal that reacts

TRY THIS

Fizzy Water

Fill a clear plastic bottle half full with vegetable oil. Add enough water to fill another quarter of the bottle. Using a pipette, place a few drops of food coloring in the mixture. Place the cap on the bottle, and shake until the food coloring is thoroughly mixed with the water. Do not disturb the bottle for a couple of minutes until the oil and water layers completely separate. Drop an effervescent tablet into the bottle and observe. Did a reaction occur? How do you know?

BIOGRAPHY

John Edward Hodge

Chemist John Edward Hodge was born in Kansas City, Kansas, on October 12, 1914. As a boy, he participated in model airplane contests and became an expert at billiards. He graduated Phi Beta Kappa from the University of Kansas, earned his master's degree, and pursued postgraduate studies at Bradley University and the Federal Executive Institute. He began his career as a chemist for the oil industry before becoming a professor of chemistry at Western University in Quindaro, Kansas. In 1941, Hodge began his 40-year career for the USDA at the Northern Regional Research Center, in Peoria, Illinois. Hodge analyzed the Mailliard reaction, the browning that takes place in foods such as bread or marshmallows when they are heated. Hodge correctly pointed out the connection between non-enzymatic browning and flavor intensity in foods, which changed cooking methods and the food industry. His paper on food browning remains the most cited work in the fields of food science and food chemistry. Hodge was passionate about encouraging black students to pursue careers in chemistry, and he recruited many minority students as summer interns.

explosively with water. When these substances interact and undergo a chemical reaction, the product is table salt. This product is neither explosive in water nor poisonous. Under ordinary circumstances, table salt will not separate into chlorine and sodium. The behavior of the products of chemical reactions truly shows God's wisdom. Imagine what would happen if the salt sprinkled on food suddenly took on the properties of chlorine and sodium!

There are five indicators that signify a reaction has occurred. The first indicator that a reaction has occurred is a change in color. Iron is a silver-white metal. When it reacts with oxygen, a brownish-red layer of rust is produced. A second indicator of a reaction is the formation of a precipitate. When two liquid substances combine, they may form a new solid substance called *a precipitate*. Typically, these liquid substances are ionic compounds. When the ionic compounds dissolve, their ions separate. Some of the ions remain dissolved in the solution. Others combine with different ions to form new solid compounds (precipitates) that do not dissolve in water. A third indicator of a reaction is the formation of a gas. When baking soda and vinegar are combined, the appearance of bubbles (carbon dioxide) indicates a reaction has taken place. A noticeable odor is the fourth indicator of a reaction. When an egg begins to rot, a pungent odor is produced. This means that a reaction inside the egg has taken place. The fifth indicator that a reaction has occurred is the release of energy in the form of light or sound. Colorful fireworks release both forms of energy when they explode.

Chemical Reaction Indicators

Color change

Formation of a precipitate

Formation of a gas

Noticeable odor

Release of light or sound energy

LESSON REVIEW

1. Describe photosynthesis as a reaction.

2. What are the reactants in the following cake recipe: 2 cups sugar, 1¾ cups flour, ¾ cup cocoa, 1½ tsp baking powder, 1½ tsp baking soda, 1 tsp salt, 2 eggs, 1 cup milk, ½ cup vegetable oil, 2 tsp vanilla extract, and 1 cup boiling water?

3. Does a reaction occur when berries are mashed and made into jam? Explain.

4. What are the five indicators that signify a reaction has occurred? Give one example of a reaction that includes one of these indicators.

OBJECTIVES

- Analyze factors that affect reaction rates.
- Compare catalysts and inhibitors.

VOCABULARY

- **catalyst** a substance that increases the reaction rate without being permanently changed
- **inhibitor** a substance used to slow down or stop a chemical reaction
- **reaction rate** a measure of the change in concentration of the reactants over a period of time

Chemical reactions occur in different parts of daily life such as food production and cooking. Browning meat, frying bacon, and baking bread and pizza all begin with a chemical reaction. Ripening of fruits and vegetables also involves many chemical reactions. Oxygen is one of the reactants in most of these reactions. The preservatives and flavorings in foods are all a result of chemical reactions. In fact, 135 reactants are involved in the chemical reactions that produce artificial raspberry flavoring!

Understanding how fast chemical reactions happen is important. For example, food chemists must know how long preservatives are able to keep certain foods safe and fresh. The **reaction rate** is a measure of the change in concentration of the reactants over a period of time. It is calculated by conducting experiments and measuring the amount of reactants and products that exist at different time intervals during the reaction.

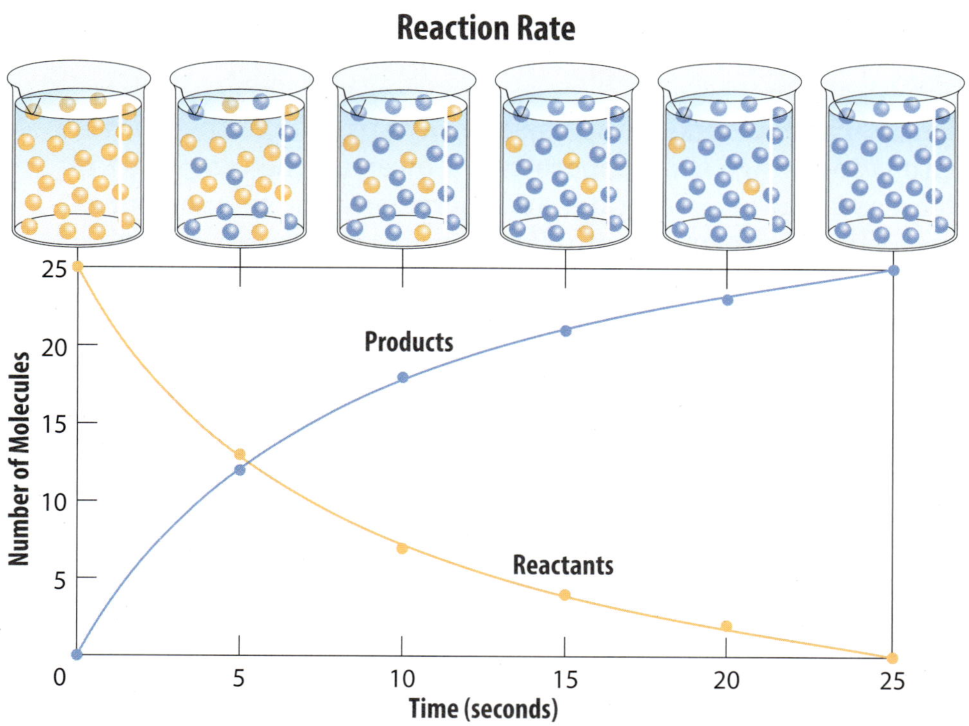

Reaction rates vary widely. Some reactions, such as burning, happen rapidly. Others may take years. The formation of rust is an example of a very slow reaction. To understand the speed of chemical reactions, it is necessary to recognize the two conditions that must exist before a reaction can take place. First, the reacting molecules must collide with enough force to break the intermolecular bonds. Second, these collisions must occur in such a way that the unbound atoms are in the right position to bond

Collision Theory

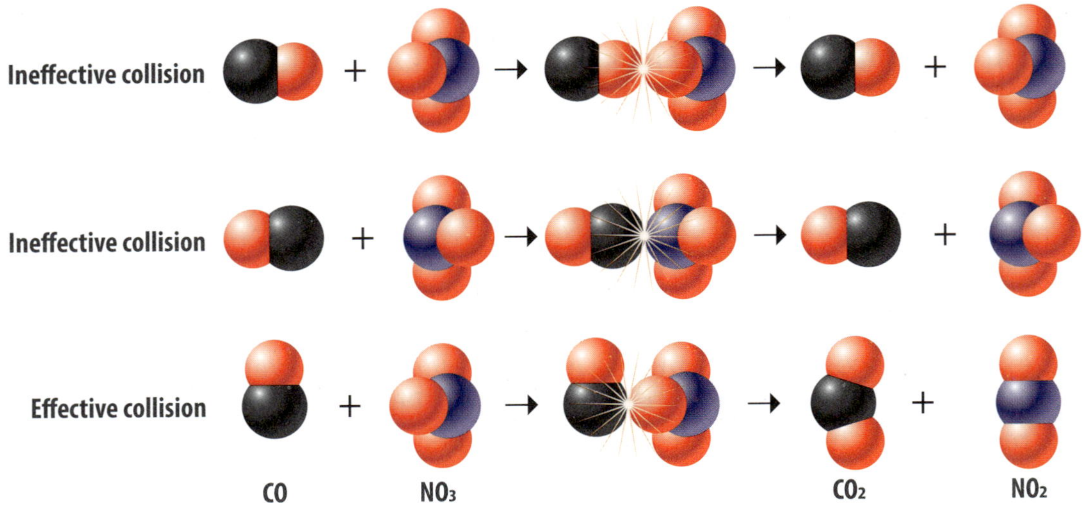

with each other. Altering the reaction conditions can change the speed with which the reactants collide with each other and form new chemical bonds. By changing the conditions, the reaction rate changes. Five factors affect reaction rates.

Temperature. One of the easiest ways to change the reaction rate is to change the temperature. Lowering the temperature decreases the reaction rate, and increasing the temperature speeds up the reaction rate. For example, frying an egg is a chemical reaction. Frying an egg on a low stove setting takes several minutes, but frying it on the highest setting takes only a minute or two. The increase in temperature causes the molecules to move faster and increases the frequency with which the molecules collide with each other. At higher temperatures, particles also have more energy. This allows molecules to bond with each other more readily.

An increase in temperature speeds up the reaction.

Surface area. Increasing the surface area of a solid reactant increases the reaction rate. A reactant crushed or ground into smaller fragments has more exposed surface area than one large piece. When more surface area is exposed, reactant atoms come in contact and bond with each other more frequently. For example, solid zinc reacts with hydrochloric acid to form zinc chloride and hydrogen gas. Cubes of zinc that have a total surface area of only 6 cm^2 react relatively slowly with hydrochloric acid. However, if the cubes are ground into a powder, the surface area increases 10,000 times. The larger surface area greatly speeds up the reaction rate.

Reaction Rate Factors

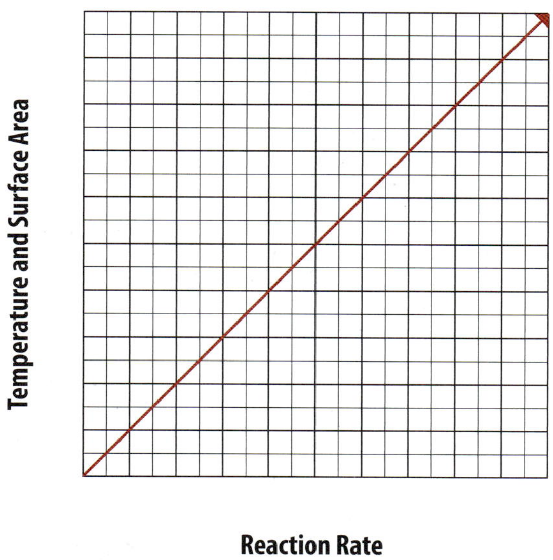

TRY THIS

Reactant Concentration

Obtain a 6 cm length of magnesium ribbon. Use a piece of sandpaper to remove any visible oxide layer from the magnesium ribbon. Cut the magnesium ribbon into two equal pieces. Using a graduated cylinder, measure approximately 10 mL of 1 M hydrochloric acid and pour it into one test tube. Place the test tube in the test tube rack. Measure approximately 10 mL of 3 M hydrochloric acid and pour it into the second test tube. Place the test tube in the test tube rack. Simultaneously drop the pieces of magnesium ribbon into the two test tubes. Observe the reactions. Which solution produced the most noticeable reaction? Why?

Concentration of reactants. The reaction rate generally increases when the concentration of the reactants is increased because there are more molecules available to collide and bond. For example, coal burns more rapidly in 100% pure oxygen than in air because air contains only 20% oxygen. Since the combustion reaction requires oxygen, extra oxygen increases the speed of the reaction.

$$\text{Rate of reaction} = \frac{\text{change in the concentration of reactants or products}}{\text{change in time}}$$

Consider the following experiment: 20 units of substance A remain after 30 units reacted with substance B to produce AB in 30 seconds. If there were 50 units of substance A at the beginning of the experiment, what is the reaction rate?

$$\text{Rate of reaction} = \frac{50\text{ units} - 20\text{ units}}{30\text{ seconds}} = \frac{1\text{ unit}}{\text{second}}$$

However, the reaction rate does not always increase with an increase in the concentration of a reactant. Chemical reactions frequently take place in a series of simple steps called *a reaction pathway*. The slowest reaction along this pathway, called *the rate-determining step*, controls the speed of the entire reaction pathway. Adding more of a reactant to one of the other pathway steps will not speed up the overall reaction. Picture the rate-determining step this way. Imagine you and your friends are making sandwiches for a fund-raiser. Three people are spreading mustard on the bread, three people are putting meat and cheese on the bread, and one person is wrapping the sandwiches in plastic wrap. Wrapping takes a long time, and only one person is assigned to this task. This means that the person who is wrapping the sandwiches determines the pace of the sandwich-making process. Wrapping the sandwiches is the rate-determining step. Recruiting more people to spread mustard or put meat and cheese on the bread will not speed up the overall sandwich-making process.

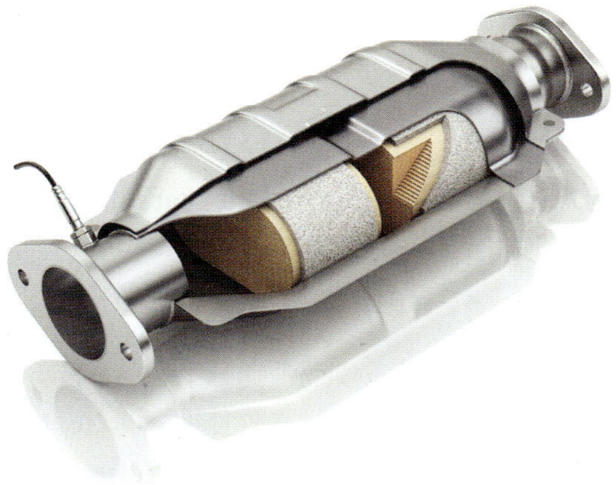

Cars have catalytic converters that utilize platinum and rhodium catalysts to increase the rate of reaction of carbon monoxide and unburnt fuel with oxygen from the air.

Catalysts. Some chemical reactions happen quite slowly and require the presence of another substance to speed up the reaction rate. A substance that increases the reaction rate without being permanently changed is called a **catalyst**. A catalyst is not part of the final product of a reaction. Sometimes, chemists can even recover and reuse catalysts. They are common in the human body. Biological catalysts are made of proteins and are called *enzymes*. Enzymes can speed up chemical reactions in the human body by a factor of 10,000. For example, an enzyme in human saliva speeds up the conversion of starch into glucose so the starch can be digested immediately instead of over the course of several weeks.

ACE inhibitors are medications that help lower people's blood pressure.

Inhibitors. Sometimes, it is important that a chemical reaction happens slower than it naturally would. In these cases, an **inhibitor**—a substance used to slow down or stop a chemical reaction—is used. Inhibitors interfere with the surface chemistry of two reactants, preventing them from rapidly bonding together. Preservatives added to food are inhibitors that slow down reactions in the bacteria or fungus that can spoil food.

LESSON REVIEW

1. What conditions must exist before a chemical reaction can take place?
2. Describe how a change in surface area might affect the reaction rate.
3. What are biological catalysts called? Provide an example of a biological catalyst.
4. How are catalysts and inhibitors similar?

Chemical Equations 3.2.3

It is important that chemists describe what happens during a chemical reaction. Chemists may describe a chemical reaction by writing a word equation. For example, the following word equation describes what happens when sodium reacts with chlorine: two solid sodium atoms react with one gaseous chlorine molecule to produce two solid sodium chloride compounds. This word equation can be simplified by using chemical equations. A **chemical equation** is a description using chemical symbols and formulas to show the reactants and products in a chemical reaction. These equations are shorthand versions of chemical word equations. The chemical equation for the sodium chloride word equation is written as follows:

$$2Na(s) + Cl_2(g) \longrightarrow 2NaCl(s)$$

Many of the symbols that are included in chemical equations represent the physical state of the reactants and products. Atoms, molecules, and compounds can exist in the solid, liquid, or gaseous state. These states are represented by the following symbols: (s), (l), and (g). Some substances also dissolve in water. They are aqueous and have the symbol (aq). There are other symbols that are included in equations. They are listed in the table below.

Common Chemical Equation Symbols	
Symbol	**Meaning**
+	This is used to separate reactants and products.
$\longrightarrow$	Produces or yields
$\rightleftharpoons$	A reversible reaction; The reaction can proceed in both directions.
(g)	Substance is in the gaseous state.
(l)	Substance is in the liquid state.
(s)	Substance is in the solid state.
(aq)	Substance is dissolved in water.
$\xrightarrow{\Delta}$	The reaction requires the addition of heat.
$\xrightarrow{cat}$	The reaction requires the addition of a catalyst.

OBJECTIVES

- Identify parts of a chemical equation.
- Demonstrate the law of conservation of mass.
- Practice balancing chemical equations.

VOCABULARY

- **chemical equation** a description using chemical symbols and formulas to show the reactants and products in a chemical reaction
- **coefficient** a whole number written in front of a chemical formula in a chemical equation

BIBLE CONNECTION

Mass Conservation, Loaves, and Fish

Jesus fed thousands of people more than once with only a few loaves of bread and a couple of fish. How can 5,000 people be fed with five loaves of bread and two fish as recorded in Matthew 14? Is this even possible if all reactions are governed by the law of conservation of mass? Jesus performed a miracle after giving thanks to the Father. Laws are created by men in an attempt to describe God's creation. Given that God is the Creator, He has the power and authority to surpass human-made laws.

Chemical equations are valuable only if they accurately represent what happens in a chemical reaction. To do this, chemical equations must fulfill three requirements:

1. The equation must represent the known facts about a reaction. All of the reactants and products must be identified.
2. An equation must contain the correct chemical formulas of the reactants and products.
3. Every chemical reaction must obey the law of conservation of mass, which states that mass is neither created nor destroyed.

Notice in the sodium chloride equation that there are two sodium atoms on the reactant side and two sodium atoms on the product side. The same is true for chlorine. This reaction adheres to the law of conservation of mass. An equal amount of elements on each side of the equation shows that atoms can be rearranged, but they are never lost or gained during the chemical reaction.

An equation in which the number of atoms of each element on the reactant side is equal to the number of atoms on the product side is a balanced equation. Chemists balance these equations by placing coefficients in front of the formulas of the reactants and products. Subscripts cannot be altered. A **coefficient** is a whole number written in front of a chemical formula in a chemical equation. When you count atoms in a formula, the coefficient is multiplied by the subscript for each element in the formula. For example, $2CO_2$ represents two carbon dioxide molecules. The coefficient 2 at the beginning of the formula indicates there are two carbon atoms and four oxygen atoms. Coefficients are used to balance equations because the subscripts cannot be changed. If the subscripts were changed, the chemical formula would no longer represent the correct substance. For example, H_2O is the chemical formula of water. Adding a subscript to the oxygen atom changes the identity of the molecule to hydrogen peroxide (H_2O_2)—a substance with drastically different properties.

Balancing an equation is a matter of counting the atoms in the reactants and products and adding coefficients until each side has the same number of each type of atom. Consider the following chemical equation:

$$H_2(g) + O_2(g) \longrightarrow H_2O(l) \text{ (Not Balanced)}$$

Element	# of Reactants	# of Products
H	2	2
O	2	1

There are two hydrogen atoms on both the reactant side and product side. Two oxygen atoms are on the reactant side, but only one oxygen atom is on the product side. This equation is not balanced. Add the coefficient 2 in front of the water formula. Now, there are two oxygen atoms on the product side.

$H_2(g) + O_2(g) \longrightarrow 2H_2O(l)$ (Not Balanced)

Element	# of Reactants	# of Products
H	2	4
O	2	2

When you added the coefficient, the oxygen atoms became balanced but the hydrogen atoms became unbalanced. Now, the product indicates there are four hydrogen atoms when the reactant only shows two hydrogen atoms. Another coefficient needs to be added in front of the hydrogen reactant molecule.

$2H_2(g) + O_2(g) \longrightarrow 2H_2O(l)$ (Balanced)

Element	# of Reactants	# of Products
H	4	4
O	2	2

It is important to count the number of reactant and product atoms after a coefficient has been added to the equation. There is an equal number of hydrogen atoms and oxygen atoms on both sides of the equation. This chemical equation is balanced.

LESSON REVIEW

1. What three requirements must all chemical equations fulfill?
2. Balance the following equation: solid copper(II) hydroxide and aqueous sodium chloride are produced when aqueous sodium hydroxide reacts with aqueous copper(II) chloride.
3. Why can subscripts not be changed when balancing a chemical equation?
4. Are the equations below balanced? If not, balance them.
   $Zn(s) + HCl(aq) \longrightarrow ZnCl_2(aq) + H_2(g)$;
   $CH_4(g) + 2O_2(g) \longrightarrow CO_2(g) + 2H_2O(l)$;
   $2CH_4(g) + S_8(s) \longrightarrow 2CS_2(l) + H_2S(g)$

LAWS & PRINCIPLES

Law of Conservation of Mass

The law of conservation of mass states that mass is neither created nor destroyed in ordinary chemical and physical changes. This is an important part of God's creation. The amount of mass that is present before a reaction must be equal to the amount of mass following a reaction. This law is demonstrated in every chemical equation. The following reaction shows that there is an equal number of reactant elements and product elements: $3K + Al(NO_3)_3 \longrightarrow Al + 3KNO_3$. There are three potassium atoms, one aluminum atom, three nitrogen atoms, and nine oxygen atoms on both sides of the equation.

3.2.4 Energy and Chemical Reactions

OBJECTIVES

- Contrast exergonic and endergonic reactions.
- Relate the exchange of heat during exothermic and endothermic reactions.

VOCABULARY

- **activation energy** the minimum amount of energy that is required to start a chemical reaction
- **endergonic reaction** a reaction that absorbs more energy at the beginning of the reaction than is released
- **endothermic reaction** a chemical reaction that absorbs heat
- **exergonic reaction** a reaction that releases more energy during bond formation than the activation energy used to begin the reaction
- **exothermic reaction** a chemical reaction that releases heat

All chemical reactions involve a change in energy in which the energy is either released or absorbed as light, sound, thermal, or electrical energy. Some reactions, such as rusting, take so long that the change in energy is not noticeable. In other reactions, such as the explosion of fireworks, the change in energy is impossible to miss.

The absorption or release of energy occurs within the chemical bonds. Energy is needed to start all chemical reactions. Bonds that hold reactant atoms and molecules together must be broken for a reaction to occur. The minimum amount of energy that is required to start a chemical reaction is called the **activation energy**. When products form new bonds, energy is released. Heat provides the activation energy for many reactions. For example, think about the reaction that occurs in a sparkler. Magnesium and oxygen react to form magnesium oxide. Without heat supplying the necessary amount of activation energy, the sparkler would not ignite.

There are two main types of reactions that describe the exchange of energy—exergonic and endergonic reactions. **Exergonic reactions** release more energy during bond formation than the activation energy used to begin the reaction. These reactions are also called *spontaneous reactions* because they do not require

Energy is released in the form of light and sound when chemical reactions occur in the atmosphere.

additional energy to proceed. The chemical reaction that releases electrical energy in a battery is an example of a spontaneous reaction. Chemical reactions in a battery happen quickly without the added energy from an external source.

Light energy is emitted when glow sticks are snapped, which allows the hydrogen peroxide inside the glass vial to react with the diphenyl oxalate inside the plastic tube.

Endergonic reactions absorb more energy at the beginning of the reaction than is released. A large amount of activation energy is required for these types of reactions. One example of an endergonic reaction is the combination of nitrogen and oxygen to form nitric oxide—a toxic gas. Nitrogen and oxygen are found in the atmosphere. These two molecules do not normally react unless sparked by a tremendous amount of energy. Nitric oxide can be produced as a result of a lightning strike. In God's wisdom, He designed this reaction to be nonspontaneous.

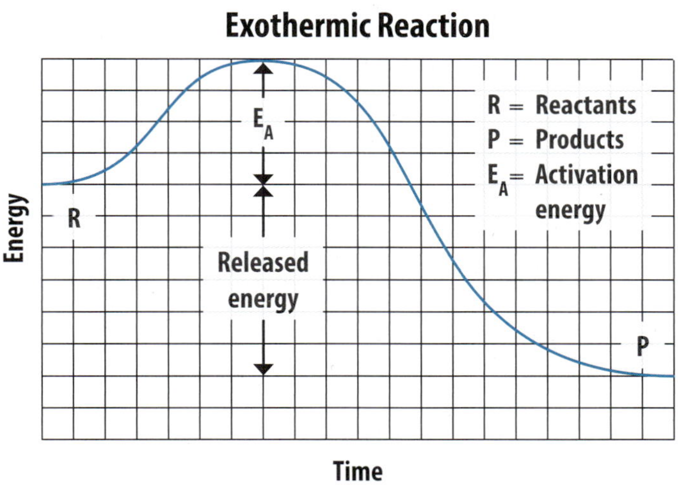

The energy that is released or absorbed during a reaction is often in the form of heat. A chemical reaction that releases heat is called an **exothermic reaction**. The root word *therm–* means "heat." *Exo–* means "outside." Exothermic literally means *the release of heat outside the system*. Combustion is an example of an exothermic reaction. It releases both heat and light. Another exothermic reaction occurs when hydrogen gas and oxygen gas react to form water. This reaction produces a lot of heat. The energy released in an exothermic reaction is often written as a product in a chemical equation.

$$\text{Reactants} \longrightarrow \text{Products} + \text{Heat}$$

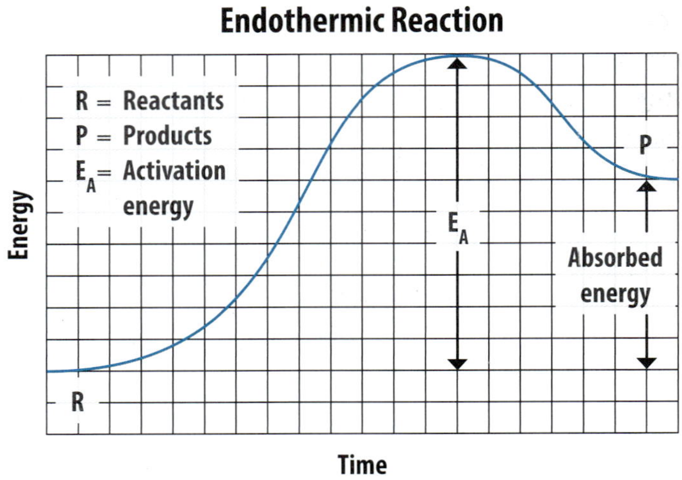

Chemical reactions that absorb heat are called **endothermic reactions**. *Endo–* means "within," so the term *endothermic* means "heat is added to the system." During an endothermic reaction, heat energy is stored in the product bonds. For example, many reactions do not occur until the reactants absorb heat from their surroundings. Limestone, which is composed of the chemical compound calcium carbonate, does not usually react with its surroundings in nature. When limestone is heated above 840°C, a chemical reaction converts the limestone into calcium

A direct electric current partially dissolved the gold metal and created chemical bonds between the gold elements and the surface metals to adhere the gold on the roof of this cathedral in Russia. This process is called *electroplating*.

TRY THIS

Hand Warmers

Fill half of a plastic cup with distilled water. Using a measuring spoon, add ¼ tsp of dry water jelly crystals to the cup of water. Allow the crystals to remain undisturbed in the cup for a couple of hours or overnight. Scoop 4 tbsp of the saturated crystals and place them in a ziplock plastic bag. Using a weighing dish and spatula, measure 20 g of iron filings and 20 g of calcium chloride. Add both substances to the plastic bag. Mix the contents of the unsealed bag. When the contents are thoroughly mixed, press out as much air as possible and seal the bag. Continue to knead the contents of the sealed bag. Note any observations. What type of reaction is occurring?

oxide (lime). Calcium oxide is a chemical used in water and waste treatment plants and in the manufacturing of paper products. Converting table salt into sodium and chlorine for manufacturing purposes is another example of an endothermic reaction.

$$\text{Reactants} + \text{Heat} \longrightarrow \text{Products}$$

In each of these different types of reactions, the total amount of energy in a closed system does not change. A closed system is isolated from and not affected by its surroundings. The energy in the system does not change, but it may be converted from one form to another. These reactions adhere to the law of conservation of energy, which states that energy cannot be created or destroyed. The potential energy that is stored in chemical bonds is converted into light, sound, or thermal energy.

LESSON REVIEW

1. Explain when energy is absorbed and released in a chemical reaction.
2. Compare the activation energy of exergonic and endergonic reactions.
3. How are exergonic reactions related to exothermic reactions?
4. Is heat considered a product or reactant in an endothermic reaction?
5. Does the total amount of energy in a closed system ever change? Explain.

LAWS & PRINCIPLES

Law of Conservation of Energy

The law of conservation of energy states that energy cannot be created or destroyed. This law applies to endergonic and exergonic reactions. The total amount of energy does not change, but it can be converted from one form to another. A certain amount of energy is stored in the chemical bonds of the reactant compounds in an exergonic reaction. That same amount of energy is released when new product bonds are formed. The energy that is released may be in the form of light, sound, or heat.

3.2.5 Types of Reactions

OBJECTIVES

- Classify chemical reactions.
- Determine whether reactions will occur according to the metal activity series.

VOCABULARY

- **combustion reaction** a reaction that occurs when a substance reacts with oxygen
- **decomposition reaction** a reaction where a substance is broken down into two or more products
- **synthesis reaction** a reaction where two or more substances combine to form a product

Some reactions happen quickly, others slowly. Some reactions produce precipitates. Water or a gas may be the product in other reactions. Chemists created a system to organize or classify these and other reactions into at least five different types, depending on how the atoms rearrange during the reaction. The five types of reactions include synthesis, decomposition, combustion, single replacement, and double replacement. Some reactions may belong to more than one type.

Synthesis reactions are also called *combination reactions* because two or more substances combine to form a product. The generic equation for a synthesis reaction is A + B $\longrightarrow$ AB. Reactant A combines with reactant B to create product AB. Magnesium combines with oxygen gas to produce magnesium oxide in a synthesis reaction.

$$2Mg(s) + O_2(g) \longrightarrow 2MgO(s)$$

Magnesium oxide is a mineral supplement for healthy muscles and bones. This compound may also relieve heartburn and indigestion.

The opposite of a synthesis reaction is a **decomposition reaction**. As the name implies, a substance is broken down into two or more products. The generic equation for a decomposition reaction is AB $\longrightarrow$ A + B. Reactant AB is broken down into product A and product B. Most reactions that belong to this type require the addition of light, heat, or electricity to proceed. An example of a decomposition reaction is the breaking down of hydrogen carbonate into water and carbon dioxide.

$$H_2CO_3(aq) \longrightarrow H_2O(l) + CO_2(g)$$

The synthesis reaction that occurs between magnesium and oxygen produces a layer of magnesium oxide.

Another example of a decomposition reaction is $2Ni_2O_3(s) \longrightarrow 4Ni(s) + 3O_2(g)$. Normally, nickel(III) oxide does not readily decompose. A tremendous amount of energy is needed to produce pure nickel and oxygen gas.

One of the most common types of reactions is a combustion reaction. A **combustion reaction** occurs when a substance reacts with oxygen. Heat or light energy is produced. Two other common products in a combustion reaction are carbon dioxide and water. These types of reactions are very common because oxygen can react with a variety of elements and compounds. The biological process of cellular respiration is an example of a combustion reaction.

$$C_6H_{12}O_6(s) + 6O_2(g) \longrightarrow 6CO_2(g) + 6H_2O(l) + \text{Energy}$$

The energy from this reaction supplies the necessary fuel for all your cells. Combustion reactions are also responsible for extracting the energy from hydrocarbons to propel vehicles and fuel gas grills. The combustion equations for methane and ethane are as follows:

The bubbles in soft drinks that tickle your nose are caused by the decomposition of hydrogen carbonate.

$$CH_4(g) + 2O_2(g) \longrightarrow CO_2(g) + 2H_2O(l)$$

$$2C_2H_6\ (g) + 7O_2\ (g) \longrightarrow 4CO_2(g) + 6H_2O(l)$$

There are two types of reactions that replace one element with another. The first type of reaction that belongs to this category is a single replacement reaction. Some individual elements have the capacity to replace another element in a compound. The generic equation for a single replacement reaction is as follows: $A + BC \longrightarrow AC + B$. Reactant BC is separated and reactant A swaps places with B to produce AC and B. A and B switch places because A is more active than B. An element's activity determines if a single replacement reaction can occur. Scientists use the metal activity series to predict if a reaction can occur. Elements at the top of the activity series are the most active and can replace less active elements in a compound that are farther down the series. For example, potassium (K) is more active than nickel (Ni). It can replace nickel in a solution.

Metal Activity Series

Most reactive

Li
Rb
K
Ca
Na
Mg
Al
Mn
Zn
Fe
Ni
Sn
Pb
Cu
Ag
Pt
Au

Least reactive

When solid magnesium is placed in a copper(II) sulfate solution, copper separates from the sulfate polyatomic ion and magnesium takes its place. The products are magnesium sulfate and solid copper. Magnesium sulfate is the main ingredient in Epsom salt.

$$Mg(s) + CuSO_4(aq) \longrightarrow Cu(s) + MgSO_4(aq)$$

Another example of a single replacement reaction is as follows:

$$2Al(s) + 3CuSO_4(aq) \longrightarrow 3Cu(s) + Al_2(SO_4)_3(aq)$$

Because aluminum is more active than copper, it replaces copper and bonds with sulfate to produce aluminum sulfate. Solid copper is the other product in this reaction.

SCIENCE IN ACTION

Combustion Engines

Combustion is an important reaction that provides energy inside car engines. There are two main reactants in an internal-combustion reaction: air and fuel. The air and fuel mix and are compressed. A spark is introduced to the system, which causes the mixture to burn. The hot gaseous products expand and come in contact with moving engine parts such as pistons, turbine blades, and nozzles. The energy from this system produces usable work. Carbon dioxide and water are eventually expelled through the exhaust system. There are two main types of internal-combustion engines: intermittent-combustion engines and continuous-combustion engines. Air and fuel mixtures are periodically ignited in intermittent-combustion engines. Gasoline and diesel engines are examples of intermittent-combustion engines. The constant supply of air and fuel mixtures in continuous-combustion engines is always ignited. Jet engines are examples of continuous-combustion engines.

The second type of reaction that replaces an element with another is a double replacement reaction. In a double replacement reaction, the two reactant compounds separate into their ions. The positive ions swap places to create two new products. The generic equation for a double replacement reaction is as follows: $AB + CD \longrightarrow AD + CB$. Both reactants separate into their ions. The positive ions A and C swap places to produce AD and CB.

There are three clear indicators that a double replacement reaction has occurred—the production of a precipitate, a gas, or water. An example of a double replacement reaction is the interaction between silver nitrate and sodium chloride to produce silver chloride and sodium nitrate.

$$AgNO_3(aq) + NaCl(aq) \longrightarrow AgCl(s) + NaNO_3(aq)$$

When this reaction occurs, the reactants separate into Ag^+, NO_3^-, Na^+, and Cl^-. The positive ions, Ag^+ and Na^+, swap places to form solid silver chloride. The sodium and nitrate ions remain dissolved in solution. Solid AgCl falls out of solution as a precipitate and is proof that a double replacement reaction occurred.

All reactions are oxidation-reduction reactions. When an atom loses electrons, it gains a positive charge and is oxidized. If an electron is gained, the atom gains a negative charge and is reduced. Oxidation and reduction occur at the same time, which is why they are called *redox reactions*. An example of an oxidation-reduction reaction is the interaction between aqueous

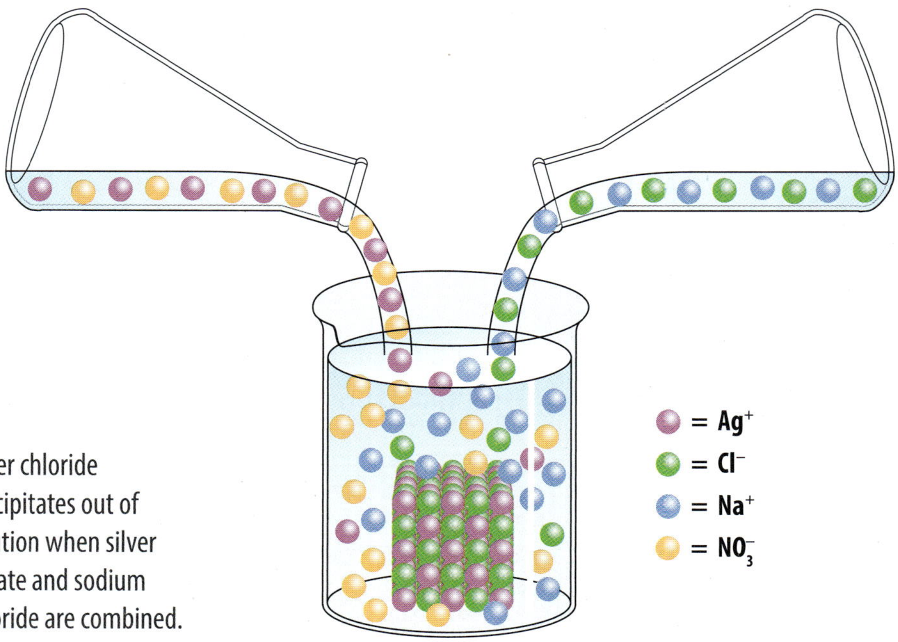

Silver chloride precipitates out of solution when silver nitrate and sodium chloride are combined.

Types of Reactions
(Generic Equations)

Synthesis:	$A + B \longrightarrow C$
Decomposition:	$AB \longrightarrow A + B$
Combustion:	$2A + O_2 \longrightarrow 2AO$
Single Replacement:	$A + BC \longrightarrow AC + B$
Double Replacement:	$AB + CD \longrightarrow AD + CB$

chlorine and aqueous potassium iodide to produce solid iodine and aqueous potassium chloride. The chlorine atom is reduced because it gains an electron when it bonds with potassium, and the iodine atom is oxidized because it loses an electron. Notice that this redox reaction is also a single replacement reaction.

$$Cl_2(aq) + 2KI(aq) \longrightarrow I_2(s) + 2KCl(aq)$$

LESSON REVIEW

1. What do most decomposition reactions require?
2. What reactant is required for a combustion reaction?
3. Classify the following reactions:
   $Mg + CuSO_4 \longrightarrow Cu + MgSO_4$
   $CaBr_2(aq) + 2AgNO_3(aq) \longrightarrow 2AgBr(s) + Ca(NO_3)_2(aq)$.
4. Predict if the reactions below can occur. Explain why some reactions do not occur. Write the full equation for the reactions that can occur.
   $Ag(s) + AlCl_3(aq)$
   $K(s) + ZnCl_2(aq)$
   $Ca(s) + CuBr_2(aq)$
5. Describe what specifically occurs when calcium bromide reacts with silver nitrate.

SCIENCE IN ACTION

Tarnish Remover
Tarnish is caused by a chemical reaction between silver and sulfur-containing materials in the air. As these sulfur particles touch an item made of silver, they combine with the silver to form silver sulfide, which is black. You can remove tarnish by using polish and rubbing it off or by dipping the item in a solution to dissolve the silver sulfide, but both methods remove bits of silver. Or, you can reverse the chemical reaction that created the silver sulfide in the first place. First, cover the bottom of a roasting pan with aluminum foil. Then, boil a gallon of water, add one cup of baking soda, and pour the solution in the foil-lined pan. Place the tarnished item in the pan and make sure it touches the foil. In seconds, the tarnish will disappear! Why? Aluminum is even more attractive to sulfur than silver, so the sulfur atoms move from the silver surface to the aluminum to form aluminum sulfide. A weak electric current is generated when the silver and aluminum touch each other, a reaction known as *an electrochemical reaction.*

3.3.1 Acids

OBJECTIVES

- Describe the products of acid dissociation.
- Identify the properties of acids.
- Compare strong and weak acids.

VOCABULARY

- **acid** a compound that releases hydrogen ions when it is dissolved in water
- **strong acid** an acid that produces many hydrogen ions when it is dissolved in water
- **weak acid** an acid that produces a small number of hydrogen ions when it is dissolved in water

You may have heard about acid rain or seen the danger labels on bottles that contain acids in the laboratory. But, what exactly is an acid? Many compounds separate into positively-charged and negatively-charged ions when they dissolve in water. This process is called *dissociation*. An **acid** is a compound that releases hydrogen ions when it is dissolved in water. Hydrogen ions (H^+) do not have any electrons and are essentially single protons. The protons attract the negative portion of water molecules to produce hydronium ions (H_3O^+). Acids are proton donors.

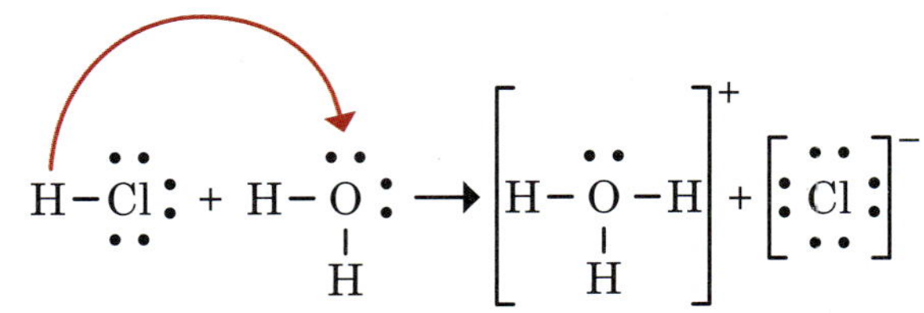

Hydrochloric acid + Water → Hydronium ion + Chloride ion

Compounds that are acids have many properties in common.

Acids taste sour. It is important to remember that it is not safe to taste any chemical in the laboratory. Lemons, oranges, and grapefruit contain citric acid. This acid gives them their sour taste. The sour taste in vinegar is produced by acetic acid. Sour milk contains lactic acid. Wine gets its sour taste from tartaric acid, and the tart flavor of many carbonated beverages comes from phosphoric acid. Aspirin dissolved in saliva tastes sour because of acetylsalicylic acid.

Acids react with carbonates, such as sodium carbonate ($NaCO_3$) or calcium carbonate ($CaCO_3$), to produce carbon dioxide.

$$Na_2CO_3(aq) + 2HCl(aq) \longrightarrow 2NaCl(aq) + CO_2(g) + H_2O(l)$$

$$CaCO_3(s) + 2HCl(aq) \longrightarrow CaCl_2(aq) + CO_2(g) + H_2O(l)$$

All acids contain hydrogen. Acids react with some metals to form hydrogen gas. Hydrochloric acid, for example, reacts with zinc metal to form zinc chloride and hydrogen gas.

Acids react with metals. They dissolve the metal and form a metal residue. The gradual wearing away of a metal as a result of a chemical reaction in which the metal is changed into a metallic compound is called *corrosion*.

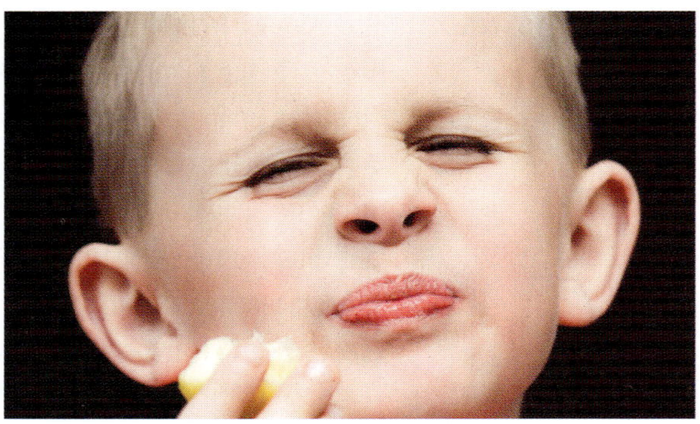

The citric acid in lemons makes some people pucker.

Common Acids

Acid Name	Uses	
Acetic acid, CH_3COOH	Plastic production, food preservation, some printing processes	
Ascorbic acid, $C_6H_8O_6$	Vitamin C, collagen synthesis, tissue growth and repair	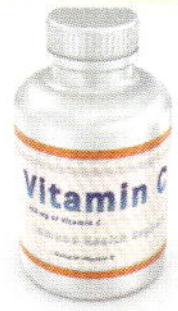
Carbonic acid, H_2CO_3	Carbonated drinks	
Hydrochloric acid, HCl	Battery production, digestion in stomach, sugar processing, metal cleanser	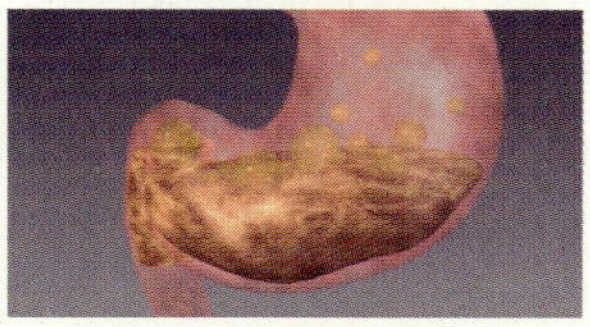
Nitric acid, HNO_3	Fertilizer production	
Phosphoric acid, H_3PO_4	Acidification of foods such as soft drinks and jams, leavening agent, detergent production, fertilizer production	
Sulfuric acid, H_2SO_4	Car batteries, manufactured fertilizers	

Because acids produce charged ions, they are electrolytes. An electrolyte is a substance that conducts an electric current when a molecule has dissociated into positively and negatively charged ions. A car battery is a good example of an acid used as an electrolyte. Car batteries are made up of a series of lead plates that are submerged in a sulfuric acid and water solution. The sulfuric acid allows charged ions to pass from one plate to another, which creates an electric current.

The pits of stone fruit, such as peaches and apricots, contain the compound amygdalin, which breaks down into hydrogen cyanide—a toxic poison. Eating these fruits is safe, as long as you do not ingest a large amount of pits.

When acids dissolve in water, some fully separate or dissociate into hydrogen ions and negatively charged ions. Others may only partially dissociate. The degree to which an acid dissociates determines the strength of the acid. When an acid completely separates in solution, a large amount of H^+ ions are produced. A **strong acid** is an acid that produces many hydrogen ions when it is dissolved in water. Nitric acid is a strong acid that completely ionizes in water. One hundred HNO_3 molecules would produce 100 hydrogen ions and 100 nitrate ions. Hydrochloric acid (HCl) and sulfuric acid (H_2SO_4) are other examples of strong acids.

$$HNO_3 + H_2O \longrightarrow H_3O^+ + NO_3^-$$

The three most common acids used in industry are all strong acids: sulfuric acid, nitric acid, and hydrochloric acid. Sulfuric acid is used to make paints, plastics, detergents, and fertilizers. Nitric acid removes tarnish from metals, and it is also important in producing explosives, plastics, dyes, and fertilizers. Hydrochloric acid is used in industrial and home cleaning agents. It is also the acid that your stomach uses to break apart foods during digestion. Strong acids are good electrolytes because they produce many ions. Strong acids are also very corrosive.

An acid that produces a small number of hydrogen ions when it is dissolved in water is called a **weak acid**. Acetic acid is a weak acid. It is used to make cellulose acetate for fibers and films. One

TRY THIS

Battery Acid

Roll two lemons on a flat surface to break the fibers inside the lemon and release the juice. Cut a small opening near one end of the lemon. Cut a second small opening parallel to the first cut, close to the other end of the lemon. Insert a penny in one of the openings. Ensure a portion of the penny is in contact with the lemon juice. Place a galvanized nail in the other opening in the lemon. Repeat this process with the second lemon. Connect the penny in one lemon to the nail in the second lemon by attaching alligator clips to each material. Attach another alligator clip to the unbound penny. Place a separate alligator clip on the unbound nail. Connect the voltmeter probes to the free alligator clips, and record the voltmeter reading. Repeat this experiment using pickles, grapefruit, and oranges. Why do some acids conduct electricity better than others?

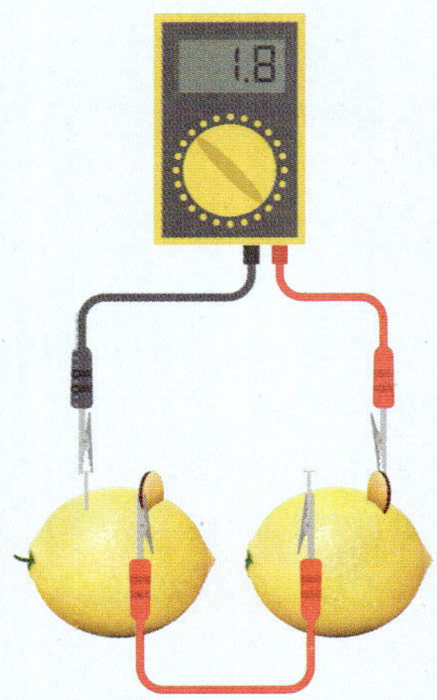

hundred of these molecules would produce about one hydrogen ion and one acetate ion. The other 99 acetic acid molecules would remain intact. Some of the acetic acid produces hydronium ions. At the same time, the hydronium ions are converted back into acetic acid molecules. Double arrows are included in this equation to show that the reaction can happen in both directions. Acetic acid never fully dissociates.

$$CH_3COOH + H_2O \rightleftharpoons H_3O^+ + CH_3COO^-$$

Acid Dissociation

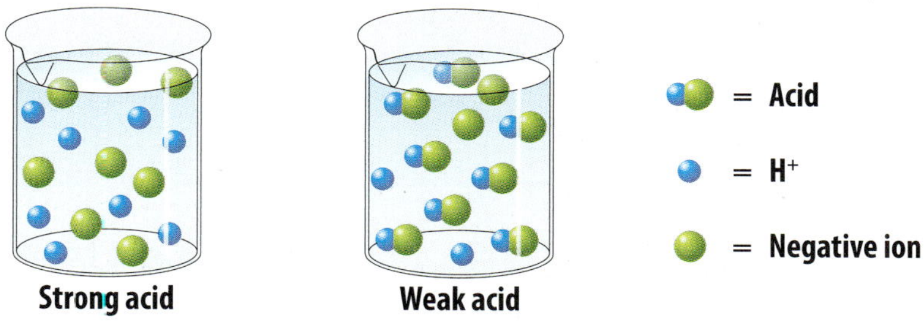

Carbonic acid, citric acid, and boric acid are also weak acids. Carbonic acid and citric acid are both used to carbonate beverages. Boric acid is used in eye drops. Weak acids are poor electrolytes because they produce fewer ions than strong acids in solution. Therefore, weak acids are not very corrosive.

Acids are useful because they dissolve in water. But this property also contributes to an environmental problem—acid rain. Rotting vegetation, industry emissions, and the burning of fossil fuels can release chemicals such as sulfur dioxide and nitrogen oxides, which react in the atmosphere with water and return to Earth as rain, fog, snow, sleet, or dew. Acid rain has a variety of effects, including damage to plants and soils, fish and other living things, buildings and monuments, and human health. Laws and regulations are in place to control the emissions that cause acid rain, but acid rain is still a concern in many regions of the world.

TRY THIS

Acid Rain

Label three 100 mL beakers: *lemon juice*, *vinegar*, and *water*. Using a graduated cylinder, measure 50 mL of lemon juice. Pour the lemon juice in the appropriate beaker. Measure and pour 50 mL of vinegar in the second beaker and 50 mL of tap water in the third beaker. Weigh each piece of chalk and record the mass. Place a piece of chalk in each beaker. Allow the chalk pieces to remain in the beakers for three days. Record your observations every 24 hours. After three days, weigh each piece of chalk and record the masses. How do the masses compare?

LESSON REVIEW

1. How do acids create hydronium ions?
2. Describe two properties of acids.
3. Citric acid is not a strong acid compared to sulfuric acid. Why?
4. Does the following equation represent a strong or weak acid: $HF + H_2O \rightleftharpoons H_3O^+ + F^-$? How do you know?
5. Acid dissociation is useful in many ways. However, what environmental problem can this process also cause? How can this impact be reduced?

3.3.2 Bases

OBJECTIVES

- Compare properties of acids and bases.
- Distinguish between strong and weak bases.

VOCABULARY

- **base** a compound that releases hydroxide ions when it is dissolved in water
- **strong base** a base that produces many hydroxide ions when it is dissolved in water
- **weak base** a base that produces few hydroxide ions when it is dissolved in water

Acids and bases are complimentary substances. The majority of their characteristics and properties are in direct contrast to one another. Unlike an acid that releases hydrogen ions, a **base** is a compound that releases hydroxide ions (OH^-) when it is dissolved in water. When bases dissociate, they produce positive ions and negative hydroxide ions. Bases are proton acceptors. They accept hydrogen ions from acids.

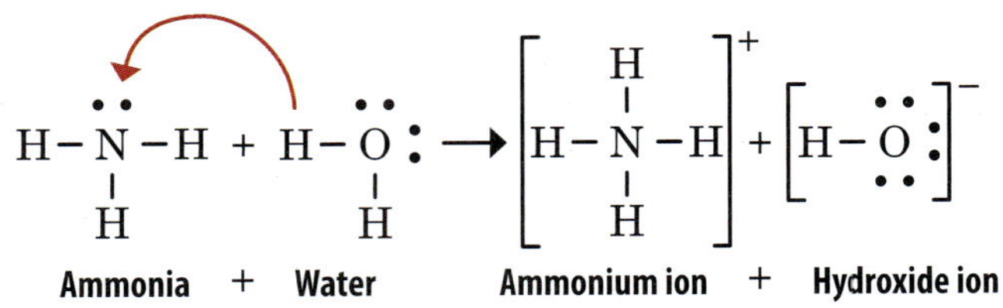

Like acids, bases are useful and have identifying properties.

Bases taste bitter. Have you ever accidentally tasted baking soda? It is a base with a very bitter taste. Bases stimulate the tongue's taste buds that react to bitterness. Some bitter foods are coffee, the peel of fruits, and cocoa.

Strong bases are caustic. They can burn or erode substances by chemical action. The labels on strong cleaning agents, such as oven cleansers, typically suggest that the user wear gloves and avoid skin contact because of the caustic property.

Bases feel slippery when they dissolve in water and form an aqueous solution. Soap is an example of a base that feels slippery when wet. Bases are used to make household cleansers and soaps.

Bases can conduct electricity in solution. They are electrolytes. Table salt and water is an electrolyte solution.

Similar to acids, bases can be classified as strong or weak. A **strong base** is a base that produces many hydroxide ions when it is dissolved in water. Metals that belong to groups IA and IIA of the periodic table usually form strong bases because the ionic bond is not strong enough to hold the metal ion and the hydroxide ion together. As a result, the compound completely dissociates in water. Examples of strong bases are sodium hydroxide (NaOH) and calcium hydroxide ($Ca(OH)_2$). Because strong bases produce large amounts of ions, they are good electrolytes. A solution containing OH^- ions from a soluble base is called *alkaline*. When strong bases dissolve, large amounts of heat are produced. It is

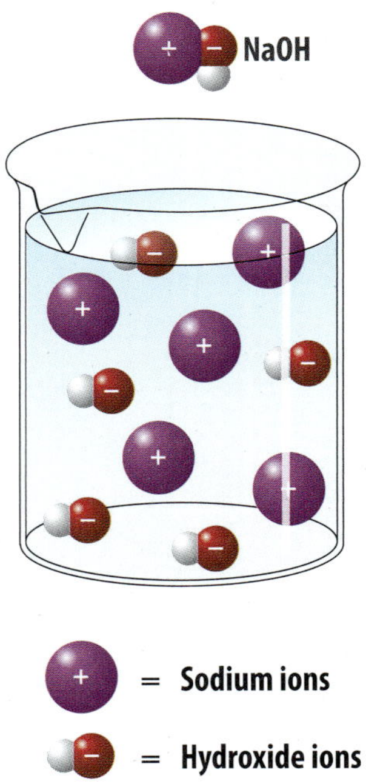

Sodium hydroxide is a strong base because it completely dissociates into sodium ions and hydroxide ions.

Common Bases		
Base Name	**Uses**	**Example**
Aluminum hydroxide, $Al(OH)_3$	Antacid, water purification	
Ammonia, NH_3	Fertilizer manufacturing, home cleansers	
Calcium hydroxide, $Ca(OH)_2$	Cement, reducing soil acidity	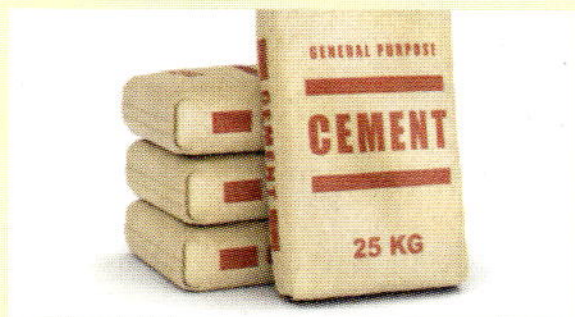
Magnesium hydroxide, $Mg(OH)_2$	Antacid, laxative	
Potassium hydroxide, KOH	Soap, cotton manufacturing, electroplating	
Sodium bicarbonate, $NaHCO_3$	Cooking, antacid	
Sodium hydroxide, NaOH	Soap production, oven cleanser, textiles	

important to practice safe laboratory techniques when working with strong bases.

$$NaOH \longrightarrow Na^+ + OH^-$$

A **weak base** is a base that produces few hydroxide ions when it is dissolved in water. Because weak bases produce very few OH^- ions, they are poor conductors of electricity. Examples of weak bases are ammonia (NH_3) and aluminum hydroxide ($Al(OH)_3$).

Basic Soap–Cleaning Action

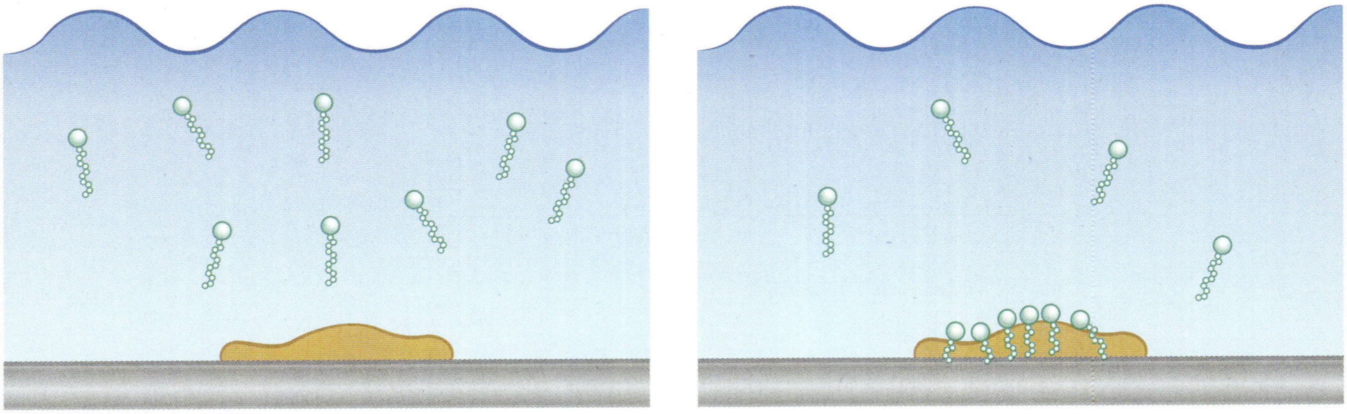

Grease is nonpolar, which is why it does not dissolve in polar water. Basic soaps have a polar and nonpolar end. The nonpolar end is attracted to grease molecules. Bases break down the grease molecules and form a micelle around the substance. The micelle, which is soluble in water, lifts the grease from the surface and washes away with the water.

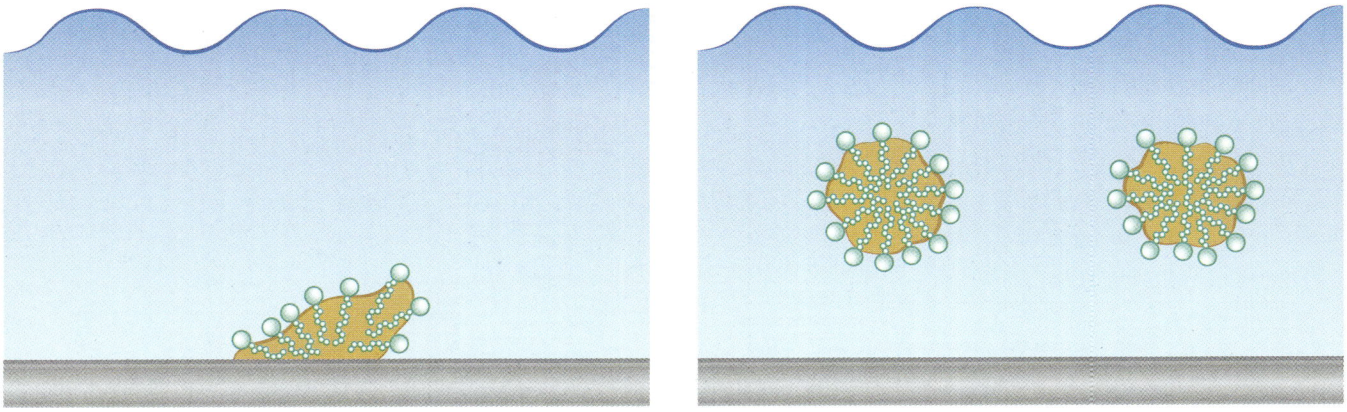

Some of the ammonia molecules react with water to produce ammonium ions (NH_4^+) and hydroxide ions. At the same time, the ammonium ions and hydroxide ions convert back into ammonia and water molecules. Double arrows are included in this equation to show that the reaction can happen in both directions. Ammonia never fully dissociates.

$$NH_3 + H_2O \rightleftharpoons NH_4^+ + OH^-$$

Bases can neutralize or "cancel out" acids because the hydronium ions from the acid and the hydroxide ions from the base combine to form water.

Many common household products are made from bases. Bases break down fat and oil molecules to make them water soluble. This is why bases make good cleaning agents. For example, household ammonia is made from ammonium hydroxide.

TRY THIS

Squeaky Clean

Put on gloves and goggles. Using a 250 mL graduated cylinder, measure 176 mL of cool water. Pour the water into a 400 mL beaker. Carefully, measure 60 mL of lye (NaOH). Slowly pour the lye into the 400 mL beaker, and gently stir with a small silicone spatula. Do not hover over the beaker as this reaction produces some fumes. The water will begin to become translucent. When this occurs, allow the solution to rest while you continue on to the next step. In a 600 mL beaker, combine 158 mL each of coconut oil, olive oil, and grapeseed oil. Heat the beaker and oils in a microwave for about one minute. Using a digital thermometer, measure the temperature of the oil mixture. Continue measuring the temperature until the mixture is between 35°C and 40°C. Also, measure the temperature of the lye mixture. It should have a temperature reading between 35°C and 40°C as well. Do not allow either mixture to cool below 35°C. When both mixtures have reached the appropriate temperature, pour the oil mixture into a stainless steel mixing bowl. Slowly pour the lye mixture into the bowl, and gently stir with the silicone spatula for approximately 5 minutes to ensure thorough mixing. Continue stirring until the mixture's color lightens and the consistency thickens. You may stir in herbs or essential oils to the mixture. Pour the mixture into soap molds, and cover the molds with plastic wrap. Wrap the molds in a towel, and do not disturb the molds for at least 24 hours. If the soap is still warm or soft after 24 hours, allow the soap to continue resting for an additional 12 to 24 hours. Remove the soap from the molds and place on a drying rack. Describe the appearance of the soap. The soap will need to cure for four weeks. After four weeks, wrap the soap in wax paper or store in an airtight container. What is the property of NaOH that makes it useful in soap making?

Sodium hydroxide, also called *lye*, is the key ingredient in products used to unclog sink drains and clean ovens. Magnesium hydroxide, which is only slightly soluble in water, is used in milk of magnesia products as an antacid to neutralize excess hydrochloric acid in the stomach. Aluminum hydroxide is another base used as an antacid and in deodorants.

LESSON REVIEW

1. Contrast acids and bases.
2. Which groups of metals form strong bases? Why?
3. Why is ammonia considered a weak base?
4. Explain how bases clean fat and oil deposits.

3.3.3 *Indicators*

OBJECTIVES

- Relate hydrogen ion concentration in solution to the pH scale.
- Identify acidic, basic, and neutral solutions using the pH scale.
- Explain how indicators are used to identify acidic and basic solutions.

VOCABULARY

- **indicator** a substance that changes color in the presence of an acid or a base
- **pH scale** a scale that measures the concentration of hydrogen ions in solution

Solutions can be classified as acidic, basic, or neutral. The classification system is determined by the concentration of hydrogen ions and hydroxide ions present in solution. If there are a greater number of hydrogen ions (H^+) compared to hydroxide ions, the solution is acidic. Neutral solutions have an equal number of hydrogen ions and hydroxide ions. If a solution has a greater number of hydroxide ions (OH^-) compared to hydrogen ions, it is basic. Scientists developed a scale to compare the strength of acids and bases using ion concentrations. This **pH scale** measures the concentration of hydrogen ions in solution. A higher concentration of hydrogen ions equals a lower pH. Higher pH values signify a lower concentration of hydrogen ions.

The pH scale is a series of numbers from 0 to 14. Neutral solutions have a pH of 7. Acidic solutions have pH values less than 7 and greater than 0. Very acidic solutions have low pH numbers. Sulfuric acid found in batteries has a pH of 0. Basic solutions have pH values greater than 7 and less than or equal to 14. Very basic solutions have extremely high pH numbers. For example, concentrated sodium hydroxide has a pH of 14. These numbers are derived from common logarithms. This means that a change of one pH unit (for example, pH 1 to pH 2) represents an exponential change in the number of hydrogen ions. An acid that has a pH of 4 has ten times the number of hydrogen ions

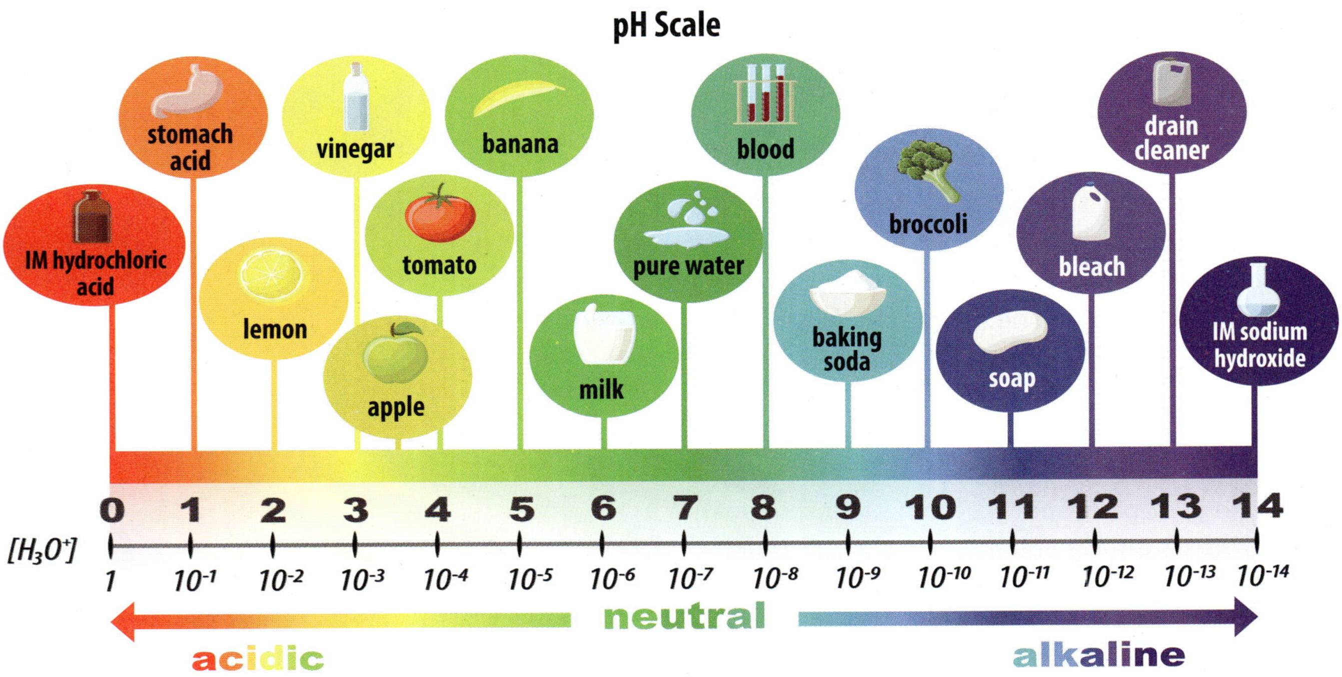

than an acid with a pH of 5. A base with a pH of 12 has ten times fewer hydrogen ions than a base with a pH of 11.

Litmus dyes are extracted from lichen.

How can the pH of a solution be tested? Chemists use indicators to determine whether a solution is acidic or basic. An **indicator** is a substance that changes color in the presence of an acid or a base. Indicators may be synthetic or found in nature. A common indicator is litmus paper that is derived from certain species of lichen. Red litmus paper turns blue in the presence of bases, and blue litmus paper turns red in the presence of acids. Phenolphthalein is another indicator used by chemists in the laboratory. This indicator is colorless in the presence of an acid, but it turns pink when it is exposed to a base. Grape juice, red cabbage water, and tea are all examples of natural indicators. Tea, for example, becomes lighter in color when mixed with the citric acid in lemons or oranges. Red cabbage water is purple on its own. However, when acids are added to the water, it turns a shade of red. With the addition of a basic solution, the cabbage water turns a shade of green.

All these indicators identify whether solutions are acidic or basic, but they do not determine the specific strength of solutions. The most accurate piece of equipment that tests the pH of a solution is a pH meter. These devices are calibrated using standard acid and base solutions with known pH values. The electrodes of the

Indicator strips change color in the presence of acids or bases. A color chart of known pH colors helps scientists determine the pH of the unknown solution.

TRY THIS

The Litmus Test

Create a data table for each sample that includes the following column titles: *sample name*, *red litmus color*, *blue litmus color*, and *acid or base*. At each acid-base station, place a red litmus test strip and a blue litmus test strip side by side on a watch glass. With the pipette, put one or two drops of solution on both test strips. Observe and record the color of the test strips. If the red litmus paper turns blue, the sample is a base. If the blue litmus paper turns red, the sample is an acid. Dispose of the test strips in the waste receptacle at each station. Compare the data table to another student's. Are the results the same? What claim can be made based on the evidence?

The color of hydrangea plants can be changed by adjusting the soil pH.

BIBLE CONNECTION

True Indicators of Faith

Chemical indicators identify if acids or bases are present in solution. The words that you speak and the way you treat others are true indicators of your faith (John 13:35). Luke 6:45 states, "A good man brings good things out of the good stored up in his heart, and an evil man brings evil things out of the evil stored up in his heart. For the mouth speaks what the heart is full of." Christians are instructed to examine their hearts to see if they are in line with God's will. Many people walk through life without ever examining their hearts. This is not wise because it is very easy to go off course and start following the deceptive paths of the world. By regularly applying God's litmus test, you ensure that you are growing in the faith, staying connected to God, and reaching others to further His kingdom.

pH meter are immersed in solution, and a digital pH reading is produced.

Maintaining particular pH values is important for many organisms. If the pH found in aquatic environments or in soil changes even slightly, fish and plants do not function as well and many do not survive. The same is true for the blood that courses through your veins and arteries. Normally, blood has a narrow pH range of 7.1 to 7.7. If blood pH becomes more acidic or basic, proteins and enzymes in the blood begin to unravel and no

TRY THIS

Cabbage Water

God created many natural indicators that change color in the presence of acids or bases. Red cabbage is an example of a natural indicator. It is red in the presence of an acid, and green in the presence of a base. Place five large test tubes in a test tube rack. Add a pipette bulb full of vinegar to the first test tube. Using separate pipettes, add a bulb full of lemon juice, apple juice, dilute shampoo, and dilute liquid soap to the remaining four test tubes. Using a dropper, add five drops of cabbage water to each test tube. Observe and record the color change of each substance. Identify which substances are acids and which substances are bases.

BIOGRAPHY

Sir William Henry Perkin

During the late 19th century, many chemists were looking for ways to produce new chemicals that could be used in industry. In 1853, Sir William Henry Perkin experimented with various substances in an effort to synthesize quinine. During this time, the spread of malaria was of great concern and the only remedy available was found from the bark of cinchona trees that were predominately grown on plantations in southeast Asia. A synthetic antimalarial compound would advance the treatment of this disease. However, instead of creating quinine, Perkin accidently made a new dye called *aniline purple*, or *mauve*. Perkin opened his own factory to manufacture mauve on a large scale. Years later, he went on to synthesize glycine (the building block of amino acids), tartaric acid, unsaturated acids, other dyes, salicyl alcohol, flavorings, and coumarin (the first artificial perfume).

longer function properly. As a result, the transport of oxygen, the clotting of blood, and the ability to fight infections are impaired. God designed people's bodies to have a natural maintenance system. This system includes special chemicals called *buffers* that regulate pH. Buffers do not allow the pH to change with the addition of acids or bases.

LESSON REVIEW

1. If a solution has a higher concentration of hydrogen ions than hydroxide ions, is it an acidic, basic, or neutral solution? Is the pH of the solution greater than 7, equal to 7, or less than 7?
2. How many more hydrogen ions does a solution with a pH of 3 have compared to a solution with a pH of 5?
3. How can scientists identify an acid or a base in solution? Include two examples in your answer.
4. Explain how indicators work.
5. How is pH regulated in blood vessels?

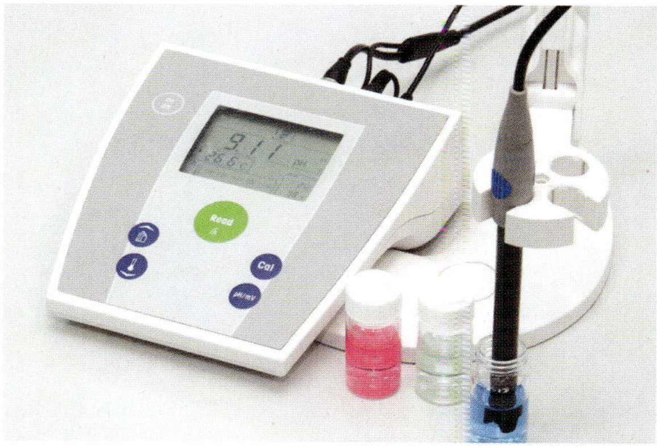

pH meters provide precise, accurate readings.

3.3.4 *Reactions of Acids and Bases*

OBJECTIVES

- Describe how the products form during a neutralization reaction.
- Outline the titration method.

VOCABULARY

- **end point** the time when an indicator changes color during a titration
- **equivalence point** the moment when the concentration of hydrogen ions equals the concentration of hydroxide ions
- **neutralization reaction** the reaction between an acid and a base in an aqueous solution that yields water and a salt
- **salt** a compound formed from the positive ions of a base and the negative ions of an acid
- **titration** a method by which a solution of known concentration is used to determine the unknown concentration of a second solution

Acidic solutions contain more hydrogen ions than hydroxide ions. Basic solutions contain more OH^- ions than H^+ ions. What do you think happens when an acid and a base are mixed together? The positive hydrogen ions produced by the acid are attracted to the negative hydroxide ions produced by the base. When these ions combine, water is formed. The mixed acids and bases also produce salts. The reaction between an acid and a base in an aqueous solution that yields water and a salt is called a **neutralization reaction**.

$$\text{Acid} + \text{Base} \longrightarrow \text{Salt} + \text{Water}$$

If the number of H^+ ions produced by the acid is equal to the number of OH^- ions produced by the base, the neutralization reaction is complete. The result is a neutral solution. There are no leftover hydrogen or hydroxide ions. If there are more H^+ ions present than OH^- ions, the solution remains acidic because of the excess H^+ ions. If OH^- ions are left, the solution remains alkaline, or basic.

Acids and bases form a salt when they react together. A **salt** is a compound formed from the positive ions of a base and the negative ions of an acid. Neutralization reactions are also double replacement reactions. Salts are neutral because they combine an equal amount of negative and positive ions. For example, table

Acids and bases react to form salt and water.

salt (NaCl) is made when hydrochloric acid and sodium hydroxide react. When the reactants dissociate in solution, the metal sodium ions from sodium hydroxide combine with the nonmetal chloride ions from hydrochloric acid.

$$HCl(aq) + NaOH(aq) \longrightarrow NaCl(aq) + H_2O(l)$$

Potassium bromide (KBr) is a salt that is made of a metal potassium ion and a nonmetal bromide ion. Salts are usually formed from positive metal ions and negative nonmetal ions. But, sometimes, salts are made of polyatomic ions, ions that contain more than one atom. Examples of salts that contain polyatomic ions include ammonium chloride (NH_4Cl), calcium sulfate ($CaSO_4$), and ammonium nitrate (NH_4NO_3).

Acids and bases are important compounds used in industry. Exact concentrations of a specific acid or base are often required. To determine the concentration of acids and bases, a process called *titration* can be used. **Titration** is a method by which a solution of known concentration is used to determine the unknown concentration of a second solution. This method works for reactions that are completed quickly, such as the mixing of an acid and a base. During a titration, a basic solution with a known concentration is placed in a burette—a glass tube used to accurately measure and dispense liquids—and added to a

Phenolphthalein indicator turns pink when the titration end point is reached.

TRY THIS

Antacid Effectiveness

Collect three different varieties of antacids. Hypothesize what characteristics make a particular type of antacid more effective than other antacid varieties. Using a universal indicator solution and a dilute acid, design and conduct an experiment that compares the pH achieved when the acid is neutralized.

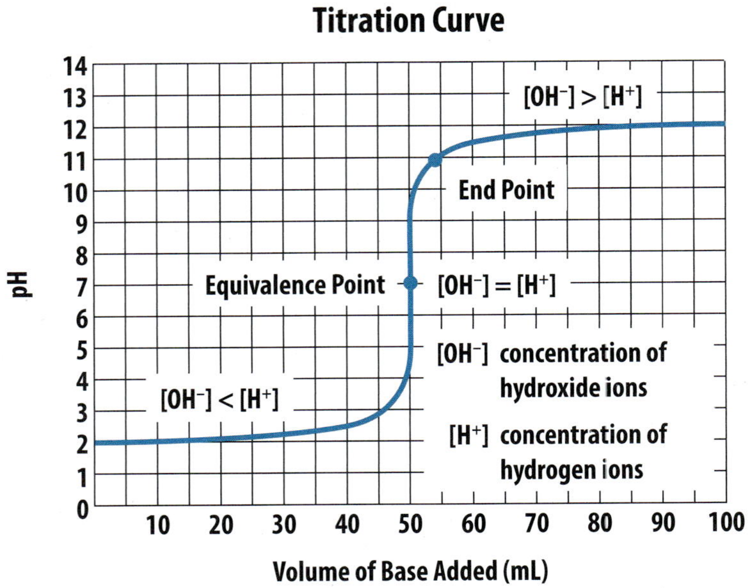

known volume of an acidic solution. The basic solution is added to the acidic solution until just enough base has been added to neutralize the acid. The **equivalence point** is reached when the concentration of hydrogen ions equals the concentration of hydroxide ions. A pH meter may be used to monitor the change in pH over time, or an indicator could be used to identify the equivalence point.

One of the most common indicators used during a titration is phenolphthalein. A small amount of phenolphthalein is added to the acid solution in the beaker. When phenolphthalein is in the presence of an acid, it appears clear. Small amounts of the base are carefully added until phenolphthalein changes to a pink color. The change in color appears just after the hydrogen ion and hydroxide ions become equal. This means that the solution went past the equivalence point and there are slightly more hydroxide ions in solution. The time when an indicator changes color during a titration is called the **end point**.

LESSON REVIEW

1. Describe the products that result from an acid and a base reaction.
2. Is a neutralization reaction complete if there are excess hydrogen ions after the reaction? Explain.
3. How are neutralization reactions also double replacement reactions?
4. Describe the titration method.
5. If phenolphthalein is the indicator used in a titration, how do you know that the end point has been reached?

UNIT 4

Matter in Motion

Chapter 1: *Forces and Motion*
Chapter 2: *Work and Energy*
Chapter 3: *Types of Machines*

Vocabulary

acceleration
buoyant force
centripetal acceleration
centripetal force
compressional force
drag
efficiency
effort force
energy
fluid
force
friction
fulcrum
hydraulic device
inclined plane
inertia
kinetic energy
lever
lift
machine
mechanical advantage
momentum
net force
newton
nonrenewable resource
pneumatic device
potential energy
power
projectile motion
pulley
reference point
renewable resource
resistance force
screw
speed
tensile force
terminal velocity
thrust
velocity
wedge
wheel and axle
work

Key Ideas

- Evidence, models, and explanation
- Change, constancy, and measurement
- Form and function
- Abilities necessary to do scientific inquiry
- Understandings about scientific inquiry
- Motion and forces
- Transfer of energy
- Conservation of energy
- Abilities of technological design
- Understandings about science and technology
- Populations, resources, and environments
- Science and technology in society
- Science as a human endeavor
- History of science

SCRIPTURE

They hammered out thin sheets of gold and cut strands to be worked into the blue, purple and scarlet yarn and fine linen—the work of skilled hands.

Exodus 39:3

4.1.1 Forces

OBJECTIVES

- Compare types of forces.
- Identify the four fundamental forces.
- Calculate gravitational force.

VOCABULARY

- **compressional force** a force that results from pushing or squeezing an object
- **force** a push or pull exerted on an object
- **friction** a force that resists motion in two objects that are moving against each other
- **newton** the unit of force that equals 1 kg × m/sec^2
- **tensile force** a force that results from a pull exerted on an object

Have you ever played baseball, football, or hockey? Have you ever pulled weeds, swept the floor, or shoveled snow? Have you ever tugged on the leash of an unwilling dog, trying to get it to follow you? These actions all involve forces. Other examples include picking up a book, throwing a Frisbee, riding a bike, and shooting a rubber band across the room.

A **force** is a push or pull exerted on an object. For example, an engine pushes a rocket through space. A magnet pulls on metal filings. The moon pulls on the oceans to create tides. A negatively charged particle repels another negatively charged particle. A potter pushes and pulls a lump of clay to form a bowl.

The amount of force applied to an object is measured in units called **newtons** (N), in honor of Sir Isaac Newton. One newton is the force exerted by Earth's gravity on an object that has a mass of 100 g, the mass of a typical apple. A Newton spring scale is commonly used to measure the force exerted on something.

When you think about pushing or pulling an object, you usually think of applying force by touching something. When you push a door open, for example, your body has to come in contact with the door. Forces that require contact between two objects or substances are called *contact forces.*

Several kinds of contact forces are at work on Earth. Each one affects how people, animals, and objects interact with the world. For instance, buoyant force is the upward force exerted on objects

Force: a push or pull

in a fluid. This force keeps ice cubes floating at the top of the water in a glass. Elastic force pulls a stretched object back to its regular shape. It is at work in a bungee cord that straps a bicycle to a car. The cord stretches when the car goes over a bump but immediately returns to its original shape and keeps your bicycle from falling off the car. **Tensile force** results from a pull exerted on an object. For example, a tensile force is exerted during a tug-of-war game.

Buoyant force

Other contact forces include **compressional force**, shearing force, and **friction**. Compressional force results from pushing or squeezing an object. Gas is compressed to fit it into a propane tank. Shearing force is a sliding force. When it is great enough, the force can break an object apart. Scissors, also known as *shears*, exert this kind of force.

Friction is a force that resists motion in two objects that are moving against each other. Friction keeps a car from sliding across dry pavement the way it slides across ice. It also prevents an object from sliding easily across carpet.

Not all forces in creation require contact between two objects. The forces that do not require contact are called *noncontact forces*. These forces are also called *the fundamental forces* because

Tensile force

TRY THIS

Tensile Force

Gather three types of lightweight string such as cotton yarn, thread, and dental floss. Cut a 75 cm length of each. Tie one end of one of the strings to a stationary object. Attach the hook on a spring scale to the other end of the string. Pull slowly, and observe the force on the scale at the moment the spring breaks. Record the data. Repeat the procedure with the remaining two strings. You may need to attach two spring scales together and add the forces. Record the data. Which string could withstand the greatest tensile force? Which could withstand the least? Give an example of a use for a string or rope that would require high tensile force.

LAWS & PRINCIPLES

Weight Versus Mass

Often science experiments call for measuring a few grams of this and a few grams of that. A gram is a unit of mass. Weight is a measure of the force of gravity acting on that mass. The unit for weight is the newton, not the gram. The formula $F = m \times g$ is used to convert mass to weight. F is the force of gravity in newtons. The letter m is the mass of the object in kilograms, and g is the acceleration of gravity (9.8 m/sec^2). If an object has a mass of 30 kg, the gravitational force, or weight, would be solved as follows: $30\text{ kg} \times 9.8\text{m/sec}^2 = 294\text{ N}$.

all forces are the result of these fundamental forces. They play a very important part in the universe. The four fundamental forces are: gravitational force, or gravity; electromagnetic force; weak nuclear force; and strong nuclear force.

Gravitational force is an obvious part of life. When a ball is thrown straight up in the air, arm muscles provide a push that sends the ball into the sky. Eventually, however, the ball changes direction as the force of gravity pulls it back to Earth. The ball does not need to be in contact with anything to be affected by gravity.

Gravity not only pulls things to Earth's surface, but it is also a force of attraction between any two objects. All objects with mass, not just Earth, exert a gravitational force on the objects around them. In fact, you are actually exerting a gravitational force on the earth even as it is exerting a gravitational force on you. The amount of gravity that an object exerts depends on the mass of the object and its distance from the other object. The more massive an object, the greater its gravitational pull, and the smaller the distance between the objects, the greater the gravitational pull. The force that a desk exerts on its user is small compared with the force that the earth exerts on you, because the desk has much less mass than the earth.

Friction

SCIENCE IN ACTION

Change in Force

When determining the gravitational force between two objects, mass and distance need to be considered. The greater the mass, the greater the force. The smaller the distance, the greater the force. These facts can be used to determine a change in force in a given system. Suppose two objects are 12 m apart and then are moved to only 3 m apart. How does the force change? To solve, take the factor needed to change 12 to 3: ¼. Then invert the factor and square it: $4^2 = 16$. The gravitational force has increased 16 times or by a factor of 16.

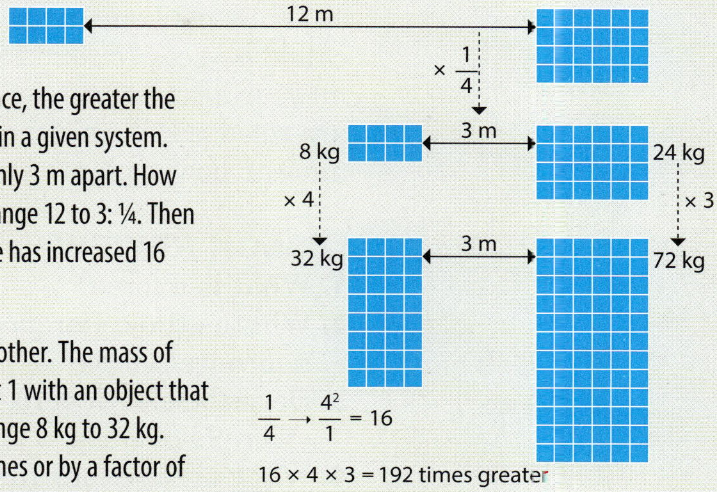

Consider the mass of the two objects exerting force on each other. The mass of Object 1 is 8 kg. The mass of Object 2 is 24 kg. Replace Object 1 with an object that has a mass of 32 kg. Determine what factor is needed to change 8 kg to 32 kg. The change means the gravitational force has increased 4 times or by a factor of 4. What happens if the second object is changed as well? If the second object is replaced with a 72 kg object, the force will increase by a factor of 3. What happens if all three changes are made? All the factors are multiplied together: $16 \times 4 \times 3 = 192$. With all three changes, the gravitational force increases by a factor of 192.

The second fundamental force is electromagnetic force. Electric and magnetic forces interact with each other so closely that physicists consider these forces as a single force. The electromagnetic noncontact force is much stronger than gravity. It determines the ways in which electrically charged particles interact with each other and with magnetic fields. It can attract or repel. The electromagnetic force binds atoms together by holding electrons in their orbits around the positively charged nucleus.

The strength of the magnetic force, like that of gravity, also depends on the distance between the objects. The closer together the objects, the stronger the force; the farther apart, the weaker the force.

Two more fundamental forces work inside the atom to contribute to its stability. One is called *the weak nuclear force*, which is the force within the atomic nucleus that is responsible for certain types of radioactivity. It is what prompts the nuclear fusion reaction that fuels the sun. The other, a powerful attractive force, is called *the strong nuclear force*, and it is the strongest of the four basic noncontact forces. It holds the protons and neutrons in

TRY THIS

Gravitational Force

Suspend a 50 g mass from the hook on a spring scale. Record the force of gravity acting on the mass. Repeat the process with a 1 kg mass and a 1.5 kg mass. How does gravitational force relate to an object's mass?

the nucleus of the atom and overcomes the repulsion between the protons as a result of the electromagnetic force.

Under certain conditions, the nucleus of an atom can be blown apart by neutrons fired at it. Splitting the nucleus of an atom is called *nuclear fission*. During nuclear fission, the very powerful nuclear force is overcome, and incredibly huge amounts of energy are released. Physicists have used that energy constructively in nuclear power plants and destructively in nuclear weapons.

LESSON REVIEW

1. What is a force?
2. What is the difference between a contact force and a noncontact force?
3. Describe elastic force and compressional force. Give an example of each.
4. What are the four fundamental forces? Are these contact or noncontact forces?
5. What affects the strength of a gravitational force?
6. Calculate the gravitational force on an object that has a mass of 82.5 kg.

All forces have two components: magnitude, which is size or strength; and direction. For instance, the force of gravity depends on the mass of the objects involved. The force of gravity is greater on an elephant than it is on a house cat. That is why the elephant weighs more. When discussing gravity, direction is always down toward the center of the earth. But what about other forces? What about contact forces?

When a force is applied to an object, the purpose is to create change. When a wagon is pulled, it will move. When a lump of clay is pressed, its shape will change. Sometimes, when a large object is pushed, however, it does not move. Why? The forces applied are balanced, so no change takes place. At all times, several forces are acting on an object. Consider a book on a table. Nothing appears to be happening. However, at least two forces are working on it. Gravity is holding it down and friction is keeping it in place. The forces are balanced so no change is taking place.

When a pushing force is applied, the book can move across the table. Pushing it produces a greater force than the frictional force working against the motion. All the forces added together is called the **net force**. For example, if a box is pushed in one direction with 95 N of force, and the weight of the box pushes back with 80 N and the frictional force also pushes back with 15 N, there is a balanced system and a net force of 0. The box does not move. But suppose two people push. Now one is pushing with 95 N, and the other is pushing with 80 N, and the box and

OBJECTIVES

- Define motion.
- Analyze a system to determine net force.
- Compare speed, velocity, and acceleration.
- Calculate speed, velocity, and acceleration.

VOCABULARY

- **acceleration** the rate of change in velocity over time
- **net force** the total of all the forces acting on an object
- **reference point** a point against which the position of an object is measured
- **speed** the rate at which an object moves
- **velocity** the speed of an object in a particular direction

friction are still pushing back with only 95 N of force. One side of the equation indicates 175 N, and the other side shows 95 N. The force is unbalanced; the net force is 80 N, and the box moves.

Motion is the result of unbalanced forces where the net force is not equal to zero. If the net force in a system is not known, how do we know something is in motion? Imagine riding in a car and traveling next to a train with the same velocity. The train appears to not be moving. To determine its motion, look for a **reference point**. Choose a part of the train to focus on and then watch to see if it passes a reference point. If it does, then the train is moving. The reference point is a stationary object against which the position of an object is measured. All motion is relative.

Here is another example. Suppose you look out the window and see the garbage truck in front of your house. A few minutes later you look out the window again. This time the truck is two houses down the street.

How do you know it moved if you did not see it move? Its position is compared to a reference point. In this case, the front of your house is the reference point. The garbage truck is no longer next to your house. If the position of an object is changing in relation to a reference point, it is in motion. What about the two people riding inside the truck? In relation to each other, they are not moving. In relation to your house, they are.

For the garbage truck to move, it has to create an unbalanced force. The engine has to move the wheels with enough force to overcome the weight of the truck and the friction on the road. When the workers are collecting garbage, the truck moves slowly. They move at a slow rate of **speed**, which is determined by measuring the distance traveled and dividing it by the time it takes to travel that distance. *Speed = d/t*. If the garbage truck travels 4 km in 2 hour, the average speed can be calculated: Speed = 4 km/2 hours, or 2 kph. Average speed is the rate at which the truck moves a certain distance over a set amount of time.

Instantaneous speed, on the other hand, measures speed at a given instance. For example, if it takes 30 minutes to drive through a city that is 10 km across, the average speed is 20 kph. For part of the drive, the car travels at 40 kph, but at other times, it is stopped at a traffic light. At that moment, the instantaneous speed is 0 kph. The speed of 20 kph is simply an average.

Besides knowing the speed of the object, it is helpful to know the direction it is traveling. **Velocity** is speed with direction. A school bus travels at an average speed of 40 kph. It benefits you to know if the bus is moving toward or away from the bus stop. If the bus is heading toward the bus stop, the velocity of the bus is 40 kph toward the bus stop. Speed only gives the distance something travels over an amount of time. Velocity includes the direction of travel.

TRY THIS

Relative Motion

The direction of travel is important in determining the impact of a crash. Make two stacks of books of equal height. Create two ramps facing each other using the stacks of books and the boards. Leave enough space between the boards for an egg to fit. Read all the directions before starting and hypothesize which trial will cause the least damage and which trial will cause the most damage. *Trial One*: Roll two eggs, one in front of the other, down the same ramp at the same time. *Trial Two*: Set one egg at the bottom of the ramp and roll another egg down a ramp into it. *Trial Three*: Simultaneously roll one egg down each ramp. Record what happened in each trial. Why is the damage different in each trial? Which trial caused the least damage? Which trial caused the most damage? How do the results compare to the hypothesis?

Another instance where velocity is important includes the motion of two objects. If two cars are traveling at 40 kph on the same road, it is important to know where they are in relation to each other. If the cars are both moving east, the passengers in the cars should enjoy the trip without incident. They can travel next to each other or one in front of the other. Relative to each other, they are not even moving. But what if one is moving east and one is moving west? If the two cars are moving toward each other, they could collide. When the two cars are moving in opposing directions, add the speeds to determine the relative speed, or in this case, the speed of the collision. This head-on collision would occur at 80 kph. If the cars are moving in opposite directions and away from each other, they are moving away from each other at 80 kph.

Could two cars traveling in the same direction ever collide? Yes. If the first car, or the one in front, is moving at a velocity of 40 kph north, and the one behind is moving at a velocity of 80 kph north, the second car could hit the first car. What would be the speed of the collision? Because the cars are moving in the same direction, the velocities are subtracted.

$$80 \text{ kph} - 40 \text{ kph} = 40 \text{ kph north}$$

A crash that occurs when both cars are moving in the same direction has a smaller magnitude than when the cars are heading toward each other.

If the driver of the first car looks in the rearview mirror and sees the car coming quickly, she may decide to change her speed

Velocity and acceleration are determinant factors in the amount of damage in a car crash.

LAWS & PRINCIPLES

Speed, Velocity, and Acceleration

Speed is the distance an object travels divided by the time it takes to travel that distance.

$$s = d/t$$

A car travels north 240 km in 4 hours: 240 km/4 hours = 60 kph. The car's speed is 60 kph.

Velocity is speed with direction. The car's velocity is 60 kph north.

Acceleration measures the rate of change in velocity. An acceleration of 9.8 m/sec^2 means that an object is changing its speed by 9.8 m/sec every second. To determine the acceleration of an object, use the final velocity, the initial velocity, and the time over which the change took place.

$$a = \frac{v_f - v_i}{t_f - t_i}$$

Suppose a skateboarder starts at the top of a hill with a velocity of zero. Three seconds later he is traveling down the hill at 12 m/sec. Use the equation to determine the acceleration:

$$\frac{12\text{ m/sec} - 0\text{ m/sec}}{3\text{ sec} - 0\text{ sec}} = \frac{12\text{ m/sec}}{3\text{ sec}} = 4\text{ m/sec}^2$$

The skateboarder increased his speed by 4 m/sec every second. His acceleration is 4 m/sec^2 down the hill.

to avoid getting hit. The rate of change in velocity is called **acceleration**. A change in velocity can be an increase in speed, a decrease in speed, or a change in direction. The first driver was originally traveling with a velocity of 40 kph north. She sees the car behind her getting closer, so she starts accelerating. Her speed increases from 40 kph to 50 kph to 60 kph to 70 kph to 80 kph over 10 seconds. The rate of acceleration can be calculated:

$$a = \frac{v_f - v_i}{t_f - t_i}$$

In this formula, a is acceleration, v_f is the final velocity, v_i is the initial velocity, t_f is the final time, and t_i is the initial time.

$$\frac{80\text{ kph} - 40\text{ kph}}{10\text{ sec} - 0\text{ sec}} = 40\text{ kph/10 sec}$$

Something is wrong with the equation. The units used to measure time are different. In order for the statement to make sense, the units must be the same. Acceleration is normally measured in meters per second per second. To calculate, convert the hours to seconds:

$$\frac{1\text{ hr}}{1} \times \frac{60\text{ min}}{1\text{ hr}} \times \frac{60\text{ sec}}{1\text{ min}} = 3{,}600\text{ sec}$$

Solve:

$$\frac{40\text{ km}/3{,}600\text{ sec}}{10\text{ sec}}$$

Reduce the numerator to 1 km/90 sec. The seconds reduce and now the acceleration is

$$\frac{1\text{ km}/9\text{ sec}}{\text{sec}} \text{ or } 1\text{ km}/9\text{ sec}^2$$

One more conversion is needed—converting kilometers to meters.

$$\frac{1{,}000\text{ m}/9\text{ sec}}{\text{sec}} = 11.11\text{ m/sec}^2$$

This means that the driver increased her speed by 11.11 m/sec every second.

In the example, the acceleration was positive, or the change in velocity was an increase. Acceleration can also be negative, meaning the object slows down. If a rabbit is running at a velocity of 3 m/sec north for 20 seconds and then takes 2 seconds to stop, the rabbit's acceleration is

$$\frac{0\text{ m/sec} - 3\text{ m/sec}}{2\text{ sec}} = \frac{-3\text{ m/sec}}{2\text{ sec}} = -1.5\text{ m/sec}^2$$

The negative sign signifies an acceleration in the opposite direction of the original motion. The rabbit slowed down, or decreased velocity, at a rate of 1.5 m/sec per second south. Also, acceleration can simply be a change in direction. A plane can travel with a velocity of 450 kph west and then accelerate to a velocity of 450 kph north.

LESSON REVIEW

1. Describe motion.
2. How do you know whether a system has a balanced force?
3. If one car travels 60 kph west and another car travels 80 kph west, what is their relative velocity? Will they collide?
4. If you travel 150 km in 4 hours, what is your average speed?
5. If it takes you 9 seconds to increase velocity from 0 kph to 40 kph east, what is your acceleration?

Newton's First and Second Laws of Motion 4.1.3

Isaac Newton carefully observed how motion works in the universe and developed three laws to describe motion. One of these, Newton's first law of motion, is also called *the law of inertia*. This law states that a body or an object at rest remains at rest unless acted on by an unbalanced force, and a body in motion continues to move at a constant velocity unless acted on by an unbalanced force. **Inertia** is the tendency of an object to resist a change in its movement, regardless of whether it is at rest or in motion. If it is still, it remains still; if it is moving, it keeps moving in a straight line and at a constant speed. In one sense, all objects in the universe are lazy. The word *inertia* comes from the Latin word meaning "lazy." Objects change their fixed state only if a force acts on them. An object that is still does not randomly move; a book on a table will remain in that position until someone moves it. A moving object speeds up, slows down, or changes direction only when a force, such as friction, acts on it. A bowling ball continues moving until something, the bowling pins or the gutter, stops its movement.

The inertia of an object is directly related to its mass. The more massive the object, the more force is needed to get it moving with a certain acceleration. For example, compare how much force you need to move a pebble with how much force you need to move a large boulder with the same acceleration. The pebble is not very massive so it has little inertia, and it takes little force to move it. The boulder is massive so it has more inertia and requires more force to move it. Newton's second law of motion applies here, which states that the net force applied to an object equals the

OBJECTIVES

- Apply Newton's first and second laws to the motion of objects.
- Calculate the distance an object falls using acceleration of gravity.
- Explain projectile motion.

VOCABULARY

- **inertia** the tendency of an object to resist a change in its movement
- **projectile motion** the curved path an object follows when it is thrown or propelled near Earth's surface
- **terminal velocity** the constant velocity after which an object stops accelerating

HISTORY

Aristotle vs. Newton

Everywhere you look, you can see objects moving—birds are flying, leaves are fluttering, cars are rolling, and people are walking. Eventually, though, everything stops moving. Why? Aristotle, an ancient Greek scientist and philosopher, believed that objects at rest stayed at rest and all objects wanted to rest. Looking at the motion of objects in the world, Aristotle's statements appeared valid. Anything that moves does eventually stop. For centuries, Aristotle's theories about motion were widely accepted. Sir Isaac Newton also believed that an object at rest stays at rest. On this point, they agreed. But on the second point, they did not. Newton did not believe that all objects "wanted" to rest. He realized that there were forces, such as gravity and friction, acting on objects to stop motion. He discovered that all objects have inertia, and once moving they would continue to move if no other forces acted on them. Therefore, instead of all objects "wanting" to rest, all objects continue doing what they are doing and will change only if acted on by an unbalanced force.

object's mass multiplied by the acceleration. It takes more force to produce the same acceleration in objects that have more mass. This law can be stated with the formula $F = m \times a$ (sometimes rearranged to $a = F/m$). In this formula, F is the net force acting on the object, m is the mass of the object, and a is the acceleration of the object. The metric units of mass are kilograms (kg), the units of acceleration are meters per second squared (m/sec^2), and the units of force are newtons (one newton = 1 kg × m/sec^2).

If a 60 kg object is pushed and it accelerates at 3 m/sec^2, the resulting net force (F) is 180 N. Net force is the sum of all forces on an object, so it includes the opposing force of friction. Suppose the frictional force (F_f) equals 580 N, and the force exerted is F_p. To find out how much force is actually exerted, use the following equation: $F_p - F_f = F$. When you solve for F_p, you find that you actually had to apply 760 N of force to gain a net force of 180 N.

LAWS & PRINCIPLES

Newton's Laws

Newton's first law, also called *the law of inertia*, states that an object at rest remains at rest unless acted on by an unbalanced force, and an object in motion stays in motion unless acted on by an unbalanced force.

Newton's second law states that force and acceleration are in the same direction, or more easily stated as the following:

$$F = m \times a$$

BIOGRAPHY

Sir Isaac Newton

Sir Isaac Newton was born in a small English village in 1643 and is regarded as one of the greatest intellects of the Scientific Revolution in the 17th century. His father, who was a farmer, died before Newton was born prematurely. His mother remarried when he was three, leaving him in the care of his grandmother. Newton attended school until his mother, widowed a second time, took him from school at age 12 to learn farming. He failed at farming and was sent back to school. Eventually, he entered Trinity College in Cambridge, where he worked on campus waiting tables and cleaning to pay for his education. Although much of Europe had embraced new scientific philosophies such as heliocentrism, Cambridge had not. Newton was taught the traditional science curriculum in class, but he read the modern philosophers and scientists in his spare time. He graduated without receiving any honors just as the Great Plague forced Cambridge to close. For 18 months, Newton studied alone at home. During that time, he developed infinitesimal calculus, began his theory of light and color, and pondered gravity and planetary motion. In 1667, he returned to Cambridge, where he earned his Master of Arts and became a professor. In 1668, Newton invented the first reflecting telescope to help prove his theory of light and color. A bitter rivalry developed between Newton and Robert Hooke, the foremost authority on optics at the time. Unable to bear criticism, Newton withdrew from society for six years. During this time, he formulated his laws of motion, which were published in *Philosophiae Naturalis Principia Mathematica* in 1687, a book that is still regarded as the single most influential work in physics. Isaac Newton was elected president of the Royal Society in 1703 and was knighted by Queen Anne in 1705.

TRY THIS

Newton's First Law

Tape a table-tennis ball to a string that is 45 cm long. Stoop down, and swing the ball in a flat circle about 3 cm above the floor. Let go of the string, and observe the direction that the ball rolls. Draw a picture of the path the ball took when you let it go. How does this demonstrate Newton's first law?

A parachute reduces terminal velocity.

TRY THIS

Inertia

Place an index card on top of a small beaker. Set a coin on the center of the card. What forces are acting on the coin? How can you get the coin in the beaker without lifting the card? Try your method. When you succeed, explain how it worked.

Gravity is the force that causes objects that are dropped to accelerate toward Earth. All objects near the earth's surface accelerate at a rate of 9.8 m/sec^2; this is called *the acceleration of gravity*. This means that for every second an object falls, its downward velocity increases by 9.8 m/sec^2. This acceleration is the same for all objects, regardless of their mass. If a 2 kg box falls from a top shelf and lands on your foot, what would be the net force of impact? F = 2 kg × 9.8 m/sec^2; F = 19.6 N. That would hurt, but it probably would not cause any damage. What would be the net force if a 20 kg box fell on your foot? Your foot would receive 196 N of force. Even though the acceleration rate is the same for all objects, regardless of mass, the force definitely changes. These examples did not take any frictional force into consideration, but frictional forces exist and must be considered.

When a tennis ball is dropped from a very tall building, it is assumed that the velocity of the tennis ball will continuously increase as it accelerates. Actually, the tennis ball will not gain velocity quite as predicted because a frictional force is also pushing up on the ball. That force, *air resistance*, acts against the free fall of the ball. The net force acting on the ball is the force of gravity minus the force of air resistance. This resistance increases as the velocity of the ball increases until the point is reached where the upward force of air resistance equals the downward force of gravity. At that point, the object stops accelerating and starts falling at a constant velocity. The constant velocity after which an object stops accelerating is called its **terminal velocity**. Terminal velocity varies according to the mass, size, and shape of the object falling.

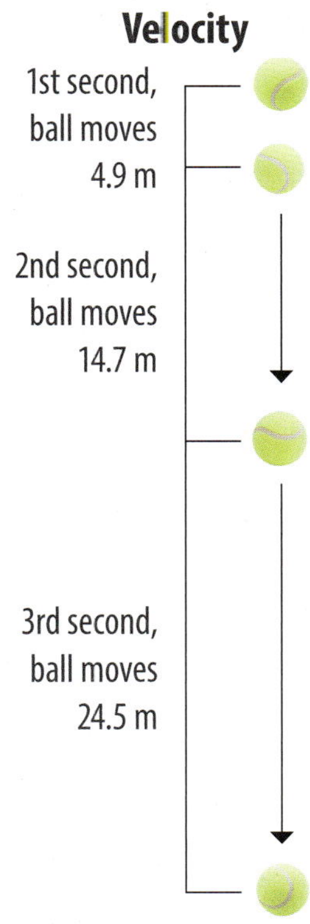

A falling object accelerates at a constant rate. Each second, the object travels faster and farther than it did the second before.

FYI

Free Fall

When objects fall, they are subject to the constant acceleration of gravity and air resistance. Because of air resistance, objects may fall at different rates. But if the only force acting on an object is gravity, it is said to be in free fall. When objects of different masses free fall, they will land at the same time. If you drop a feather and a billiard ball inside a vacuum, a chamber without air, they will fall at the exact same rate.

Terminal Velocity

A skydiver wears a parachute to reduce his or her terminal velocity. The parachute does not change the diver's mass but increases the surface area so more air can resist that fall and decrease the terminal velocity.

Because the acceleration of gravity is constant, the distance an object travels in free fall can be calculated if the time it falls is known. The equation $d = 1/2 \times a \times t^2$ shows the relationship between acceleration, the distance traveled, and the time that something falls. In this equation, d is the distance traveled, t is the time, and a is the acceleration.

If you drop an object and measure the time of the fall from the moment you let the object go until it hits the floor, you can calculate the distance of the fall. What distance did an object fall if it took 0.75 seconds to reach the ground?

$$d = 1/2 \times (9.8 \text{ m/sec}^2) \times (0.75 \text{ sec})^2 = 2.75 \text{ m}$$

The acceleration of gravity also affects the motion of projectiles in a predictable manner. **Projectile motion** is the curved path an object follows when it is thrown or propelled near Earth's surface. The motions of leaping dancers, thrown balls, hopping frogs, and arrows shot from a bow are all examples of projectile motion. Projectile motion has two components—horizontal and vertical. The two components are independent; that is, they have no effect on each other. When the two motions are combined, they form a curved path.

Think about shooting an arrow. The bow's force on the arrow makes the arrow move forward. This force results in horizontal motion. After the arrow is released, no horizontal forces, except air resistance, are acting on the arrow, so the horizontal velocity of the arrow is constant. After the arrow is released, gravity pulls it downward, giving it vertical motion. Projectiles accelerate downward with the acceleration of 9.8 m/sec^2, as do all falling objects. Because objects in projectile motion accelerate downward, the archer must aim above the target instead of directly at it.

LAWS & PRINCIPLES

Calculating a Free Fall

$$d = 1/2 \times a \times t^2$$

d = distance
a = acceleration
t = time

Projectile Motion

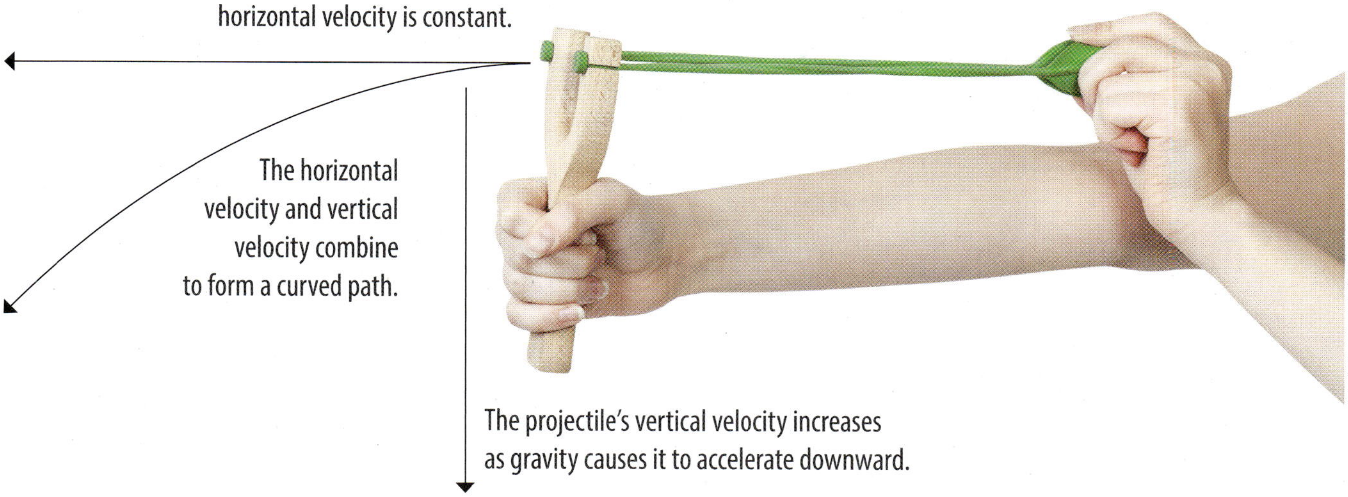

LESSON REVIEW

1. What is the law of inertia? When a ball rolls across the floor, why does it stop moving?
2. Imagine that two objects of different masses have the same acceleration. Which has the greater force acting on it? Why?
3. If a man pushes an object with 210 N of force, and the object resists with 150 N of force, what is the net force? If the object accelerates at 2 m/sec^2, what is the object's mass?
4. If the acceleration of gravity is a constant, why is terminal velocity different for different objects?
5. If a planter is knocked off a ledge and falls for 1.2 seconds, how far does it fall?
6. What two components create the curved path of projectile motion?

TRY THIS

Newton's Second Law

Tie one end of a string around a tennis ball and the other end to a ring stand. The ball should hang just above the floor. Set a table-tennis ball on the floor in the path of the swinging tennis ball. Pull back the tennis ball to the height of the ring stand and release it so it strikes the other ball. Measure the distance the table-tennis ball traveled. Now, replace the table-tennis ball with a tennis ball. Pull the hanging tennis ball to the height of the ring stand and release it so it strikes the second tennis ball. Measure the distance the second tennis ball traveled. Did the two balls travel the same distance? Why? How are mass and acceleration related to force?

4.1.4 Circular Motion

OBJECTIVES

- Identify force and motion of a circular path.
- Correlate centripetal acceleration to Newton's laws.

VOCABULARY

- **centripetal acceleration** the change in velocity of an object traveling in a circular path
- **centripetal force** the force that continuously pulls objects traveling in a circular path toward the center of the circle

Newton's first law of motion, or the law of inertia, states that an object travels at a constant velocity unless an unbalanced force acts on it. Velocity includes both speed and direction. Since maintaining a constant velocity includes traveling in a straight line, how can Earth orbit the sun? Why does Earth not travel in a straight line? Does circular motion violate Newton's first law of motion? The answer is no. An object's inertia maintains the object's speed, but a force acts on the object to create a circular path and to keep it from traveling in a straight line. The force that continuously pulls objects into a circular path toward the center of the circle is called **centripetal force**. It is always perpendicular to the velocity of the orbiting object.

If you tied a string around a tennis ball and swung it in a circle around your head, the tension on the string would provide the centripetal force that pulls the ball toward the center. But if the string snapped, the tennis ball would not keep traveling in a circle around you; it would fly off in a straight line in the direction that its velocity was headed at the time the string broke. A force must act on the object to keep it moving in a circle. For instance, the force of gravity becomes the centripetal force that holds the planets in their orbits.

If an object is moving in a straight line, its changes in velocity are the result of a change in speed rather than direction. But this is not true of an object moving in a

A windmill with uniform circular motion needs a centripetal force toward the center to maintain its circular path of rotation.

circular path. Even if the object is moving in a circular path at a constant speed, its velocity changes continuously because its direction is constantly changing. This change in velocity of an orbiting object is called **centripetal acceleration**, and it always points to the center of the circle. If the object is traveling at a constant speed around a circle, the total acceleration also points to the center of the circle. If an object is speeding up, the total acceleration points in the direction the object is moving. If an object is slowing down, the total acceleration points back, away from the direction the object is moving.

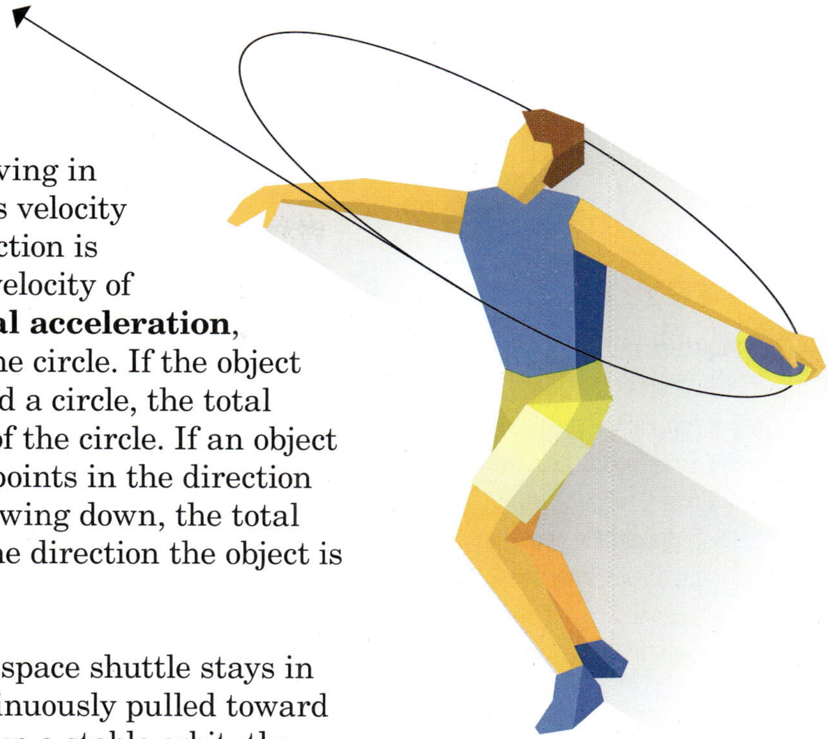

The muscles of a discus thrower apply the centripetal force needed to make the discus move in a circle until the moment it is released. When the discus thrower releases his grip, the discus flies off in a straight line.

Consider how an artificial satellite or space shuttle stays in orbit around Earth. A satellite is continuously pulled toward the center of Earth by gravity. To set up a stable orbit, the centripetal acceleration needed to maintain the orbit must be exactly equal to the attraction of gravity between the satellite and Earth. Through careful calculations, scientists can launch a satellite to orbit Earth at exactly the same rate that Earth rotates. Although the satellite is orbiting, it seems at all times to hover over the same point above Earth.

Why do astronauts in space appear to be weightless as they float inside their spaceship? It might be assumed that astronauts are so far away from Earth that there is no gravity pulling on them, but that is not accurate. Gravity is still acting as a centripetal force. The astronauts appear weightless because they are in free fall—the condition of an object when gravity is the only force acting on it—along with everything else in the spacecraft. If the cables of an elevator in a very tall building broke and the elevator

Velocity and Acceleration

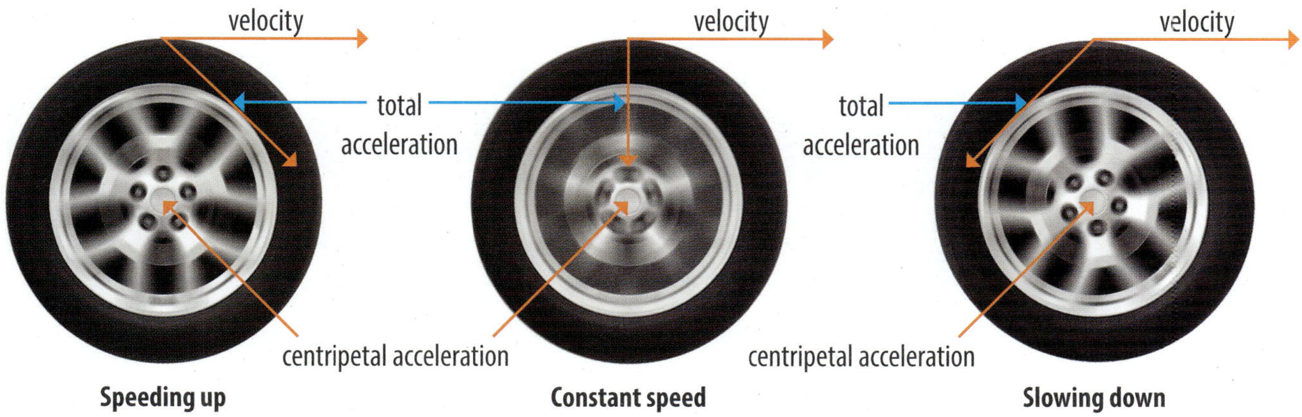

TRY THIS

Weightlessness

Punch a hole in the side of a paper cup. Fill the cup with water, covering the hole with your fingers. Hold the cup high off the ground or over a sink, and drop it. Why does the water not flow out of the hole when the cup is falling?

BIBLE CONNECTION

David and Goliath

Before David faced Goliath, Scripture says that he chose five smooth stones from a brook. He then headed to the battlefield with the stones and his slingshot to meet the giant. King Saul offered David his armor for protection, but David refused it. David, the shepherd boy, was familiar with the power he could generate with centripetal force. He had used his slingshot many times to kill or ward off the lions and bears that threatened his flock, and he knew the Lord was with him. On the day of battle, he confidently took his slingshot, loaded it with a stone, and began to swing it around and around, over his head. As he continued to accelerate the stone, he was increasing the potential force with which he would eventually knock down and kill Goliath. When he had achieved his maximum velocity, he released the stone, sending it in a straight line toward Goliath's head. The aggressor fell dead; David had defeated Israel's enemy.

accelerated at the same rate as gravity, the cargo inside the elevator would seem to float inside the elevator as it was falling.

LESSON REVIEW

1. Define centripetal force.
2. What does the law of inertia say about the motion of a moving object?
3. What direction is the centripetal force in relation to the velocity of an orbiting body?
4. What keeps an orbiting satellite from falling to Earth?
5. What is centripetal acceleration?
6. What keeps an orbiting satellite from flying off into space?

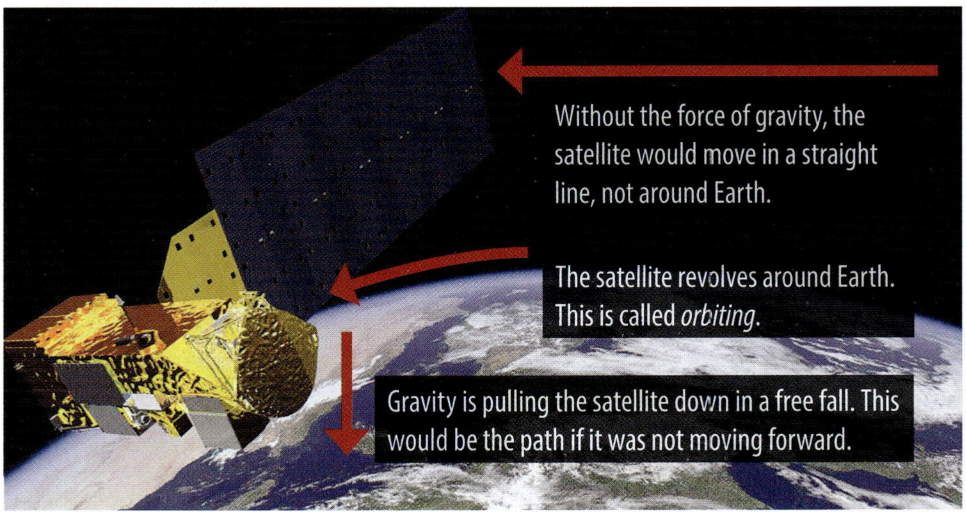

Newton's Third Law of Motion 4.1.5

Newton's third law of motion states that for every action there is an equal and opposite reaction. In other words, forces always come in pairs. When one object exerts a force on another object, the second object exerts a force of the same size on the first object—but in the opposite direction. Consider the chair you are sitting on. Are you exerting a downward force on the chair? Or is it exerting an upward force on you? Both! You cannot touch the chair without the chair touching you. When you walk, your feet push back against the ground, and the ground pushes forward on your feet, moving you forward.

Newton's third law of motion can be used to explain how a rocket is launched. When the rocket is ignited, the hot gases escaping from the engine push in a downward direction. Just as Newton's third law predicts, an opposite force thrusts the rocket upward. This third law of motion also explains why rocket engines can be used to adjust the course of a spacecraft: when an engine is fired to the right side of a spacecraft, the opposite force turns the spacecraft to the left.

A force on an object changes its momentum. **Momentum** is the property of a moving object that depends on its mass and velocity. This can be expressed with the formula $p = m \times v$. In this formula p is the momentum, m is the mass of the object, and v is the velocity of the object. A moving bowling ball has more momentum than a marble moving with the same velocity, and a bicycle moving at 10 kph has less momentum than a bicycle of equal size going 30 kph.

The more momentum an object has, the harder it is to stop the object or change its direction. It takes more force to stop a heavy truck than it takes to stop a light compact car. It also takes more force to stop a truck going 70 kph than the same truck traveling at 30 kph.

When a moving object hits another object, momentum can be transferred from one object to the other, but the total momentum

LAWS & PRINCIPLES

Newton's Third Law of Motion

For every action, there is an equal and opposite reaction.

When a tennis ball hits a tennis racket, the ball exerts force on the racket, causing it to bow, and the racket exerts force on the ball, causing it to fly through the air.

OBJECTIVES

- Explain Newton's third law of motion.
- Define momentum.
- Apply the law of conservation of momentum.

VOCABULARY

- **momentum** the property of a moving object that depends on its mass and velocity

Rocket launch

LAWS & PRINCIPLES

Law of Conservation of Momentum

Momentum is the property of a moving object that depends on its mass and velocity.

The greater the mass, the greater the momentum. The greater the velocity, the greater the momentum.

Momentum (p) equals mass (m) times velocity (v).

$$p = m \times v$$

The law of conservation of momentum states that momentum can be transferred from one object to another, but the total momentum is the same before and after the collision.

HISTORY

Roller Coasters

Extreme heights, stomach-dropping plunges, death-defying loops, and high speeds define today's roller coasters, which can be found in almost every amusement park around the globe. Coaster designers compete to make rides higher, faster, and steeper, and with more inversions—generally more exciting than the last one. But where and when did the roller-coaster mania start? In the 17th century, Russians built giant ice slides that people would ride down on large sleds made from wood or ice. A pile of sand at the bottom of the hill would stop the sled. Later, a Frenchman tried to follow suit, but the climate in France was too warm to support an entire hill made of ice. Then a "centrifuge railway" was built in the 1850s. The rail car would travel through a loop with only centripetal acceleration to keep it on the track. People loved it. The craze made its way to America near the end of the 19th century, and a new era of entertainment began. La Marcus Thompson built the switchback railway at Coney Island in 1884. The coaster was driven by gravity alone. Passengers would climb stairs to board the car, and then it was pushed to travel down a hill and over a few bumps. At the bottom, passengers would disembark and climb another set of stairs as workers pulled the car to the top of the next hill. Passengers got back in and rode down another hill. The Flip-Flap Railway, made in 1898, included a vertical loop. The force generated by the circular motion was so strong that many riders' necks were snapped. This did not deter the forward momentum of the roller-coaster industry. As time went on, bigger, faster, and safer roller coasters were made. The year 1959 marked the construction of the first tubular steel roller coaster, The Matterhorn, at Disneyland in Anaheim, California. Designers and architects all over the globe continue the race to push the limits of physics and provide the most thrilling coaster ride ever.

TRY THIS

Pushy, Pushy

Sit on a skateboard. Have another person of different mass sit on a skateboard facing you. Push each other. Which one of you moves? Why?

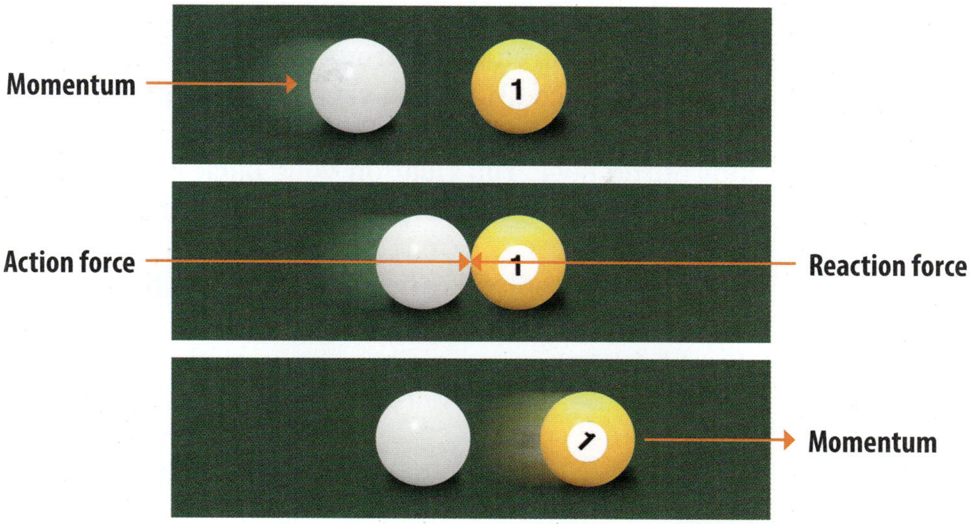

stays the same. This is called *the law of conservation of momentum*. Consider the example of a cue ball hitting a billiard ball. The combined momentum of the cue ball and the billiard ball before the collision is the same as the combined momentum after the collision.

Imagine hitting the billiard ball with the cue ball with so much force that the billiard ball starts moving, and the cue ball stops. The cue ball has a certain amount of momentum before the collision. During the collision, the cue ball's momentum transfers to the billiard ball. The billiard ball then moves away with the same amount of momentum that the cue ball had before the collision. The total momentum remains the same.

Newton's third law of motion provides understanding about what happens. The cue ball hits the billiard ball with a certain amount of force—the action force. The billiard ball exerts an equal but opposite force on the cue ball—the reaction force. The action force causes the billiard ball to start moving; the reaction force makes the cue ball stop.

LESSON REVIEW

1. What is Newton's third law of motion?
2. Explain how rockets depend on Newton's third law of motion.
3. What is momentum?
4. What factors determine the amount of momentum of a moving object?
5. What is the law of conservation of momentum? Give an example.
6. Explain why a bowling ball is better than a foam ball for knocking down bowling pins.

TRY THIS

Mass Times Velocity

Drop a table-tennis ball and a tennis ball from a height of about one meter. Note the height to which both balls return. Then, with the table-tennis ball just above the tennis ball, drop the two balls at the same time. Note the height to which both balls return. What causes the difference?

4.1.6 *Forces in Fluids*

OBJECTIVES

- Compare the effects of compression on a solid, a liquid, and a gas.
- Identify the relationship of force, pressure, and area in a fluid.
- Define Pascal's principle.
- Explain buoyant force.

VOCABULARY

- **buoyant force** the upward force exerted on objects in a fluid
- **fluid** any substance that flows and takes the shape of its container
- **hydraulic device** a device that uses liquids to transfer pressure from one point to another

Imagine spending an afternoon at the pool. You dive into the water, cleanly breaking through the surface. You challenge your friends to a race and swim several pool lengths. Then you all have a water fight, splashing water at each other and over the sides of the pool. You float on your back for a time, and then you inflate a raft so you can float with less effort.

You can swim, splash, and float in a pool because of the way water molecules are created. Although the molecules in solids are bound together in a rigid structure so the object does not readily change shape, that is not true of a liquid like water.

Any substance that flows and takes the shape of its container is called a **fluid**. Both liquids and gases are considered fluids because their molecules move around more freely than molecules in solids. Forces acting on fluids have different effects than the same forces acting on solids. Compressing a solid changes the volume of that solid, but the change is so small that it is generally not noticeable because the molecules in solids are already so close together that they cannot be forced much closer together. When a liquid is compressed, the effects are transmitted to all parts of the liquid. You can squeeze a water bottle, and all the water is affected. Most liquids, however, are not easily compressed into smaller volumes.

Gases, on the other hand, can be compressed into a smaller space because the molecules are usually so far apart. The volume of a compressed gas is less than before it was compressed. Spray

Pascal's Principle

The force of the water on the bottle is the same at any given depth in the bottle. When the bottle is squeezed, the increased pressure is evenly distributed throughout the liquid.

cans contain compressed gas and are pressurized to maintain the smaller volume. When the valve is opened on the can, the air rushes out because fluids move from areas of high pressure to low pressure. When the pressurized air is released, the air carries the paint, sunscreen, cooking oil, or other added can contents and disperses it quickly.

Pressure is defined as force per area. The molecules in fluids are constantly bouncing around. When the molecules hit the walls of the container or hit an object in the fluid, they push against it. Imagine a fluid in a cylinder with a tight-fitting piston as a lid. The pressure being exerted on the underside of the piston is a measure of how many times molecules are bouncing against a surface. This can be expressed in the formula $P = F/A$. P is the pressure being exerted by the molecules bumping against the piston, F is the total force on the piston, and A is the area of the piston. The units used to measure pressure in the metric system are pascals (Pa).

Hydraulic devices use liquids to transfer pressure from one point to another. A device called *a hydraulic press* is used to lift dentist chairs, hospital operating tables, and cars at repair shops. How does it work? A small force, such as a person's foot stepping on a pedal, is applied to a small piston at one point in the system. According to Pascal's principle, this small force affects all the fluid in the system and translates into a large force (F) at another point in the system, where there is a piston with a much larger area. The larger force is used to move the larger area, like the dentist chair or an auto lift.

LAWS & PRINCIPLES

Laws & Principles

The pressure on a fluid is equal to the force divided by the area.

$$P = F/A$$

Force on a fluid can be increased by increasing the pressure or the area over which it is distributed.

$$F = P \times A$$

Density of a solid or fluid is the mass divided by the volume.

$$d = m/V$$

Forces in Fluids

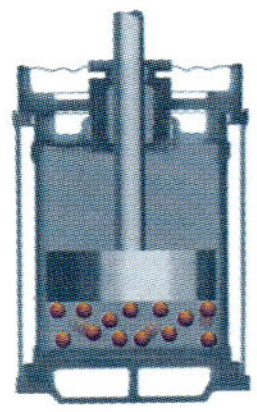

The molecules of liquid or gas are compressed inside the cylinder of this piston.

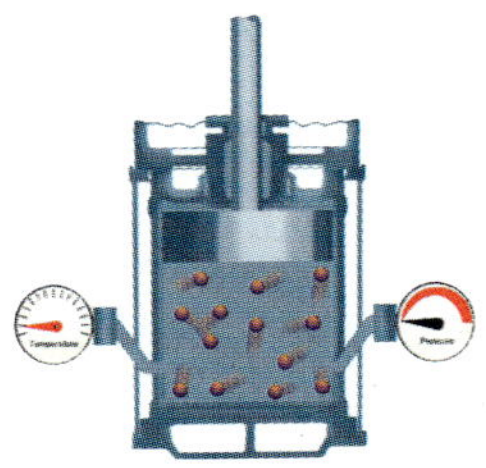

The molecules in the cylinder are not under much pressure.

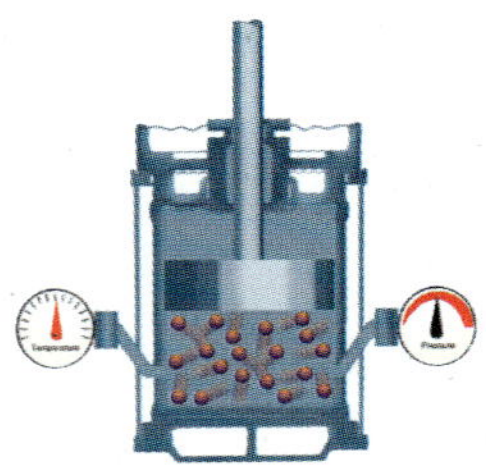

The increase in molecules and temperature and the compression of the piston raise the pressure inside the cylinder.

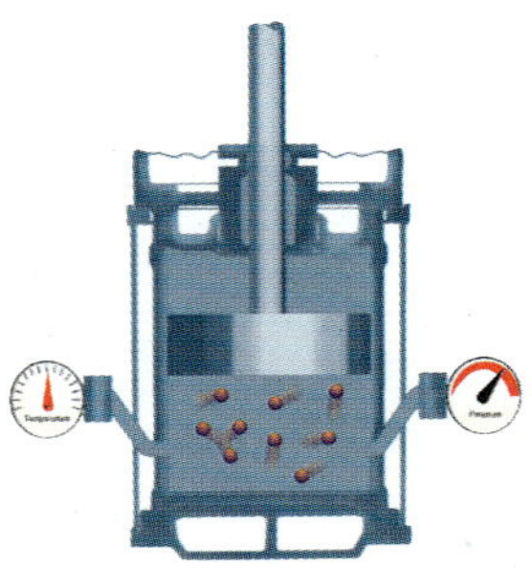

The larger cylinder has more room for more molecules to bounce around, adding to the pressure in the piston.

Piston

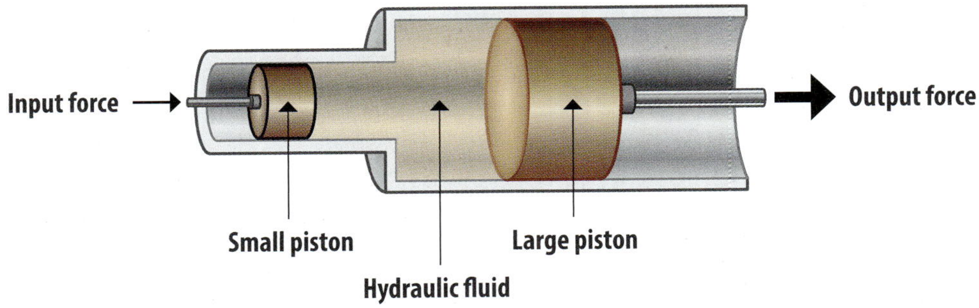

How could the force on the piston be increased? The total force exerted depends not only on the pressure but also on the size of the area being hit by the molecules. Using the equation $P = F/A$, the variables can be manipulated to solve for F. For example, $F = P \times A$. This means that total force can be increased by increasing the pressure, the area, or both. In the example, force can increase by increasing the surface area of the piston or by increasing the pressure. Increasing the surface area can be achieved by using a wider cylinder.

To increase the pressure, the molecules must hit the piston more often. This can happen in three ways. The first way is to increase the density of the fluid in the cylinder by adding more molecules. More molecules bouncing against the piston increases the pressure. The second way is to increase the speed of the molecules. The faster molecules travel, the more often they will bounce off the piston. Since the speed of the molecules in an object is directly related to temperature, the hotter the molecules, the faster they move around. So increasing the pressure with this method means heating the molecules. The third way is to compress the fluid in the cylinder. This is done by pushing the piston down and squeezing the molecules together into a smaller volume. Because the piston is moving as the molecules hit it, they bounce off faster than before. The same thing happens when a moving tennis racket contacts a tennis ball. The ball bounces off a moving racket with a faster speed than it bounces off a racket that is not moving. The moving racket has a greater velocity, which provides more momentum.

A greater number of molecules in a given space results in a greater density. Consider the density of two common fluids. Water has a density of 1 g/mL or 1,000 g/L; air has a much lower density, 1.29 g/L. Since water is so much more dense than air, the pressure exerted on an object as it travels down below the surface of the water is much greater than the pressure under the

blanket of air in the atmosphere. For example, the deeper a diver goes in the ocean, the greater the pressure.

What happens when additional pressure is exerted on a fluid? French scientist and mathematician Blaise Pascal determined that a change in pressure at any point in an enclosed fluid results in an equal change in pressure at all other points in the fluid. In other words, any additional pressure applied to a fluid is transferred in all directions throughout the fluid and even against the walls of the fluid's container. This fact is known as *Pascal's principle*. If a person jumps off a high diving board into a pool, the pressure from the landing will be felt by all the people in the pool around the diver, in every direction.

Fluids exert a buoyant force on objects that are totally or partially submerged in them. **Buoyant force** is the upward force exerted on objects in a fluid. Since pressure is greater farther down in a fluid, the force on the lower part of the object is greater than the force on the upper part. The result is an upward

HISTORY

Archimedes' Principle
Ancient Greek scientist Archimedes discovered how to measure the buoyant force on an object by immersing objects of the same mass but different volumes in water. He determined that the buoyant force acting on an object in a fluid is equal to the weight of the fluid that it displaces. This is called *Archimedes' principle*.

buoyant force. For example, an inflatable ball becomes more difficult to push down the deeper in water it goes. At some point, the buoyant force becomes so strong that the ball flies up out of the water. The buoyant force pushes back with an equal and opposite force to the push.

How is density related to buoyancy? First, the pressure exerted by a fluid (and therefore the force exerted by that pressure) depends on the density of the fluid. An object that is less dense than the fluid it is in will float. For example, Styrofoam, which has a density less than that of water, floats on water. An object with the same density as the fluid it is in hangs suspended within the fluid. An object that is more dense than the fluid it is in will sink. Furthermore, a fluid that is more dense than water, such as corn syrup, has a greater buoyant force. However, on the moon where the gravitational force is less than that on Earth, the buoyant force exerted by a fluid would also be less.

Buoyant force is the reason why some people with arthritis exercise in a swimming pool. Because the buoyant force of the water cancels out much of the force of gravity on their joints, they can exercise with less pain.

Steel alloys have a density of 7.75 g/cm^3 to 8.05 g/cm^3, which is eight times more dense than water. And yet ships are made of steel. The buoyant force is strong enough to keep them afloat

TRY THIS

Floating Egg

Fill a tall glass 3/4 full with water. Add a raw egg. What happened? Remove the egg and add 1 tsp of salt. Gently drop the egg in the water again. What happened? Repeat the procedure, adding 1 tsp of salt at a time until the egg actually floats in the water. How many teaspoons of salt were added? Why does the egg float?

BIOGRAPHY

Blaise Pascal

Blaise Pascal was born in France and lived from 1623 to 1662. After his mother died when he was three, he was raised by his father and sister. By the time he was 12, Pascal had mastered Euclid's Elements, a geometry text written by Euclid, a brilliant Greek mathematician. At 19, Pascal invented the adding machine. By the time he was 31, he was famous for his mathematical and scientific contributions.

During the night of November 23, 1654, Pascal had an experience that changed his life, which he did not mention to anyone. After his death, a servant discovered a piece of paper with these words sewn into the lining of his coat: "Fire. God of Abraham, God of Isaac, God of Jacob.... Certainty, certainty, heartfelt joy, peace. God of Jesus Christ.... Joy, joy, joy, tears of joy!" This experience led him to write down his thoughts in defense of Christianity. After he died at age 39, his thoughts were gathered into the book *Pensées*.

because the overall volume of the steel changes when it is formed into a ship. A large part of the volume of the ship is made of air, so the total density is much less than the solid steel block of the same volume. The big ship displaces large amounts of water, and the weight of that water equals the buoyant force exerted on the ship.

Swimming pool exercise

LESSON REVIEW

1. Explain how compression affects a solid, a liquid, and a gas.
2. How can the pressure of a certain gas be determined?
3. What factors increase the force in a hydraulic device?
4. Name three ways the pressure can be increased in a hydraulic device.
5. What is Pascal's principle?
6. Describe buoyant force.

Would Archimedes have believed this ship could float?

4.1.7 *Flight*

OBJECTIVES

- Apply Bernoulli's principle to flight.
- Describe how thrust, drag, lift, and weight work in flight.
- Explain how a plane accelerates.

VOCABULARY

- **drag** the frictional force of air molecules exerted on a flying aircraft
- **lift** the force that pushes an aircraft up
- **pneumatic device** a device that uses compressed gas to transfer pressure from one point to another
- **thrust** the force that pushes an aircraft forward

Interesting things happen to the pressure of fluids in motion. Swiss physicist Daniel Bernoulli (1700–1782) studied steadily flowing fluids. He determined the relationship of factors such as the pressure, speed, and elevation in a moving fluid, and the relationship between compressibility and viscosity as they contribute to the total energy of the system. Bernoulli discovered that a fluid's velocity increases when the flow is restricted. When you place your thumb over the end of a garden hose and restrict the flow of water, the water actually moves much faster. Plumbers use large pipes to bring large quantities of water into a house and then increase the velocity of the water for the faucets by connecting the larger pipes to smaller pipes.

Bernoulli also found that the faster moving fluid created an area of low pressure. Bernoulli's principle states that the pressure that fluids exert on their surroundings decreases as their speed increases. Liquids flowing through pipes will exert less pressure the faster they flow. As gases flow over and around objects, they exert less pressure on the objects the faster they flow. Have you ever noticed how the shower curtain moves closer to you while

BIOGRAPHY

Amelia Earhart

Perhaps you have heard about the disappearance of aviator Amelia Earhart and her navigator, Fred Noonan, in the South Pacific in June of 1937. Had Earhart completed the flight, she would have been the first woman to fly around the world. Earhart was used to setting such records; in 1928, she became the first woman to fly across the Atlantic, a feat that cost three other female aviators and their crews their lives earlier that year. In 1932, she was the second person and first woman to fly solo across the Atlantic, and in 1935, she became the first person to fly solo across the Pacific. Born June 24, 1897, Earhart was a self-professed tomboy who preferred outdoor activities like climbing trees and hunting to the conventional activities for girls in the early 1900s. She left her girls' finishing school midway through her second year to work as a nurse's aide at a military hospital during World War I. Following the war, she attended college and became a social worker in Boston. She began taking flying lessons in 1921 and bought her first plane six months later. Seven years later, she flew across the Atlantic for the first time.

Since her disappearance, many scientists have searched for clues to what happened to Earhart. In 2015, an expedition attempted to search near Nikumaroro Island for Earhart's plane using an ROV, a remotely operated vehicle, but was unsuccessful. However, as undersea search and forensic science technology improve, scientists may be able to finally determine what happened to Earhart and her navigator.

LAWS & PRINCIPLES

Bernoulli's Principle

The pressure that fluids exert on their surroundings decreases as their speed increases.

Bernoulli's Principle and Water

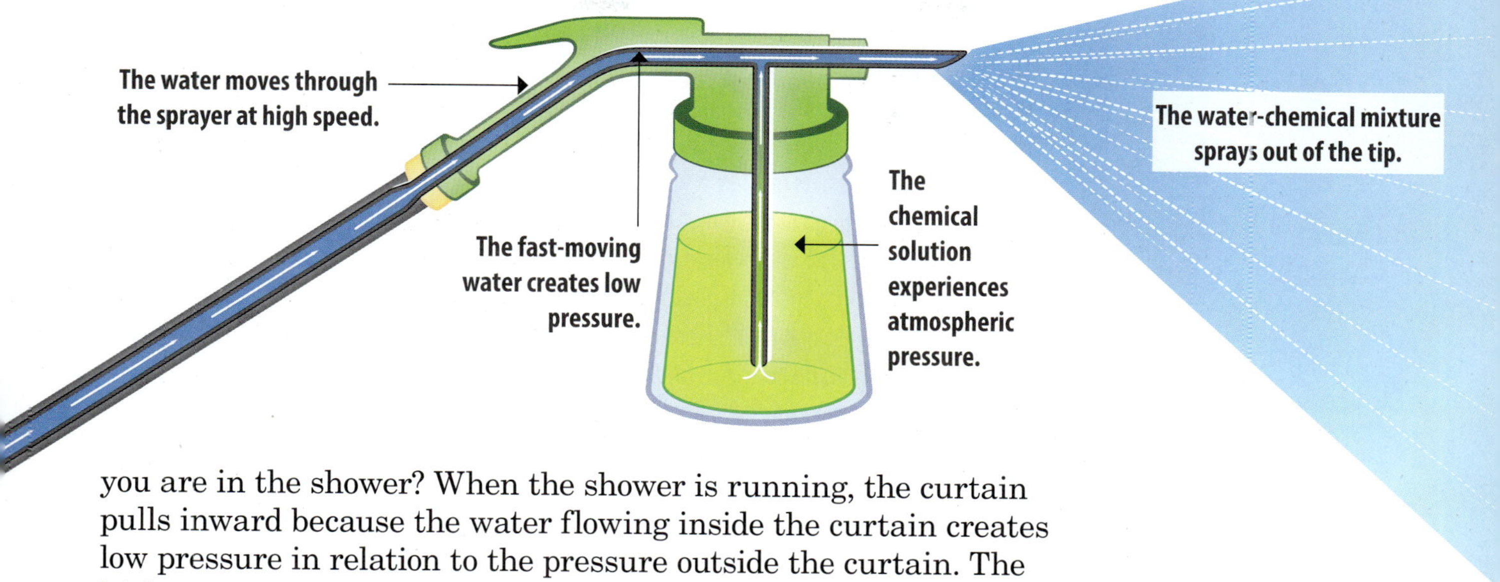

you are in the shower? When the shower is running, the curtain pulls inward because the water flowing inside the curtain creates low pressure in relation to the pressure outside the curtain. The higher pressure on the outside pushes the curtain in.

Bernoulli's principle is used in designing aircraft wings. The specific shape is called *an airfoil.* The air flowing over the upper curved surface of an aircraft wing moves faster than the air beneath the wing. This makes the pressure underneath greater than that on the top of the wing. The difference in pressure causes **lift**, a force that pushes upward on the wings and makes flight possible. Fluid flow principles are also used in designing ships; submarines; and streamlined, fuel-efficient cars.

Another factor that causes lift is the slant of the wings. According to Newton's third law, the slant of the wings forces air down, and the equal and opposite reaction force is exerted up on the wing.

Bernoulli's Principle and Air

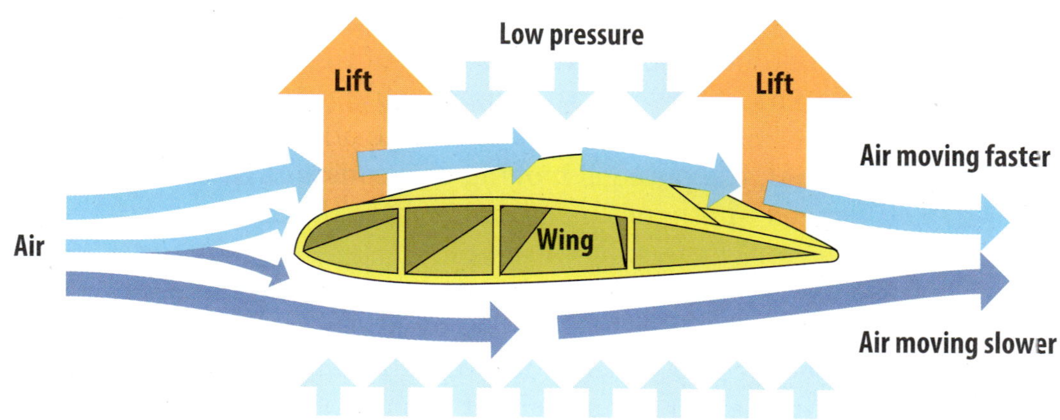

SCIENCE IN ACTION

Physics of Sailing

Sailboats also depend on Bernoulli's principle. Boats of all kinds are subject to it while traveling through the water, but the sail on a sailboat acts just like a wing, creating lift. If a sailor only wanted to go in the direction of the wind, he or she would not need lift. The wind would just push the boat along. But, how is sailing into the wind possible? The boom, or pole, at the bottom of the sail is maneuvered to "catch" the wind so it will actually pull the boat. It is important for the sailor to maintain the curve in the sail. Remember, the wind moves faster over the rounded part of the sail, creating low pressure. What does that do on the other side? It creates higher pressure that pushes the sail and moves the boat. Since there is almost always a sideways force, the keel keeps the boat from moving sideways. This fin provides drag to resist sideways motion and keep the boat sailing forward and into the wind.

In addition to lift, three other forces are working during flight: gravity, or weight; thrust; and drag. Weight works in opposition to lift. If it is greater than the lift, the plane will descend. If the plane begins its flight with too much weight, it will not be able to create enough lift to take off. How does the plane create enough lift to take off? Consider Bernoulli's principle. The speed of the air over the top of the wing has to create extremely low pressure. To increase the speed of air over the wing, the plane has to reach a certain velocity. This is why an airplane speeds down a runway. It continues to accelerate until it creates enough lift to get off the ground.

Thrust is generated by the engines and is the force that pushes the plane forward. **Drag** works in opposition to thrust and is the friction created by air molecules. Imagine how much drag would be generated if the front of the airplane were a flat square. Too much drag on the plane would prevent its ability to create lift. That plane would never take off. When thrust and drag are equal, the plane is moving at a constant rate. For the plane to speed up, thrust has to be greater than drag.

The motion of an airplane relies on the relationship of these four forces. When all are balanced, the plane cruises at a constant speed. Suppose the pilot wants to maneuver the plane while it is in the air. Consider that when driving on the ground, a driver can move a car side to side by turning the wheel, or front to

HISTORY

History of Flight

Louis Blériot's first flight across the English Channel, 1909

The thought of flying is exhilarating, and the desire to fly has probably existed since the beginning of time. Watching the path of a soaring bird delights the heart and inspires the mind. Many have dabbled in the possibilities and contemplated what it would take to make human flight a reality. Leonardo da Vinci drew plans for a flying machine as far back as 1490. For centuries, man attempted to fly by emulating the flapping wings of a bird, but the attempts were futile. Finally, in the late 19th century and early 20th century the dream of human flight became a reality.

But those who hope in the LORD will renew their strength. They will soar on wings like eagles; they will run and not grow weary. Is. 40:31a

1884 Horatio Phillips of England designs the airfoil.

1896 Samuel P. Langley launches a steam-powered model aircraft over the Potomac River, with flights reaching 0.75 of a mile.

1901 In a speech Wilbur Wright makes to a group of Chicago engineers, he notes three general difficulties in sustaining flight:

1. The construction of sustaining wings
2. The generation and application of power to drive the machine through the air
3. The balancing and steering of the machine after it is actually in flight

The Wright brothers have much of what they need to be successful, building on over four centuries of research from Leonardo da Vinci and Galileo Galilei of Italy, Christiaan Huygens from the Netherlands, Sir Isaac Newton in England, Daniel Bernoulli and Leonhard Euler from Switzerland, John Smeaton from England, and many others.

Wilbur and Orville Wright make their famous flight on December 17, 1903, at Kitty Hawk, North Carolina. They actually make four flights that day and only consider the last one a true flight. They fly for 59 seconds and travel 260 meters.

Their accomplishment ushers in the age of flight.

1906 A patent is issued for the brothers' flying machine.

1908 Madame Thérèse Peltier is the first woman to fly solo.

1908 US War Department purchases a flying machine from the Wright brothers.

1918 The US post office inaugurates airmail service.

1924 US Army Service planes avert a flood in Nebraska by dropping bombs to clear an ice jam.

1932 Amelia Earhart flies solo across the Atlantic.

1947 The US Air Force is established as an equal element of the US armed forces, no longer a division of the US Army.

1950s The Space Race begins.

1954 The first Air Force One, private plane for the president of the United States, is inaugurated for service.

1964 Geraldine Mock becomes the first woman to fly around the world.

1994 Test flight of Boeing 777 occurs, the first aircraft to be designed entirely on a computer.

2010 Bertrand Piccard and Solar impulse team fly more than 24 consecutive hours nonstop in a solar airplane.

2016 FAA predicts that 7 million UAVs (unmanned aircraft vehicles) will swarm the skies by the year 2020.

Lift
Thrust
Drag
Weight

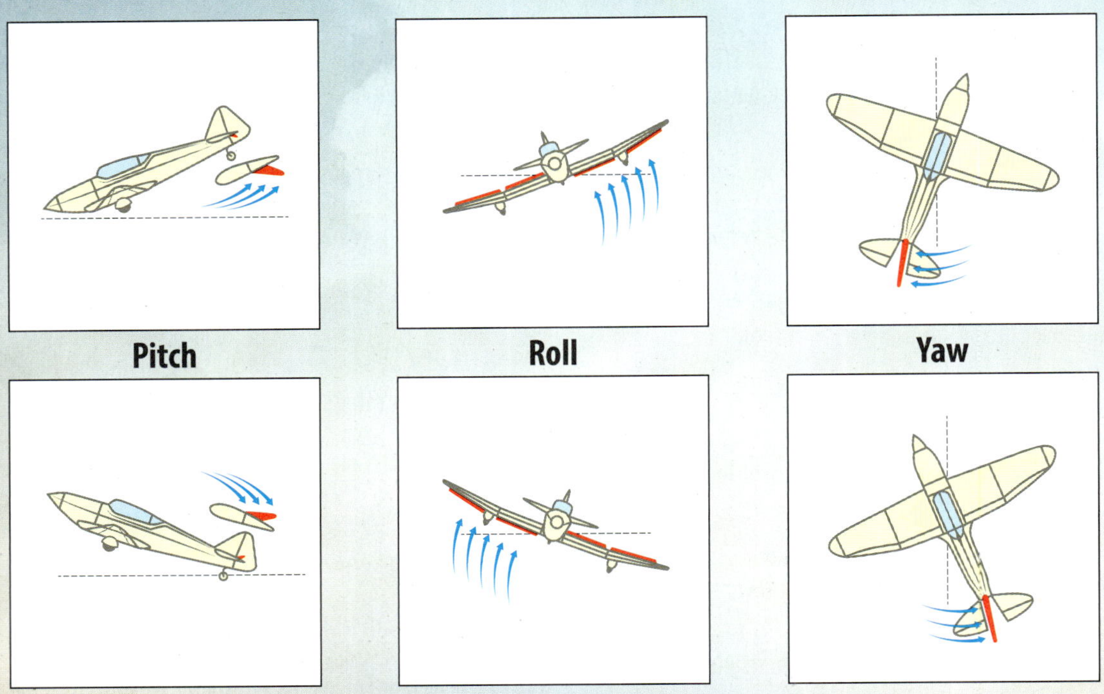
Pitch
Roll
Yaw

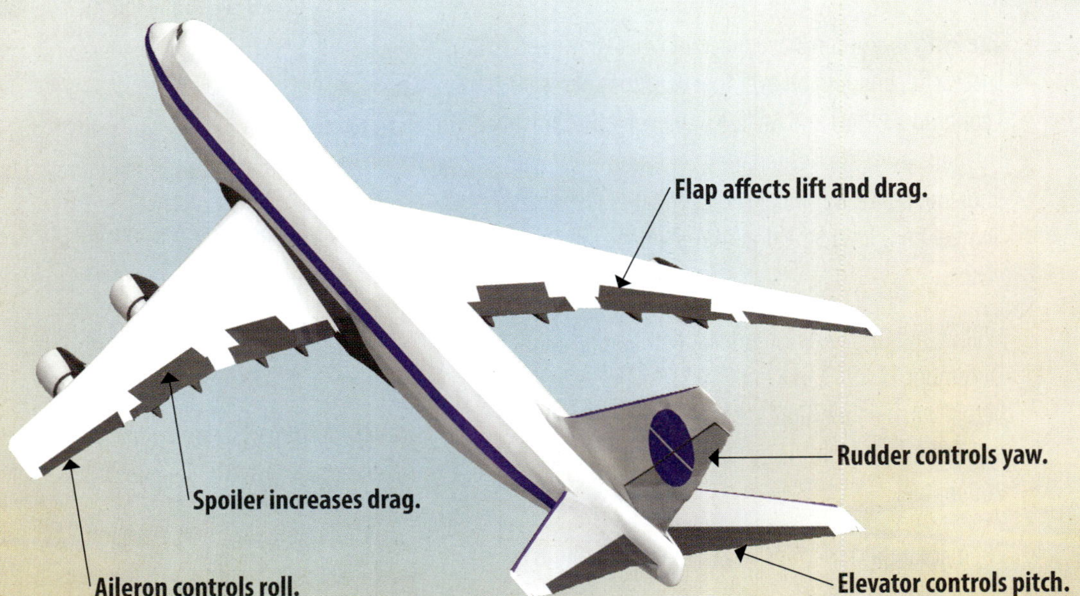
Flap affects lift and drag.
Rudder controls yaw.
Spoiler increases drag.
Elevator controls pitch.
Aileron controls roll.

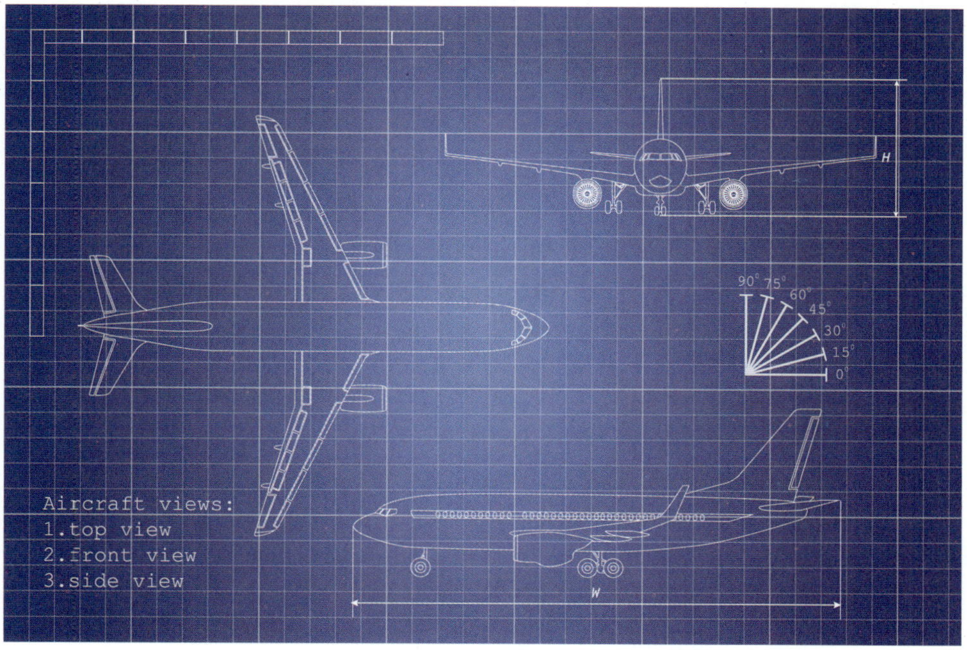

back by placing the car in drive or reverse. A pilot has different methods to create an unbalanced force and to accelerate. To change the yaw, or left-and right-movement, the pilot moves a rudder on the tail. To change the pitch, the up-and-down movement, the pilot adjusts the elevators on the tail of the plane. To roll the plane, the pilot moves a part of one wing up and a part of the other wing down. These parts are called *ailerons.*

Pneumatic devices also make use of Pascal's principle and Bernoulli's principle. Pneumatic devices are devices that use compressed gas to transfer pressure from one point to another. Pneumatic tools are connected with hoses to compressed air tanks. When a valve is opened at the end of the hose, the force of the compressed air moving from high pressure to low pressure provides the energy needed to make the tools work. Examples include rock drills, pavement breakers, and blast cleaners.

LESSON REVIEW

1. What is Bernoulli's principle?
2. How does Bernoulli's principle apply to flight?
3. Explain how lift, weight, thrust, and drag exert force on an airplane.
4. What creates acceleration, or a change in velocity?
5. Describe how a plane accelerates, or changes direction, using the terms *yaw*, *pitch*, and *roll*.

4.2.1 *Work and Energy Defined*

OBJECTIVES

- Correlate the terms *force*, *motion*, *work*, and *energy*.
- Calculate the amount of work done when given force and distance.
- Distinguish between potential and kinetic energy.

VOCABULARY

- **energy** the ability to do work
- **kinetic energy** the energy contained in moving objects
- **potential energy** the energy that is stored by way of an object's position, chemical makeup, or elasticity
- **work** the measure of the energy transferred when a force is applied to an object that causes it to move, and at least part of that force is in the direction of the movement

Climbing a ladder is work.

Words used in science are oftentimes the same words used in everyday language, but the definitions can be vastly different. The word *work* is one of these words. You may work hard to study for a test or to plan a science project. When you work as a babysitter, you might watch or check on a sleeping baby. A counselor's work includes talking and listening to people all day. Someone might work long and hard but unsuccessfully to push open a stuck door. The word *work* is used to describe these different scenarios, but according to scientists, these situations are ones where no actual work is being accomplished.

In science, **work** is the measure of the energy transferred when a force is applied to an object that causes it to move, and at least part of that force is in the direction of the movement. You may do lots of thinking when you do your homework, but according to scientists, you have not done any work because you have not moved an object. A person may exert great force trying to push a broken-down car, but unless the car actually moves, no work has been done. If you lift your backpack two meters, you have done twice as much work than if you lifted it one meter; but if you hold your backpack for 20 minutes, you have not done any work,

Doing Work

Example	Direction of force	Direction of motion	Doing work?
	↑	→	No
	→	→	Yes
	↘ (components: → and ↓)	→	Yes
	↑	↑	Yes

no matter how heavy the backpack is! When you walk down the hallway while wearing the backpack, you are not doing work on the backpack because the force on the backpack is upward but the motion is forward.

The amount of work can be calculated by using the following formula, assuming that the force is constant:

$$\boldsymbol{W = F \times d}$$

In this formula, W is the work done, F is the force in newtons, and d is the distance the object moved in meters. Only the distance traveled in the direction of the force counts. The unit for the product of the newtons and meters is *the newton meter* (Nm), or *the joule* (J). A kilojoule (kJ) is 1,000 joules.

Doing work requires energy. In fact, **energy** is defined as the ability to do work. You cannot do work without using energy, and you cannot give anything energy without doing work. The two go hand in hand. A car gets its energy from gasoline; when it runs out of gas, the car stops and no longer has the energy to do the work of overcoming friction. Animals get energy from the food they eat. When they have used all the energy from the food they have eaten, they must eat again.

Energy can be either kinetic or potential. **Kinetic energy** is contained in moving objects. Objects with kinetic energy can do

A bowling ball knocking down bowling pins is doing work.

TRY THIS

Tabletop Bowling

Set up dominoes in a triangular configuration, similar to bowling pins. Create a bowling lane with the edges of books. Take turns rolling a marble down the lane to knock over the dominoes. Is any work being done? Explain. Identify all instances of potential and kinetic energy.

LAWS & PRINCIPLES

Work

If a girl with a mass of 40 kg climbs a vertical ladder that is 5 m high, how much work does she do?

Find the weight of the girl in newtons.

$$40 \text{ kg} \times 9.8 \text{ m/sec}^2 = 392 \text{ N}$$

Find the amount of work the girl did.

$$\boldsymbol{F \times d = W}$$

$$392 \text{ N} \times 5 \text{ m} = 1{,}960 \text{ J}$$

The girl used 1,960 J of energy to climb the ladder, 1,960 J of work.

work. When you roll a bowling ball down the lane, you do work on the ball to cause motion; the moving ball has kinetic energy. As a result, the ball can do work on the pins. **Potential energy** is energy that is stored by way of an object's chemical makeup, elasticity, or position. The food you eat, a stretched rubber band, and books on a bookshelf all have potential energy.

Consider the position of two books in a bookcase. They have equal mass, but one is on the top shelf and the other is on the bottom shelf. According to the equation $F = m \times a$, each book would exert the same force if it were to fall to the ground. However, the potential energy stored in each book is different because of the distance each would fall. Experience would suggest that the amount of work gravity exerts on the book from the top shelf would be greater than the amount of work exerted on the book from the bottom shelf. And if both books landed on your foot, you would know which one fell from the higher shelf!

Even though energy is invisible and untouchable and cannot be measured directly, energy can be measured, with great accuracy, indirectly. It is not a substance, but it is real. Like work, energy is measured in joules.

LESSON REVIEW

1. What is work?
2. According to the scientific definition, which of these is work: picking up a pencil or holding a heavy box? Why?
3. How much work is required to push a 29 N object over a distance of 6 m?
4. What is energy?
5. Explain the difference between kinetic and potential energy.

FYI

Get the Ball Rolling

Have you ever heard someone say, "Just get the ball rolling?" People understand that the hardest part of doing a project is starting it. This phrase implies that it is easier to keep something in motion than it is to get it moving in the first place. Imagine the amount of energy needed to get a large boulder moving. It may take two or three people. But once it is moving, it may only take one person to keep it going. A ball has potential energy when it is just sitting, and it is also subject to static friction. As soon as enough force is added to overcome static friction, the ball has kinetic energy, and the kinetic friction is much less resistant to motion. The work involved to start the ball rolling requires a large amount of energy. Once the ball is rolling, however, the work is much easier.

Books on a higher shelf store more potential energy than those on a lower shelf.

The whole stadium feels tense with the strain of the crowd, watching and waiting, as the timer counts down the final moments of the game. The teams have been tied at 2-2 since the first quarter. With 10 seconds left on the clock, a player in red blasts through the defense and scores! The crowd goes wild and the stadium erupts with deafening cheers! The home team has triumphed! You, however, missed the excitement. You have just returned from the concession stand, and all you see is the ball rolling out of the goal.

OBJECTIVES

- Identify various forms of energy.
- Explain the transfer of energy in a given system.
- Calculate kinetic and gravitational potential energy.

What actually happened before that? What did it take for the ball to roll out of the net? Where did the ball get the energy to move? The net transferred its elastic potential energy to the ball. Before that, the ball flew into the net and transferred its kinetic energy to the net. The soccer player kicked the ball, using his mechanical energy to move the ball. The player ate a big bowl of pasta before the game, taking the chemical energy of his food and converting it to thermal energy to heat his body, electrical energy to control his nervous system, and mechanical energy to move his muscles. Where did his food get the energy? The pasta is made from grains that get energy from the sun. The meat on the pasta came from animals that eat the grain that gets energy from the sun. Nuclear fusion creates the radiant energy released from the sun.

Energy has many forms and is useful not only because of the many forms in which it is found but also because these original forms can be converted. This means that energy has many practical uses.

FYI

Wind-Up Toys

What do a robot, a bird, and a monkey all have in common? They can be made into wind-up toys along with hundreds of other mobile objects and animals. Wind-up toys get the energy to move by a spring being wound up, which builds up elastic potential energy. Mechanical energy is used to wind the spring. The more it is wound, the more potential energy it has. When the key is released, the spring unwinds and the elastic potential energy is converted into kinetic energy so the toy can move.

Chemical potential energy is stored in the chemical bonds that hold atoms together in molecules. For example, energy stored in gasoline can be converted by combustion (burning) into kinetic energy to move a car. Burning wood provides thermal energy to cook food. The chemical energy in food changes into thermal, electrical, and kinetic energy for the body to use.

Elastic potential energy is stored in a stretched spring or other elastic object. Elastic bands and diving boards have this type of energy when they are stretched or deflected. When the diver bounces on the board, the elastic energy of the board is returned to the diver, giving her the kinetic energy she needs for the dive.

Gravitational potential energy is stored in an object that has been lifted to some height. The greater the height, the greater the potential energy. Waterfalls convert gravitational potential energy to kinetic energy as the water falls. Hydroelectric facilities change the kinetic energy into electrical energy.

Electrical energy is the kinetic energy of electrons. The attraction and repulsion of electrical charges causes electrons to move. Their energy can be transferred to kinetic energy for motors in DVD players, blenders, and power tools.

Radiant energy from the sun allows plants to create chemical energy through the process of photosynthesis, which they store in starch. The energy from the sun also provides light and warmth

A campfire is an example of chemical potential energy.

SCIENCE IN ACTION

Gravitational Potential Energy

To calculate the potential energy of gravity, use the weight of the object in newtons and the position of the object in relation to the ground. For example, if a rock with a mass of 400 kg sits on a cliff that is 300 m high, what is its gravitational potential energy?

$$PE = m \times g \times h$$

m = mass of the object, g = acceleration rate of gravity (9.8 m/sec^2), h = height

PE = (400 kg × 9.8 m/sec^2) × 300 m = 3,920 N × 300 m = 1,176,000 Nm = 1,176,000 J = 1,176 kJ

Some energy can be converted into sound. Energy vibrates the air and creates sound.

and can be converted to electrical energy. Solar energy provides a renewable resource.

Thermal energy is the total kinetic energy of the atoms that make up an object. The hotter the object, the faster the atoms move, and the more kinetic energy the atoms have to use.

Mechanical energy is the total of all potential and kinetic energy within an object or body. Anything, or anyone, that moves has mechanical energy.

God designed your body to be maintained through a constant transfer of energy. The chemical potential energy in your food becomes the heat to keep your body at a constant temperature.

Hydroelectric dam

TRY THIS

Warming Your Hands

Rub your hands together quickly. Continue rubbing for as long as you can. Have another student time you. Then switch roles. What happened to the temperature of your hands? Why did that happen? Why did you finally stop rubbing your hands together? Trace the energy conversions from the food you ate to the heat energy from the hand rubbing.

It also becomes electrical energy that sends messages along your nerves, and mechanical, or muscular, energy that allows your muscles to move.

How does energy change from one form into another? Consider a pendulum. When the weight on the end of the string hangs straight down, the pendulum has no kinetic energy because it is not moving. It also has the minimum possible potential energy. Lifting a pendulum is work, and the pendulum gains potential energy. When the pendulum is released and begins to fall, the potential energy changes into kinetic energy. By the time the pendulum is swinging through its lowest point, all the potential energy has been converted into kinetic energy—no potential energy is left. As the pendulum swings out toward the other side, however, it slows down. As it slows down, it loses some of its kinetic energy. At this point, the kinetic energy is being transferred into potential energy. At some point, the pendulum has converted all its kinetic energy into potential energy, and it stops moving. All the energy is again potential energy, and the pendulum begins to swing back once more.

The gravitational potential energy decreases as it moves lower and increases as it moves higher.

LESSON REVIEW

1. Give an example of chemical potential energy and how it is used.
2. What is thermal energy?
3. Why does a 2 kg book that is 3 m off the ground hurt more when it lands on your foot than a 2 kg book that is 1 m off the ground?
4. Use the example of the pendulum to explain how energy is transferred.
5. If an arrow has a mass of 0.02 kg and travels north at a velocity of 65 m/s, what is its kinetic energy?
6. How much gravitational potential energy does a bucket with a mass of 17 kg have if it is sitting atop a 2 m wall?

LAWS & PRINCIPLES

Kinetic Energy

The kinetic energy of a moving object equals one-half its mass times its velocity squared.

$$KE = 1/2 \times m \times v^2$$

What is the kinetic energy of a 300 kg object traveling at an average velocity of 10 m/sec to the south?

$$KE = 1/2 \times 300 \text{ kg} \times (10 \text{ m/sec})^2 = 150 \text{ kg} \times 100 \text{ m}^2/\text{sec}^2 = 15{,}000 \text{ J}$$

Energy is never really lost; it is simply converted into a different form. To better understand this, think about a roller coaster and the energy required to push it over the ride's many hills. A mechanical device often pushes the cars over the first hill, but after that the cars usually do not get any assistance. Therefore, each hill must be smaller than the previous one. Why?

At the top of the first hill, the roller-coaster cars have a lot of potential energy. The sum of this energy must carry it through the whole ride. As the cars go up and down the hills, the potential energy converts to kinetic energy and back into potential energy again.

The cars never return to the same height they started from because of friction. On the way down the hill, some of the cars' potential energy changes to kinetic energy, but some converts into thermal energy from friction. This leaves less kinetic energy to climb the second hill, so the second hill has to be shorter than the first hill. On the way down the second hill, more potential energy becomes thermal energy; the rest changes into kinetic energy that the cars need to climb the third hill. Although none of the original energy is lost, some of the mechanical energy is always converted into thermal energy through friction. The sum of the energy of the entire system—kinetic energy at the bottom of the first hill, sound energy, and thermal energy—equals the amount of the original potential energy at the top of the first hill. If all the energy remained as mechanical energy, the ride could

OBJECTIVES

- Identify forms of energy in a given system.
- Compare renewable and nonrenewable resources.
- Determine methods for conserving energy.

VOCABULARY

- **nonrenewable resource** a resource that cannot be replaced once it is used, or can be replaced only over long periods of time
- **renewable resource** a resource that is constantly available, or that can be replaced in a relatively short period of time

Roller coaster

CAREER

Roller-Coaster Designer

Imagine getting to design the fastest, the tallest, and the most extreme roller coasters ever, with tunnels, complete inversions, and death-defying drops and speeds. Very few people have this opportunity. About 100 companies in the United States design and build roller coasters for amusement parks. A roller-coaster designer has a degree in mechanical, electrical, or drafting engineering. Usually a team of several different engineers works together to design and assemble a roller coaster. Oftentimes, an amusement park will hire a company to build a roller coaster to fit in a certain space, carry a certain number of people, and have specific design features, such as loops, drops, tunnels, or water. The designers have to create a roller coaster that meets all the criteria, is fun to ride, and is safe for passengers. When the roller coaster is complete, the designers get to test-ride it for free!

go on forever. But some of the energy is converted into non-useful forms, so the ride must eventually come to an end.

According to the law of conservation of energy, energy cannot be created or destroyed. In physics when something is conserved, the amount does not change. Although it cannot be created or destroyed, energy can be wasted. Leaving the door open in the wintertime when the furnace is running to keep the house warm or sitting in a long drive-thru line is using energy unnecessarily. Taking a longer route than necessary to get somewhere or turning on more lights than needed is using energy inefficiently.

Some energy sources, such as solar energy, a type of radiant energy, and wind energy, a type of kinetic energy, constantly replenish. These are **renewable resources** because they are continuously available or can be replaced in a relatively short period of time. The radiant energy of the sun is constantly created through nuclear fusion. As air masses move across the earth, temperatures change and produce wind. As long as the sun continues to burn in space, shining light and radiant energy

Nonrenewable Energy

Petroleum

Natural gas

Coal

Nuclear

on the planet, these two types of energy will be available. Energy from waterfalls and dams is also a type of renewable energy, one that is often converted into electrical energy.

Other types of energy come from **nonrenewable resources** that cannot be replaced once they are used or can be replaced but only over long periods of time. These are mostly chemical forms of energy such as coal, gasoline, and other petroleum products. Old-earth scientists believe that these products form from organic matter over long periods of time and under intense pressure. Although God has placed great quantities of these resources on Earth, in the past 100–150 years they have been used in large quantities. Scientists speculate as to how long these resources will last.

Much of industrialized society depends on nonrenewable forms of energy, so energy conservation is a very important topic. Using public transportation, such as buses, subways, and trains, is more efficient than using a personal car. Also, the heating systems of most North American houses rely on petroleum products. Although one alternative is to use electric heat generated at nuclear power stations, these stations create radioactive waste, an environmental hazard. Wood is not a good alternative because forests take decades to renew, and the ash and carbon dioxide produced by burning wood contribute to air pollution.

Electrical energy is needed for computers, lights, and hundreds of appliances, including washing machines, hair dryers, televisions, toasters, and microwaves. Waterfalls and dams provide a strong renewable resource for generating electricity. However, there are not enough waterfalls to supply the large quantities of electricity that are needed. This means that electricity must be generated in other ways, including by using nuclear power plants and by burning coal or natural gas, all of which are nonrenewable.

TRY THIS

Conserving Energy

Efficiency is a key to conserving energy. Design a pendulum system that uses string, weights, marbles, paper cups, dominoes, and a metric ruler to dump 20 g of rice out of a paper cup into a container. The goal is to use just enough energy to get the job done. Use each of the supplies in the smallest amount possible. The energy must travel 100 cm and start with the gravitational potential energy of the pendulum and finish with the rice pouring into the container. Illustrate and label the different forms of energy in your system. Write a summary of your process, detailing how you conserved energy. List any additional materials used. Compare your system to other systems in the class. Which system used the least amount of energy? Which system used the most?

Renewable Energy

Hydropower

Wind

Solar

Bioenergy

Oil Production by Region

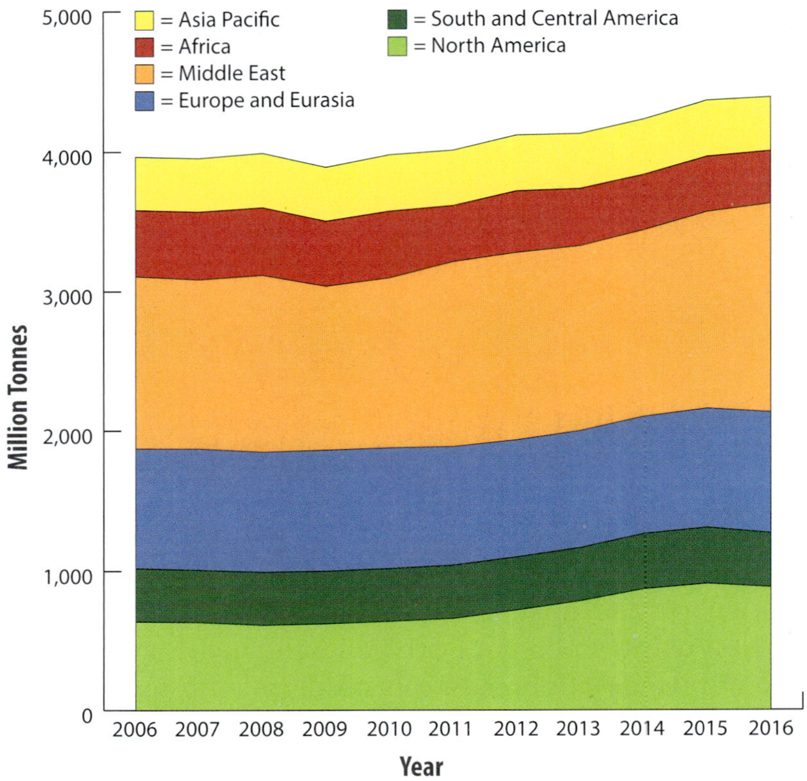

Oil Consumption by Region

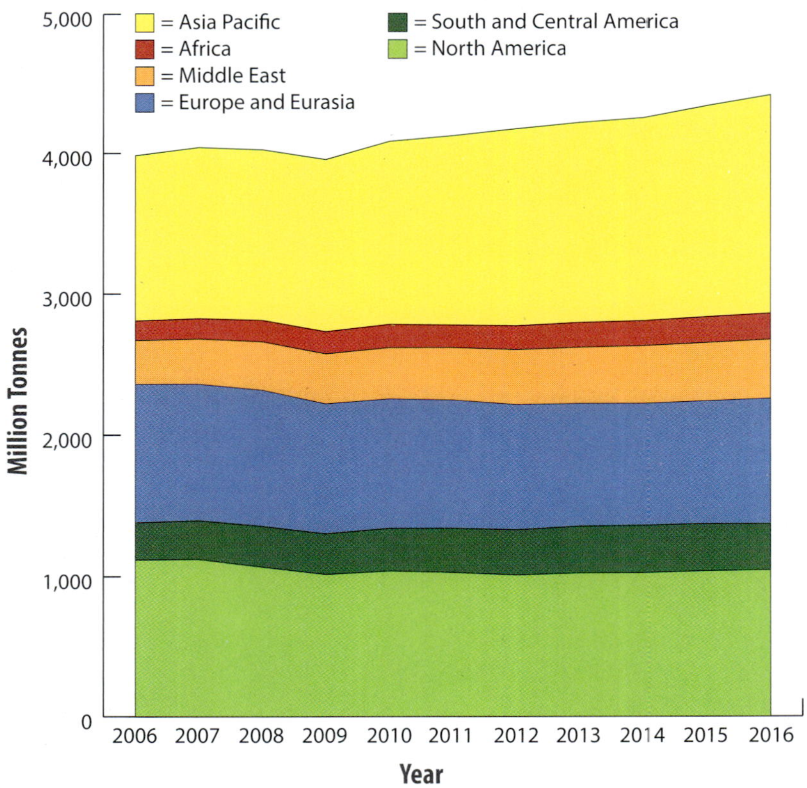

A tonne is one metric ton. A tonne of oil is equivalent to about seven 42-gallon (159-liter) barrels of oil.

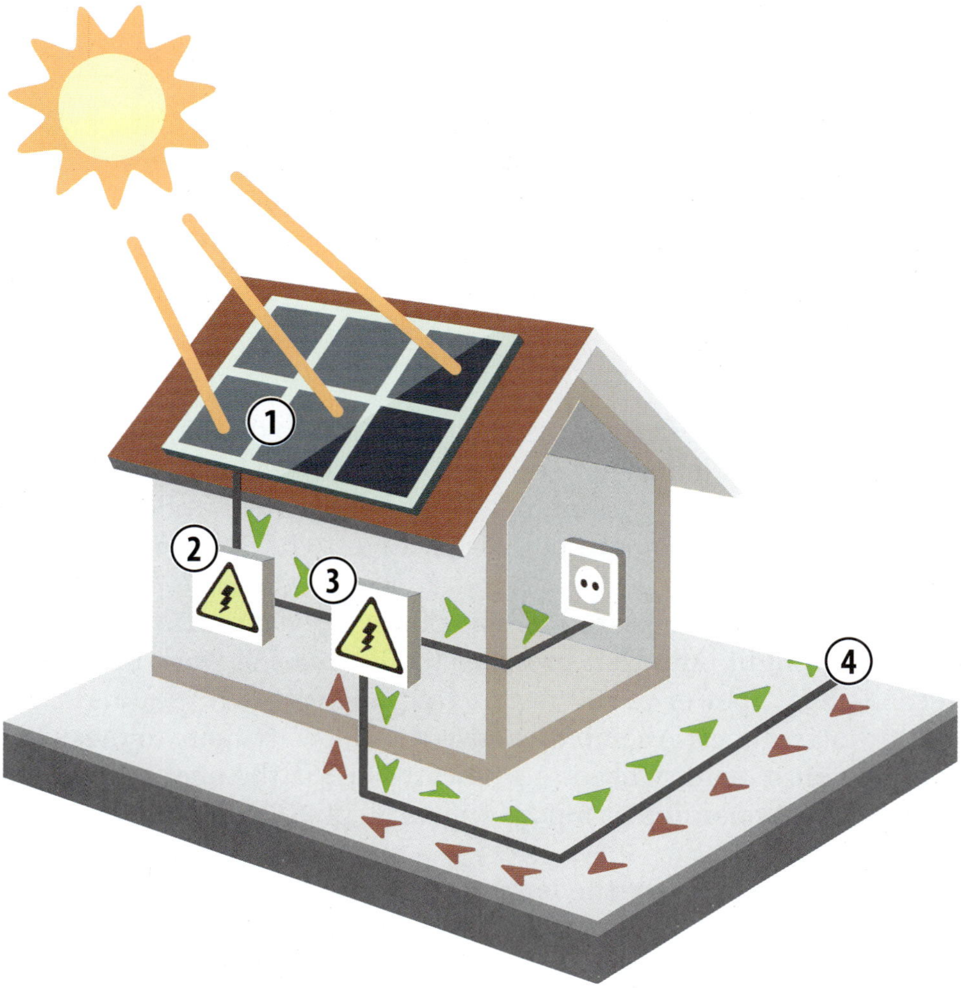

1. Solar panels convert sunlight to DC electricity.
2. Inverter converts DC into AC for household use.
3. Electricity passes through the breaker box and is distributed throughout the house. If the solar panels do not produce enough electricity, power is taken from the grid.
4. If solar panels produce more energy than needed, the excess can be sent to the grid.

More people and industries are turning to solar power and wind power to generate electricity. What deters increased use of these renewable energy sources? Changing requires large amounts of capital—monetary funding—which makes it cost prohibitive for many. Researchers and technicians continue to develop more cost-efficient technologies for harnessing solar and wind energy. Research and development often has high costs, but the future benefits make it worthwhile.

LESSON REVIEW

1. What is the law of conservation of energy?
2. Identify the energy at different points in a roller coaster's path and explain why each hill is smaller than the previous hill.
3. If energy cannot be created or destroyed, explain how it can be wasted.
4. Describe the differences between renewable and nonrenewable energy sources, and give an example of each.
5. Why are more nonrenewable than renewable energy sources used?
6. List several ways to conserve energy in a home.

4.2.4 Power, Machines, and Force

OBJECTIVES

- Define power in terms of work and time.
- Describe the benefits of machines.
- Calculate mechanical advantage.

VOCABULARY

- **effort force** the force needed to do the same work with a machine
- **machine** a device that makes work easier by changing the size or direction (or both) of a force
- **mechanical advantage** a measure of how much a machine increases a force applied to it
- **power** the amount of work done or energy provided in a period of time
- **resistance force** the force needed to do work without a machine

Fluorescent bulbs and LED bulbs produce more light with less power than incandescent bulbs.

The word *power* is used in many ways. It is said that a storm packs a lot of power, that someone has a lot of muscle power, and that a government official is in power. In physics, however, **power** is the amount of work done or energy provided in a period of time. Average power can be calculated by using the formula:

$$P = W/t$$

P is power, W is work, and t is the time used to do the work. The unit for power is joules per second. One joule/sec is defined as one watt (W). The watt is named after James Watt, a Scottish engineer who made steam power practical to use in 1769.

Electrical devices are rated for the power they can deliver. For example, a light bulb that provides 6,000 joules of work in 60 seconds has a power of 100 watts. One kilowatt (kW) equals 1,000 watts. Electric companies use the kilowatt-hour to measure the amount of energy used. A kilowatt-hour is the energy used by a device that operates at a power level of 1,000 W for one hour.

Some tasks, because of the force of direction involved in the work, require a lot of power. A **machine** makes work easier by changing the size or direction (or both) of a force. For example, lifting a treasure chest straight up out of a hole would be more difficult than tying a rope around it and using a pulley suspended from a tree branch to lift it. The pulley, which makes the work easier, is a machine. Machines lessen the workload in one of four ways.

Pulleys change the direction of a force. Pulling down is easier than pushing up because the work is in the same direction as the force of gravity.

Gears on a bicycle reduce the size of the force that must be applied to move an object over a distance.

Levers transfer forces from one place to another or increase the size of the force. A shovel, acting as a lever, transfers the force of your arm pushing down on the handle end to the blade end. Car jacks increase the size of a force. A small force on the handle of the jack translates into a large force on the body of a car, raising the car off the ground.

The force needed to do work without a machine is called the **resistance force** or the load force. The force needed to do the same work with a machine is called the **effort force**. If a person putting a 20 kg box into a truck lifts the box straight up from the ground and places it in the truck, the force of gravity acting on that box is 200 N. (Remember: $F = m \times a$ and the acceleration of gravity is approximately 10 m/sec^2.) Therefore, the resistance force is about 200 N. If the person pushes the box up a ramp into the truck, the effort force is less than 200 N because the box is not being pulled straight up. The effort force might only be around 40 N. Pushing 40 N is much easier than lifting 200 N, and the ramp acts as a machine to make the work easier. Keep in mind that the 40 N is exerted over a longer distance. The ramp, like all machines, offers a trade-off—a smaller force can be applied, but it must be applied over a longer distance. Although the work is easier, the same amount of work is done.

Bicycle gears reduce force.

A machine can increase the force that is applied to it. By using a machine, a load of 200 N could ideally be moved up a ramp with an effort force of 40 N. A measure of how much a machine increases a force applied to it is called the **mechanical advantage** (*MA*). The mechanical advantage can be calculated by dividing the resistance force by the effort force.

MA* = *resistance force / effort force

In this case, the mechanical advantage of using the ramp is 5 (200 N/40 N). What does that mean? It means that the machine magnifies the effort by a factor of 5. But this does not take into account the friction that is acting on the box as it is pushed up the ramp. The calculated mechanical advantage that does not take friction into account is called *ideal mechanical advantage* (*IMA*). To truly know the mechanical advantage, the resistance of friction needs to be subtracted. Friction always works in opposition to the force.

TRY THIS

Bulb Swap

Take an inventory of the incandescent bulbs in your school. Record the wattage for each. Research fluorescent or LED replacements for those bulbs and record what the new wattage would be if all the incandescent bulbs were replaced. Calculate and compare one hour of power usage of the old bulbs to the new bulbs. Could you save energy by swapping out bulbs?

BIBLE CONNECTION

God's Power

God chose Gideon to lead the Israelites against the Midianites (Judges 6–8). But Gideon did not feel powerful enough to do what God asked. God consistently reassured Gideon that He had a plan that included Gideon leading the attack. The only work God required of Gideon and his 300 men was to break vases, blow trumpets, and yell. The actual defeat came through God's power as He set the Midianites against themselves and sent them running into the night. When God calls His people to action, He provides the power to do the work.

FYI

Item	Wattage
cellphone charger	10 W
Blu-ray or DVD player	15 W
tablet	10–20 W
32" LED/LCD TV	50 W
laptop computer	20–75 W
video game console	40–140 W
42" plasma TV	240 W
refrigerator	500–750 W
washing machine	500–1,000 W
coffee maker	600–1,200 W
vacuum cleaner	300–1,500 W
toaster	800–1,500 W
hair dryer	1,000–1,875 W
microwave	1,000–2,000 W

LESSON REVIEW

1. What is power?
2. What is the power of a machine that does 3,000 joules of work in 60 seconds?
3. Compare resistance force and effort force.
4. How do machines make work easier?
5. What is the mechanical advantage of a machine that uses an effort force of 25 N to lift a 125 N load? What does that mean?

LAWS & PRINCIPLES

Power

Power is the amount of work done or energy used in a given amount of time. The unit for power is called *a watt*, which is defined as the number of joules used in one second.

$$P = W/t$$

If a drill runs for 10 seconds and uses 500 J, the amount of power is 500 J/10 sec = 50 W. How many watts would the drill need if it used 1,000 J in 10 seconds? 1,000 J/10 sec = 100 W. What if you used a 200 W drill for 10 seconds? How many joules of energy would it need? 200 W × 10 sec = 2,000 J

Efficiency and Friction 4.2.5

Suppose you and your classmates have chosen to give gifts to the children at a local shelter. You have 100 shoe boxes to fill and only two hours to get the job done. How do you set up all the items to go in the boxes? What order should the gifts be placed in each box? Should one person be responsible for one box, or should one person be responsible for one thing that goes into every box? All these questions have to do with efficiency. Someone who gets a lot of work done in a small amount of time is efficient. Machines also have efficiency. The **efficiency** of a machine measures the amount of useful work done by the machine, *work output*, compared to the amount of work done to make the machine operate, *work input*. When your class becomes a "human machine" assembling 100 gift boxes, as much of the work input as possible should go into the making of the gift and not into moving things around and walking from one place to another.

OBJECTIVES

- Evaluate the efficiency of a machine.
- Identify factors that contribute to friction.
- Distinguish between ideal and actual mechanical advantage.

VOCABULARY

- **efficiency** a measure of the amount of useful work done by the machine compared with the amount of work done to make the machine operate

Think about lifting a 20 kg box 1.5 m into the back of a truck. The resistance force is about 200 N (9.8 m/sec^2 × 20 kg). The work output can be calculated by multiplying the resistance force times the distance (200 N × 1.5 m). In this example, the work output is 300 J.

work output** = **resistance force** × **distance

Calculating the work input requires knowing how much work is accomplished by a machine. Suppose you pushed the same box with 40 N of effort force up a 10-meter ramp to the truck. The work input would then be calculated by multiplying the effort force (40 N) by the length of the ramp (10 m). The work input equals 400 J.

work input** = **effort force** × **distance

The efficiency of the machine (the ramp) is then calculated as the ratio of the work output to the work input and expressed as a percentage.

efficiency** = **(work output / work input)** × **100%

The ramp's efficiency equals 75% (300 J/400 J × 100%). What does this mean? It means that 75% of the work put into moving the box actually went into moving the box. The work (400 J) that you had to do to get the box up the ramp was actually more than that required to lift the box straight up (300 J). Even though the amount of work increased, it was easier for you to do the work because only 40 N of force had to be exerted instead of 200 N.

Machines are designed with efficiency in mind.

Robots can do repetitive tasks very efficiently.

No machine is 100% efficient. Forces cause some of the energy to be converted into waste forms. For example, in moving the box up the ramp, some of the energy is converted into heat instead of kinetic energy. Friction converts kinetic energy into heat because it resists motion in two objects that are moving against each other, and it is the biggest factor contributing to the inefficiency of machines. Increased friction converts even more of the kinetic energy of the moving object into heat. And in mechanical systems, heat is usually considered a waste of energy.

Deep treaded tires add friction in the snow.

Two things cause friction: the attraction of molecules on the two surfaces that are in contact with each other and the physical interlocking of irregularities of the two surfaces. The surface of ice is very smooth with few irregularities; people slip on ice because ice has little friction.

The amount of friction that occurs when two surfaces rub together depends on several factors. First, it depends on the characteristics of the surfaces themselves: whether they are rough or smooth, soft or hard, slippery or sticky. Second, it depends on how much force is pushing the surfaces together. Objects with more mass naturally provide a greater force as gravity pushes that mass down. You can increase the friction of a piece of sandpaper on wood if you press down harder on it. Heavy

FYI

Hovercraft

Would you take a boat to travel across land? Would you try to drive a car over a lake? Are these questions a little absurd? Of course they are. Boats and cars are designed to move over a specific surface. A boat is designed to stay afloat and move against the friction of the water. Car tires provide just enough friction to push against the road and propel the car forward. A hovercraft can do both. The air cushion created by a hovercraft reduces so much friction that it can skim across land and water. Fans on the back of a hovercraft, or air cushion vehicle, force air beneath the vehicle to create an almost frictionless environment, and they force the air behind the vehicle, creating thrust. How does a hovercraft stop? What is the purpose of the skirt?

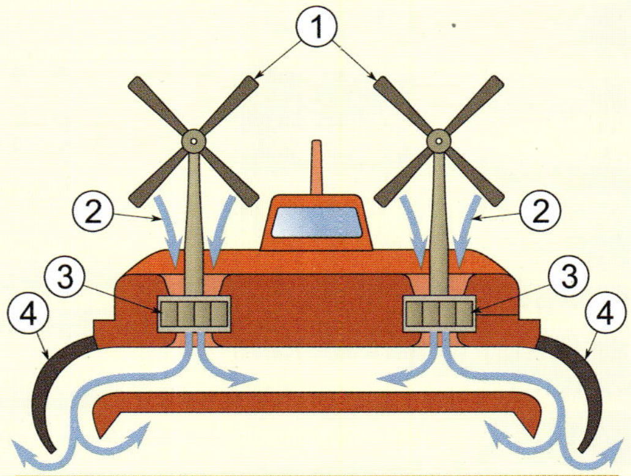

1) Propellers
2) Air currents
3) Fans
4) Flexible skirts

vehicles and deep treads on tires add the necessary friction to get through slippery snow.

Ideal mechanical advantage assumes that no friction is working against the operation of a machine. In reality, however, friction is always at work. So the mechanical advantage of a real working machine must take friction and other forces into account. The actual mechanical advantage (*AMA*) includes friction and any other forces working against the efficient operation of a machine. The efficiency of the machine measures how much friction reduces the ideal mechanical advantage. The actual mechanical advantage can be calculated by multiplying efficiency times the ideal mechanical advantage (*IMA*).

$$AMA = efficiency \times IMA$$

If the efficiency of a ramp equals 75% and the *IMA* is 5, the *AMA* is 3.75 (75% × 5 = 3.75). The actual mechanical advantage is always less than the ideal mechanical advantage.

LESSON REVIEW

1. What is efficiency?
2. What is the efficiency of a machine with a work input of 375 J and with a work output of 125 J?
3. What causes friction?
4. Why are roller blades not worn on an ice rink or ice skates on the sidewalk?
5. How might oil increase the efficiency of a machine?
6. What is the difference between ideal mechanical advantage (*IMA*) and actual mechanical advantage (*AMA*)?

TRY THIS

Friction Race

Friction occurs when two surfaces come into contact with each other. The more bumps or grooves the surfaces have, the more friction prevents motion. With the rough side of a board facing up, place an eraser, an ice cube, and a rock in a row at the short end of the board. Place that end on top of a book. Do any of the items start to slide? If yes, measure the height of the board and note which item started moving. Stack a book on top of the first one and raise the board to that height. If any items start sliding, record the height and the item. Continue to add books or increase the height until all items slide. Which item experienced the least friction? Which item experienced the most? Flip the board over so the smooth surface is facing up. At what height will each item start to slide? Write a hypothesis and test for answers. Did the objects experience more or less friction in the second trial? Why?

4.3.1 *Simple Machines and Inclined Planes*

OBJECTIVES

- Determine the effect of an inclined plane on work and force.
- Calculate mechanical advantage and efficiency.

VOCABULARY

- **inclined plane** a simple machine that has a slanted surface

How have machines affected people's lives throughout history? How have they changed society? The invention of the printing press, for example, permitted books to be printed quickly and made it possible for people to have their own Bibles. Before the the printing press existed, books were written by hand. The train and automobile ushered in a mobile lifestyle, and today the computer provides instant access to information and people all over the world.

History provides many other examples of machines. The plow is a machine that improved agriculture. The ramp helped the Egyptians build the pyramids. The wheel and axle led to carts and chariots. All these things are machines—devices that make work easier by changing the force and distance or by changing the direction of a force. Whenever you use a broom, a key, a door, a screwdriver, or a knife, you are using a machine.

Since God created people in His image, they are able to be creative. Many inventive people have used their creative abilities to give the world a variety of devices. Some of these devices make life easier, such as cars and bicycles. Others just make life more fun, like roller coasters and Ferris wheels.

An exact replica of a Gutenberg printing press with its wooden screw. (Museum of the Bible)

The world is full of machines—machines of all shapes and sizes, machines that perform all sorts of tasks. Most of these machines are made of a combination of only six simple machines. Some of these are so simple that they are not often thought of as machines. The list includes the inclined plane, the lever, the pulley, the screw, the wedge, and the wheel and axle.

An **inclined plane** is a simple machine that has a slanted surface; a ramp is an inclined plane. Ideally, a ramp does not change the amount of work that you have to do, but it changes the way to do the work. A ramp increases the distance over which you must do the work, but it decreases the amount of force that you need to exert. If you needed to put a piano into a truck bed that is 2 meters off the ground, would it be easier to push the piano up a 10-meter ramp or to hoist it 2 meters straight up? Even though you would have to exert a force over a distance of 10 meters, pushing the piano up the ramp would be easier than hoisting it straight up because you would not have to exert as much force over that 10 meters.

BIOGRAPHY

Jan Ernst Matzeliger

Have you ever found yourself thinking, "There has got to be a better way to do this job," or "I need to get more done in less time"? Jan Ernst Matzeliger did. He was born in 1852 in the Republic of Suriname, to a Dutch engineer and a Surinamese mother. As a young boy, he worked with his father, but he went to work as a sailor during his teenage years. He settled in the US when he was 19 years old and found work as an apprentice to a shoe cobbler in Philadelphia. His career opportunities were limited, so he left.

In 1877, Matzeliger was working for a shoemaker in Lynn, Massachusetts. At that time, the Harney Brothers factory could finish 50 pairs of shoes in a 10-hour work day. Matzeliger believed he could make a machine to do the work faster and more efficiently. In his spare time, he would take scraps of metal, wood, nails, and cigar boxes to try different configurations for a machine. Five years later, he had finally developed a full scrap-metal model of a machine to put the lasts, or the bottoms, on shoes. He took his model to Washington, DC, demonstrated how it worked, and in March of 1883, he received US Patent #274207. His machine could produce more than 150 pairs of shoes a day. This new mass producing machine cut prices drastically, to the point that almost everyone could afford a good pair of shoes. After several refinements were made to his machine, it could produce 700 pairs of shoes in a day. Matzeliger did not get to see his machine reach its full potential because he died of tuberculosis in 1889. In 1991 during Black American History Month, a US postage stamp was issued with Matzeliger's portrait to commemorate his important contributions to the shoe industry.

If the piano weighs 300 N, the resistance force is also 300 N. The work needed to lift the piano 2 meters would require 600 J of energy. If you pushed it up a 10-meter ramp instead, there would be an ideal mechanical advantage (*IMA*) of 5. Using that advantage (300 N/5), only 60 N of force would be needed. How much work would that require? 60 N × 10 m = 600 J. The same amount of work would be needed to move the piano that distance, but the force would be reasonable. The efficiency of the system is 100% because the output force equals the input force. An efficiency of 100% is ideal; but, this is really not a perfect machine because the resistant force of friction is missing. Friction always reduces the advantage and because it works against motion, the force necessary to push the piano would need to be increased. If friction is taken into account and input force increases to 80 N, the work required to push the piano would then equal 800 J. The actual efficiency would be 75% (600 J/800 J × 100%). With an *IMA* of 5 and an efficiency of 75%, the *AMA* is 3.75. The ramp actually increases the force by a factor of 3.75.

The plow revolutionized agriculture.

Simple Machines

Inclined plane

Lever

Pulley

Screw

Wedge

Wheel and axle

Many mountain roads are modified inclined planes.

The use of machines necessitates a trade-off between distance and force. The inclined plane requires a greater distance, but the advantage is the reduction in input force.

The less slanted the inclined plane, the longer the distance over which the force is exerted. It requires more work but less force, which in turn increases the mechanical advantage. The Egyptians relied on ramps to provide the mechanical advantage they needed to move enormous blocks of stone to build their pyramids. Escalators, launching ramps for boats, and mountain roads are examples of inclined planes. What other examples of inclined planes do you use?

LESSON REVIEW

1. How is an inclined plane helpful in moving a large load?
2. If a 10 m ramp that has a height of 2 m were replaced with a 15 m ramp of the same height, how would that change the mechanical advantage? What would the *IMA* be?
3. Would the longer ramp increase or decrease the work? Why?
4. If an inclined plane has an *IMA* of 8, an output force of 160 N, and an input force of 320 N, what is the *AMA*?
5. How could the mechanical advantage in the previous question be increased?

SCIENCE IN ACTION

Mechanical Advantage of an Inclined Plane

Finding the mechanical advantage of an inclined plane requires two measurements: length of the incline (l) and height of the plane (h).

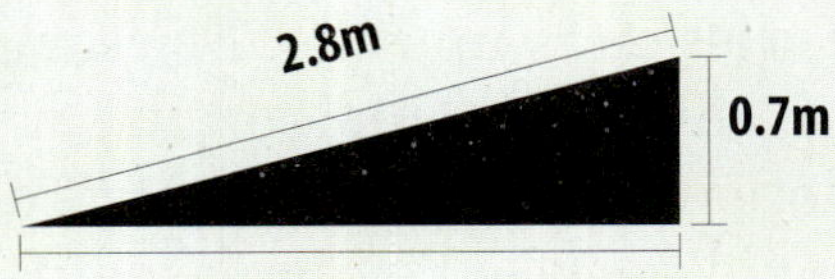

$$MA = l/h$$

If $l = 2.8$ m and $h = 0.7$ m, then 2.8 m/0.7 m = 4.

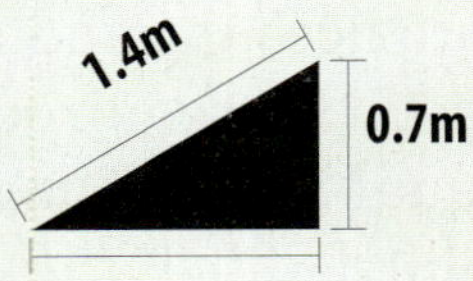

This ramp can ideally increase force by a factor of 4. What happens to the mechanical advantage if the ramp is shortened to a length of 1.4 m? What happens to the angle?

Levers 4.3.2

Have you ever used a screwdriver to pry the lid from a paint can, or played on a seesaw, or used a wheelbarrow? If so, you have used a lever.

A **lever** is a simple machine made of a rigid bar that pivots at a fixed point. A lever consists of two parts: a stiff bar that rotates and a **fulcrum**, which is the point at which a lever pivots. When a force is applied on a part of the bar, the lever pivots at the fulcrum and overcomes a resistance force. Levers help move heavy objects. The Greek scientist Archimedes claimed that if he had a lever long enough and a place to stand, he could move the whole world!

Levers are sorted into three classes according to the relative positions of the fulcrum, the effort force (the force a person applies somewhere along the arm to make the lever pivot), and the resistance force, or the load to be moved.

In a first-class lever, the fulcrum is somewhere in the middle; the effort force is applied on one side and the resistance force on the other. Seesaws, crowbars, and the claw of a hammer used to remove nails are all classified as first-class levers. By shifting slightly forward or backward, children on each end of a seesaw can raise or lower the opposite end—even if one child is smaller than the other. Similarly, to remove a nail from a piece of wood with a crowbar, a person pushes down on one end of the crowbar and the nail moves in the other direction.

A seesaw is an example of a first-class lever.

OBJECTIVES

- Classify and compare different types of levers.
- Calculate mechanical advantage of a lever.

VOCABULARY

- **fulcrum** the point at which a lever pivots
- **lever** a simple machine made of a rigid bar that pivots at a fixed point

First-Class Lever

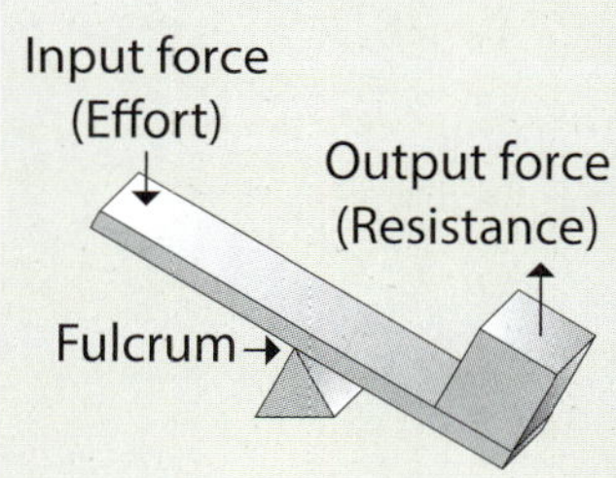

SCIENCE IN ACTION

Mechanical Advantage of a Lever

***MA* = Effort Arm Length / Resistance Arm Length**

First-Class Lever

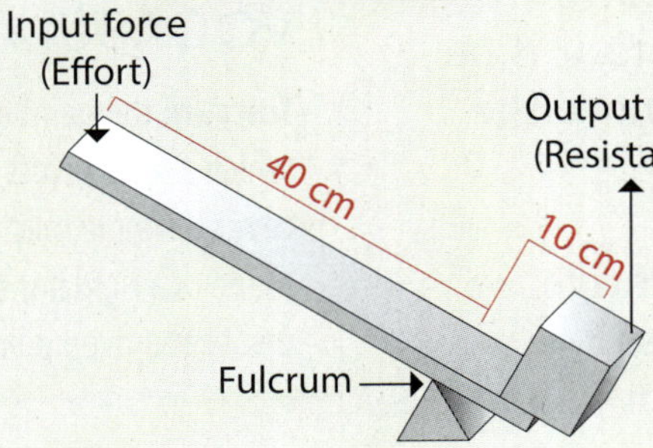

Second-Class Lever

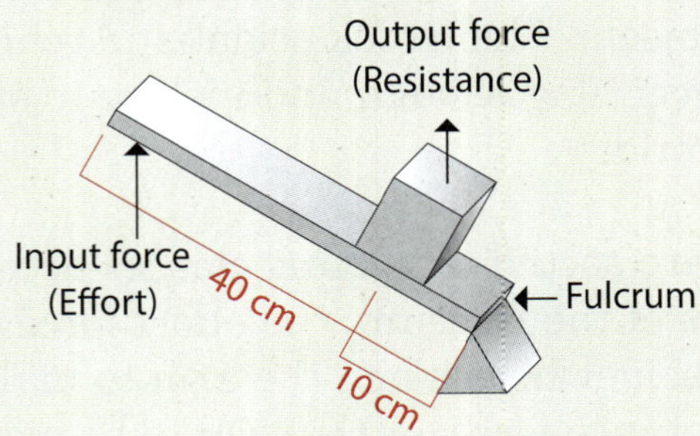

For both the first- and second-class levers:

Effort arm length = 40 cm **Resistance arm length = 10 cm**

$$MA = 40\text{ cm}/10\text{ cm} = 4$$

Third-Class Lever

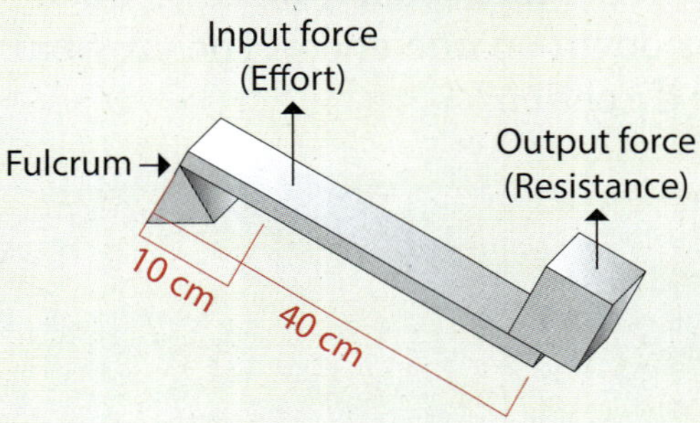

For this third-class lever:

Effort arm length = 10 cm **Resistance arm length = 40 cm**

$$MA = 10\text{ cm}/40\text{ cm} = 1/4$$

Invert the answer to find how many times the speed of the effort is magnified.
Speed is magnified by a factor of 4.

In a second-class lever, the resistance force lies between the fulcrum and the effort force. Doors, bottle openers, and wheelbarrows are examples of second-class levers. On a wheelbarrow, the wheel acts as the fulcrum. The resistance force, in the middle, is the weight of the load. The effort force is applied to lift the wheelbarrow. It is exerted over the distance that the handles are lifted. For example, when a wheelbarrow is used to move a bag of dirt, a smaller force is exerted than would be if the bag were moved directly—but the force is exerted over a longer distance.

A third-class lever is very similar to a second-class lever, but there is one difference. The effort force and the resistance force change positions. In a third-class lever, the effort force is between the fulcrum and the resistance force. A third-class lever does not change the direction of the force; the resistance and effort go in the same direction. With this type of lever, there is always a gain in speed and distance with a loss in force. Examples of third-class levers include hockey sticks, tweezers, baseball bats, and tennis rackets.

Pushing a dolly with boxes is an example of using a second-class lever.

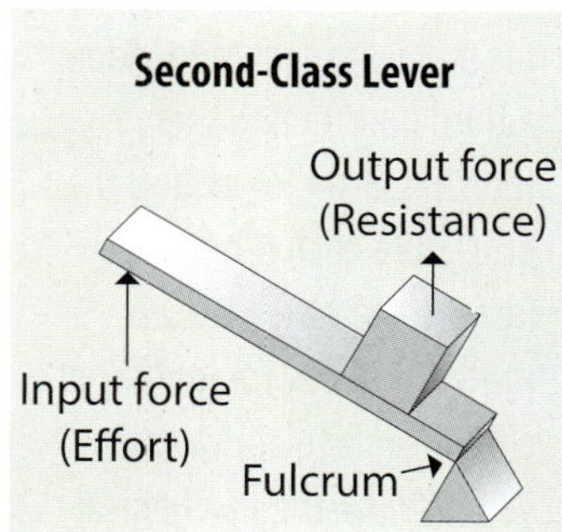

You have learned that the mechanical advantage (*MA*) of a machine can be calculated by dividing the resistance force by the effort force. For example, if the resistance force = 400 N and the effort force = 200 N, then *MA* = resistance force/effort force = 2. But there is another way to calculate the mechanical advantage for levers. Because the distance from the forces to the fulcrum also depends on the size of those forces if things are in balance, *MA* can be calculated by dividing the length of the effort arm (distance from the effort force to the fulcrum) by the length of the resistance arm (distance from the resistance force to the fulcrum). If the length

A hockey player with a hockey stick is an example of a third-class lever.

TRY THIS

The Strength of a Quarter
Place a wooden pencil on the table. Balance a metric ruler on the pencil. Place a quarter on each end of the pencil. If necessary, move them slightly to keep the ruler balanced. Stack another quarter at one end. This is now the resistance arm of the lever. The other end that has one quarter is the effort arm. Move the stack of two quarters until the one quarter can lift the two quarters. Measure the resistance arm and the effort arm. Calculate the mechanical advantage. Add a third quarter and then a fourth quarter, repeating the procedure each time. Which instance had the greatest mechanical advantage? What class lever is this? How did you increase the mechanical advantage?

What class of lever does a hammer removing a nail illustrate?

of the effort arm = 2 m and the length of the resistance arm = 1 m, then MA = 2 m/1 m = 2.

The mechanical advantage for first- and second-class levers indicates how many times the input force is magnified when using the machine. When calculating mechanical advantage for a third-class lever, however, the advantage will be a fraction. Why? Because it does not multiply force. It actually requires more force than it puts out. The advantage is an increase in speed. Inverting the fraction of the mechanical advantage determines how many times faster the resistance moves than the effort.

LESSON REVIEW

1. What is a lever?
2. Give an example of a first-class and a second-class lever.
3. What makes a third-class lever different from the first- and second-class levers?
4. What is the mechanical advantage of a lever that uses an effort force of 90 N to lift a load, or resistance force, of 30 N? What does it mean?
5. If the effort arm on a second-class lever is 2 m long and the resistance arm is 50 cm, what is the mechanical advantage of the lever?

OBJECTIVES

- Distinguish between different pulleys.
- Determine the mechanical advantage of pulleys.

VOCABULARY

- **pulley** a simple machine consisting of a rope or cable that turns around a wheel

Have you ever been to the top of a skyscraper? How did you get there? You probably rode in an elevator. Without elevators, skyscrapers would be very impractical. Elevators depend on a simple machine, the pulley, to lift a load of passengers. The **pulley** is a simple machine consisting of a rope or cable that turns around a wheel. Often the wheel is grooved to keep the rope from slipping. There are two types of pulleys—fixed and movable.

A fixed pulley attaches to something that does not move, such as a door frame, a ceiling, or a tree branch. A fixed pulley changes the direction of the force that must be exerted, but it does not decrease force. Because the distance through which the effort force and the resistance force are exerted is the same, a single fixed pulley has an ideal mechanical advantage of 1. By using a fixed pulley, you can pull down on a rope in order to pull an object up, which is often easier than lifting an object straight up.

A movable pulley connects to the object being moved. The pulley sits at the lowest point of the rope with the object attached to its axle. Half of the object is supported by the rope attached above it, and the other half is supported by the other end of the rope, the place where the effort force is exerted. The mechanical advantage of a movable pulley is 2 because the effort force is only half of the resistance force.

Combinations of pulleys can be strung together to further reduce the force to move an object. By combining pulleys, heavy objects can be hauled or lifted. An arrangement of pulleys designed to move heavy objects is called *a block and tackle*. To figure the mechanical advantage of complex block and tackle arrangements, count the number of ropes that bear the load. In a single movable pulley, the mechanical advantage equals 2 because two ropes share the load. In a block and tackle in which three ropes share

Pulleys

Fixed pulley

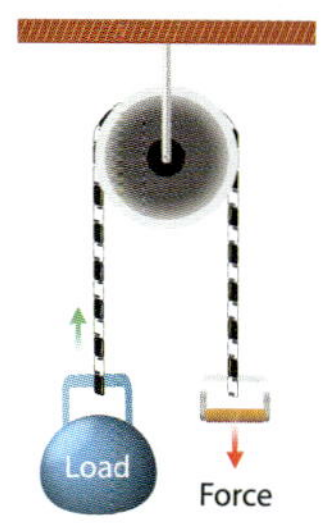

Movable pulley

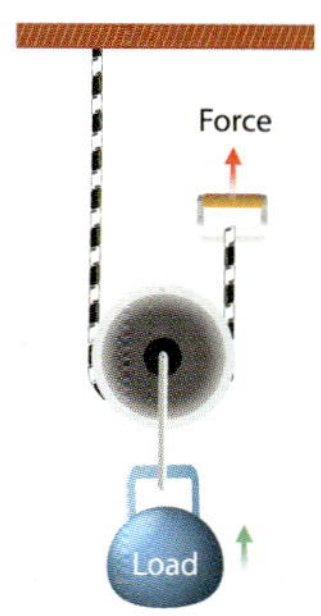

System of two pulleys

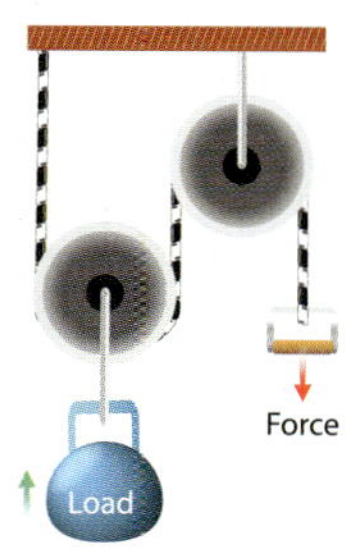

Crane hook with fixed and movable pulley

Examples of Pulleys Used in Various Applications

Ski lift pulley

Elevator motor and pulley

Glass elevator and pulleys

Window-washing pulleys

the load, the mechanical advantage is 3. The load-supporting ropes pull up. The rope that is being pulled down does not count; it is sharing the load of the adjacent rope.

A machine that needs less force requires more distance. How does this fact affect pulleys? In a pulley system with a mechanical advantage of 2, the rope must be pulled twice as far as in a system with a mechanical advantage of 1. In a pulley system with a mechanical advantage of 3, the rope must be pulled three times as far. Of course, friction has an effect in real-life situations, which results in a need for more force to move an object. Friction reduces the mechanical advantage.

LESSON REVIEW

1. What is a pulley? Which pulley gives a greater mechanical advantage—fixed or movable?
2. How does adding more pulleys to a system affect the mechanical advantage? Why?
3. What is the mechanical advantage of a block and tackle system that contains four pulleys?
4. If you add more pulleys to gain more mechanical advantage, what else do you have to add?
5. What happens to the frictional force when more pulleys are added? Why? How does that affect the mechanical advantage?

CAREER

Vertical Dancer

Maybe you have considered mountain climbing as a hobby, or dancing as a career. Did you know there are people who do both? Vertical dancers gracefully and fearlessly leap and swoop alongside skyscrapers, mountains, and cliffs. Every move requires the combined skills of dancing and rock climbing, with constant attention to risk management. The choreography is beautiful and breathtaking, a delightful combination of fine art and architecture. How do the dancers manage to move vertically while defying the laws of physics? They are secured with harnesses, carabiners, and climbing rope. Simply put, they are attached to movable pulley systems, which allow them to move up and down along a variety of vertical and irregular surfaces. The choreography and challenges change for each new venue, often pushing the dancers to their limits. Hours and hours of practice and preparation result in a unique and spectacular work of art.

Screws, Wedges, Wheels and Axles 4.3.4

The **screw** is a simple machine that consists of an inclined plane wrapped around a cylinder. Think of the threads of a screw as a very long inclined plane. An inclined plane increases the distance but decreases the amount of force needed. When a screw turns, it moves a certain distance up or down with each rotation. The screw exerts effort force over a long distance.

A jar lid is an example of a screw. It would be hard to seal a jar tightly by pushing down on the lid, so the jar lid is threaded. The effort force is small when the lid is turned but it is multiplied because it is exerted over a long distance with each turn of the lid.

The pitch of a screw spans the distance between two adjacent threads along the length of the screw. This measures the depth that the screw will advance with one complete rotation. The closer together the threads sit along the screw, the longer the distance over which the effort force is exerted and the more the force is multiplied. Screws with threads that are closer together, then, have a greater mechanical advantage than screws with threads that are far apart. They also have more holding power.

The mechanical advantage of a screw is determined by the circumference of its head divided by its pitch. Circumference equals the value of π (approximately 3.1416) times the diameter. If a screwhead has a diameter of 0.4 cm, multiply that by π to find the circumference: 1.25 cm. If the pitch is 0.1 cm, then MA = 1.25 cm/0.1 cm = 12.5.

screw's *MA* = *screwhead's circumference / pitch of threads*

OBJECTIVES

- Describe the advantages of using different simple machines.
- Calculate the mechanical advantage of screws, wedges, and wheels and axles.

VOCABULARY

- **screw** a simple machine that consists of an inclined plane wrapped around a cylinder
- **wedge** a simple machine that consists of two back-to-back inclined planes
- **wheel and axle** a simple machine that consists of a wheel attached to an axle that has a smaller diameter

Jar lids are examples of screws.

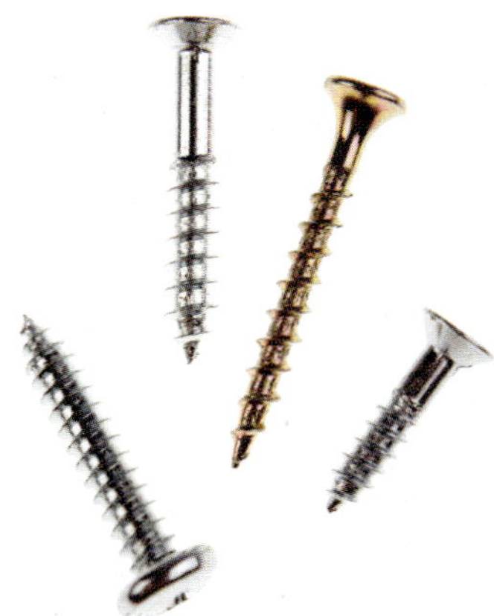

The screw is a simple machine made of an inclined plane wrapped around a cylinder.

SCIENCE IN ACTION

Mechanical Advantage

Screw: *MA = circumference / pitch*

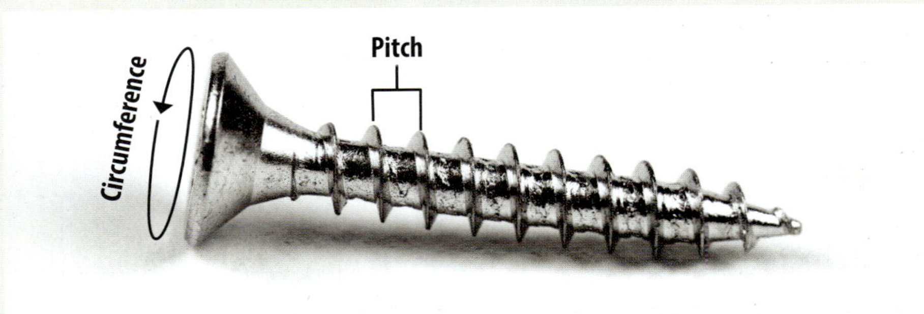

Find the circumference of the screwhead and the pitch of the threads.

Wedge: *MA = length of slope / height*

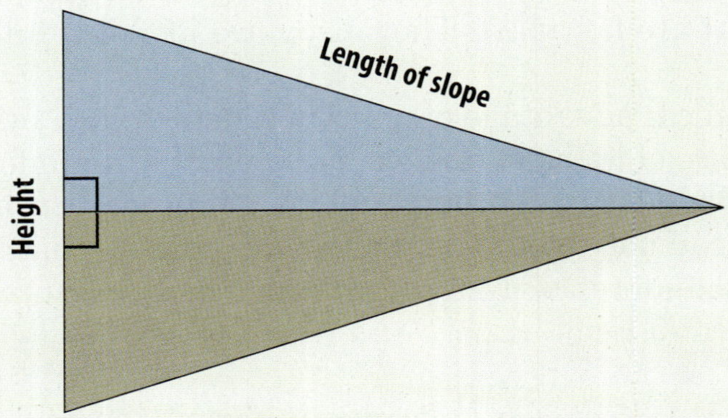

Find the height and the length of the slope.

Wheel and Axle:
MA = diameter of the wheel / diameter of the axle

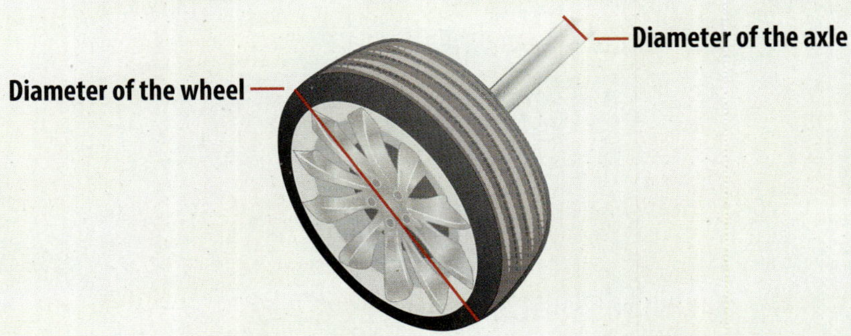

If the effort is applied to the wheel, the force is magnified. If the effort is applied to the axle, the speed is magnified.

It may seem that an *MA* of 12.5 is very beneficial, but experience says that fingers do not have the strength to turn a screw with bare hands. A screwdriver is needed to increase strength. Using a screwdriver increases the diameter because the handle's diameter is larger than that of the screwhead. Consider the *MA* of the screw noted earlier. Suppose it is turned by a screwdriver that has a handle with a 3 cm diameter. Find the circumference: 3.1416 × 3 cm = 9.4 cm. Next, divide 9.4 cm by the pitch of 0.1 cm to get the *MA* of using the screwdriver with the screw. 9.4 cm/0.1 cm = 94. Now that is an advantage!

An ax-head is an example of a wedge.

A **wedge** consists of two back-to-back inclined planes. Instead of the force being magnified by pushing an object along the slope, the wedge actually separates an object or pushes it apart. The greater the distance a wedge moves, the greater force it applies, which then produces a greater mechanical advantage. Imagine splitting a log with an ax. What shape of ax-head would be best

SCIENCE IN ACTION

Gearing on a Bicycle

Suppose you are about to ride your new bike to your friend's house. But, you are not sure which gears to use because the bike has so many. You have two chainrings in the front and seven different-sized gears on the rear axle. The best thing to do is to start in the middle of the gear range. This will require a moderate amount of force and result in a moderate speed. Start with the inner ring on the front of the bike and a middle gear on the back. After you start riding, if you want to go faster, shift to a smaller ring in the rear. This will turn the rear wheel faster or with more rotations in relation to one rotation of the pedals. It will require more force but result in greater speed.

If you come to a steep incline, you may need to quickly shift to the largest gear on the rear wheel. Now the chain will travel a much greater distance to turn the rear wheel one rotation, but your effort force will decrease and the output will increase.

Enjoy your new bike, and find what works best for you. Choose the front chainring first, and then fine-tune your ride by changing gears in the back. Be careful not to overstretch the chain by matching the largest chainring with the largest gear, or the smallest chainring with the smallest gear. Your bike will not work as smoothly and you could derail or snap the chain.

TRY THIS

Wheel Around

Create a wheel and axle, using a cardboard circle for the wheel and a pencil for the axle. Compare the distance of one rotation of the axle with the distance of one rotation of the wheel. Find the circumference of both the wheel and the axle. Calculate the mechanical advantage. What is the advantage if you apply force to the wheel? What is the advantage if you apply force to the axle?

to split a log? A long, thin wedge would take much less effort than a short, thick one.

Knives, chisels, plows, and ax-heads are all examples of wedges. Tremendous force is exerted on the wedge as well as on whatever material it is being pushed into or through. Although the efficiency of wedges is low because their friction is so high, they are very effective at converting a large impulse force into a large force perpendicular to the direction of the moving wedge. The advantage of a wedge is more about changing the direction of the force than magnifying it. The MA of a wedge is calculated by dividing the length of its slope by its height.

wedge's MA = length of slope / height

The **wheel and axle** is the most common simple machine used to transfer motion and force from one place to another. A wheel and axle is made of two circular pieces—a wheel, or disk, attached to an axle, or shaft. The wheel always has a larger diameter than the axle. The wheel and axle are firmly attached to each other; when one moves, so does the other.

Screwdrivers, wrenches, steering wheels, and doorknobs are all examples of wheels and axles. A wheel and axle makes work easier by magnifying the force that can be exerted on something. But like other machines, a wheel and axle does not change the amount of work that has to be done. If the force decreases, the distance over which the work is done must increase. In the wheel and axle, the wheel always moves over a greater distance than the axle because the wheel is larger. The force applied to the wheel is multiplied when it transfers to the axle because the axle travels a shorter distance. If the radius of the wheel is six times greater than the radius of the axle, the force exerted on the wheel is multiplied by six.

Steering wheels vary in size depending on the size of the vehicle and the force needed to move the tires.

The mechanical advantage of a wheel and axle is calculated as follows:

wheel and axle's MA =
wheel's diameter/axle's diameter

The force of a fan is transferred from the axle to the wheel, the hub to the fan blades.

If the wheel's diameter is 25 cm and the axle's diameter is 5 cm, then MA = 25 cm/5 cm = 5. Applying the effort to the wheel magnifies the force on the axle 5 times. Consider the steering wheel of a car versus that of a large truck. The car is not as heavy, so the steering wheel is not that big. A semitruck has a very large steering wheel because a great amount of force is needed to move the tires. A larger steering wheel gives greater force. The trade-off is the distance the wheel must be turned. The steering wheel must be turned a greater distance around to provide the greater force.

Some other devices use the wheel and axle in an opposite way. On a fan, for example, the effort is exerted around the axle. The force is reduced as it is transferred to the wheel (the blades), which has a larger diameter. At the same time, the distance becomes greater. The force travels a greater distance, allowing the blades to travel farther and faster. A wheel and axle is similar to a lever. Apply force to the wheel, and the force is magnified in the same way force is boosted in a first- or second-class lever. If force is applied to the axle, it is similar to a third-class lever because more force is required to achieve greater speed.

A wheel and axle can be used to transfer rotary motion to rotary motion or to transfer rotary motion to linear motion. For example, a crank uses rotary motion to raise and lower a bucket

Gears are a type of wheel and axle. The wheels of gears have small projections, called *teeth*, along their edges. The teeth of one gear fit into the teeth of a neighboring gear. As a result, when one gear turns, so does the other. Sometimes gears are connected by a chain, as seen here. The wheel and gears on a bicycle work together to make getting places easier.

from a well. The crank turns around and around as the rope and bucket travel in a straight line (linear motion).

LESSON REVIEW

1. Explain how the following simple machines make work easier: screw, wedge, and wheel and axle.
2. Why is it that screws with threads that are closer together have a greater mechanical advantage than those with threads that are farther apart? Which would give a greater mechanical advantage, a screwdriver with a wide handle or one with a longer shaft?
3. Find the mechanical advantage of a screw with a pitch of 3 mm and a head with a diameter of 8 mm.
4. What is the mechanical advantage of a wheel and axle with a wheel diameter of 300 cm and an axle diameter of 15 cm?
5. How is a wheel and axle like a lever?
6. What is the mechanical advantage of a wedge with a slope of 400 mm and a height of 20 cm?

FYI

Gear Ratios

Bicycle gears are often named by the number of teeth they have. A chainring is the teethed ring that pulls the chain around the wheel. Standard-sized chainrings range in size from 36 teeth to 53 teeth. The rear wheel gears range from 10 teeth to 36 teeth. The ratio indicates how many times the rear wheel turns for every one revolution of the chainring.

Rear Wheel Ring	Chainring 53	Chainring 39
12	4.42	3.25
14	3.79	2.79
16	3.31	2.44
18	2.94	2.17
21	2.52	1.86
26	2.04	1.50
32	1.66	1.22

Compound Machines 4.3.5

It is said that "necessity is the mother of invention." Many simple machines have been created and used to accomplish important tasks. Over time, more sophisticated machines have been made using a combination of machines. A compound machine results from a combination of two or more simple machines. A compound machine does work that a simple machine alone cannot do. Like a simple machine, a compound machine cannot produce more work than is put into it.

The total mechanical advantage of a compound machine is the product of the individual mechanical advantage of the simple machines that comprise it. For example, the mechanical advantage of a bicycle is the product of the mechanical advantage of each gear, wheel, and other simple machines that causes its motion. For compound machines to work efficiently, friction must be reduced as much as possible through lubrication and other techniques.

People have used their God-given gifts of intelligence and creativity to invent machines that bring blessings to the world. Mechanized farming allows relatively few farmers to feed billions. Amazing new medical equipment such as MRI machines help diagnose illnesses. Surgeons use advanced surgical tools to perform microsurgery on infants or on tiny areas of the body. Cars and other vehicles allow faster travel to farther distances.

But technological developments also have the potential of causing harmful effects. For example, many communities in North

OBJECTIVES

- Identify simple machines within complex machines.
- Evaluate the advantages and disadvantages of compound machines.

TRY THIS

From Simple to Complex

Examine the following compound machines, and determine what simple machines they are made of:

- scissors
- ax
- piano keyboard
- bicycle
- hand-powered drill
- zipper
- manual pencil sharpener
- nail clippers

Wood block press

Movable type press

Letterpress

Sheetfed offset press

Web press

Digital press

Printing presses were first designed to be made of only simple machines, and then quickly progressed to be comprised of compound machines.

HISTORY

Bicycle

The bicycle you ride, today, differs greatly from the original bicycles. The first bicycles were actually called *hobby horses* and did not have pedals. The rider would push on the ground with his feet to move the hobby horse. In 1839, Kirkpatrick Macmillan created a hobby horse with pedals attached to rods that moved back and forth and extended to the rear wheel to make it move. The velocipede was introduced in the early 1860s, constructed with two steel wheels, a wooden frame, pedals, and a simple gear system. It became known as *the bone shaker* for its bumpy ride. Later, James Starley created the famous penny-farthing bicycle. This looked more like the modern day tricycle than a bicycle. The giant wheel helped the rider gain great distance with each crank of the pedals. The name comes from the smallest and largest coins in England at the time, the penny and the farthing. Today, bikes have springs, shock absorbers, and more gearing to give the best mechanical advantage in any situation.

CAREER

Mechanical Engineer

Mechanical engineers solve problems and design equipment related to the production, transmission, and use of mechanical power and heat. They design and develop machines that use or produce power: engines, turbines, machine tools, industrial production equipment, and more. Mechanical engineers may research and develop specifications for new products; supervise the assembly, installation, and inspection of products; and test the durability, quality, and strength of products. These engineers may work in laboratories, offices, or on production sites.

America have been designed around the automobile—from the size of the streets to the location of shopping centers and other community services. Few people walk to the library or the grocery store because it is necessary or it is more convenient to take the car. More cars can lower the quality of life by reducing the need for human activity and polluting the air with exhaust.

If wisdom is combined with the intelligence and creativity that God has given, scientists and researchers can find solutions to the problems brought about by technology. Thoughtful choices made about the use of machines demonstrate good stewardship of God's creation.

LESSON REVIEW

1. What is a compound machine? What is the advantage of a compound machine?
2. A bicycle is a compound machine. Name another compound machine, and list the simple machines that compose it.
3. What is the *IMA* of a machine that uses a wheel and axle with an *IMA* of 6 and a pulley system with an *IMA* of 4?
4. Name two machines that have made life easier in the past 200 years. What could be considered disadvantages of using these machines?

UNIT 5

Energy at Work

Chapter 1: *Thermal Energy*
Chapter 2: *Nuclear Energy*
Chapter 3: *Energy Resources*

Vocabulary

alpha particle
beta particle
calorie
carbon dating
conduction
convection
critical mass
fossil fuel
gamma ray
half-life
heat
heat engine
irradiation
meltdown
nuclear chain reaction
nuclear radiation
radiation
radioactive decay
radioactivity
specific heat
stable nucleus
temperature
thermal conductor
thermal insulator
tracer
transmutation
unstable nucleus

Key Ideas

- Evidence, models, and explanation
- Change, constancy, and measurement
- Form and function
- Abilities necessary to do scientific inquiry
- Understandings about scientific inquiry
- Properties and changes of properties in matter
- Motion and forces
- Transfer of energy
- Structure of atoms
- Structure and properties of matter
- Conservation of energy
- Abilities of technological design
- Understandings about science and technology
- Personal health
- Populations, resources, and environments
- Natural and human-induced hazards
- Risks and benefits
- Science and technology in society
- Science as a human endeavor
- History of science

SCRIPTURE

If we are thrown into the blazing furnace, the God we serve is able to deliver us from it, and He will deliver us from your majesty's hand.
Daniel 3:17

5.1.1 *Heat*

OBJECTIVES

- Correlate kinetic theory and temperature.
- Describe the transfer of energy in a system.

VOCABULARY

- **heat** the energy that is transferred between substances that are at different temperatures
- **temperature** the measure of the average kinetic energy of the particles in an object

Many places on the planet record extreme temperatures every year. These are places with very few permanent residents. On July 10, 1913, Death Valley, California, baked under a temperature of 57°C. Very high temperatures are dangerous, but so are very cold temperatures. Most of the universe experiences temperatures at the opposite extreme, with bone-chilling average temperatures of –269°C. But God has blessed Earth with warmer temperatures that make life possible.

Heat has puzzled scientists for thousands of years. What is heat? Where does it come from? How does it move from place to place? Why are some places hot and others cold? Two hundred years ago, scientists believed that heat was an invisible substance flowing through the air from one object to another. Today, ideas about heat have advanced. By understanding the orderly ways that heat behaves, people have learned not only how to channel heat energy to keep warm in frigid temperatures but also how to power modern society. Heat is an amazing gift.

Why does a roaring fire in a fireplace warm your skin and dry your wet clothes? How does the sun scorch your skin and make you sweat? How does a stove cook food? To understand heat, the study must begin with the atom. All objects consist of atoms that constantly move. Even the atoms of solids—such as you, your desk, and your sandwich—are in motion. The movement of these structures gives them energy. The energy contained in moving objects is called *kinetic energy*. Any object that has mass and is moving has kinetic energy. The kinetic energy of an object is determined by using the formula $KE = 1/2 \times m \times v^2$. In this formula, m is the mass and v is the velocity. The unit of kinetic energy is the joule (J). Because all matter moves, all matter has

Desert temperatures can reach dangerous highs, but most of the universe experiences extremely cold temperatures.

Fire creates radiant energy.

kinetic energy. The atoms of all objects move even if the object itself is not moving. Atoms always have kinetic energy, but objects do not.

Atoms and molecules are too small to observe, so how can scientists measure the motion of an atom of gold or a molecule of oxygen? They do this by observing the average behavior of large groups of atoms. The measure of the average kinetic energy of the atoms in an object is proportional to its **temperature**. Physical scientists describe the behavior of moving particles by using the kinetic theory, a model developed to explain the states of matter according to the random movement of particles.

TRY THIS

Heat by Radiation

Refrigerate a beaker of water. When it is chilled, take it out of the refrigerator and measure the water's temperature. Place it in direct sunlight on a windowsill. Keep the thermometer in the water and observe the heat added by the sun's radiant energy as measured by the temperature change on the thermometer. What is heat? What direction did the heat travel? How much did the temperature of the water change? What happened to the molecules in the water?

SCIENCE IN ACTION

Kinetic Energy

To calculate kinetic energy, use the mass and velocity of the moving object. If a 0.4 kg bird is flying south at a velocity of 6 m/s, what is the bird's kinetic energy?

$$KE = 1/2 \times m \times v^2$$

$$KE = 1/2 \times 0.4 \text{ kg} \times (6 \text{ m/s})^2$$

$$KE = (1/2 \times 0.4 \text{ kg}) \times 36 \text{ m}^2/\text{sec}^2$$

$$KE = 0.2 \text{ kg} \times 36 \text{ m}^2/\text{sec}^2 = 7.2 \text{ J}$$

TRY THIS

Speed up Molecules
Hold a thick rubber band close to your mouth, stretch the band tightly, and quickly press it on your lips. How does it feel? Then relax the rubber band and hold it to your lips again. How does it feel? What do you think happened to the molecules of the rubber band as it stretched?

FYI

Shivering

The average internal temperature of the human body is 37°C. Most of the time, the air temperature is cooler than your body temperature and you lose energy as the heat travels from your body to the air. Usually, you lose heat slowly enough that your body can make up for it by burning more calories from the foods eaten. Sometimes, however, the air temperature is so much cooler than the body temperature that you lose heat faster than you can warm it up, so the body starts to shiver. The movement of those shivering muscles creates kinetic energy that transfers to the cells throughout your body to warm you up.

The kinetic theory has three basic assumptions.

- All matter is made of tiny particles that are in constant, random motion.
- The speed of the particles directly corresponds to their temperature.
- At a given temperature, the total amount of kinetic energy of the colliding particles remains constant.

Imagine how difficult it was for scientists to begin to identify and explain the nature of heat. Before the 18th century, scientists believed that heat was an invisible, weightless fluid capable of flowing from hotter objects to colder objects. They called this substance "caloric." In 1798, American scientist Benjamin Thompson demonstrated that heat is related to the motion of particles. Forty years later, in 1838, British scientist James Prescott Joule showed that heat, mechanical energy, and other forms of energy can be changed into one another. The kinetic theory was derived from the work of these two scientists.

When you look at a substance, you cannot see its molecules moving because the molecules are too small. How fast the molecules move relates to their kinetic energy. The water molecules in ice have very little kinetic energy because they are locked rigidly in place; they only vibrate back and forth weakly. The molecules of liquid water, though they are still attached to each other, have more kinetic energy. The increased kinetic energy allows water molecules to roll over one another; this is why liquids flow. The molecules of water vapor have even higher kinetic energy because they are free-moving particles. Water vapor is a gas, and the molecules of a gas are widely separated.

The high kinetic energy of gas molecules allows them to move away from each other rapidly and scatter throughout a room. These three phases of water, ice, water, and vapor, have varying amounts of kinetic energy.

Kinetic theory supports an understanding of **heat**, which is energy transferred between substances of different temperatures. When you place your cold hand on someone's warm neck, the person feels the cold as the heat leaves their neck and is transferred to your hand. When you have a fever and you place a cold cloth on your head, the heat from your head is transferred

BIOGRAPHY

James Joule

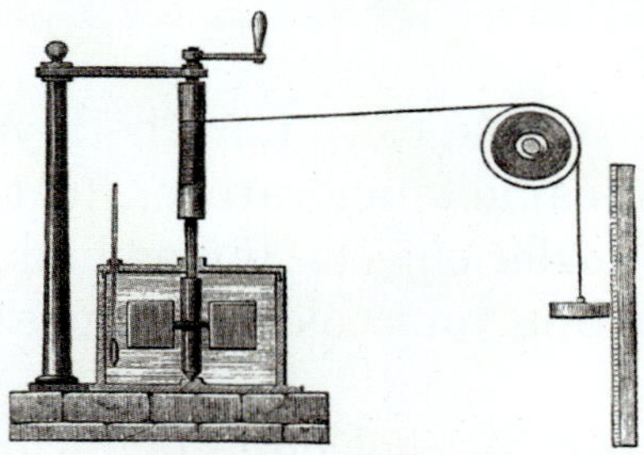

Machine used to measure mechanical equivalent of heat

When you read about a famous scientist making a breakthrough in scientific understanding, you rarely think about what it took for that breakthrough to occur. Oftentimes, a scientist who finds something new encounters great opposition from the current scientific community. James Joule was not a university professor or a prestigious scientist. He was homeschooled until the age of 15 and then was tutored by John Dalton, a famous chemist, in physics, chemistry, and the scientific method. He was not able to attend university because he had to work at the family business. But in his off time, he conducted scientific research, motivated to find a replacement for the steam engines at the family business.

He pursued the idea that electricity would be a good replacement. In this pursuit, he discovered a mathematical relationship between the heat generated by an electric current, the current itself, and the resistance of the current-carrying wire. He presented his work to the Royal Society in 1838, but he was not well received. He lacked notoriety and his findings went against the current caloric theory, which suggested that heat was a fluid substance. Even without support, Joule continued his research. His own writings suggest that his relationship with God was part of his motivation. Joule wrote, "It is evident that an acquaintance with natural laws means no less than an acquaintance with the mind of God therein expressed."

As his research continued, he gained the endorsements of William Thomson (also known as Lord Kelvin), Michael Faraday, and George Stokes. History recognizes Joule as the founder of thermodynamics. His work demonstrated that energy is neither created nor destroyed but simply changes into different forms. He believed this idea was completely consistent with the teachings of Scripture. His words are recorded in Reynolds's *Memoir of James Prescott Joule*: "It is manifestly absurd to suppose that the powers with which God has endowed matter can be destroyed any more than they can be created by man's agency."

TRY THIS

Heat by Mechanical Energy

Feel a large nail and the head of a hammer. Hammer the nail rapidly into a piece of wood until the nailhead is just sticking out. Then feel the nail again. Record and explain your observations.

TRY THIS

Heat by Electricity

Blow a hair dryer over a thermometer. Observe the temperature change caused by the electrical generation of heat. Why does the temperature change?

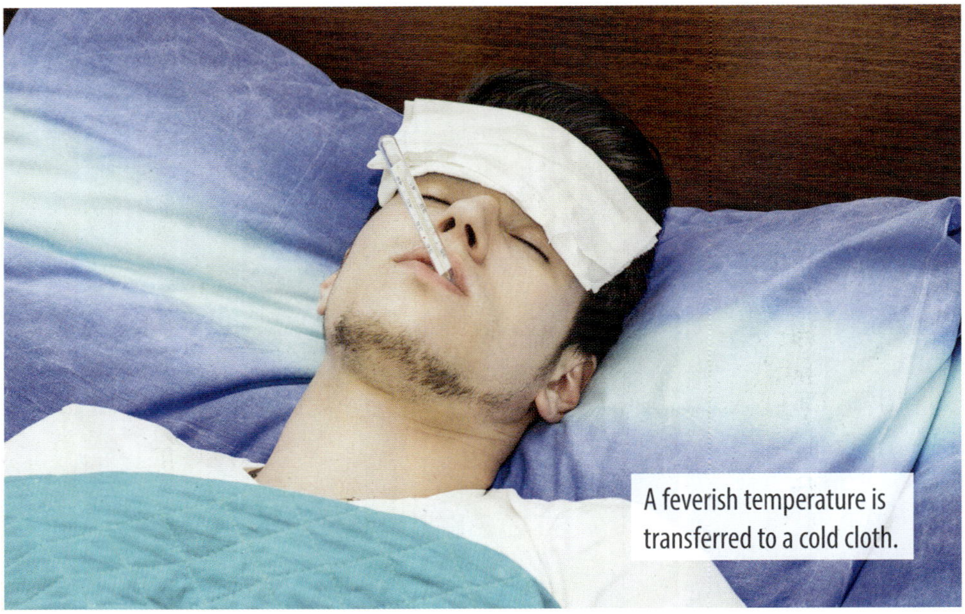

A feverish temperature is transferred to a cold cloth.

to the cold cloth until both your forehead and the cloth reach the same temperature. Heat always flows from warmer objects to colder objects. If God had not designed heat to work this way, holding your cold hands near a campfire would not warm them.

When the molecules of a warm object collide with the molecules of a cold object, energy transfers from the warm object to the cold object. The average kinetic energy of the warmer object decreases, and that object becomes cooler. The average kinetic energy of the cooler object increases, and that object becomes warmer. To understand how energy is transferred, picture what would happen if a marble was shot quickly against a slowly moving marble; the faster marble would slow down, transferring energy into the slower marble, which would then speed up. Now imagine what would happen if a piece of hot metal was dipped into cold water. The water would warm up, and the metal would cool down. And heat would continue to flow from the metal into the water until the metal and water reached the same temperature.

LESSON REVIEW

1. What is kinetic energy? What objects have kinetic energy?
2. List the three basic assumptions of kinetic theory. Why is it called a theory?
3. What does temperature measure?
4. In order for a system to have heat, what must be true?
5. If two bricks are placed next to each other and one has a temperature of –5°C and the other has a temperature of –10°C, is there heat in the system? Why?

Specific Heat 5.1.2

Adding or removing heat is a way to change the state of a substance. For example, whether water is a solid, liquid, or gas depends on heat. Heat also changes the amount of kinetic energy in a substance's molecules. How do you measure change in heat? How do you calculate the amount of heat needed to create these differences?

Heat cannot be measured directly by watching moving molecules, but heat can be determined by calculating temperature changes. Temperature, the measure of an object's kinetic energy, indicates how fast an object's molecules are moving. An increase in temperature signifies heat being added; a decrease in temperature identifies heat being taken away.

When heat is added to a solid, the atoms of the solid vibrate more quickly, and the object becomes warmer. As the solid's atoms receive more energy, they begin to loosen from their rigid positions and their bonds break. At this point, the solid melts into a liquid. The molecules of liquids and solids are held together by strong cohesive forces. But when heat is added to a liquid, the liquid's molecules gain kinetic energy and can overcome these forces, breaking free. At this point, the liquid evaporates into a gas. If the gas's molecules are heated, they also receive more kinetic energy and move faster—flying farther apart from each other. But if kinetic energy is removed from the gas molecules, the molecules connect again to form a liquid in the process of condensation.

The unit for measuring energy is the joule, but heat energy is often expressed in units called *calories*. A **calorie** represents the amount of heat needed to raise the temperature of 1 g of water by 1°C. One calorie of heat must be added to a gram of water at

OBJECTIVES

- Explain what specific heat measures.
- Calculate the amount of energy needed to raise the temperature of a given substance.
- Compare the heat of fusion with the heat of vaporization.

VOCABULARY

- **calorie** the amount of heat needed to raise the temperature of 1 g of water by 1°C
- **specific heat** the amount of heat needed to raise the temperature of 1 g of that substance 1°C

Sugar boils at 110°C.

Water boils at 100°C.

A calorimeter is used to measure heat capacity.

Substance	Specific Heat (cal/g°C)	Specific Heat (J/g°C)
gold	0.0301	0.126
lead	0.0305	0.128
mercury	0.033	0.14
silver	0.06	0.23
copper	0.09	0.39
zinc	0.09	0.39
aluminum	0.22	0.90
ice (–10°C)	0.49	2.05
alcohol (ethyl)	0.58	2.4
water	1.00	4.18

10°C to raise its temperature to 11°C. To lower the temperature of a gram of water from 11°C to 10°C, one calorie of heat must be taken away. Five calories will raise the temperature of a gram of water from 25°C to 30°C, but five calories will only raise the temperature of five grams of water from 25°C to 26°C. The word *calorie* originates from the original caloric theory about heat. Even though scientists know that the theory is false, they have held on to the unit. The scientific calorie is different from the calories used to measure the energy potential of food. Food calories are literally kilo-calories. One kilo-calorie equals 1,000 science calories. The US food industry does not use the term *kilo-calorie*. To distinguish between the two, the science calorie is written as *calorie* and the food calorie is written with a capital letter as *Calorie*.

A substance's ability to absorb the kinetic energy of heat is called its **specific heat**. The specific heat of a substance is the amount of heat needed to raise the temperature of 1 g of that substance 1°C. Sometimes the specific heat of a substance is expressed

TRY THIS

Thermometer

To make a model of a thermometer, fill the plastic bottle halfway with isopropyl alcohol. Add a couple of drops of food coloring and mix thoroughly. Place the straw in the bottle. Use the clay to suspend the straw just above the bottom of the bottle and to seal the mouth of the bottle. Pour ice water in the bowl and place the bottle in the center of it. What happens to the level of the alcohol in the straw? Why? Remove the bottle and pour out the ice water. Add hot water to the bowl and set the bottle in the hot water. What happens to the level of the alcohol in the straw? Why?

in calories per gram degrees Celsius (cal/g°C). Different substances have different specific heats. This explains why a lake can be cold on a day when the air is very hot. Water has a very high specific heat—raising the temperature of water requires more heat than raising the temperature of air. Imagine what would happen to sea life if water had a low specific heat. Water temperatures would change more quickly and would have greater extremes, all of which could be life-threatening to the organisms living in the water.

The potential energy of food is measured in Calories.

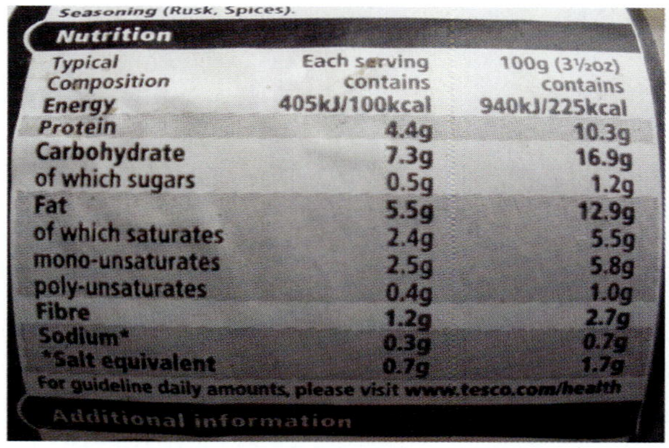

The potential energy of food is measured in kilo-calories.

A change in the phase (or state) of a substance requires heat. Ice absorbs heat to become liquid water. The heat of fusion of a substance is the amount of heat required to melt a certain amount of a particular substance—to change it from a solid to a liquid. The unit used to express the heat of fusion is calories/gram. Ice, for example, has a heat of fusion of 80 calories/gram. This means that one gram of ice must absorb 80 calories of heat before melting into liquid water. Also, one gram of liquid water must lose 80 calories of heat to change into ice. How many calories of heat must be removed from 25 grams of water at 0°C to turn it into ice? (25 g × 80 cal/g = 2,000 cal)

Heat is also required to turn liquids into gases. If you put a pot of water on a stove burner, it will take several minutes for the water to start boiling because the liquid water must absorb enough heat. Absorbing enough heat is a two-step process. Heat

LAWS & PRINCIPLES

Heat

To find the amount of heat needed to raise the temperature of a specific substance, use the following equation:

$$Q = m \times c \times \Delta T$$

where Q = calories, m = mass in grams, c = specific heat, and ΔT = change in temperature. How many calories of heat are needed to raise 30 g of silver from 20°C to 30°C? Solve for Q.

$$Q = 30\text{ g} \times 0.06\text{ cal/g°C} \times 10\text{°C}$$

$$Q = 30\,\cancel{\text{g}} \times 0.06\text{ cal}/\cancel{\text{g°C}} \times 10\cancel{\text{°C}} = 18\text{ calories}$$

causes the temperature of the liquid to rise to 100°C; more heat causes the liquid to turn into vapor.

> **TRY THIS**
>
> **Calculate Energy**
>
> Using the formula $Q = m \times c \times \Delta T$, determine how many calories are needed to raise the temperature of 10 g each of gold wire, copper wire, and aluminum wire from a temperature of 30°C to 40°C.

The heat of vaporization is the amount of heat required to vaporize a certain amount of a substance—to change it from a liquid to a gas. The unit used to express the heat of vaporization is also calories/gram. The heat of vaporization for water is 540 cal/g. The heat of vaporization of a substance is higher than the heat of fusion because it requires more energy to pull a liquid's molecules apart to form a gas than to pull a solid's molecules apart to form a liquid. How much heat is required to turn a liter of water at 100°C into a liter of steam? (1 liter = 1,000 mL = 1,000 g; 1,000 g × 540 cal/g = 540,000 cal)

LESSON REVIEW

1. What does specific heat measure?
2. Why would it take less energy to raise the temperature of a gold wire to a specific temperature than to raise the temperature of a wire made of copper or aluminum?
3. In the past, why did people with fevers take alcohol baths to cool down?
4. How much heat is needed to convert 150 g of ice into water and 150 g of water into steam? Compare the two values.
5. Why does it take more energy to convert water to steam than to convert ice to water?

Water in geysers can come from deep in the ground. The deeper the water, the greater the heat of vaporization. Why?

Heat Transfer 5.1.3

OBJECTIVES

- Classify methods of heat transfer.
- Discriminate between thermal conductors and thermal insulators.

VOCABULARY

- **conduction** the transfer of heat from one substance to another through direct contact of the substances' molecules
- **convection** the heat transfer that occurs in moving fluids (liquids or gases) caused by the circulation of currents from one region to another
- **radiation** the transfer of energy through space or matter by waves or rays
- **thermal conductor** a substance that conducts heat well
- **thermal insulator** a substance that does not conduct heat well

You may not realize it, but you think about heat transfer several times a day. If your soup is too hot, you let it cool a little before you take a sip—or risk burning your mouth. If the bathwater is too cold, you add some hot water. If you are too hot in the sun, you move to the shade. You do these things naturally, without thinking about why they work. But why do they work? How can a spoon in boiling water burn your hand? How does the heat from the sun move millions of kilometers through space to warm the earth? These are questions about the transfer of heat that scientists have struggled for centuries to explain.

Heat always travels from an object of higher temperature to an object of lower temperature. If you come inside from shoveling snow, heat travels from the warm room into your body instead of the other way around. The movement of heat from a warmer object to a cooler object is called *heat transfer*. Heat may be transferred from one object or location to another through conduction, convection, or radiation.

Conduction transfers heat from one substance to another through direct contact of the substances' molecules. When you warm your cold hands by holding a mug of hot chocolate or cool your hot face by holding an ice pack against it, you are benefiting from conduction. All molecules move, but warmer molecules move faster than colder molecules. When fast-moving, warmer molecules come in contact with slower, cooler molecules, some of the kinetic energy of the warmer molecules transfers to the cooler molecules. That is why two substances of different temperatures that stay in contact with each other eventually reach the same temperature. If you put a small piece of frozen meat into warm water, after a while the water and the meat will be the same temperature.

The sun provides radiant heat.

Warming hands through conduction

Some substances transfer heat better than others. A substance that conducts heat well is called a **thermal conductor**. Metals such as silver and copper are excellent heat conductors. A substance that does not conduct heat well is called a **thermal insulator**. Glass, wood, rubber, and plastic are all insulators. Air is also an important insulator. Storm windows use air to reduce heat loss. Because air insulates, the layer of air trapped between the two panes of glass helps to keep the house warm.

Psalm 139:14 says, "I praise You because I am fearfully and wonderfully made; Your works are wonderful, I know that full well." Part of this wonder comes from consideration of human body tissue, which maintains body temperature. Because your body tissues insulate well, the inner core of your body stays warm in a cold environment. What would happen if God had not designed the human body this way?

Convection is the second method of heat transfer. Convection occurs in moving fluids (gases and liquids) through the circulation of currents from one region to another. Have you ever watched a hawk or eagle glide in circles searching for prey? These birds are riding air currents. In other words, they take advantage of convection currents to conserve their energy.

Most heat transfer in gases and liquids occurs through convection. The warmer part of the fluid becomes less dense and rises, carrying heat with it. When a warmer fluid arrives in cooler regions, it cools, becomes more dense, and sinks again.

FYI

Coober Pedy

Coober Pedy is an opal mining town in South Australia. The desert climate creates extremely hot and cold temperatures, so many residents live underground in homes called *dugouts*. Many businesses are hewn from the rock as well. The rock and dirt provide great insulation from the daytime heat and nighttime cold. The temperature in the dugouts stays at about 24°C year-round.

Opal

Rising warm air and sinking cool air form convection currents that transfer heat around the world. These convection currents in the atmosphere cause Earth's weather patterns. Convection currents occur in the ocean as well, transporting nutrients to organisms living at different ocean depths.

One benefit of convection currents is the use of a hot-water or steam radiator. Air near the radiator heats up and rises. Air near the walls and windows away from the radiator cools and falls. These convection currents establish a pattern of flow to distribute heat throughout the room.

Radiation is the third method of heat transfer. Conduction and convection move heat through substances, but radiation does not need a medium. Radiation transfers energy through space or matter by waves or rays. The sun's heat reaches Earth and the rest of the solar system through radiation. The heat given off by a stove, fire, or electric heater also travels by radiation. Because of radiation, a marshmallow can be toasted over a fire even though the marshmallow never touches the flames.

About one-billionth of the sun's radiation falls on the earth, providing renewable energy for life and work. Fossil fuels actually store solar radiation, and the energy released from burning wood also comes from stored solar radiation. Eco-hotels in the tropics use solar energy to heat black drums full of rain water to supply guests with hot showers. Around the world, people take advantage of solar radiation by drying their clothes on clotheslines. Solar cells and conductors trap solar energy,

SCIENCE IN ACTION

Induction Cooker

The induction cooker transmits an electromagnetic signal through a copper coil to an iron pan, or a steel pan with a high percentage of iron. Iron is not a good conductor of electricity, so the pan actually produces the heat. The stove top does not get hot because the burner does not produce heat. The electromagnetic waves travel from the copper coil into the iron pan. The magnetic field generated in the iron creates kinetic energy and ultimately heat in the pan, which is then transferred to the food. Induction cookers can cook faster than conventional stove tops. Would an aluminum pan work on an induction cooker? Why?

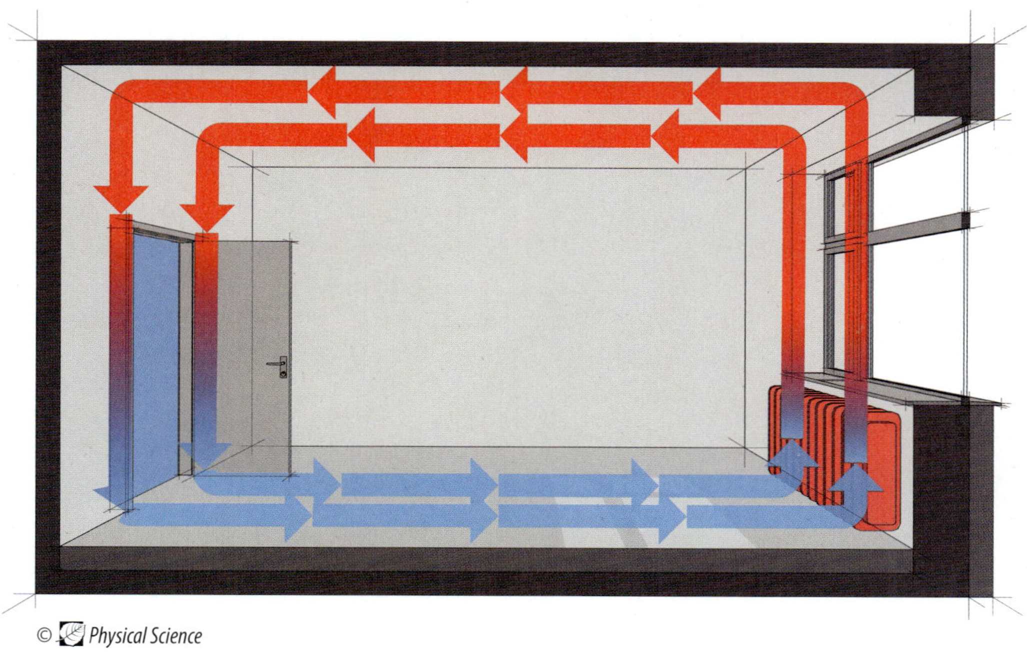

A radiator creates convection currents.

FYI

Zeroth Law of Thermodynamics

The first law of thermodynamics, which is also called *the law of conservation of energy,* states that energy can neither be created nor destroyed. The zeroth law of thermodynamics was instituted after the first law, but seemed more foundational than the first, so it has been given the name *zeroth law.* The zeroth law states that if two systems are in thermal equilibrium with a third system, then they are in thermal equilibrium with each other. In other words, If the temperature of object A (T_A) equals the temperature of object B (T_B), and the temperature of object B (T_B) equals the temperature of object C (T_C), then the temperature of object A (T_A) equals the temperature of object C (T_C).

If $T_A = T_B$ and $T_B = T_C$, then $T_A = T_C$.

converting it into electricity. In the past, solar power has been an expensive form of energy, but new technology promises to reduce costs.

LESSON REVIEW

1. Name and describe the three ways heat can be transferred.
2. What is the difference between a thermal conductor and a thermal insulator?
3. What materials are used to make pots and pans? Why?
4. Explain how conduction works.
5. Explain how convection works.
6. Give some examples of radiation.

The earth benefits from solar radiation.

BIBLE CONNECTION

God's Heat

"How many are Your works, Lord! In wisdom You made them all" (Psalm 104:24a). Heat is one of God's works. In wisdom, God designed heat and molded its behavior to sustain creation. Think about what life would be like without convection, conduction, radiation, or insulation. What would you have to do to survive? Could you survive?

How do you get to school, music lessons, or soccer practice? How do you travel to your favorite vacation spot? How do the clothes, groceries, books, sports equipment, electronics, and art supplies you buy get to the stores where you shop? Life would certainly be different without the planes, cars, trucks, and buses that carry people and materials across the country and around the world.

OBJECTIVES

- Compare types of engines.
- Describe the heat transfer in a heat pump.

VOCABULARY

- **heat engine** a machine that converts heat energy into mechanical energy in order to do work

None of these forms of transportation would be possible without heat engines. A **heat engine** is a machine that converts heat energy into mechanical energy in order to do work. Many heat engines use the process of combustion, the burning of substances, to produce heat. During combustion, a fuel is burned—heated to a temperature at which it will combine with oxygen to give off heat. Some heat engines use solar energy for power.

There are three types of heat engines. An external combustion engine is a heat engine that burns fuel outside the engine. Steam engines, once widely used in train locomotives and steamships, are external combustion engines. Coal, wood, or other combustible materials burn to heat the water in a boiler, outside the engine. The steam produced is then routed through a valve in the engine. Early steam engines used the steam to move a piston, which moved back and forth in a cylinder. This moving piston was attached to a rod; the rod, in turn, moved a wheel or propeller. But modern steam engines no longer use

Diesel truck

HISTORY

Engines

- **Hero of Alexandria** created a steam engine in the first century AD. The aeolipile was a hollow sphere attached to two metal tubes that carried steam into it. Two other bent tubes were attached to the tube in such a way that when the steam blew out of the tubes, the sphere would spin. This is the first known steam-powered machine.
- **Thomas Newcomen**, a British engineer, designed the steam pump and built the steam engine with Thomas Savery in 1712.
- **James Watts**, a Scottish inventor, improved Newcomen's engine by lessening the consumption of steam and in 1782 patented the double-acting engine. The piston pushed and pulled.
- **Nikolaus Otto**, a German engineer, was the first to build a gasoline-powered engine. In 1876, he constructed an internal-combustion engine with a four-stroke cycle. The piston moved four times for each ignition. Otto had received a patent for his engine but it was revoked in 1886, because French engineer Alphonse Beau de Rochas had his four-stroke engine patented in 1862. Because Otto was the first to build an engine using this design, and because more than 30,000 of them were built, this four-stroke cycle is referred to as *the Otto cycle*.

Steam Turbine

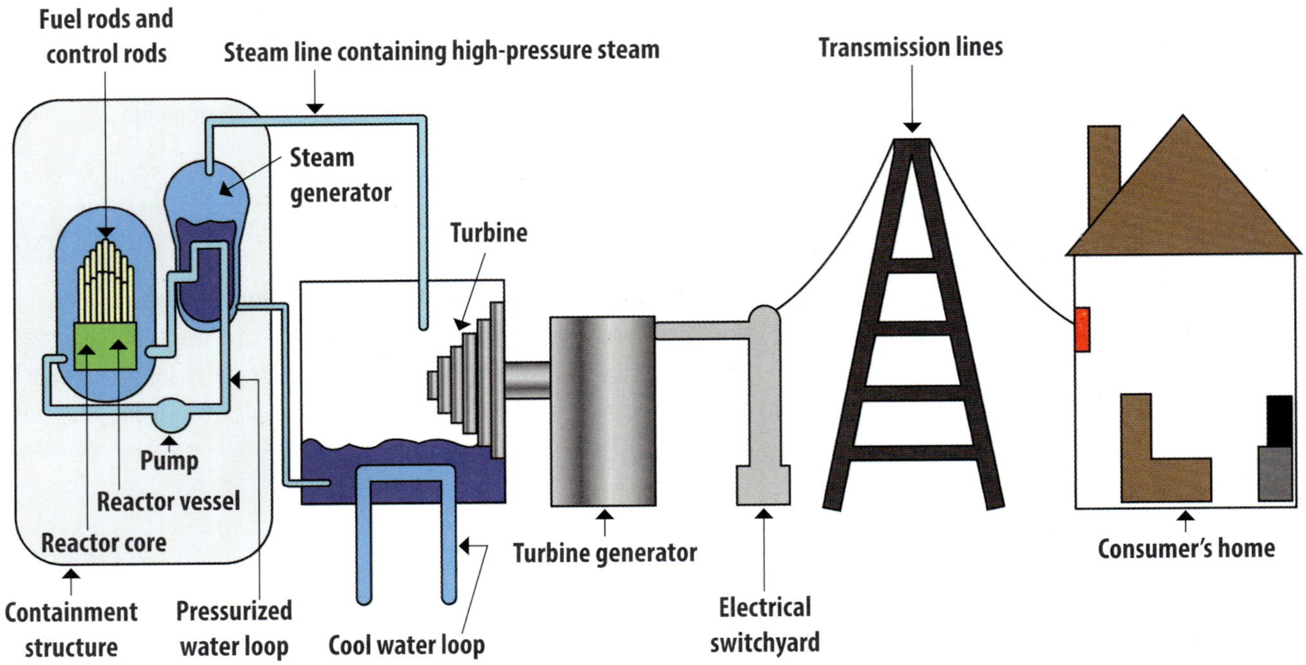

pistons and cylinders. In newer engines, steam is placed under great pressure and passed through holes onto wheels called *turbines*. Turbines, which waste less energy than pistons, rotate at high speeds and produce kinetic energy. The kinetic energy is transferred to mechanical energy, which is used to produce electrical energy.

Another method for producing steam uses the heat created from nuclear fission. Nuclear power plants use the radioactivity of uranium to heat water and produce steam. The steam powers the turbines in the same way the external combustion engines do.

An internal combustion engine is a heat engine that burns fuel inside the engine. A gasoline engine is a familiar type of internal combustion engine. This type of engine may work in a four-step or two-step process. A gasoline engine is called *a four-stroke engine* and it works in the following ways:

Intake stroke. During the first stroke, gasoline turns to vapor and mixes with air in a carburetor or fuel injector. The air-gasoline mixture moves to a cylinder that has a piston inside.

Compression stroke. The piston rises in the cylinder and compresses the vapor mixture to a fraction of its original volume.

CHALLENGE

Newton's Steam Wagon
Construct a working model of a steam engine or aeolipile and use it to convert thermal energy to mechanical energy.

Four-Stroke Engine

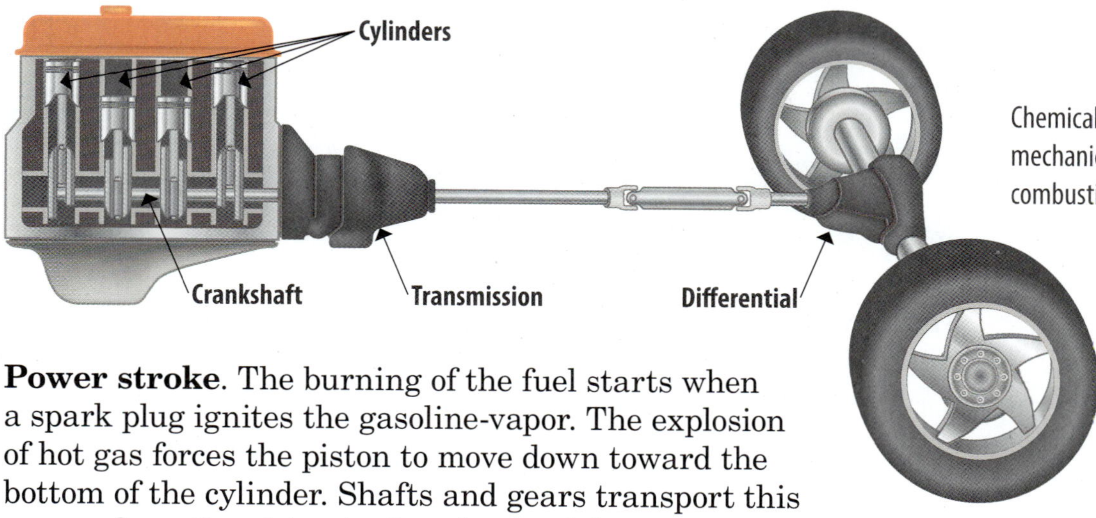

Chemical energy is converted to mechanical energy in an internal combustion engine.

Power stroke. The burning of the fuel starts when a spark plug ignites the gasoline-vapor. The explosion of hot gas forces the piston to move down toward the bottom of the cylinder. Shafts and gears transport this energy from the piston to the wheels of the car.

Exhaust stroke. The piston moves back to the top of the cylinder to get rid of excess gases through the exhaust valve.

As the piston falls back down in the cylinder afterward, the carburetor, or fuel injector, produces more vapor to start the cycle over again.

In a diesel engine, only air is added during the intake stroke. The fuel is injected into the air at the end of the compression stroke. The act of compression raises the temperature enough to ignite diesel fuel, which has properties that allow ignition to take place without a spark plug. That is why diesel engines do not need spark plugs.

Four Stages of the Four-Stroke Engine

A motorboat uses an internal combustion engine.

Most lawn mowers, jet skis, chain saws, and motorboats have internal combustion engines that work in a two-step process. Two-stroke engines combine the exhaust and intake steps near the end of the power stroke. Two-stroke engines are cheaper than four-stroke engines, but they use more fuel.

A combustion engine burns fuel and converts chemical energy into mechanical energy. A heat pump, on the other hand, has a different job. It transfers heat energy from one location to another. A heat pump circulates liquids that evaporate easily at low temperatures in order to move the heat in the air out of or into a building. The liquid picks up heat from the air or ground.

Diesel trucks do not use spark plugs.

The heat turns the liquid into a gas. The gas travels into a compressor that applies pressure to raise the temperature of the gas. This hot gas then passes through a coil inside the building and releases the heat. Once the hot gas gives off its heat, it changes back into a hot liquid. The liquid is then cooled and returned to the outside to collect more heat.

BIOGRAPHY

Margaret (Mattie) Knight

When you think of important inventions, things like light bulbs, phonographs, computers, and cell phones may come to mind. But what about the paper bag? Margaret Knight, born in 1838 in the state of Maine, began inventing at the age of 12. She was visiting a textile mill when she saw a shuttle fly off a loom and stab a boy. Bothered by this unsafe environment, she invented a device that would shut down the machine if something were caught in it.

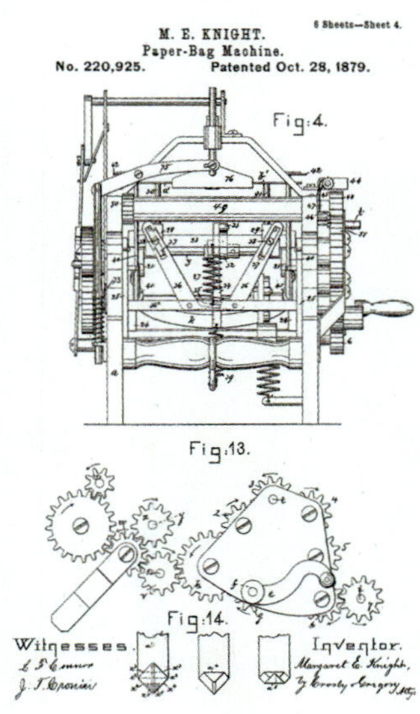

As an adult, she worked for a paper bag company in Boston, MA. At the time, paper bags did not have flat bottoms and were difficult to fill. Knight spent several years designing and building a model of an attachment for the paper bag machines to create the flat bottom. In 1870, she received a patent for her invention. Today, grocery bags, lunch bags, and gift bags still use Margaret Knight's design. Knight went on to invent several more machines including a skirt shield, a clasp for robes, and a rotary engine. She earned 27 patents in her lifetime.

FYI

Perpetual Motion Machine

Through the centuries, inventors and scientists have struggled to develop a machine that can perpetually generate its own energy to keep it running. Many attempts have been made and many machines have maintained motion for a length of time, but physics dictates it is not possible. The law of conservation of energy states that energy can neither be created nor destroyed. In order for a machine to stay in motion indefinitely, three things would need to be true: 1. The machine could not have any rubbing parts, so no friction, 2. The machine would have to operate inside a vacuum to avoid air resistance, and 3. The machine could not produce any sound, because sound is energy. The reality is, energy would have to be added to the machine to start it, and even if it kept itself going, it could not produce any usable energy to make anything happen. All the energy would have to stay in the machine to keep it going. A perpetual motion machine, then, amuses the mind and tickles the imagination, but it cannot be created.

LESSON REVIEW

1. What is the difference between a modern steam engine and a gasoline engine?
2. How does the fuel in an engine become hot enough to burn?
3. In which of the strokes of the gasoline engine does the piston do work that can be used by a car?
4. What is the difference between a combustion engine and a heat pump?
5. Why must the liquid in a heat pump be cooler than the air around it before it can collect more heat?

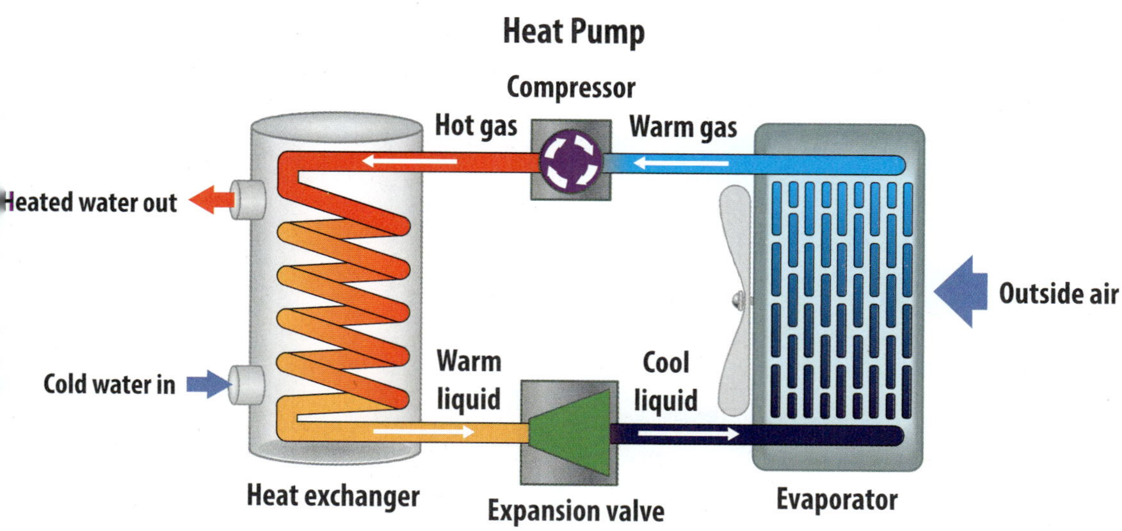

5.2.1 *Radioactive Elements*

OBJECTIVES

- Describe Becquerel's discovery of radioactivity.
- Compare stable and unstable isotopes.

VOCABULARY

- **radioactivity** the ability of some elements to give off nuclear radiation as a result of a change in the atom's nucleus
- **stable nucleus** a nucleus that does not decay and fall apart
- **unstable nucleus** a nucleus that naturally decays and falls apart

Nuclear energy is sometimes referred to as "atomic energy" because it is the energy stored in the nucleus of atoms. Through the fusion and fission of atomic particles, nuclear energy is created to produce electricity and to provide diagnostic and curative medicine. Nuclear fission was first used in the early 1940s with the Manhattan Project. The project produced a military weapon in the form of a bomb that released a fast neutron chain reaction of nuclear fission. The war against Japan ended with the bombing of Hiroshima and Nagasaki. Nuclear fission was not used to produce electricity until 1953. Nuclear power potentially offers large amounts of energy. But, nuclear energy also creates the burden of tons of toxic waste that will remain radioactive for millions of years. Radioactive substances can be very dangerous as well as beneficial. It is important to weigh the costs, benefits, and advantages associated with nuclear energy.

The source of nuclear energy, or radiation, was discovered by accident. French scientist Henri Becquerel (1852–1908) knew that fluorescent substances glow in sunlight. In 1896, he wrapped a photographic plate in lightproof paper to determine if fluorescent substances give off X-rays. He placed uranium salt on top of the paper and left these materials in direct sunlight to cause the uranium to become fluorescent. Becquerel hypothesized that if the fluorescent uranium gave off X-rays, the X-rays would pass through the paper to form an image on the plate. When he developed the plate, he saw an image. Becquerel wanted to try the experiment again, so he prepared

Nuclear power station

the uranium salt, lightproof paper, and plate. Then the weather turned cloudy, so he stored the materials in a drawer. Becquerel decided to develop the plate even though it had not been exposed to sunlight. He was astounded to find an image as clear and strong as the image produced in the sunlight. Becquerel's discovery was much more exciting than he had intended! Sunlight had nothing to do with the image on the plate, and something other than X-rays had passed through the paper to produce an image. This invisible "something" was later named **radioactivity**.

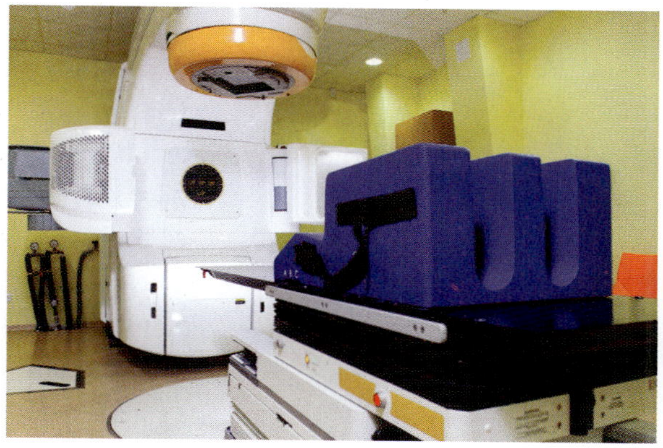

Radiation technology used in cancer treatments

After testing many other elements, Becquerel found that the source of the radioactivity was uranium. Radioactivity is the ability of some elements to give off nuclear radiation as a result of a change in the atom's nucleus. During the same era, other scientists were testing radioactivity. While Polish scientist Marie Curie (1867–1934) and her husband were working with uranium, they discovered two other radioactive elements: polonium (named after Curie's native country) and radium.

Many people are afraid of radiation, but most do not know the details of what it is or where it comes from. Small amounts of

BIOGRAPHY

Marie Curie

Maria Sklodowska was born in Warsaw, Poland. In 1891, Maria went to Paris to study mathematics, chemistry, and physics at the Sorbonne. In 1895, she married Pierre Curie, a physics teacher at the University of Paris. Maria, now called Marie, and her husband worked together, researching radioactive substances. The Curies discovered two highly radioactive elements, radium and polonium. Marie and Pierre, along with Henri Becquerel, won the 1903 Nobel Prize in physics for their discovery. After Pierre's death in 1906, Marie became the first woman to teach at the Sorbonne. She continued to research radioactive elements, and in 1911 she won the Nobel Prize in chemistry for isolating radium and studying its chemical properties. She was the first woman to win a Nobel Prize.

In 1914, Madame Curie helped found the Radium Institute in Paris and served as its first director. She died in 1934 of leukemia, which may have been caused by exposure to radiation during her research. After her death, the Radium Institute was renamed the Curie Institute.

TRY THIS

Design an Experiment

Create an experiment to measure the effects of radiation on irradiated radish seeds. Include a hypothesis, materials list, procedure, and type of data to be gathered. Describe what variable(s) would indicate a difference between irradiated seeds and regular seeds.

FYI

Radon

Radon is a radioactive gas that results from the natural, radioactive decay of uranium in soil, rocks, and water. Airborne radon is considered a pollutant because high levels, or concentrations, can contribute to lung cancer. The gases can seep into buildings and collect in large quantities. It is colorless and odorless, so special equipment is required to detect the presence of the gas. Consumers can buy test equipment or hire a company to test for the presence of radon. If radon is detected, a vent pipe system and fan can be installed to mitigate the gas. Sealing cracks in foundations improves the efficiency of the mitigation system. These systems are inexpensive and very effective.

radiation are everywhere. Radiation comes to Earth from space constantly and there are many Earth-based natural radioactive elements.

To understand how radiation works, think back to the intricate design of atoms, the foundation of matter. Most elements are not radioactive. These substances have a **stable nucleus**—a nucleus that does not decay and fall apart. In contrast, an **unstable nucleus** naturally decays and falls apart. This means

BIOGRAPHY

Lise Meitner

Lise Meitner (1878–1968) was an Austrian nuclear physicist in an era when most women could not go to college. Her brothers were able to go to a grammar school to prepare for university, but she was not. Lise stayed home and studied on her own but eventually asked her father if she could go to the University of Vienna. He paid for a private tutor to prepare her for the entrance exam. She entered the university in October of 1901, and by February of 1906, Meitner had passed her doctoral oral exam with the highest honor, and had earned a PhD in physics.

Meitner spent the next few decades researching and teaching. In the mid-1930s, Meitner and two chemists, Otto Hahn and Fritz Strassmann, engaged in research that led to the discovery of fission in their lab in Berlin. Because Meitner was Jewish, she was forced to flee from the Nazis in 1938. She continued her work on nuclear fission at the Bohr Institute in Copenhagen, Denmark, and then at the Nobel Institute in Stockholm, Sweden. In 1944, Hahn alone was awarded the Nobel Prize for the discovery; the Nobel Prize committee overlooked the contributions of both Strassmann and Meitner.

Isotopes of Carbon

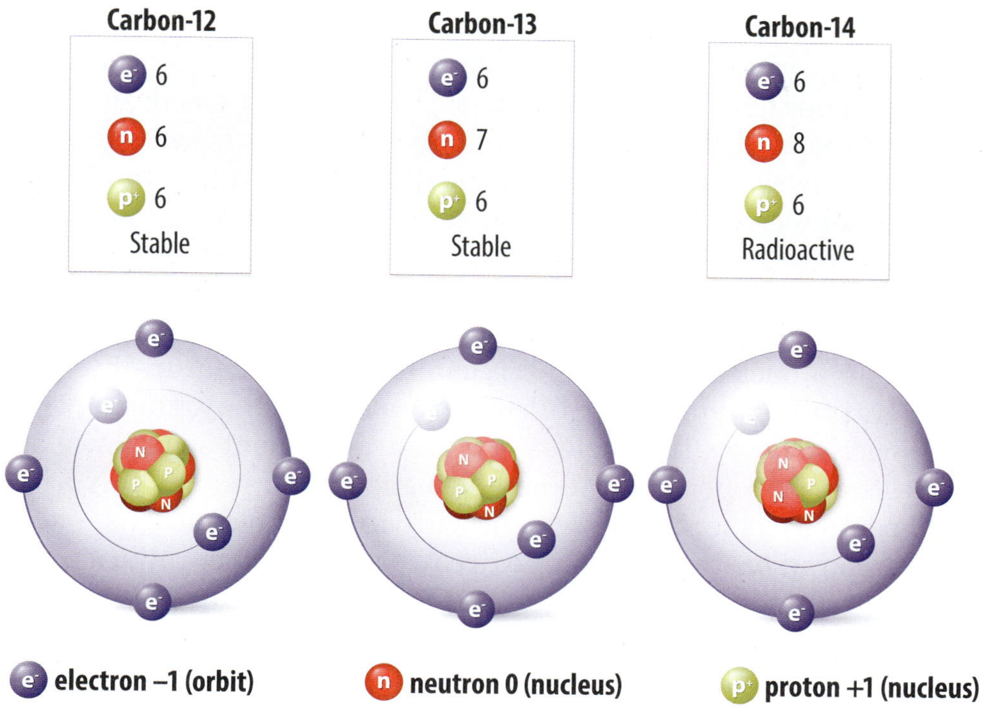

that it gives off particles. An element with an unstable nucleus is called *a radioactive element.*

All elements with an atomic number greater than 83 are radioactive. Many elements with atomic numbers less than 83 have at least one radioactive isotope, an atom of the same element that has the same number of protons but different numbers of neutrons. A radioactive isotope has an unstable nucleus. Carbon-14 (used to date organic substances), hydrogen-3, and oxygen-14 are examples of radioactive isotopes. The number that follows an isotope refers to its mass number, which is the total number of protons and neutrons.

LESSON REVIEW

1. What was Becquerel's hypothesis?
2. Describe Becquerel's experiment.
3. How did Becquerel's experiment demonstrate that sunlight and fluorescence did not form the image on the film?
4. What is a radioactive isotope?
5. What is the difference between a stable and an unstable nucleus?
6. Explain the difference between the carbon-12 and carbon-14 isotopes.

5.2.2 *Nuclear Decay*

OBJECTIVES

- Describe three types of radioactive decay.
- Calculate the amount of decay in a sample of radioactive isotope.

VOCABULARY

- **alpha particle** a particle with two protons and two neutrons released during alpha decay
- **beta particle** an electron or a positron released from a nucleus during beta decay
- **gamma ray** an electromagnetic wave with the shortest wavelength and highest frequency released during gamma decay
- **half-life** the time it takes for one-half of the atoms of an isotope in a given sample to decay into atoms of another element
- **nuclear radiation** the particles released during radioactive decay
- **radioactive decay** natural transmutation
- **transmutation** the process of change within an atom's nucleus that produces a different element

Imagine what life would be like if common elements in the world could suddenly turn into other elements. What if most of the oxygen in the room could turn into nitrogen? Or what if the iron in a bridge could turn into soft aluminum? There is no need to worry about disasters such as these, because God has designed most elements to be stable and to not change into other elements. He did, however, create a few elements with the ability to change into other elements through a process called *transmutation*.

Transmutation is the change within an atom's nucleus that produces a different element. This process can happen in a lab or in a nuclear reactor, but it can also happen naturally. Natural transmutation is called **radioactive decay**. During radioactive decay, the nucleus of a natural element spontaneously breaks apart to form a different element that is stable. Only a radioactive element—an element with an unstable nucleus—can undergo radioactive decay. The decay produces a nonradioactive element in a sequence of steps—a decay series. Uranium-238, for example, goes through 15 steps of radioactive decay before forming the stable element lead-206.

During radioactive decay, particles are released. The particles released during radioactive decay are called **nuclear radiation**. There are three different types of radioactive decay: alpha, beta, and gamma. Each releases a different kind of particle.

In alpha decay, a nucleus releases an **alpha particle,** which contains two protons and two neutrons. (An alpha particle is the equivalent of a helium nucleus.) An element's atomic number is the number of protons in its nucleus; this number identifies the element. The mass number is the total number of neutrons and protons in a nucleus. Because protons are lost from the nucleus in alpha decay, the atomic number changes. This changes the element itself. If polonium-214 goes through alpha decay, it loses 2 protons and 2 neutrons. What does it become? To determine that, look at the periodic table to find the element with an atomic number that is 2 less than polonium. It is lead—Pb. Then, subtract 4 from the original mass number, because the alpha particle contains 4 particles. The new mass number of the isotope is 210. Therefore, the new isotope is lead-210. Alpha particles cannot travel very far. They travel only a few centimeters in the air and have so little energy that a sheet of paper can stop them.

The second type of radiative decay is beta decay. Beta decay is the release of a beta particle from a nucleus. A **beta particle** is an electron (with a negative charge) or a positron (with a positive

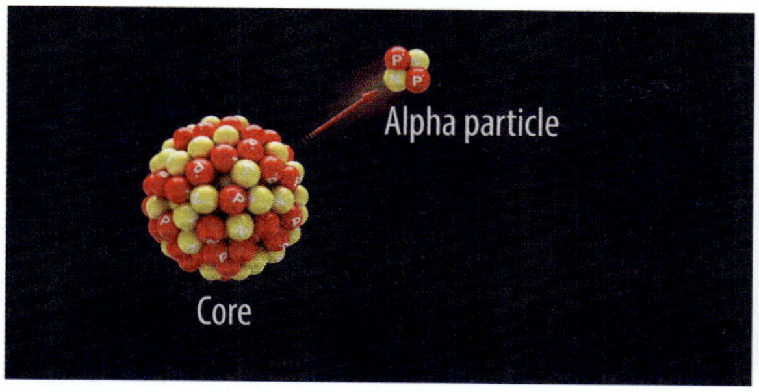

Alpha decay

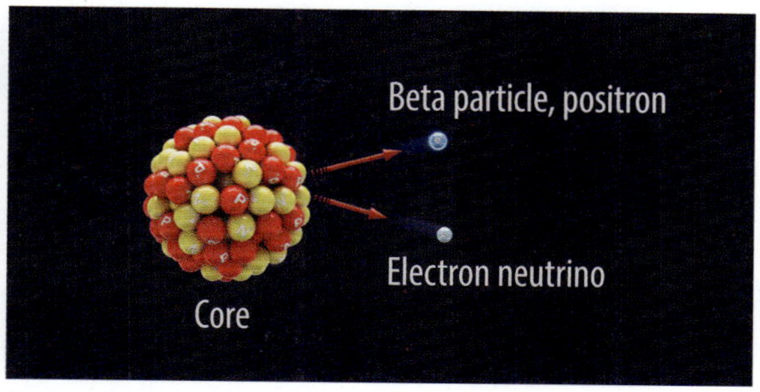

Beta-plus decay

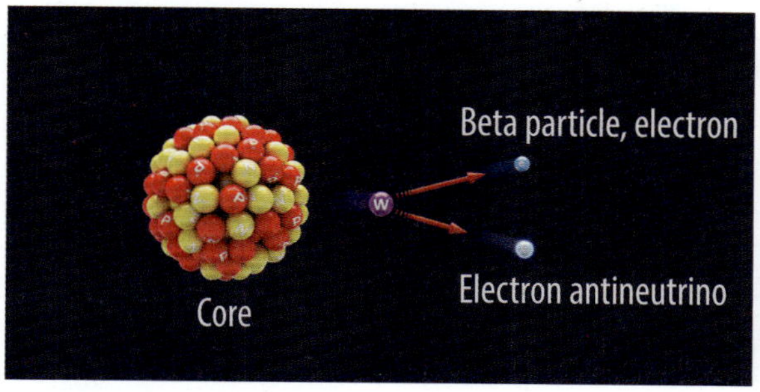

Beta-minus decay

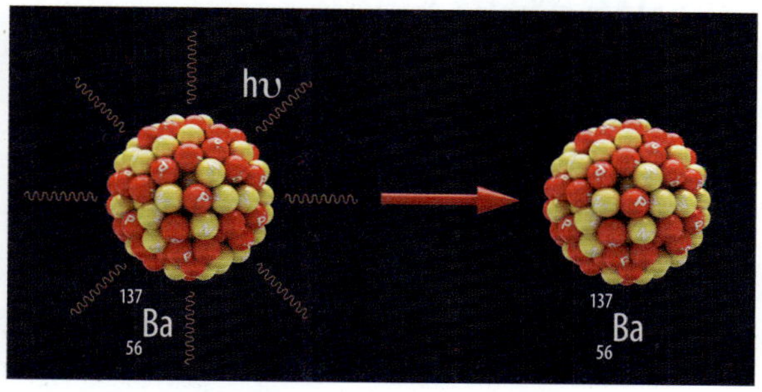

Gamma radiation

charge). During beta decay, a neutron in the nucleus splits into a proton and an electron. The electron is released from the nucleus. During beta decay the neutron is lost because of the split, but a proton is gained; the mass number of the atom remains the same, but a new element is formed. For instance, the atomic number for bismuth is 83. When bismuth-214 goes through beta decay, its atomic number increases and its mass number remains the same. The next element on the periodic table is polonium, with an atomic number of 84. When bismuth-214 goes through beta decay, it becomes polonium-214. Beta particles have more energy than alpha particles and they are able to travel several

Penetrating Power of Various Types of Radiation

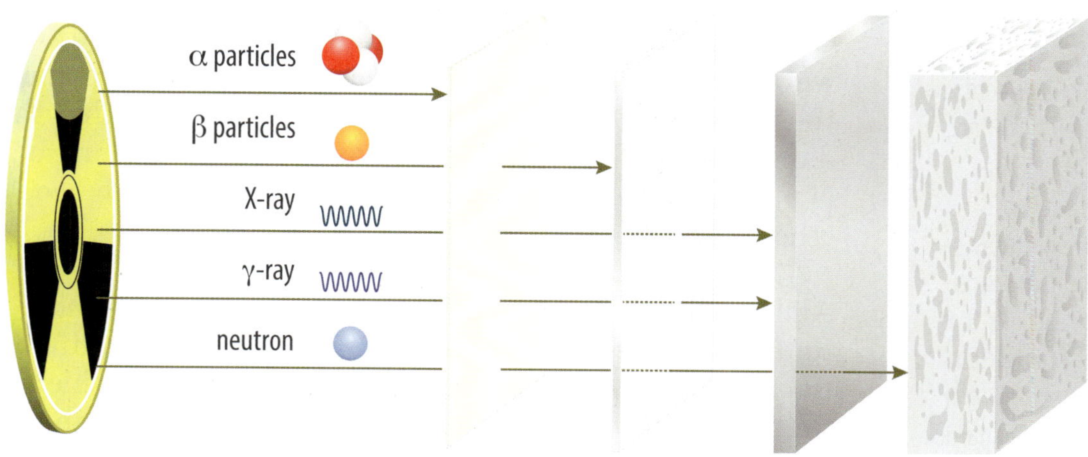

FYI

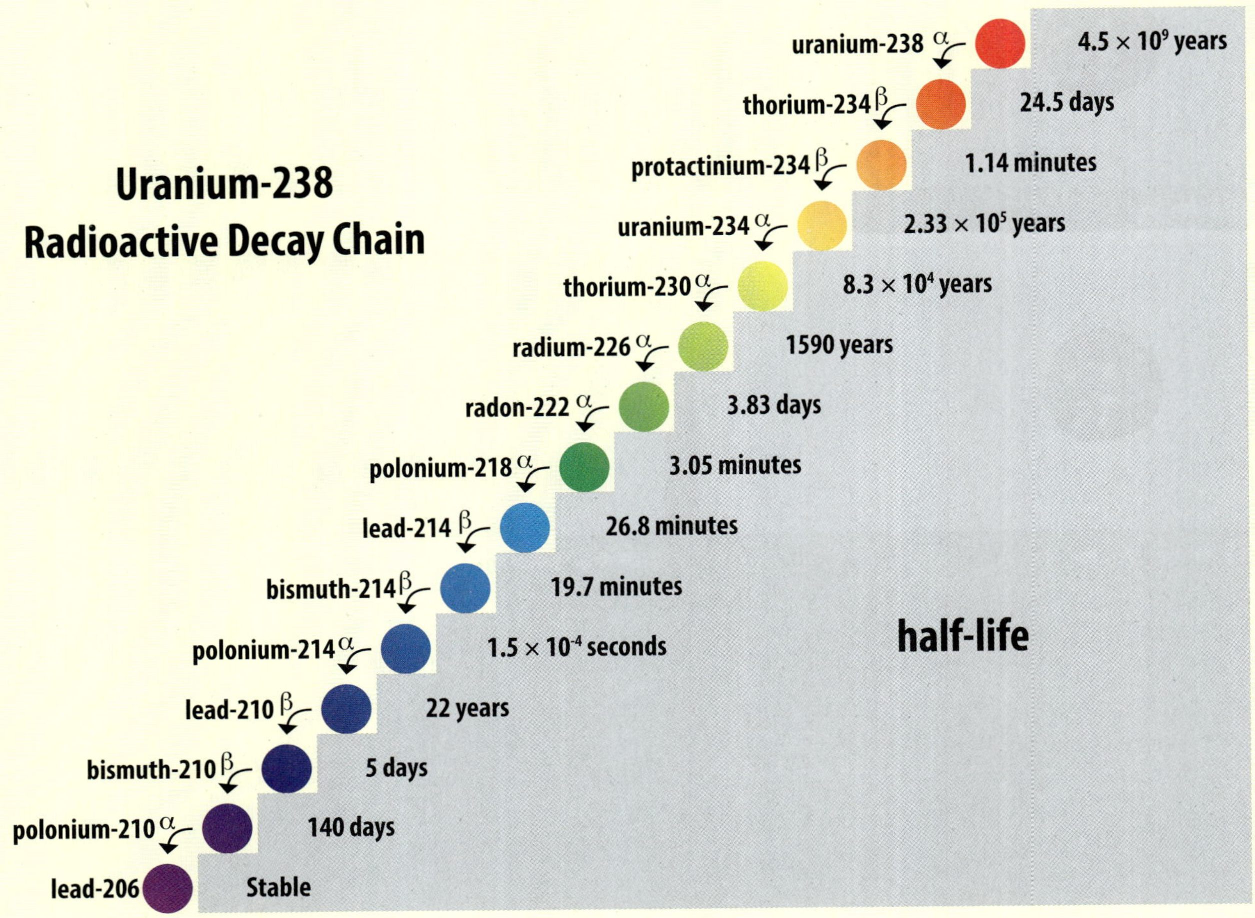

TRY THIS

Cloud Chamber

Substances everywhere give off microscopic radiation. A small cloud chamber can provide a glimpse of radioactive particles rocketing through space. Follow your teacher's directions to build a cloud chamber and record what types of radiation or particle movement are observed.

meters in the air. They can travel through paper, but they can be stopped by thin layers of metal or plastic.

Alpha and beta decay are sometimes accompanied by gamma decay. Gamma decay releases **gamma rays** from a nucleus. Gamma rays are electromagnetic waves with the shortest wavelengths and highest frequencies. Their high frequency gives them extremely high amounts of energy—so high that they are dangerous to all living things. Gamma rays travel up to hundreds of meters in the air and can easily go right through people and other living organisms.

Alpha, beta, and gamma decay do not happen instantly. Radioactive elements decay into other elements at different rates. Each radioactive element has a unique **half-life**, which is the time it takes for one-half of the atoms of an isotope in a given sample to decay into atoms of another element. Usually the new

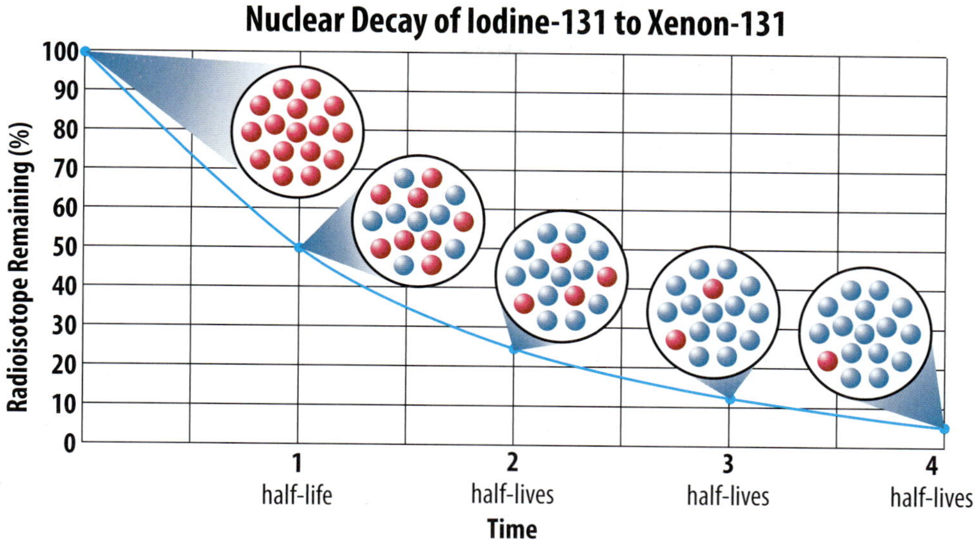

element, or daughter product, is also an isotope. For example, bismuth-210 has a half-life of five days. This means that every five days, half of a quantity of bismuth-210 will go through beta decay and become polonium-210. If given a 400 g sample of bismuth-210, in five days there will be 200 g of bismuth-210 and 200 g of polonium-210. Half of the sample decays in one half-life. In five more days, half of the 200 g of bismuth-210 will become polonium-210, leaving 100 g of bismuth-210 and 300 g of polonium-210.

High amounts of nuclear radiation are dangerous to living things. Radiation levels are monitored to keep people safe. Nuclear radiation can be detected and measured by an instrument called *a Geiger counter.* A Geiger counter contains a tube filled with a gas such as argon or helium. When the gas comes in contact with radiation, some of the gas is separated into ions, or charged particles, by the radiation. The ions are detected by an electrical pulse. The current produces a clicking sound or a flashing light. The faster the clicks or flashes, the more radiation is present. Geiger counters detect alpha, beta, and gamma radiation.

LESSON REVIEW

1. What is transmutation?
2. What happens during alpha decay?
3. What happens during beta decay and gamma decay?
4. What is the half-life of a radioactive element?
5. Do stable isotopes have half-lives? Why?
6. How much bismuth-210 will be left in 20 days if there is an 800 g sample initially?

FYI

Half-Lives of Radioactive Substances

carbon-14	beta
5,730 years	
chlorine-36	beta
301,000 years	
cobalt-60	beta
5.27 years	
iodine-131	beta
8.03 days	
phosphorus-32	beta
14.26 days	
polonium-215	alpha
1.78 milliseconds	
radium-226	alpha
1,600 years	
radon-222	alpha
3.8 days	
sodium-24	beta
15 hours	
thorium-232	alpha
14.05 billion years	
uranium-235	alpha
7.04 billion years	
uranium-238	alpha
4.5 billion years	

5.2.3 *Nuclear Power*

OBJECTIVES

- Explain how Einstein's formula relates to the energy derived from nuclear power.
- Distinguish between nuclear fusion and nuclear fission.

VOCABULARY

- **critical mass** the minimum amount of isotope needed to sustain a chain reaction
- **meltdown** what occurs when the fuel in a reactor melts and radiation leaks out of the reactor
- **nuclear chain reaction** a continuous series of nuclear fission reactions

Gazing at the sky on a clear summer night is a wonderful experience. The distant stars glow in the blackness. The scene is peaceful. Yet, each of the trillions of stars in the universe blazes like an inferno similar to the sun. Imagine hiking through the desert at noon on a hot summer day. That kind of heat pales in comparison with the energy released from every star.

This immense heat is created by nuclear energy. The two types of nuclear energy are fusion and fission. Nuclear fusion is the joining of two light atomic nuclei to form a single heavier nucleus. Nuclear fission is the splitting of a heavy atom's nucleus into two lighter nuclei. In both of these processes, a small amount of matter is converted into energy—as explained by Albert Einstein's famous equation, $E = mc^2$. The equation states that if a small amount of mass is destroyed, a tremendous amount of energy is produced. The E stands for energy, the m is the mass destroyed, and the c equals the speed of light, which is about 300,000,000 m/sec, or the rate at which the energy is released.

The process of nuclear fusion requires heat. For nuclear fusion to take place, the atoms must be over a million degrees Celsius. Nuclear fusion occurs in the sun and stars. God uses this process to create the sun's energy and to make the sun and other stars burn so brightly. The sun's core temperature can reach 20,000,000°C, and this tremendous amount of heat sustains its fusion reactions. In the fusion reaction, four hydrogen nuclei fuse together to become one helium nucleus. If the fusion continued unchecked, the sun would eventually lose so much

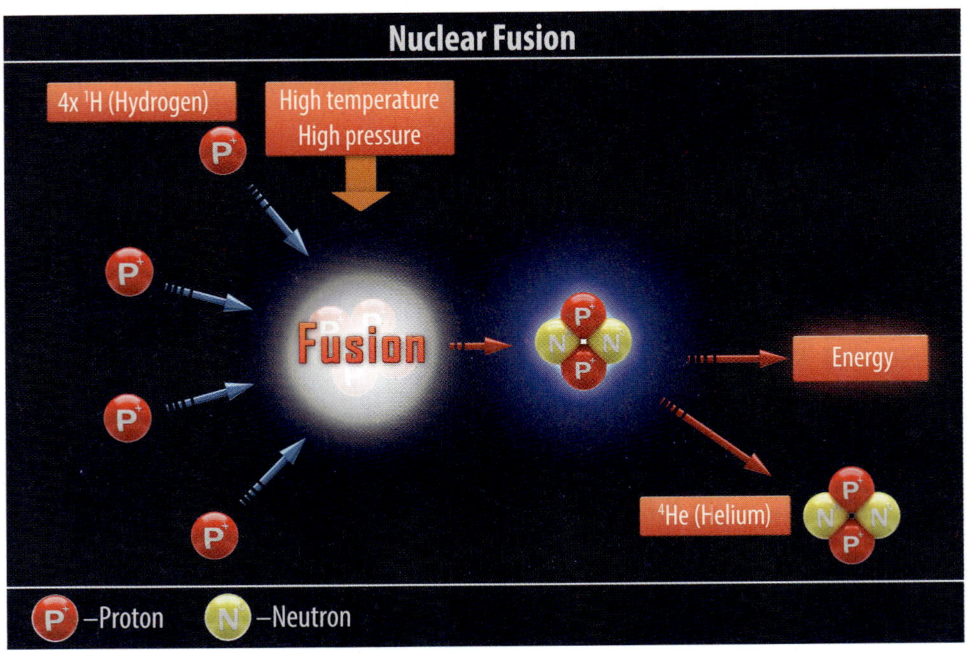

density that fusion would stop. But gravity maintains a viable level of density to keep the fusion reactions going. God designed the sun with a regulating system that changes the temperature of the core to slow down or increase the rate of fusion as needed. And He created the perfect balance of gravity and fusion. With too much gravity, the sun would implode. With too much fusion, the sun would explode.

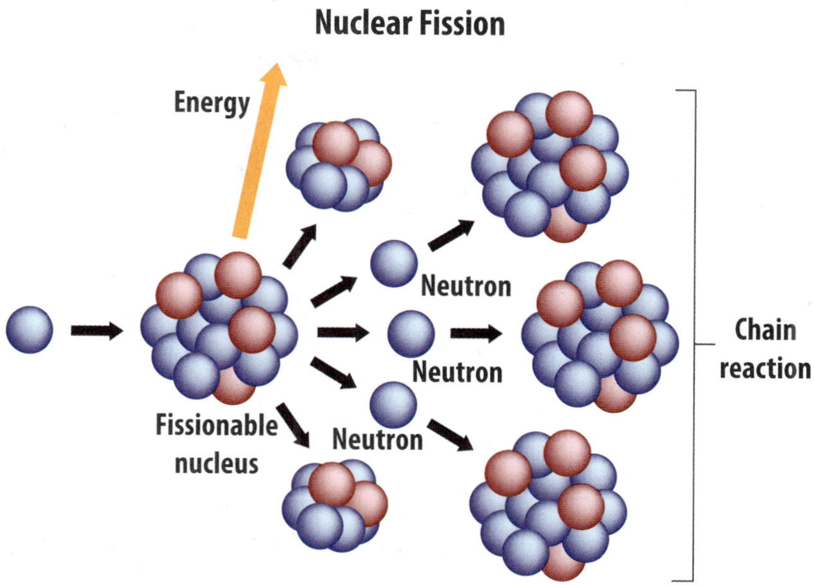

When the sun runs out of hydrogen, it will stop producing heat. Scientists estimate that this will happen in approximately five billion years. Although it is not known how or when God will choose to end life on Earth, it is known that at the end of time, "The sun will no more be your light by day ... for the Lord will be your everlasting light, and your God will be your glory" (Isaiah 60:19). There is no need to worry about what will happen because God has a plan for that time.

In a nuclear fusion reaction, some of the matter is turned into energy. Nuclear fusion is very powerful. Think of it: the sun resides millions of kilometers away yet it provides enough heat to warm the earth. On the sun, nuclear fusion is self-sustaining; on Earth, it does not happen naturally. Creating nuclear fusion in a lab is still very expensive, so at this time it is not a practical energy source.

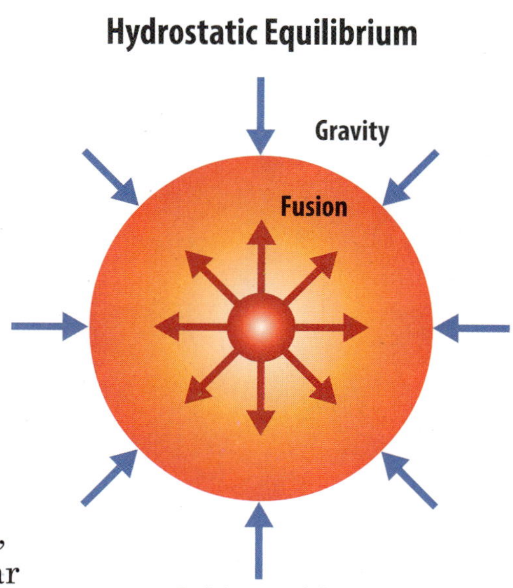

God designed the sun to have a balance of gravity and fusion.

Nuclear fission, on the other hand, is a practical energy source. Fission occurs most often inside a nuclear reactor. Nuclear fission is the splitting of uranium (U) or of plutonium (Pu) into two or more lighter nuclei. Generally, only atoms with an atomic number greater than 90 can undergo fission and release energy. In a nuclear fission reaction, a uranium-235 nucleus is hit by a neutron, which scientists call *a nuclear bullet*. As a result, the uranium nucleus splits. The collision is similar to a billiard ball hitting a bowling ball so hard that the bowling ball splits in half. The nucleus may split, for example, into krypton-92 and barium-141. This nuclear fission process also releases three neutrons. These three neutrons act as nuclear bullets and split three more uranium

TRY THIS

Nuclear Fission
Design a demonstration of nuclear fission using a twisting balloon to model a heavy isotope like uranium-235. Show the fission process and the release of energy. Discuss ways to dispose of the waste products.

nuclei, which also become nuclear bullets. This continuous series of nuclear fission reactions is called a **nuclear chain reaction.**

Billions of reactions can occur each second. Nuclear reactors contain only a **critical mass**, which is the amount of isotope necessary to sustain a chain reaction. There is no danger of nuclear explosion because that requires a quantity significantly greater than critical mass. When all of the uranium atoms are used up, they leave behind radioactive by-products. These leftovers are called *nuclear waste*, which is very harmful to people and other living things. Power plants have special disposal sites and containers for storing nuclear waste.

Nuclear fission takes place within the reactor of a power plant, and it is used to generate electricity. First, nuclear fission heats water into steam; then, the steam turns a turbine and generates electricity. Some of the neutrons released by fission are absorbed. The absorbed neutrons keep the reaction continuing at a steady rate. The biggest danger in a power plant is a **meltdown**, which occurs when the fuel in a reactor melts and radiation leaks out of the reactor. Meltdowns are caused by the melting of part of the nuclear reactor core because of inadequate cooling of the fuel.

Two significant meltdowns have occurred in the history of nuclear power. In 1986, the nuclear power plant in Chernobyl, Ukraine, experienced a steam explosion that triggered a meltdown. Since the devastating Chernobyl accident, scientists

FYI

Radioactive Waste
Radioactive waste is classified as high-level, intermediate-level, or low-level waste. The majority of high-level waste consists of spent fuel rods, which contain many radioactive atoms. These fuel rods are stored in water-filled, steel-lined tanks in the power plants.

Low-level waste includes nuclear medical waste, commercially produced radioactive materials, and irradiated materials that have had contact with highly radioactive substances. Low-level radioactive waste is contained in storage units appropriate to their level of hazard. Low-level wastes may be shielded with lead or concrete.

	Volume	Radioactive Content
High-level waste	3%	95%
Intermediate-level waste	7%	4%
Low-level waste	90%	1%

and governments have become more aware of the importance of safety regulations and now require safety features to be installed in nuclear reactors. Even with the protocols in place, however, safety precautions did not stop the damaging earthquake and tsunami of 2011 from destroying the nuclear power plant and causing a meltdown in Fukushima, Japan. The systems built to cool the reactors were overcome by the disaster. Even several years later, the radioactive waste kept piling up in the midst of persistent clean-up efforts.

Chernobyl is the site of one of the most significant nuclear power-plant meltdowns in history.

Scientists continue to research viable ways to initiate and maintain nuclear fusion. Currently, recreating the conditions of the sun is expensive and impractical. But nuclear fusion would produce much lower levels of radioactive waste than nuclear fission and it would be classified as a clean, renewable energy source.

Fuel rods contain fuel pellets of fissionable material.

LESSON REVIEW

1. How does Einstein's equation relate to nuclear fission?
2. Explain nuclear fusion in the sun.
3. How is nuclear fusion maintained in the sun?
4. Describe nuclear fission in a power plant.
5. Is it possible for a nuclear power plant to experience a nuclear explosion?
6. Why is fission a more practical energy source than fusion?

IAEA (International Atomic Energy Agency) experts review Japan's plans to decommission the Fukushima nuclear power plant.

5.2.4 *Uses of Radioactivity*

OBJECTIVES

- Identify uses of radioactivity.
- Examine the advantages and disadvantages of using nuclear energy.

VOCABULARY

- **carbon dating** a method used to determine the age of things of organic origin by measuring the radioactivity of their carbon-14 content
- **irradiation** the exposure of foods to radiation in order to kill harmful bacteria
- **tracer** a radioactive isotope whose pathway can be traced through the body

Although nuclear energy can be dangerous, it offers many benefits if it is used wisely. For example, nuclear power can be used to create electrical energy. Nuclear energy is considered a clean source of power. It generates high-level nuclear waste, but it does not burn fuel, pollute the air, or cause acid rain as fossil fuels do. Nuclear reactions are more efficient because they produce much more energy with a given amount of fuel than fossil fuels do.

Nuclear energy is also used in medicine. Radiation destroys cancerous tissues or tumors. The radioactive isotope cobalt-60 is often used in cancer treatments, and erbium-169 is used to relieve arthritis pain. Radioactive substances are also injected into a person to act as tracers. A **tracer** is a radioactive isotope whose pathway can be traced through the body. For example, doctors use iodine-131 tracers to diagnose problems in the thyroid gland. The thyroid naturally absorbs iodine; when the iodine-131 tracer is introduced into the body, the thyroid gland absorbs it. Then radiation scanners detect this tracer and produce an image on a screen that shows whether the thyroid is normal or enlarged. Radiation can be harmful to living tissues, so the isotopes used as tracers have short half-lives. Doctors use isotopes that last long enough to help diagnose the problem but not long enough to harm the patient.

Control room of a nuclear power plant

CAREER

Nuclear Engineer

Nuclear engineers work with nuclear power and radiation. Some use their knowledge of nuclear energy to design, develop, monitor, and operate nuclear power plants that generate electricity and that power naval vessels. Others may work on the nuclear fuel cycle, which involves producing, handling, and using nuclear fuel and safely disposing of nuclear waste. Or, they may focus on fusion energy research. Nuclear engineers may also specialize in developing nuclear power sources for spacecraft or developing medical uses for radioactive materials to diagnose and treat diseases.

Nuclear engineers are exposed to radiation, but they wear protective gear. This gear includes a badge that indicates the amount of radiation exposure and warns of overexposure.

Radiation also kills bacteria in food. Exposing foods to radiation in order to kill harmful bacteria is called **irradiation**. In irradiation, meats and vegetables are passed through radiation to kill harmful microorganisms, such as certain bacteria that cause food poisoning or even death. Irradiated food does not become radioactive. The radiation energy levels are low enough that only chemical changes are possible and not nuclear changes that could make the food itself radioactive. Using radiation to

Doctor reads X-ray results.

CHALLENGE

Conflicting Dates

Permian coalified bark found in Sydney, Australia, yielded a radiocarbon date of 33,700 years. If carbon-14 has a half-life of 5,730 years and fossils of wood found in Permian layers of rock are considered to be hundreds of millions of years old, why do these fossils have measurable amounts of carbon-14?

FYI

Some Radioactive Isotopes Used in Medicine

Isotope	Half-Life	Use
bismuth-213	46 minutes	cancer therapy
cesium-137	30 years	sterilization of blood
cobalt-60	5.27 years	brain cancer treatment
fluorine-18	110 minutes	tracer for imaging tumors
iodine-131	8 days	imaging and treatment of thyroid cancer
krypton-81m	13 seconds	diagnosis of lung disease
rhenium-186	3.8 days	pain relief in bone cancer
selenium-75	120 days	the study of digestive enzymes
thallium-201	73 hours	diagnosis of coronary disease
yttrium-90	64 hours	liver cancer therapy

kill bacteria is a common practice in Japan, China, and many European countries, but it is not as common in the United States and Canada.

Scientists use their knowledge about the half-lives of radioactive isotopes to date artifacts and rocks. **Carbon dating** determines the age of things of organic origin by measuring the radioactivity of their carbon-14 content. By using the half-life of carbon-14 isotopes (5,730 years), scientists can calculate up to 50,000 years of age in samples of bone, shells, wood, and other organic materials. Carbon-14, a radioactive isotope of carbon, is produced in the earth's atmosphere. Because plants get their carbon from carbon dioxide in the air, some of their carbon is carbon-14. This carbon-14 is also passed on to animals who eat the plants. As a result, all animals contain some carbon-14. If an organism dies and is preserved, the carbon-14 undergoes radioactive decay. Half of the carbon-14 will be gone after 5,730 years; after another 5,730 years, half of that amount will be gone. So the amount of carbon-14 in a preserved organism can help determine how long ago that organism was alive. The accuracy of this dating method has been greatly improved since its first use in the 1940s. It now takes into account different carbon-14 levels in recent decades.

In the same way, uranium and potassium are used to date rocks in Earth's crust. Uranium-238 decays into lead-206, and potassium-40 decays into argon-40. Knowing the half-life of these substances helps scientists determine the age of rocks.

BIBLE CONNECTION

Dead Sea Scrolls

In 1947, a Bedouin herdsman threw a rock into a cave at Qumran, a place near the border of the Dead Sea 13, miles east of Jerusalem. Hearing the sound of breaking pottery, he went to investigate and made one of the most important archeological discoveries of the century: the Dead Sea Scrolls. The Dead Sea Scrolls were the library of the Essenes, a Jewish community that hid away in caves during the outbreak of the Jewish-Roman War (68 AD). Several historical documents tell of the Essenes and the Jewish-Roman War. The library contains copies of the Scriptures, including the book of Isaiah and the various Psalms; copies of the apocryphal books, such as the book of Enoch; holy writings by the Essenes; and nonreligious works, such as military dispatches and legal writs. Other historical documents bear witness to the time in history of these writings. Archaeologists found related scrolls at Masada, the rock fortress taken over by Jewish zealots after the fall of Jerusalem in 70 AD and finally taken by the Romans in 73 AD. Because the scrolls are composed of organic substances, such as papyrus and linen, archaeologists used carbon dating to test the age of the historical documents. As with any good science, several scientists have conducted independent tests on the scrolls to determine the age, and they yielded varying results. Some carbon dating tests resulted in ages from 200 BC to 68 AD. Other tests have given dates as late as 585–660 AD. What does this mean? Does it mean the scrolls are fake? That historical documents and biblical records are wrong? No, it means that science is imperfect. The Word of God is the only thing that is perfect. Conflicting dates in scientific research are not a reflection on the truth of the Bible but only on the fallibility and incomplete understanding of science.

LESSON REVIEW

1. Why is nuclear energy considered a clean energy source?
2. What is the disadvantage of using nuclear energy as a power source?
3. How is nuclear energy used in medicine?
4. How is radiation used in the food industry?
5. Describe how carbon dating can help determine the age of organic substances.

5.3.1 *Renewable and Nonrenewable Resources*

OBJECTIVES

- Explain why developed countries are dependent on fossil fuels.
- Compare renewable resources to nonrenewable resources.

VOCABULARY

- **fossil fuel** a source of energy formed from the buried remains of dead plants and animals

Why is your house warm in the winter and cool in the summer? The energy used today for heating and cooling, cooking, and transportation would amaze those who lived 100 years ago. Where does the energy come from to power this standard of living?

For centuries, wood and animal fuel provided the energy needed to warm houses and cook food. With industrialization came the need for larger supplies of fuel, and coal replaced wood. A few decades later, private transportation grew into a huge industry, and oil became the primary fuel source. Modern technologies have brought great advances in medicine, science, industry, and healthy living, but all of these require fuel in some capacity. Oil and coal, the most widely used sources of fuel, are nonrenewable resources. They cannot be replaced once they are used or it will take an extremely long period of time for them to be replaced.

Oil powers the car or the bus that transports you to school each day. What heats your house and provides the electricity to run your computer and television? For many people, the answer to these questions is oil. The world runs on oil, and the supply is diminishing. Scientists believe the world's supply will last for several more decades, but whenever a person drives a car or runs a furnace, a tiny bit of that oil reserve is used up.

The three most common nonrenewable resources are oil, coal, and natural gas. These types of fuel are called **fossil fuels** and they are composed of layers of dead plants and animals that were buried beneath sediments that turned into rock. Heat

Coal mine with excavator

and pressure changed the sediments into rock and changed the organic material into fossil fuels. Fossil fuels contain hydrocarbons, organic compounds containing only hydrogen and carbon. Hydrocarbons burn easily to produce a lot of heat. A kilogram of coal, for example, produces twice the amount of heat as a kilogram of wood. Oil and natural gas produce three times the heat that wood produces. Fossil fuels are also easy to store and transport—two more reasons why they have become primary energy sources.

Because the supply of fossil fuels is finite, scientists are searching for practical renewable resources of energy. Renewable resources are constantly available or can be replaced over a relatively short period of time. Renewable resources include geothermal energy, solar energy, wind, biomass, hydroelectric power, and wood.

Many people in developing countries burn wood for heating and cooking. Since new trees can be planted and grown, wood is considered a renewable resource. But using this energy resource has contributed to the deforestation of tropical rain forests. In some countries, trees are cut into logs, which are placed in pits beneath layers of soil and other materials and slowly burned to make charcoal. The charcoal is then used to cook meals. This is the only source of cooking fuel for many people.

Although energy can never be created or destroyed, nonrenewable energy sources can be used up. And, of course, energy itself can be used wisely or unwisely. It is hard to imagine that fossil fuels

Cars and trucks use gasoline and diesel fuel.

FYI

World Coal Reserves

As of 2016, the world's proven coal reserves were estimated to be sufficient for 153 years of worldwide production and use. The graph shows reserves to production ratios, or the estimated years left of an available resource in a particular region. For example, North America's reserves are estimated to supply North America's usage for 350 more years. The Asia Pacific region is estimated to have enough reserves to supply its usage for about 100 more years.

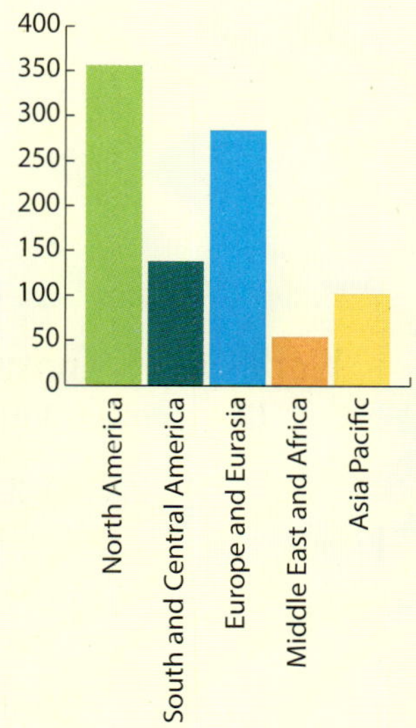

FYI

Muons

When cosmic rays enter the atmosphere, they collide with molecules of air and create elementary particles, some of which are called muons. These tiny, microscopic particles can pass through earth and rock. Scientists have been able to create detectors for these particles and measure their movements as they pass through structures. This gives them the opportunity to "see" through solid rock. Archaeologists and geologists have used this technology to map structures inside the pyramids of Giza. The technology could also be helpful to detect changes in density underground, which could indicate carbon dioxide (CO_2) deposits. Until recently, the muon detectors were too big for subsurface imaging, some the size of a small car. A new detector has been designed with a diameter of 15 cm and a length of 1 m. The new detectors fit through the boreholes made in the ground and contain optical fibers and electronics that count each muon that passes through them. The information gathered is converted to an image that allows the geologists to "see" the movement of CO_2 underground.

will eventually run out. When the car runs low on gas, you buy more. When it is dark, you turn on the lights. When it is cold, you turn up the heat. However, wise stewardship dictates that people should regularly consider how much energy is being used.

Small energy choices matter, and they matter now. How can fuel consumption be reduced? What ways can the public and private sectors of business decrease their use of electricity, which in turn will lower the use of fossil fuels burned to produce electricity?

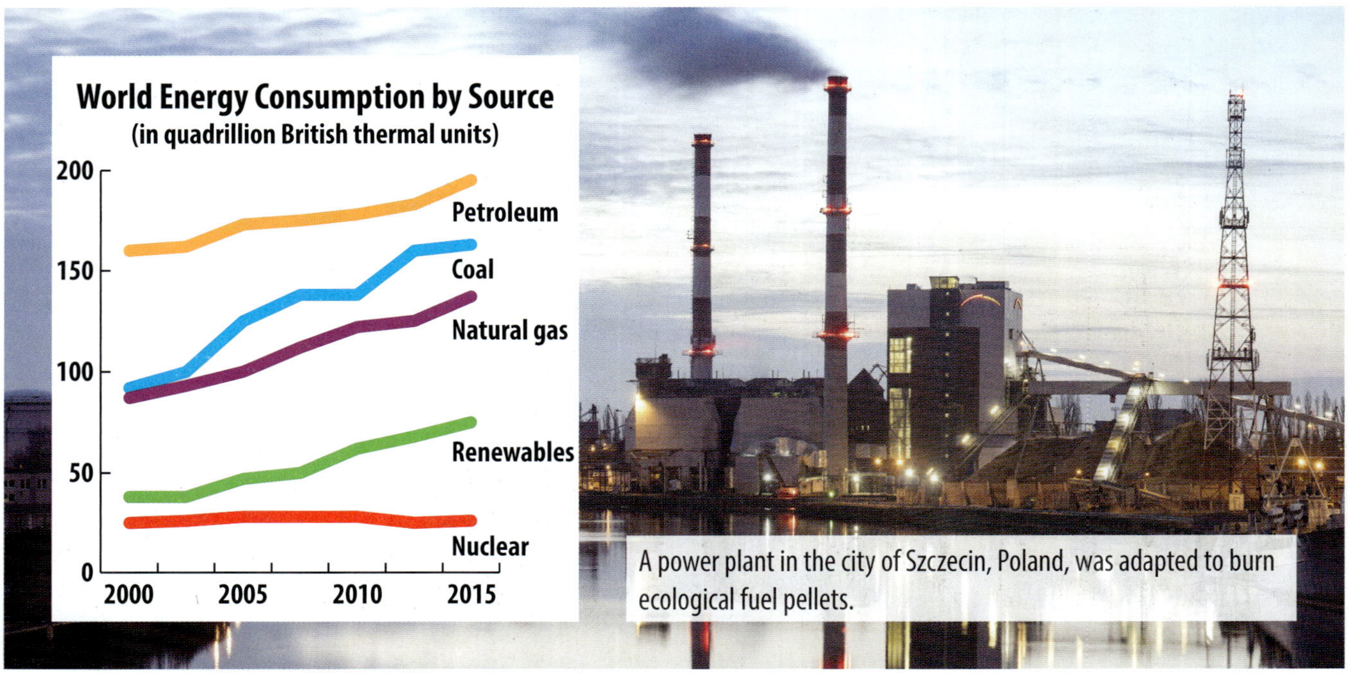

A power plant in the city of Szczecin, Poland, was adapted to burn ecological fuel pellets.

What about personal responsibility? Can you walk, bike, or carpool instead of driving?

Nonrenewable resources have in many ways been a blessing, but they have sometimes been used unwisely. Fossil fuels generate the energy needed to create all disposable items. Piles of newspapers, paper plates, napkins, plastic forks, razors, foam cups, and diapers are thrown away every day. Society's throwaway, wasteful lifestyle consumes nonrenewable resources in large amounts.

Choose to ride a bike instead of driving a car.

As the industrial revolution progressed, assembly lines grew, mass production became commonplace, and motorized transportation became a way of life. How these advances would impact the future and world fuel supplies was unforeseen. Today, the developed world depends on millions of tonnes of fossil fuels every year. The supply is limited, and the need to conserve current resources and find alternative forms of renewable energy has become imperative. Making good energy choices takes some careful thought and a lot of personal responsibility. But individuals can make a difference. What can you do to make the best possible energy choices?

Excess fuel is used when cars are stuck in traffic.

LESSON REVIEW

1. Why are developed countries so dependent on fossil fuels?
2. What is the difference between nonrenewable resources and renewable resources?
3. Name three common nonrenewable resources.
4. Why are fossil fuels used more than renewable resources?
5. Could one renewable resource supply energy for all regions of the world? Why?

Fossil fuels are used to make plastics.

5.3.2 *Fossil Fuels*

OBJECTIVES

- Compare the theories of fossil fuel formation.
- Describe four types of coal.
- Highlight the properties and uses of fossil fuels.
- Examine the advantages and disadvantages of processing fossil fuels.

In 1859, Edward Drake punched a hole 21 m into the ground outside of Titusville, Pennsylvania that released a gush of oil. This was the world's first oil well. Suddenly, the world was thirsty for oil! And, oil refined into kerosene was becoming the most popular fuel used in lamps. Within a century after its discovery, oil replaced coal as the world's leading energy source.

Energy-generating fossil fuels can be traced back to producers that harnessed energy from the sun by way of the photosynthesis process. This process models the first law of thermodynamics that states that energy cannot be created or destroyed. Plants convert energy from the sun into stored energy. As the plants decay, pressure and heat cause the matter to form layers that are eventually converted into fossil fuels. When the fossil fuels are burned, the thermal energy is released in the form of heat. Therefore, the energy in the natural gas that a furnace burns and the energy in the gasoline that fuels a car ultimately came from the sun!

Column of Coal Types

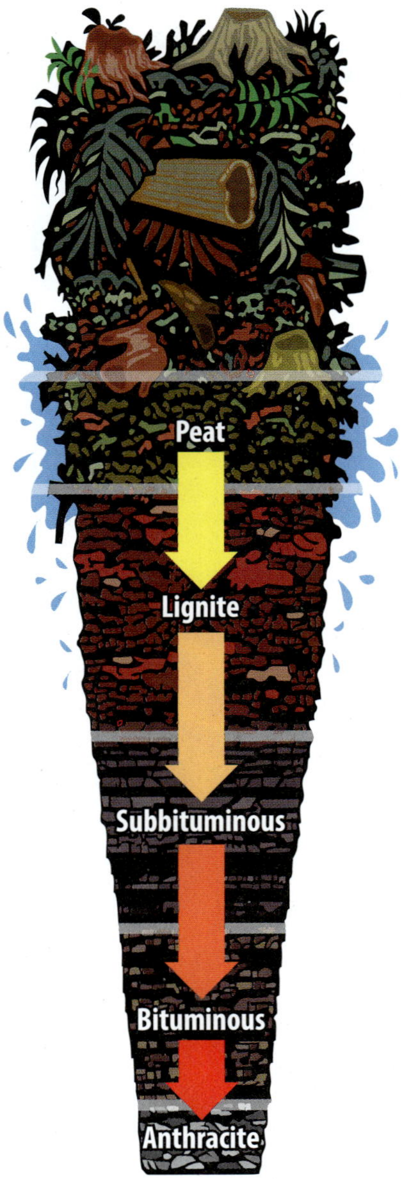

The process that forms fossil fuels takes time, but how much time? Scientific theory states that the formation of fossil fuels takes thousands or even millions of years. Old-earth scientists believe the principle of superposition and radiometric dating methods measure the age of rocks to be in the millions of years. Bituminous coal is found in layers of rock from the carboniferous period and is believed to be 290–354 million years old. Many coal

BIBLE CONNECTION

Noah's Flood

Young-earth scientists attribute the formation of many fossils, rock layers, and fossil fuels to the violent nature of the Flood. "On that day all the springs of the great deep burst forth, and the floodgates of the heavens were opened. And rain fell on the earth forty days and forty nights.... For forty days the flood kept coming on the earth, and as the waters increased they lifted the ark high above the earth. The waters rose and increased greatly on the earth.... and all the high mountains under the entire heavens were covered. The waters rose and covered the mountains to a depth of more than fifteen cubits. Every living thing that moved on land perished—birds, livestock, wild animals, all the creatures that swarm over the earth, and all mankind. Everything on dry land that had the breath of life in its nostrils died.... Only Noah was left, and those with him in the ark. The waters flooded the earth for a hundred and fifty days" (Genesis 7:11–12, 17–24). Whether or not the Flood accounts for a majority of the fossil record and the formation of fossil fuels, the truth remains that God flooded the whole earth and only those within the ark were saved.

Pumps bring crude oil from underground.

mines contain fossils of extinct trees and shrubbery. Fossils of modern species are not found.

Young-earth scientists' geologic data suggests fossil fuels can form rapidly; it does not take millions of years. Natural gas can be produced in a matter of hours by adding heat and water to brown coal. The growth rate of modern forests creates enough mass to produce the amount of coal believed to be buried in the earth within a time frame of 10,000 years. Several fossilized trees travel vertically through many sedimentary layers, including strata of coal, suggesting the layers did not take millions of years to form.

This tree fossilized vertically through layers of strata.

As you know, the three major forms of fossil fuel are coal, natural gas, and oil (petroleum). Coal, a solid fossil fuel formed from decomposed plant remains, was the first fossil fuel to be used by humans. Prior to this time, wood was the primary source of fuel throughout the world. Coal mining in Europe dates back to at least the 13th century. This fossil fuel was once the leading energy source for home heating in the United States and Canada. Many power plants still use coal to generate the steam needed to produce electrical energy.

The formation of coal can be traced to wetlands such as swamps, bogs, and moors. Peat is a substance made of partially decayed plant matter. Over time, heat and pressure turn the peat into coal. There are four kinds of coal that contain various types and amounts of carbon and other impurities. Each type produces a different

Coal power plant

1905 Peat gatherers in West Yorkshire, England

Peat is used for growing plants and for fuel, and it is mined in Northern Europe and North American wetlands.

amount of heat energy. When placed under tremendous pressure from overhead rock layers, peat is transformed into lignite. Lignite is a soft coal with a woody texture. It is very moist and crumbles when exposed to air. Of the different types of coal, lignite has the lowest heating value. Subbituminous coal forms when lignite is placed under pressure. This type of coal is less moist and has more carbon than lignite. The application of both pressure and heat to lignite produces bituminous coal. The majority of coal is bituminous coal, which has less moisture and more carbon than subbituminous coal. It also has a high sulfur content. When bituminous coal is burned, sulfur dioxide is released into the atmosphere. Sulfur dioxide is a serious form of air pollution that causes a variety of lung diseases. When sulfur dioxide mixes with the water in the atmosphere, acid rain precipitates. Bituminous coal under extreme pressure forms anthracite coal, the highest ranked coal and the hardest form of coal. Anthracite coal contains the highest energy content.

Burning coal has both advantages and disadvantages. Coal is the most abundant fossil fuel, is relatively inexpensive to produce, and is easy to transport. Devices called *scrubber systems* work like a shower to dissolve sulfur oxides before they are released into the atmosphere. This process helps to reduce pollution. However, these systems do not reduce all pollution. Burning coal produces more carbon dioxide per unit of heat energy than burning either petroleum or natural gas. Coal mines can also pollute the land, and strip-mining coal out of the ground can harm a region's environment.

Peat

Lignite coal

Bituminous coal

Anthracite coal

The other two fossil fuels are natural gas and oil (petroleum). Natural gas is a mixture of methane and other gases formed from decomposed marine organisms. When these organisms died and sank to the ocean floor, they were covered by layers of sediments and compressed into sedimentary rock. Pressure, heat, and bacterial activity transformed these organic substances into oil and natural gas. Geologists typically find oil and natural gas together. The gas layers usually rise above the oil layers because gas is less dense than oil.

Almost all natural gas provides fuel for heating and for generating electricity. People commonly burn methane to warm houses and cook on gas stoves. Butane and propane are also natural gases. Butane is used in lighters and camp stoves. People burn propane as heating fuel and as cooking fuel for outdoor grills. Some motor vehicles are powered by natural gas. These vehicles produce less pollution than gasoline and diesel vehicles. Natural gas burns cleaner than other fossil fuels and it creates less pollution. Natural gas is also plentiful and it is convenient to transport and use. Burning natural gas does have a drawback; it produces carbon dioxide, a major greenhouse gas. However, it produces less carbon dioxide than petroleum or coal for the same amount of heat energy produced.

Although the use of coal and natural gas is common, most of the world's energy needs are fulfilled through the use of oil (petroleum). Petroleum, a crude oil, is a liquid fossil fuel formed from microscopic plants, animals, and marine organisms. It can be found in underground pools or reservoirs. This oily, flammable mixture is separated into liquid fuels and other products in refineries. The liquid fuels include gasoline, jet fuel, diesel fuel,

TRY THIS

Coal

Check the labels on various appliances at home, and record how much electricity, in watts or kilowatts, they use. Estimate how long these appliances run during an average day, and determine how much electricity they use per day and per year.

About 0.45 kg of coal generates 1.25 kw of electricity. How much coal would it take to run each appliance per day? Per year? How much coal would be needed per day and year to run all of the appliances?

FYI

Fossil Fuels: a Misnomer?

According to photographs of Saturn's moon Titan, methane lakes occupy parts of this planetary satellite. The lakes are full of hydrocarbons and appear to debunk the idea that petroleum is actually a fuel formed from fossils and is in limited supply. Russian scientists have been arguing against the idea of fossil fuels for decades. They submit that the primary components of petroleum derive from highly compressed minerals, which suggests the fuel is renewable. In experiments modeling conditions about 40–95 miles deep in the earth's crust, ethane and heavier hydrocarbons were produced. The limited understanding of science has only scratched the surface of comprehending God's miraculous creation and His provision of resources.

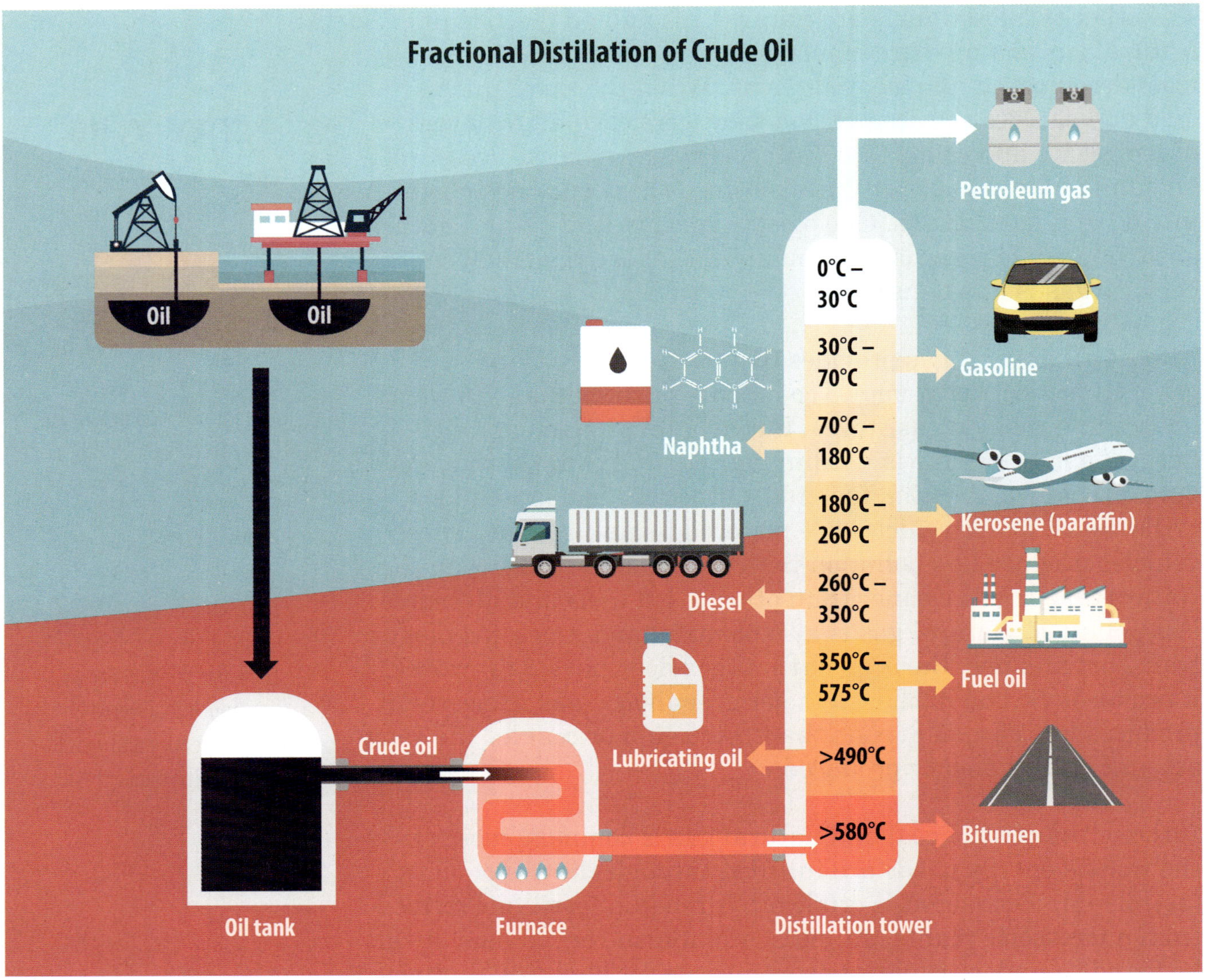

kerosene, and fuel oil. Other products separated from petroleum include lubricants, waxes, and tars such as asphalt.

Most petroleum is drawn from oil wells, but some petroleum is located near the surface of the earth in the form of oil shale and tar sands. Oil shale is a fine-grained rock containing a solid, waxy mixture of hydrocarbon compounds. Tar sands are layers of sand soaked with thick petroleum.

Pipes for pumping components of tar sand

Some of the world's largest mining projects mine oil from the tar sands in northeastern Alberta, Canada. Clay, sand, and water along with bitumen, a thick black oil, are the components of tar sand. Alberta's tar sands are the world's largest. This form of oil can be mined using

FYI

Heating Value and Components of Coal Types

Components	Lignite	Subbituminous	Bituminous	Anthracite
Heat content (Btu/lb)	4,000–8,300	8,500–13,000	11,000–15,000	13,000–15,000
Moisture	30%–60%	10%–45%	2%–15%	< 15%
Fixed carbon	25%–35%	35%–45%	45%–85%	85%–98%
Ash	10%–50%	≤ 10%	3%–12%	10%–20%
Sulfur	0.4%–1.0%	< 2%	0.7%–4.0%	0.6%–0.8%

two methods. Open-pit mining involves digging huge holes in the earth's surface. Oil miners use heat and diluents (thinning agents) to remove the oil from the tar sand. The remaining sand is put back into the open-pit mine, which is reclaimed—brought back to a condition where it can be used for other purposes. The second method involves injecting steam into the tar sand deposit. Oil miners pump out and process the heated oil. This method is useful for deep deposits.

The reliance on petroleum has advantages and disadvantages. It is an inexpensive, trusted source of energy. High-grade oil and gasoline burn relatively efficiently. Petroleum is convenient to transport, and new drilling technologies continue to improve oil production. One disadvantage of burning petroleum is the production of atmospheric pollution, such as smog and carbon dioxide. Another issue is the potential for oil spills that harm the environment. The advantages and disadvantages continue to be considered in petroleum's use.

LESSON REVIEW

1. Compare and give evidence for the theories of fossil fuel formation.
2. Name four types of coal and explain how they are formed.
3. Describe how natural gas and petroleum are formed.
4. Contrast the properties and uses of the three major fossil fuels.
5. Examine the advantages and disadvantages of processing fossil fuels.

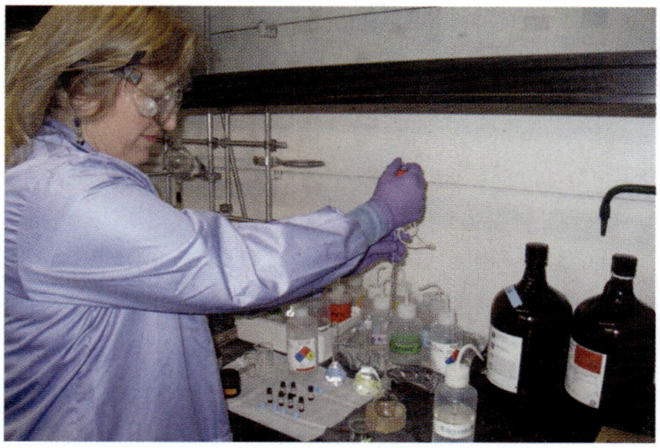

A scientist tests samples taken from water contaminated by the Deepwater Horizon Drilling Platform explosion in 2010.

5.3.3 Renewable Energy Sources

OBJECTIVES

- Appraise the value of energy resources.
- Describe how energy resources are captured and used.

Even if energy is conserved today, better access and use is needed in the future. God wove enough energy resources into Earth and space to provide all the energy needed. Scientists continue to research better ways to tap into the resources and for conserving the environment.

The sun supplies the earth with solar energy. Every year, Earth's surface receives 10 times more energy than is contained in all of the known oil, coal, natural gas, and uranium reserves combined. Every year the sun radiates enough sunlight to Earth to supply 15,000 times more energy than the world consumes.

Solar energy warms the Earth, provides plants with the means to photosynthesize, and offers limitless energy. Simple ways of using solar energy include positioning windows in a house to absorb the maximum amount of sunlight and hanging laundry outside to dry. Solar collectors and solar cells are two other ways of using solar energy. A solar collector heats water or air using solar power. The surface of the solar collector is covered with dark glass or plastic to absorb energy from the sun. Water is pumped through pipes inside the collector, and solar energy heats the water. The warm water is then stored to provide heat at a later time. Eco-hotels in Saint John, Virgin Islands, use solar collectors to provide hot showers for their guests. The sun warms the water all day and guests can have a hot shower in the evening.

TRY THIS

Turbine

Measure the precut slits in the cork from your teacher and cut four fins from a plastic lid to fit in the slits. Slip the fins into the slits. Secure the fins with dots of glue. Stick pins into the ends of the cork for axles. Glue craft sticks together to form two Xs. Let the glue dry. Set the axles on the Xs. Allow water from a faucet or hose to flow underneath the turbine. What other renewable resources could move the turbine?

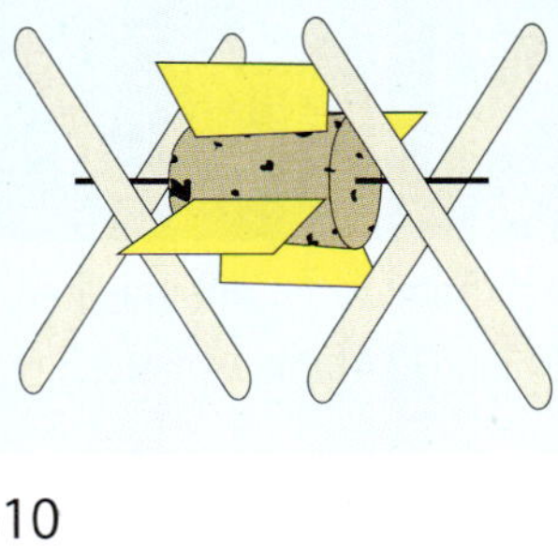

Solar panels with photovoltaic cells

A solar cell, also called *a photovoltaic cell*, is a device that converts sunlight directly into electrical energy. A solar cell is made of layers of silicon and metal. When sunlight hits the cell, electrons flow across the layers, creating an electrical current.

Wind turbine ready for installation

Uneven distribution of solar radiation produces wind. If it were possible to capture all the energy generated, the energy needs of the entire world could potentially be met. Wind machines can generate electricity. Blowing wind spins the blades on a wind turbine. The blades are attached to a hub that is mounted on a turning shaft. The shaft goes through a gear transmission box, which increases the turning speed of the blades. The transmission is attached to a high-speed shaft that turns a generator. The generator then makes electricity. Wind power is very inexpensive in areas where strong winds blow. Germany, the United States, and China share more than 50% of the total wind power capacity in the world.

Imagine using energy from plants to run a car or heat a house. This may be possible in the future by using biomass. Biomass is organic matter, such as crops or waste products, transformed to produce fuel. Biomass uses the energy that plants receive and store from the sun. Burning wood or paper can create biomass energy. Biomass includes yard clippings, crop waste, and sawdust

Wind farm

FYI

Electric Cars

Electric cars could potentially replace gasoline engine cars in the 21st century. An electric car is powered by an electric motor. Because its motor is connected directly to its wheels, it uses no energy when at rest or coasting. Electric cars offer nearly pollution-free, low-maintenance, cheap transportation. Even though the cars do not produce harmful emissions, fossil fuels are still used in the production of the vehicles and the generation of electricity needed to charge the batteries.

Electric car-charging station

from lumber mills. Using biomass not only helps conserve fossil fuels but also frees up space in landfills.

New ways of using biomass are being developed. One use converts biomass into ethanol and methanol. This process begins with the fermentation of biomass. Fermentation converts sugars and starch to ethanol. Also, certain bacteria act on biomass to make methane gas, which can be used to produce methanol. Both of these alcohols, ethanol and methanol, are blended with gasoline to make cars run more efficiently and to reduce pollution levels. Biomass is also converted into gases that can be burned to produce electricity. In Brazil, deforested land could be transformed into biomass plantations.

Geothermal energy is collected from trapped heat in Earth's crust. As geologists penetrate Earth's crust, the temperature rises about 3°C every 100 m. Rock located only a few kilometers beneath Earth's surface is very hot. Water that flows close to the hot rock becomes scalding hot or turns into steam. The hot water can reach temperatures near 150°C—much hotter than boiling water. Hot water that rises through cracks in Earth's surface can be piped through buildings to provide heat, and the steam this hot water produces turns turbines and creates electricity.

Iceland, which sits on a volcanic zone, harnesses steam from underground hot rocks to drive turbines that produce electrical energy. Geysers—hot spots where heated water and steam burst through Earth's crust—are also used to heat buildings. Geysers

Iceland's geothermal power plant in Gunnuhver, Reykjanes

Energy Source Comparison

Energy Source	Pros	Cons
Biomass (Biofuel)	fuel burns clean; efficient; grows on low-quality land, which can preserve tropical rain forests; requires low-maintenance technology; can be used in diesel and gasoline engines; renewable	requires a lot of water; produces greenhouse gases; may not work in all climates
Coal	abundant energy source; inexpensive; easy to transport; generates large amounts of power	causes air pollution and acid rain; produces greenhouse gases; harmful to ecosystem because of mining
Geothermal power	unlimited fuel source; does not produce pollution; renewable	expensive technology; limited locations available to use it
Heavy oil (from tar sands and oil shale)	1.2 trillion barrels have been found	harmful to the environment; requires water to process
Hydroelectric power	very inexpensive; 75,000 dams already in place; low maintenance technology needed; does not produce pollution; renewable	prevents rivers from flowing freely; clogged reservoirs; impact to dams during times of drought
Hydrogen fuel	limitless supply; does not produce pollution or greenhouse gases; inexpensive to transport; renewable	requires other fuels to generate
Natural gas	abundant energy source; convenient to use and transport	causes some air pollution; creates greenhouse gases; ecosystems impacted by pipelines
Nuclear fission	well-developed technology; no greenhouse gases produced; abundant uranium resources	produces radioactive waste; aquatic life harmed by heated wastewater
Nuclear fusion	could potentially be very inexpensive fuel source; does not produce greenhouse gases	costly fusion research; resolution needed for many theoretical and technological problems; produces radioactive waste
Oil	inexpensive; burns efficiently; convenient to transport; technology already in place to use in industry and transportation	causes air pollution; produces greenhouse gases; ecosystems harmed by oil spills
Solar energy	most abundant energy source; nonpolluting; renewable	high start-up costs; requires large areas for panels; not practical in areas with little sunshine; does not work at night
Wind power	inexpensive in areas with strong winds; little disruption to ecosystems; renewable	inconsistent source of power; cannot be used in many places; extensive land use; noisy; harms bats and birds

are also an important heat source in Hawaii, New Zealand, and California. A place called *the Geysers* in California creates an energy supply to meet the needs of about 1,000,000 homes.

Hydroelectric power is electricity produced from the power of moving water. A dam is built to stop the natural flow of a river. The water is channeled through a pipe and forced to push against blades in a turbine, causing them to turn. The turbine spins a generator to produce electricity. Although hydroelectric dams do not pollute the air, they can disrupt the river environment. Hydroelectric power, which supplies nearly one-fifth of the world's electrical power, is the world's most commonly-used renewable resource. This type of power is a good choice for hilly countries that receive a lot of rain.

Nuclear power plants provide electricity for many regions. This is another clean energy source. Nuclear energy is more expensive than fossil fuels because maintaining nuclear facilities is costly. But as fossil fuels become scarce, nuclear energy looks more appealing. And even though maintaining a nuclear power plant is costly, uranium, the fuel used in nuclear power plants, is inexpensive. A pellet of uranium with a mass of several grams can produce as much energy as three barrels of oil or one ton of coal. Uranium, of course, is a nonrenewable resource, but it is projected that the world's uranium supplies will last longer than its oil supplies.

LESSON REVIEW

1. Why is it important to pursue the use of renewable resources?
2. Name the renewable energy resources.
3. What are the advantages and disadvantages of each renewable energy resource?
4. Describe how two energy resources are captured and used.
5. Which resource do you think is the best? Why?

Hydroelectric power station in Upper Svaneti, Georgia

UNIT 6

Sound and Light

Chapter 1: *Waves*
Chapter 2: *Sound*
Chapter 3: *Light*
Chapter 4: *Optics*

Vocabulary

absorption
acoustics
amplitude
antinode
coherent light
compression
concave mirror
convex mirror
crest
diffraction
Doppler effect
electromagnetic spectrum
fiber optics
fluorescent light
focal length
focal point
frequency
hologram
illuminated object
image
incandescent light
incoherent light
index of refraction
infrared wave
intensity
interference
laser
lens
light filter
longitudinal wave
luminous intensity
luminous object
medium
microwave
mirror
music
neon light
node
noise pollution
opaque substance
optical axis
optical fiber
particle theory
penumbra
period
photoelectric effect
photon
pigment
pitch
plane mirror
polarized light
radiant energy
radio wave
rarefaction
real image
reflection
refraction
resonance
shadow
sonar
sound
surface wave
timbre
translucent substance
transparent substance
transverse wave
trough
ultrasonic sound
ultraviolet (UV) wave
umbra
virtual image
visible light
wave
wave height
wavelength
X-ray

Key Ideas

- Evidence, models, and explanation
- Change, constancy, and measurement
- Form and function
- Abilities necessary to do scientific inquiry
- Understandings about scientific inquiry
- Motion and forces
- Transfer of energy
- Structure and properties of matter
- Abilities of technological design
- Understandings about science and technology
- Personal health
- Natural and human-induced hazards
- Science and technology in society
- Science as a human endeavor
- Nature of science
- History of science

SCRIPTURE

The unfolding of Your words gives light; it gives understanding to the simple.

Psalm 119:130

6.1.1 *Wave Properties*

OBJECTIVES

- Discuss how waves transfer energy.
- Analyze the motion of a medium as each type of wave passes through it.
- Compare the three types of mechanical waves.

VOCABULARY

- **compression** a section in a longitudinal wave where the particles are tightly packed together
- **longitudinal wave** a wave in which the movement of the medium's particles is parallel to the direction of the wave's movement
- **medium** any substance through which a wave can travel
- **rarefaction** a section in a longitudinal wave where the particles are loosely packed together
- **surface wave** a wave in which the medium's particles move in a circular motion
- **transverse wave** a wave in which the medium's particles move in a direction that is perpendicular to the direction of the wave's movement
- **wave** a disturbance that transfers energy through matter or space

Thunderstorms are powerful parts of God's creation. During these amazing events, flashes of lightning and cracks of thunder can be seen and heard. How? Lightning can be seen because of light waves. Thunder is heard because of sound waves. A large amount of energy in creation is in the form of waves. A **wave** is a disturbance that transfers energy through matter or space. A better understanding of wave characteristics and behavior gives you the opportunity to enjoy the world of sight and sound.

Waves begin as vibrations in different materials. Consider a pebble being dropped into a pond. The falling pebble transfers kinetic energy to the water molecules. When the water molecules begin to move and vibrate, they transfer the energy to surrounding water molecules and a wave is produced. The vibrations cause waves to spread out from the source of the disturbance. Waves from the energy of the pebble do not continuously occur. They stop when all of the energy has been exchanged with the water and air. An earthquake beneath the earth's crust or an explosion in the air also sets off vibrations in solid and gaseous substances.

The waves described in the previous examples are called *mechanical waves*. They require a medium to travel through. A **medium** is any substance through which a wave can travel. Mediums can be a solid, a liquid, or a gas. Ocean waves travel through a liquid water medium. Most sound waves travel through an air medium. The solid core of the earth serves as a medium for seismic waves. But some waves, such as light waves

Vibrations may be transferred in the form of kinetic energy to create waves.

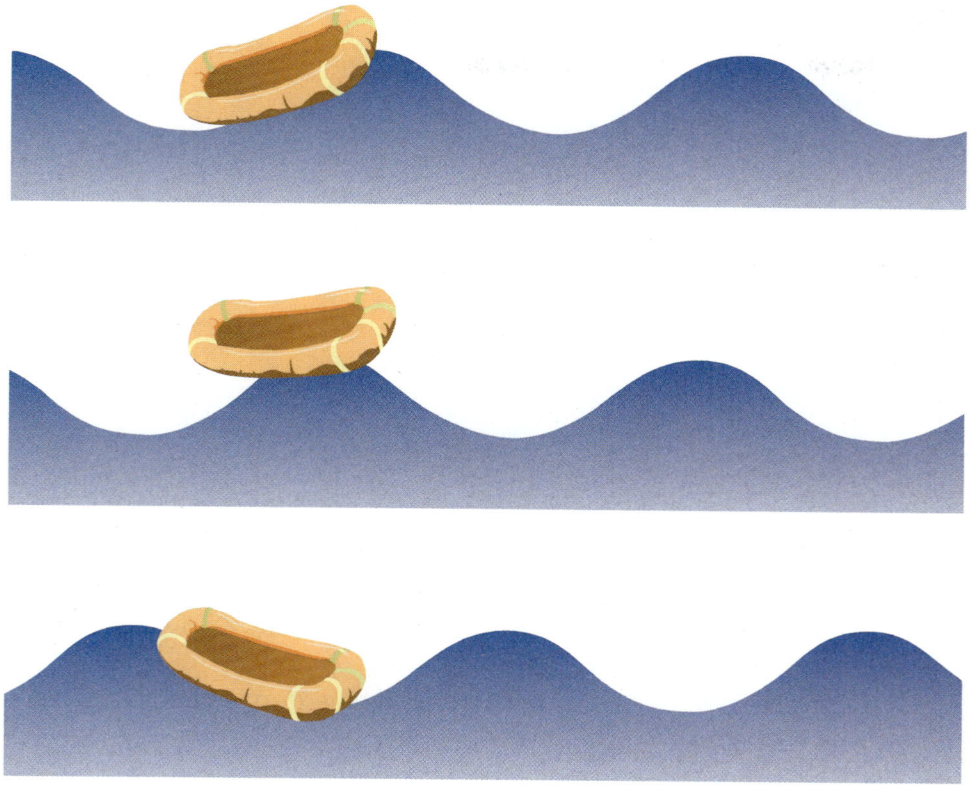

Waves transport energy, not matter. This is why the raft remains in the original position.

produced by the sun, do not need a medium. Light and other forms of radiation can travel through the vacuum of space.

A wave transfers energy, but this energy is transported without moving the medium as a whole. When energy is transferred to individual particles in the medium, the particles vibrate around a fixed point. After the energy is dispersed, the particles return to their original position. Imagine sitting on a raft beyond the breakers on the ocean. The waves pass under the raft, but do not move it. The raft may move up and down. This is because some of the energy from the wave is transferred to the raft, but the waves do not cause the raft to move toward or away from land. Wind and currents are responsible for that type of movement.

Another example that illustrates the exchange of energy in a wave is a volcanic eruption. When the volcano on the island Krakatau in Indonesia erupted in 1883, the explosion created an air wave that bounced back and forth many times. However, the atmosphere that served as the wave's medium remained in one place.

Waves travel at different speeds through different mediums. A wave's speed depends on two properties of the medium through which it travels: density and compressibility. Typically,

The eruption on the island of Krakatau in Indonesia illustrates how energy was transferred, but the atmosphere molecules did not relocate.

a mechanical wave passes quickly through a dense medium because it is easier to transfer energy between molecules that are close together. Mechanical waves travel the fastest through solids, followed by liquids. They travel the slowest through gases.

Waves also move faster in a more compressed medium. Compressibility is the ability of a medium to return quickly to its original shape. For example, steel has more compressibility than air. Steel molecules that have been disturbed move back to their original positions quicker than air molecules. This is why sound waves travel faster through steel than through air.

There are three main types of waves: transverse, longitudinal, and surface. A **transverse wave** is a wave in which the medium's particles move in a direction that is perpendicular to the direction of the wave's movement. Imagine two people each holding one end of a rope. When one person raises a portion of the rope up and down, nearby points begin to move up and down. The wave is eventually propelled in a forward direction until it reaches the person at the opposite end of the rope. The rope particles move perpendicular to the direction of the wave movement. Deep ocean waves, some earthquake waves, mechanical waves generated when a rope is flipped back and forth, and X-rays are all examples of transverse waves. Snakes model transverse waves because their bodies wind sideways across the ground or water while moving in an overall forward direction.

TRY THIS

Transverse Waves

Grasp one end of an extendable spring toy. Do not allow the toy to move. Have your partner grasp the other end of the toy and begin to move the toy up and down to produce a transverse wave. Generate a series of transverse waves and draw your observations. Move the toy up and down more vigorously and draw your observations. How does the wave's size relate to the toy's speed?

Transverse Waves

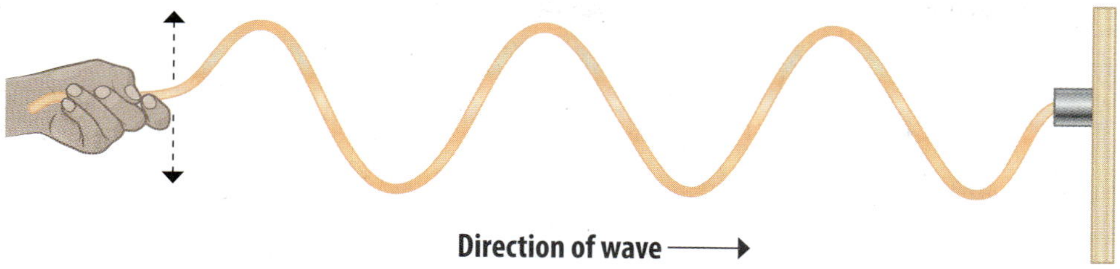

A **longitudinal wave** is a wave in which the movement of the medium's particles is parallel to the direction of the wave's movement. Longitudinal waves consist of two parts: compressions and rarefactions. A **compression** is a section in a longitudinal wave where the particles are tightly packed together. A **rarefaction** is a section in a longitudinal wave where the particles are loosely packed together. Longitudinal waves have a series of compressions and rarefactions. Picture an extendable spring toy that has been placed on a flat surface. When the toy is pushed, portions of the coil contract and other portions expand. The coil contractions represent compressions, and the coil expansions represent rarefactions. Earthworms and inchworms model longitudinal waves. They travel by shortening and lengthening their bodies across the ground.

Longitudinal Waves

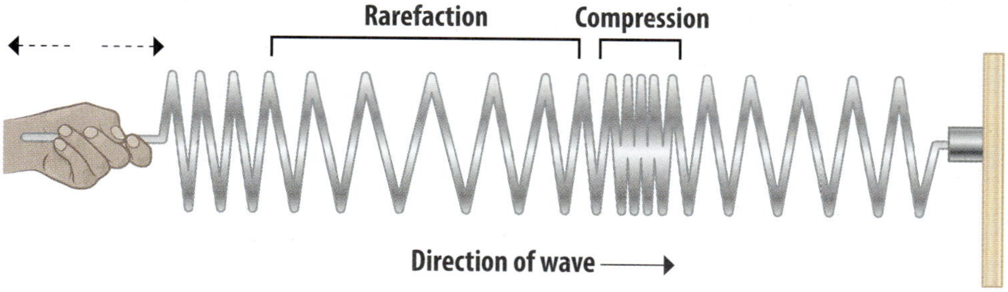

Waves that travel through solid mediums can be either transverse or longitudinal. Waves that travel though fluid mediums (liquids or gases) are longitudinal only. Earthquakes can produce both transverse and longitudinal waves through the solid portion of Earth. However, only the longitudinal waves can travel through the core of the earth. This discovery led scientists to hypothesize that Earth's outer core is made of a liquid substance.

The waves that travel along the surface of the ocean are an example of the third main type of wave, which is known as *a surface wave*. A **surface wave** is a wave in which the medium's

TRY THIS

Longitudinal Waves

Tie a piece of ribbon to the middle section of an extendable spring toy. Place the spring toy on a flat surface. Grasp one end of the toy. Do not allow it to move. Have your laboratory partner grasp the other end of the toy and push the toy toward you. After your partner pushes the toy, have them pull the toy back once to produce a longitudinal wave. Observe the compressions and rarefactions in the wave. Draw your observations. Generate a series of longitudinal waves, and draw your observations. Relate the movement of the coils to the movement of the wave.

FYI

Sound Waves Detect Sewer Clogs

Every day, utility workers in the United States are hard at work trying to unclog the more than 800,000 miles of underground sewer pipes. This can be a daunting task that often requires many hours of hard labor. A group of individuals has found relief for this issue by using sound waves. The National Science Foundation provided its support and funding for this endeavor that seeks to improve lives. The small business sends down a device, called *the Sewer Line Rapid Assessment Tool*, or *SL-RAT*, into manholes. This device sends sound waves through the pipe to a receiver. The receiver processes the signal and helps utility workers determine the severity of the clog as well as any defects in the pipe. The traditional system that uses cameras and cleaning trucks covers about 370 m per day. SL-RAT can help utility workers cover around 3,000 m per day.

Surface Waves

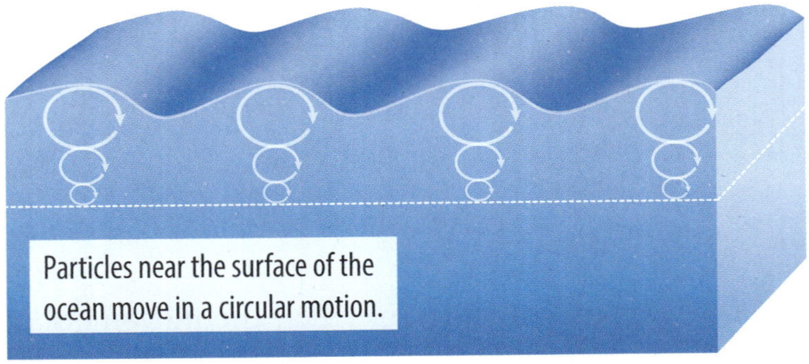

particles move in a circular motion. Unlike transverse or longitudinal waves that affect all of the medium's particles, surface waves affect only the particles at the surface of the medium. A wave produced on the surface of ocean water is an example of a surface wave.

LESSON REVIEW

1. Describe how waves are produced in water.
2. What do all mechanical waves require?
3. Are water molecules transported with energy when a wave is produced? Explain.
4. What two properties of a medium determine the wave's speed?
5. If a transverse wave is transporting energy left to right, how will the particles of the medium move?
6. What type of wave is produced when the particles of the medium are moving in the same direction as the wave? How does this type of wave compare to surface waves?

FYI

Earthquakes

Earthquake waves are divided into two categories known as *body waves* and *surface waves*. Seismic waves that travel underground are body waves. They consist of P (primary) waves and S (secondary) waves. P waves are the faster of the seismic body waves. They can travel through solids, liquids, and gases. As P waves travel, they cause consecutive sections of the earth to push together and then spread apart like a giant spring. P waves are longitudinal waves, and they usually cause little to no damage. S waves are slower than P waves, and they travel only through solids. S waves are transverse waves. They move the earth perpendicular to the direction they are traveling. S waves can cause greater damage than P waves because they are stronger than P waves and because their motion can cause buildings to sway. Seismic waves that travel on or just under the earth's surface are surface waves. They are detected after the body waves pass. Surface waves cause more damage than body waves because they have greater force and last longer.

OBJECTIVES

- Describe the parts of a wave.
- Calculate wavelength, frequency, and wave speed.

VOCABULARY

- **amplitude** the maximum distance a wave vibrates from its rest position
- **crest** the highest point of a transverse wave
- **frequency** the number of complete waves that pass a particular point in a unit of time
- **period** the time it takes two successive wave crests or troughs to pass a given point
- **trough** the lowest point of a transverse wave
- **wave height** the distance between a trough and a crest
- **wavelength** the distance between corresponding points on a wave

Waves are everywhere and in many forms. Light from the sun or a light bulb travels in waves. Radio and television transmissions move through the air as waves. Even X-rays travel through your bones and teeth in waves. God designed the many different kinds of waves—visible light, radio waves, ocean waves, X-rays, microwaves, gamma rays, and seismic waves—with certain properties. These properties help people discern one wave from another.

When an actor in a Western movie cracks a bull whip, he creates a transverse wave. There are two distinct features in transverse waves—crests and troughs. A **crest** is the highest point of a transverse wave, and a **trough** is the lowest point of a transverse wave. The distance between a trough and a crest is called the **wave height**. Longitudinal waves do not have crests or troughs. Instead, they are divided into compressions and rarefactions.

Oceanographers, surfers, and wave specialists measure different aspects of waves for various reasons. One aspect measured is wavelength. The **wavelength** is the distance between corresponding points on a wave. For a transverse wave, the wavelength can be measured between adjacent crests. Wavelengths can also be measured between adjacent troughs. The wavelength of a longitudinal wave can be measured between adjacent compressions or rarefactions. Energy is related to the wavelength. The shorter the wavelength, the greater the energy of the wave.

Wavelengths are generally measured in meters or centimeters, but some wavelengths are much smaller. For example, visible

Parts of a Wave

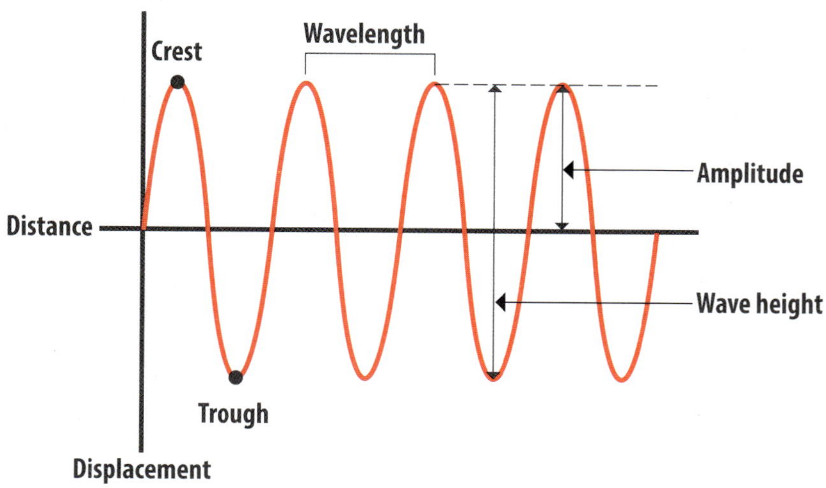

Electromagnetic Spectrum

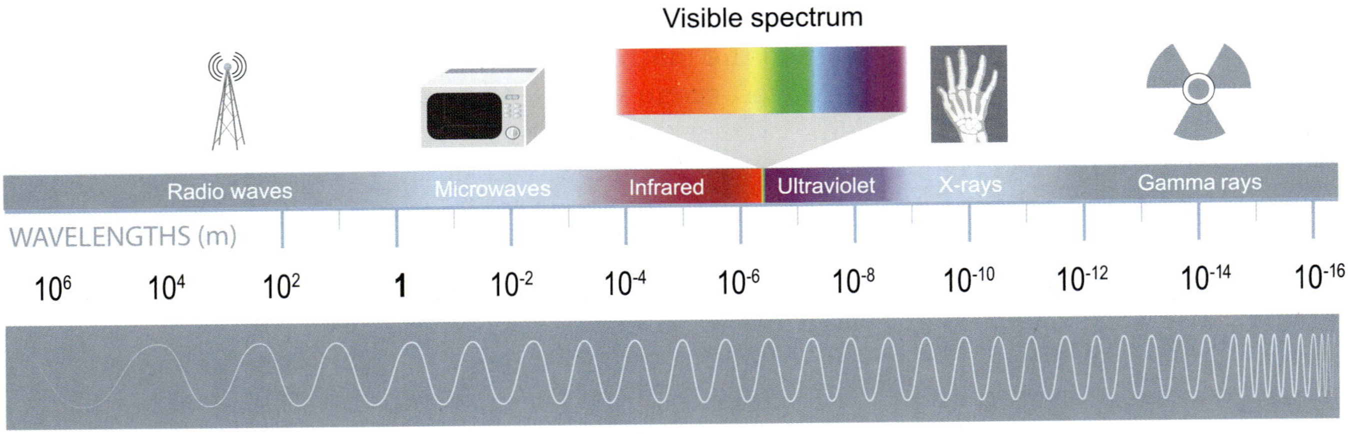

As the wavelength increases, the frequency decreases.

light waves have wavelengths that measure between 400 and 700 nm. The wavelength of X-rays is 0.001 to 10 nm. Gamma rays have wavelengths less than 0.001 nanometers. Shorter wavelengths generally produce faster, more destructive forms of radiation.

Wave **frequency** is the number of complete waves that pass a particular point in a unit of time. The frequency of a transverse wave can be measured by counting the number of crests or troughs that pass a particular point in a certain amount of time. To measure the frequency of a longitudinal wave, the number of compressions or rarefactions that pass a particular point in a certain amount of time are counted. For example, if 50 compressions pass a point in a medium in one second, the frequency of the longitudinal wave is 50 cycles per second.

Frequency is measured in hertz (Hz). This unit is named after Heinrich Hertz, one of the first scientists to study electromagnetic waves. A hertz is one wave, or cycle, per second.

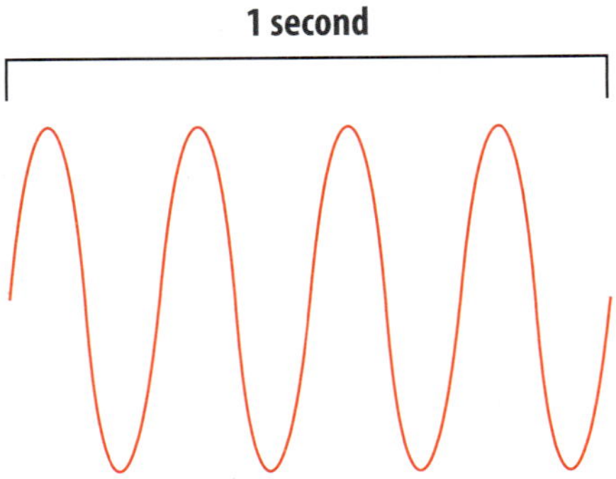

4 waves per second (4 Hz)
(Low frequency)

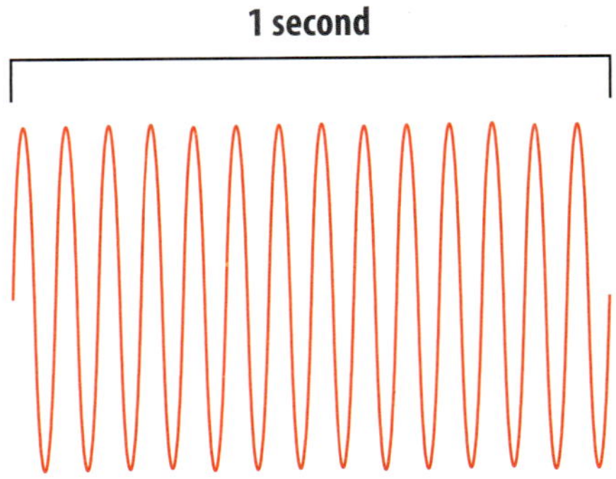

14 waves per second (14 Hz)
(High frequency)

One hertz equals 1 cycle/second. Gamma rays have a frequency of around 10^{20} Hz or greater. Visible light waves have a frequency of around 10^{15} Hz. Radio waves have frequencies up to 105 Hz.

The **period** of a wave is the time it takes two successive wave crests or troughs to pass a given point. The period (T) is the inverse of the frequency and is calculated $T = 1/f$. Tap your finger on the table. If you tap two times in one second, the frequency is 2 Hz. If each tap lasts for half a second, the period is 0.5 second. It is easy to confuse frequency and period. They are not the same, but they are related. Frequency refers to how often something happens. Period refers to the time it takes something to happen.

Wavelength and frequency are related. Consider a jump rope. When you create a transverse wave using the jump rope, you can adjust the rate at which you move the jump rope up and down. If you move the jump rope slowly, the wavelengths are long. The frequency is also small at this point. If you increase the frequency by moving the jump rope up and down quickly, the wavelengths shorten. If the waves are traveling at a constant speed, their frequency is inversely proportional to the wavelength. If the frequency decreases, the wavelength increases. When the frequency increases, the wavelength decreases.

Frequency and wavelength help physical scientists determine the overall wave speed. The wave speed is affected by the medium through which the wave moves. Most waves travel quickly through solids and liquids, but slower in gases. Sound waves often travel faster through mediums that have gained more energy with the addition of heat. Wave speed can be determined by using the following formula: $v = f \times \lambda$. Normally, the letter

Dogs can hear much higher frequency waves than humans. The approximate hearing range for humans ranges from 20 Hz to 23,000 Hz. Dogs can hear frequencies that range between 67 Hz and 45,000 Hz.

LAWS & PRINCIPLES

Wave Speed

Question: A wave has a frequency of 50 Hz and a wavelength of 10 m. What is the speed of the wave?

Step 1: Identify the known variables.
Frequency (f) = 50 Hz
Wavelength (λ) = 10 m

Step 2: Identify the unknown variables.
Wave speed (v) = ?

Step 3: Choose the formula that will help you answer the question.

$$v = f \times \lambda$$

Step 4: Solve.
$v = 50 \text{ Hz} \times 10 \text{ m}$
$v = 500 \text{ m/s}$

Step 5: Verify the units are correct.
Hz = 1 cycle/s
1/s × m = m/s

v represents velocity. However, velocity also involves direction. The speed of a wave does not depend on the direction the wave is traveling. So, v represents speed in this case. Frequency is represented by the letter f, and wavelength is represented by the Greek letter lambda, λ. Wave speed is usually expressed in meters per second.

Another characteristic that waves share is amplitude. This characteristic is related to the amount of energy in a wave. **Amplitude** is the maximum distance a wave vibrates from its resting position. As a wave's energy increases, the amplitude of the wave increases. Think of the part of a wave that a surfer rides. This is a good visual picture of amplitude. The more energy in an ocean, the bigger the wave, the larger the amplitude, and the more thrilling the ride for the surfer.

LESSON REVIEW

1. What is the difference between wave height and wavelength for a transverse wave?
2. How are wavelength and energy related?
3. What is the frequency of a 10 m longitudinal wave when 25 compressions pass a point in a medium in 5 seconds? What is the speed?
4. How does the frequency of a wave respond when the wavelength decreases?
5. Calculate the wavelength of a wave that has a frequency of 10.0 Hz and a speed of 8 m/s.

Wave Behavior 6.1.3

Sometimes two waves join to form one large wave. Other waves may hit rocks or the poles of a pier and bounce back, smacking into nearby waves. Although ocean waves look chaotic, they behave in an orderly manner. When ocean waves or other types of waves, such as light waves, meet another wave or a barrier, they react in one of five basic ways: reflection, absorption, refraction, diffraction, or interference.

Reflection is the bouncing back of a wave after it hits a barrier that does not absorb all of its energy. When you look in a mirror, you are able to see your appearance because light waves are reflected twice. The waves from an external light source bounce off your face and strike the mirror. Then, they bounce off the mirror and an image is reflected into your eyes. A similar experience occurs when you are surrounded by numerous moving swimmers in a pool. The waves that these swimmers make bounce off the sides of the pool and into other people.

Light and water waves are not the only types of waves that can be reflected. Sound waves also reflect. Echolocation is an example of sound wave reflection. When the sound wave exits a bat's mouth, the wave bounces off potential food sources and reflects back to the bat's ears. The wave that is directed toward a barrier or object is called *the incident wave*, and the returning wave is called *the reflected wave*. The line perpendicular to the barrier or object is called *the normal*. The angle formed by the incident wave and the normal is called *the angle of incidence*, and the angle formed by the reflected wave and the normal is called

OBJECTIVES

- Examine the five basic wave behaviors.
- Differentiate between refraction and diffraction.
- Apply basic wave behaviors to real-world examples.

VOCABULARY

- **absorption** the conversion of a wave's energy into another form of energy
- **antinode** the place where constructive interference causes the largest wave
- **diffraction** the bending of a wave around a barrier
- **interference** the effect produced by the combination of two or more waves
- **node** the point where destructive interference causes no displacement
- **reflection** the bouncing back of a wave after it hits a barrier that does not absorb all of its energy
- **refraction** the bending of a wave because of a change in the wave's speed as it passes from one medium to another

Light Wave Reflection

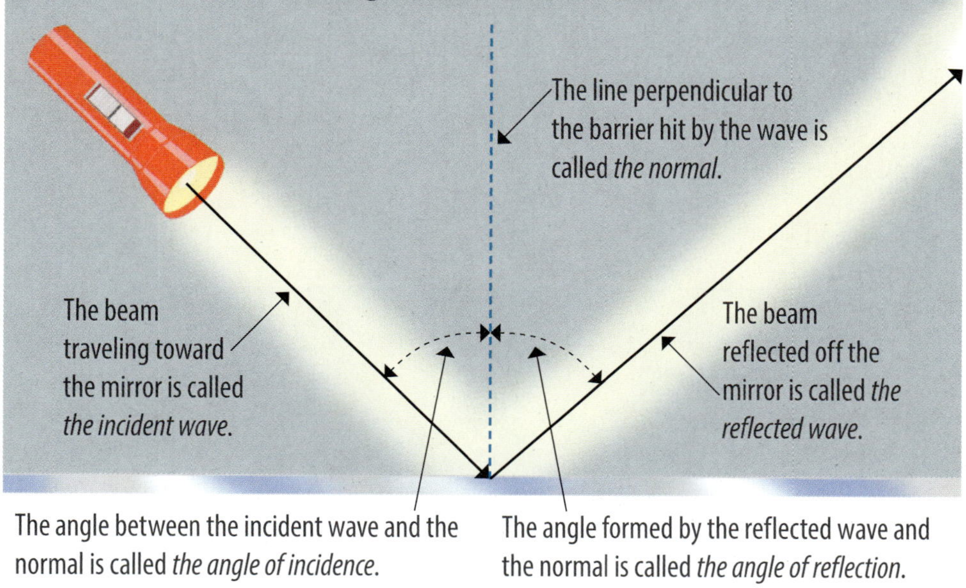

Reflection

the angle of reflection. According to the law of reflection, the angle of incidence is equal to the angle of reflection.

Absorption is the conversion of a wave's energy into another form, such as thermal energy or heat. For example, a light wave striking a piece of clear glass passes through the glass and comes out the other side. But a light wave striking a piece of heavy black fabric is absorbed by the fabric. None of the light wave comes out the other side. The fabric, however, becomes warm as the light energy is converted into heat energy.

Light waves slow down when they travel from a gaseous medium to a liquid medium. The result is a bending of light waves, called *refraction*.

Refraction is the bending of a wave because of a change in the wave's speed as it passes from one medium to another. A wave travels through a uniform medium in a straight line. But, when it enters a new medium at an angle, the speed of the wave changes and the wave bends. For example, if a straight stick is plunged into water at an angle, the stick seems to split into two pieces. A light wave traveling through the air slows down and bends when it passes into the water. If a wave slows down when it enters a different medium, the wave bends or refracts toward the normal. The wave bends away from the normal when it enters a medium, and it speeds up.

Diffraction is the bending of a wave around a barrier. For example, when an ocean wave hits a boat, the wave is forced to

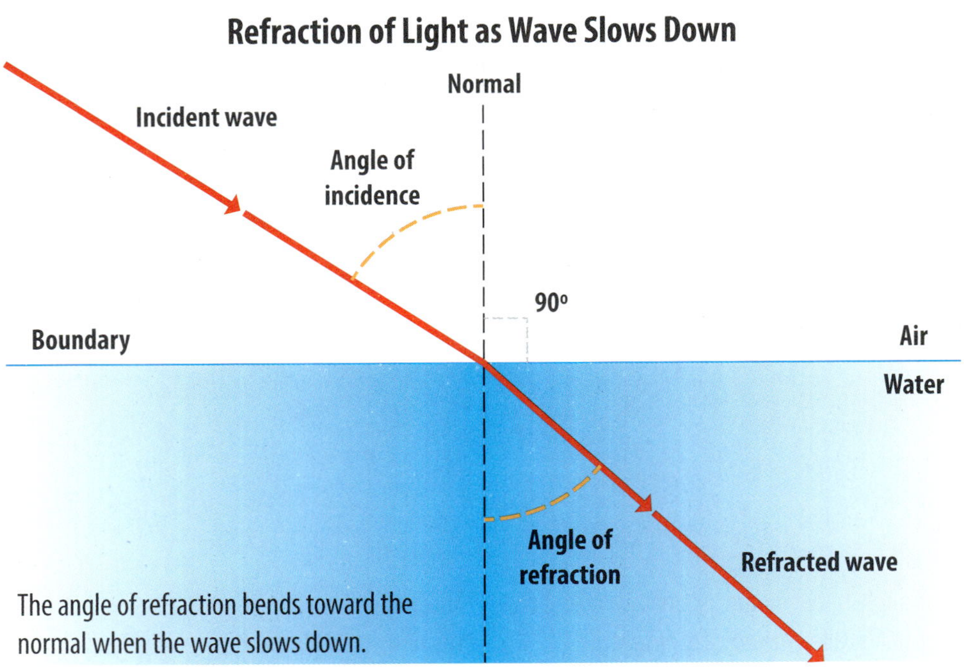

change direction and bend around the boat. Diffraction can also occur when a wave passes through a narrow opening. As the wave passes through, it bends around the edges of the opening and spreads out. Sound waves can diffract around corners. The diffraction of sound waves is what allows spectators to hear the noise of a ball game while they are in the tunnels leading to the stadium.

The degree of diffraction depends on the size of the wavelength in relation to the size of the barrier or opening. If the wavelength is smaller than the barrier or opening, the wave has little diffraction. If the wavelength is larger than the barrier or opening, there is a large amount of diffraction. Diffraction and refraction may seem the same. However, refraction occurs when a wave travels through an object. Diffraction occurs when a wave travels around an object.

Sometimes two waves travel in a direct path toward each other. When they meet, the two waves overlap and combine to form a new wave. **Interference** is the effect produced by the combination of two or more waves. This process generally has two possible results. The two waves can either add to each other or subtract from each other. When the crests from two transverse waves meet, they combine to form one large wave with an amplitude equal to the sum of the amplitudes of the two original waves. This meeting of two wave crests that form one large wave causes constructive interference. The place where constructive interference causes the largest wave is called the **antinode**.

Refraction of Light as Wave Speeds Up

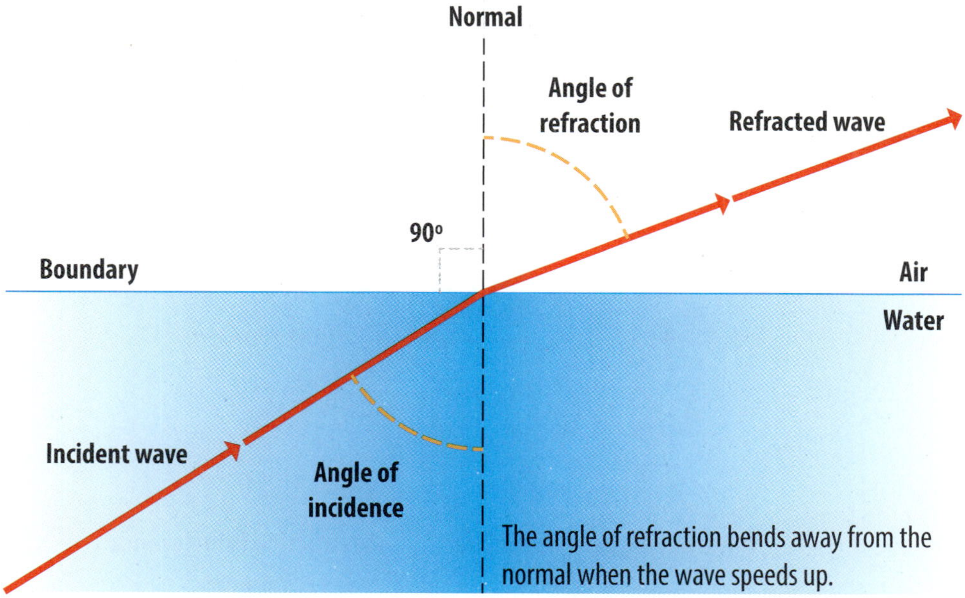

Diffraction of Waves

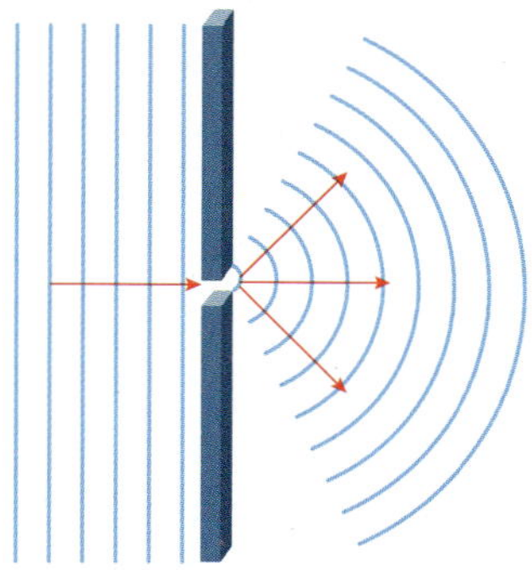

Large amount of diffraction because the wavelength is larger than the opening

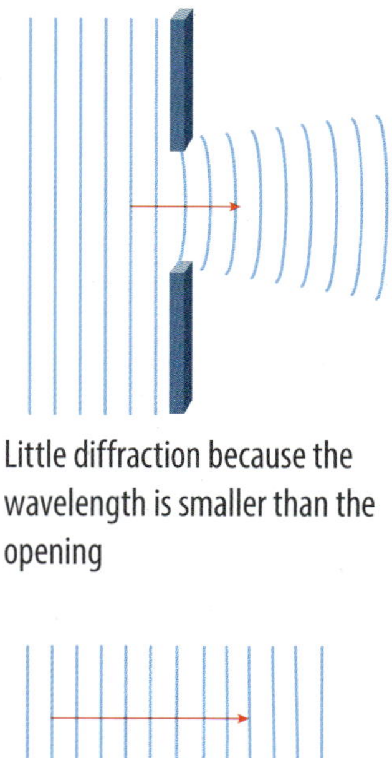

Little diffraction because the wavelength is smaller than the opening

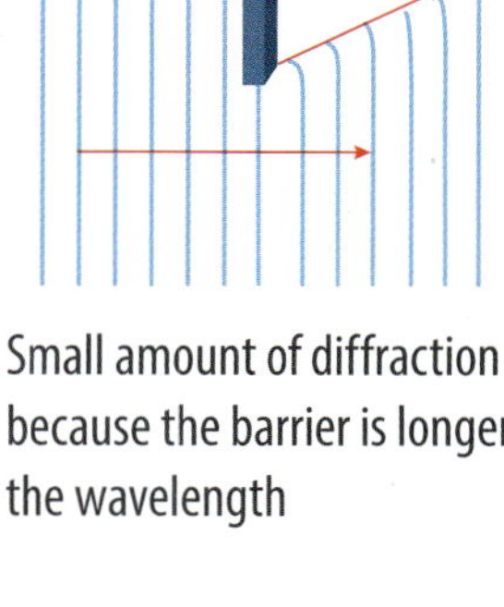

Small amount of diffraction because the barrier is longer than the wavelength

BIBLE CONNECTION

A Test of Faith

Ocean waves are powerful and often intimidating. However, they are no match for the Son of God. With just one word, Jesus quieted the storm. Mark 4:39 states, "He got up, rebuked the wind and said to the waves, 'Quiet! Be still!' Then the wind died down and it was completely calm." Winds were raging, swells were pouring in, and the disciples were terrified. Jesus knew that His Father sent Him for a purpose, and nothing would deter Him—not even a storm. It was Jesus' hope that His disciples would have that type of faith.

TRY THIS

Liquid Refraction

Using a graduated cylinder, pour 75 mL of water and 75 mL of olive oil in two separate 100 mL beakers. Place a pencil in each beaker. Using a protractor, measure the angle of refraction created between the portion of the pencil above the liquid and below the liquid. Why is the angle of refraction larger in some liquids?

Interference Patterns in Waves

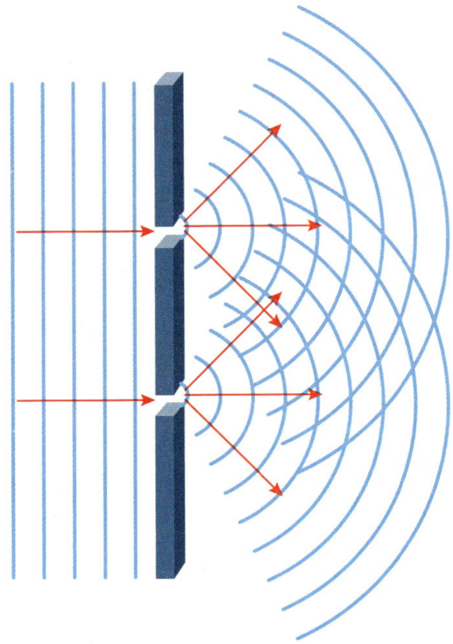

When waves pass through narrow gaps, they spread out. If two sets of waves spread out near each other, they overlap and create both constructive and destructive interference.

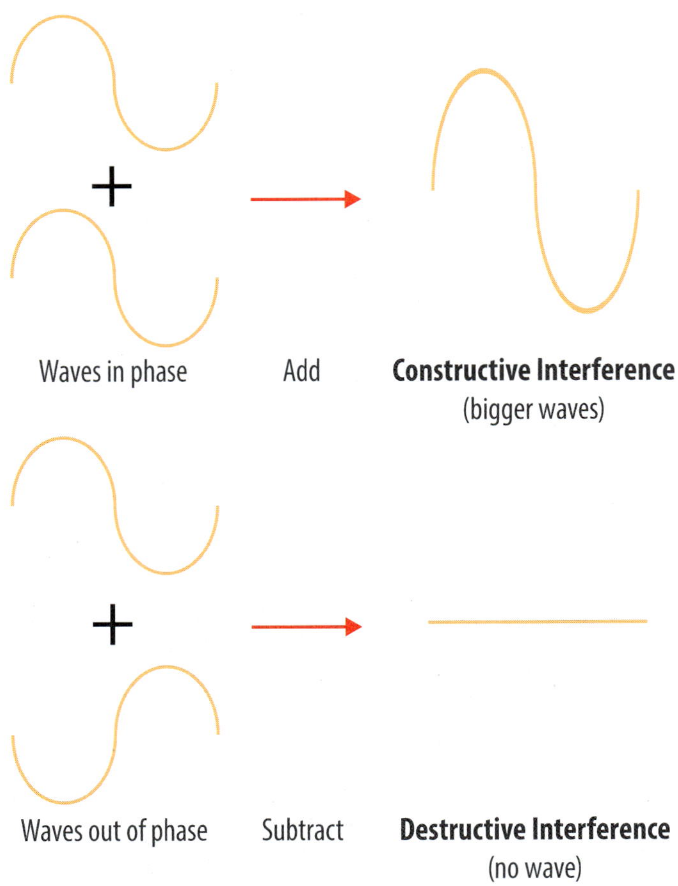

FYI

Noise-Canceling Technology

Have you ever put on a pair of headphones in anticipation of listening to your favorite music only to be distracted by some external noise? This is especially true when riding on an airplane. The noise from the engines can make it almost impossible to hear anything else. In an attempt to remedy these situations, scientists and engineers have developed noise-canceling headphones. They come in one of two varieties: passive and active. Passive noise-canceling headphones rely on high-density foam or other sound-absorbing material to reduce the impact of external noises. However, they only reduce the sound by about 15 to 20 decibels. The more effective option is active noise-canceling headphones. They are made of the same materials as passive headphones, but they can also "erase" lower-frequency sound waves. Active headphones do this by producing sound waves that are similar to the incoming sound waves. The difference is the sound waves produced by the headphones are 180 degrees out of phase with the incoming sound waves. The two sets of sound waves have the same frequency and amplitude, but their crests and troughs are in opposition to one another. Active headphones produce destructive interference, and the two waves cancel each other. They can block about 70% of the external sound.

CAREER

Sound Engineer

Why do the sounds of movie battle scenes seem so intense or the music during a worship concert sound so clear? It is most likely attributed to the knowledge and talents of skilled sound engineers. They work on the technical aspects of sound and music production. The work of sound engineers is not limited to movies and concerts. They also design and control the sound at conferences, sporting events, and theater productions. Sound engineers may even configure sound for video and computer game development. They control microphones, sound levels, and outputs using their well-developed knowledge of acoustics. The sound editing process usually involves four steps: recording, editing, mixing, and mastering. People who are interested in becoming sound engineers should investigate training in a variety of fields such as radio and television, music, broadcasting, or electrical engineering.

If the wave crest of one wave meets the wave trough of a second wave, the crest and trough subtract from each other to form one smaller wave. The amplitude of this combined wave is equal to the difference of the amplitudes of the two original waves. If both waves are the same size, the two waves completely cancel each other at the place of meeting, causing destructive interference. Destructive interference is interference that results in a wave with a smaller amplitude than that of the original waves. The point where destructive interference causes no displacement is called a **node**.

LESSON REVIEW

1. Echolocation is an example of what type of wave behavior? Explain how waves behave during this process.
2. State the law of reflection.
3. Describe how waves bend during refraction when they pass from one medium into another.
4. How are refraction and diffraction different?
5. Describe the two types of interference.
6. What type of wave reaction occurs when light waves hit a dark surface and turn into heat?

6.2.1 *Sound Waves*

OBJECTIVES

- Rank the mediums that transmit sound the best.
- Describe the two factors that affect the speed of sound.

VOCABULARY

- **sound** a mechanical disturbance produced by an object's vibrations

How is it possible for your friends to hear one another when they are talking? How does hearing discern the difference between the droning of a small airplane and the song of a bird? Why is the music of some singers more pleasing than the music of others?

Each of these scenarios involves the transfer of energy through sound waves. **Sound** is a mechanical disturbance produced by an object's vibrations. Sound travels in longitudinal waves—the medium particles vibrate back and forth along the path that the sound wave travels. When someone shouts to you across the school cafeteria, the vibrations produced by their vocal cords collide with nearby air molecules, and transfer energy. The air molecules continue colliding with other air molecules to create dense energy regions (compressions) and less dense energy regions (rarefactions). These regions eventually reach your ear as the sound waves travel across the room.

Sound waves can travel in any medium. If you are underwater, you can hear the sound of someone jumping into the pool. The knock on a wood door is amplified if you place your ear on the door. But sound waves cannot travel through a vacuum. If two astronauts on the moon could take off their helmets and talk to each other, they would hear nothing because sound waves cannot transfer energy without a medium.

Solids transmit sound waves the best because they are dense. The molecules are packed closely together, which allows for the energy to easily transfer from one molecule to the next. This is why sound will travel faster in steel, iron, and wood than it

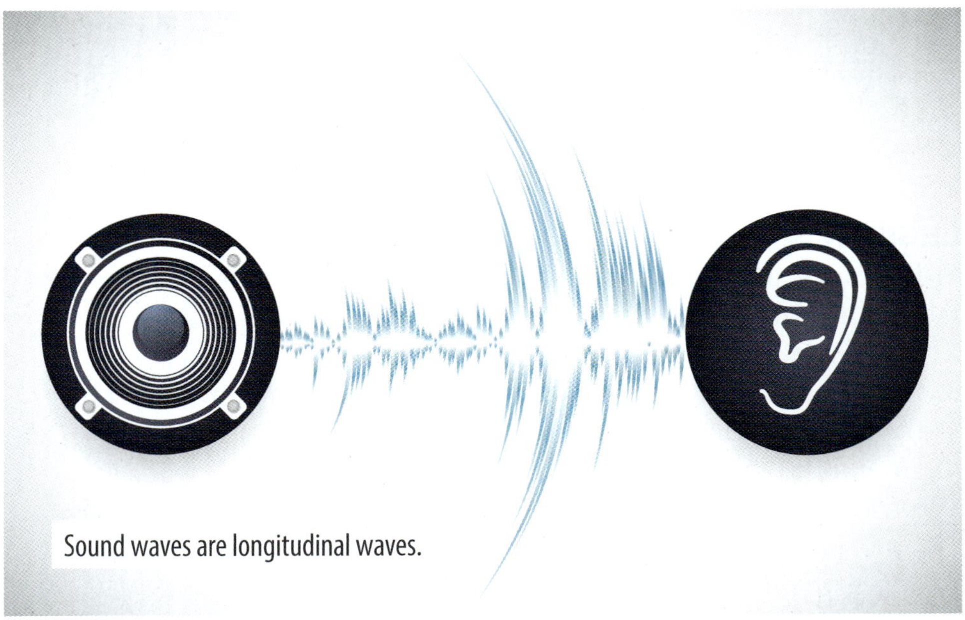

Sound waves are longitudinal waves.

will in water or air. The molecules of a solid are held rigidly in place, but they still have enough kinetic energy to vibrate back and forth. This rigid structure keeps them from moving very far, so they bounce quickly back into place after the compressions and rarefactions of the sound waves pass. Stone, nickel, and iron are some of the best mediums for transporting sound waves quickly.

Sound waves cannot travel through space because no medium is present.

Sound waves travel much slower in liquids than they do in solids because the liquid molecules are not locked into position, and roll over each other. This partial freedom of movement keeps the molecules of liquids from bouncing back into place quickly when sound waves pass through them. Still, sound waves travel nearly five times faster through water than through air. Sound waves travel slowest through gases because gas molecules are unconnected and spread apart. This total freedom of movement prevents gas molecules from passing sound waves quickly.

The speed of sound waves depends on the state of the medium but also on the temperature of the medium. Sound waves pass much quicker through mediums at elevated temperatures. The addition of heat speeds up molecules, which promotes collisions. More collisions mean sound waves can transfer energy in a shorter amount of time. For example, the speed of sound through liquid water at a temperature of 8°C is 1,439 m/s. The speed

BIOGRAPHY

Charles Yeager

Charles Elwood Yeager was born in West Virginia on February 13, 1923. After graduating from high school, Yeager enlisted in the Army Air Corps. He was a fighter pilot ace during World War II where he flew 64 missions over Europe, shot down 13 German planes, and was shot down over France. Yeager was not captured, and he escaped to Spain. When he returned to the United States, Yeager was selected to test pilot the Bell X-1 rocket. During one of the tests, he reached a top speed of 1,066 kph at an altitude of 12,000 m. Chuck Yeager was the first person to break the sound barrier (Mach 1). In 1953, Yeager set a world speed record in the air of 2,660 kph.

TRY THIS

Watching Sound

Stretch a large balloon tightly over one end of an oatmeal box or large metal can. Fasten the balloon to the container with a large rubber band. Affix a small mirror with a piece of tape or glue to the center of the surface of the balloon. Lay the cylinder on its side. Position the cylinder perpendicular to a wall or white piece of paper. Darken the room. Shine a flashlight at a 45° angle toward the mirror. Have your partner quickly shout into the open end of the box or can. Observe the light reflection on the wall or paper. What did you notice? How is sound produced?

FYI

Low-Frequency Waves

The United States Navy utilizes two extremely low-frequency (ELF) radio transmitters to communicate with their submarines. These transmitters are located in Clam Lake, Wisconsin, and Republic, Michigan. The communications remain silent and undetectable. In order to accomplish this mission, the Navy has set up 45 km of overhead signal transmission lines that serve as the antenna that radiates the ELF signals. This system is the only operational form of communications that can reach great ocean depths and is not susceptible to interference. Low-frequency wave communication allows this part of the US military to progress without having to surface and still maintain contact with the President and the Secretary of Defense. It adheres to the principle that states lower frequencies produce signals that can travel to greater depths. Very low frequencies of about 20,000 Hz can only travel below the water surface to depths less than 10 m. The Navy's ELF system that operates at about 76 Hz can transmit signals to depths greater than 30 m.

of sound through liquid water at a temperature of 50°C is 1,541 m/s.

LESSON REVIEW

1. How is the energy in sound waves transferred?
2. Do sound waves travel better through solids or gases? Explain.
3. Explain why sound waves cannot travel in a vacuum.
4. How does an increase in temperature affect the transport of sound waves?

The Speed of Sound in Different Mediums

Medium	Speed of Sound (m/s)	Medium	Speed of Sound (m/s)
carbon dioxide (0°C)	259	seawater (25°C)	1,533
oxygen (0°C)	315	rubber	1,600
air (0°C)	331	brick	3,650
air (25°C)	346	wood	3,850
helium (0°C)	972	glass	4,540
ethanol (25°C)	1,144	aluminum	5,100
lead	1,322	iron	5,130
mercury (25°C)	1,450	stainless steel	5,790
freshwater (25°C)	1,493	titanium	6,070

Temperatures are included for liquids and gases because their densities are greatly affected. The density of solids is not altered by temperatures as much.

Speed, Frequency, and Intensity 6.2.2

Suppose you hear a low rumble of thunder, so you decide to settle near a window to watch the storm roll in. Lightning zigzags across the sky in the distance. You hear the sound of thunder moments later. Then, a huge flash of lightning strikes nearby and the thunder crash is immediate and deafening. Why is the sound of thunder delayed following the strike of lightning in the distance? Sound travels slower than light—about a million times slower. Light waves travel at a speed of 299,337,984 meters per second (m/s). Sound waves travel at a speed of 346 m/s through air at 25°C. So, if a lightning bolt strikes 346 m away, you would hear the thunder one second later.

Much of the energy that is released in the form of sound during thunderstorms relates to the amplitude of the wave. Amplitude in longitudinal waves can be determined by measuring the density of the particles in the compressions and rarefactions. A wave that has extremely dense compressions and particles that are spread out in the rarefactions has higher amplitude. Energy can be transferred more easily in this type of wave. If the particles in the compressions are not densely packed and the particles in the rarefactions are less spread out, the sound wave has lower amplitude because it carries less energy.

The amplitude of a sound wave also determines the intensity of the wave. **Intensity** is the average rate of energy that flows through a defined area. If the volume on a music device is turned down, there is less energy that flows through the sound waves. The notes did not change, but the intensity was reduced. Waves of greater amplitude have greater intensity. The distance that a

OBJECTIVES

- Relate amplitude to intensity and loudness.
- Explain how frequency corresponds with pitch.
- Summarize the Doppler effect.

VOCABULARY

- **Doppler effect** the change in the frequency and pitch of a sound caused by the motion of the sound source or the listener
- **intensity** the average rate of energy that flows through a defined area
- **pitch** how high or low a sound is
- **ultrasonic sound** a sound with a frequency higher than 20,000 Hz

The airplane produces a shock wave when it moves faster than the speed of sound.

Amplitude and Air Molecules

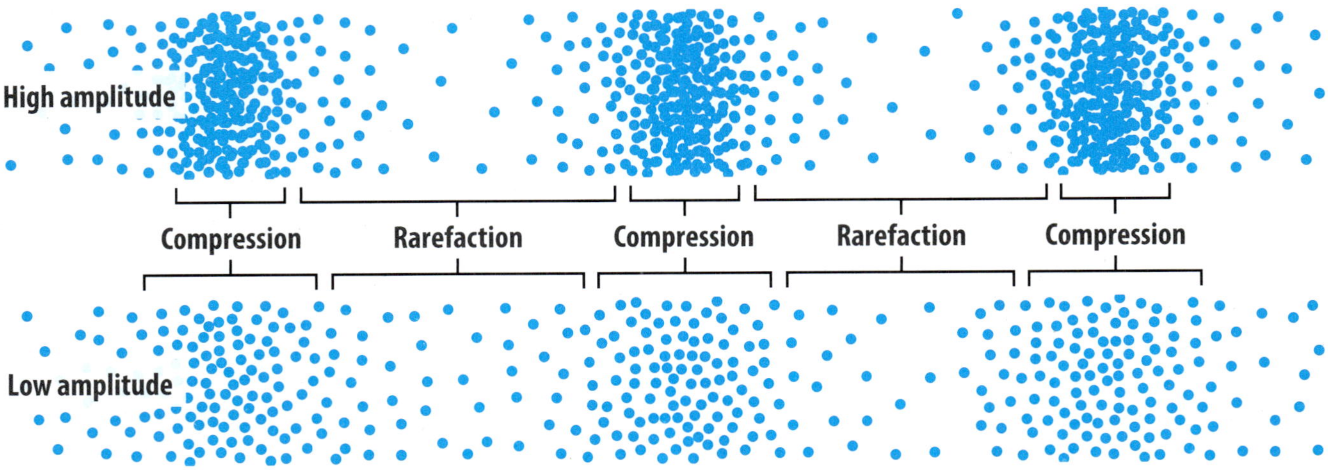

listener is from the source of the sound also affects the intensity of the wave. When cheerleaders are cheering on the sidelines at an outdoor football game, they do not normally whisper because the crowd would not be able to hear them. There is not enough energy produced for the sound to travel a far distance. The cheerleaders add energy to the sound by shouting, which allows the sound wave to travel a farther distance with more intensity. The way in which a person perceives the intensity of sound waves is called *loudness*. This is a subjective characteristic of sound. What may seem extremely loud to one person may seem only moderately loud to another person.

Have you ever heard sounds with such high intensity that they hurt your ears? Loudness is subjective, but intensity can be measured on a scale. The most common unit on the scale

BIBLE CONNECTION

Star Songs

In Job 38:6–7, God challenges Job by asking, "On what were its [the earth's] footings set, or who laid its cornerstone—while the morning stars sang together? ..." Can stars really sing? How can there be sound in space that has no medium? Astrophysicists from the University of Birmingham and NASA scientists decided to analyze this phenomenon. They discovered that stars have "starquakes" that cause them to vibrate like a musical instrument. Because sound is not audible in space, these scientists measured the flickers of light that were caused by the quakes and constructed a representation of what the star waves would sound like. The result was a harmonic hum. This information has helped scientists learn more about the internal structure and size of stars. They have determined that large stars vibrate at low frequencies, and small stars vibrate at high frequencies.

SCIENCE IN ACTION

Determining Distance of Lightning

Light travels faster than sound. To be precise, light travels at a speed of 299,337,984 m/s. Sound travels at a speed of 346 m/s. These two values are useful if you ever get stuck outside in a thunderstorm because they can help you determine how far away lightning strikes occur. To determine the distance between you and the lightning strike, practice the following steps:

1. Observe the lightning strike.
2. Start counting the number of seconds that occur between the time you see the lightning strike and when you hear the thunder.
3. Multiply the number of seconds by 346.

This value will indicate how many meters away the lightning is. For example, if you counted two seconds between the time you saw the lightning and heard the thunder, that means the lightning is 692 m away.

is a decibel. The faintest sounds an average human can hear measure 0 decibels. Nearby thunder can have an intensity of 120 decibels, and a jet engine has a level of 170 decibels. Sounds above 85 decibels can damage your ears. Pain is not always associated with damage, but sounds above 140 decibels will produce pain in the ears. Going to loud concerts or listening through headphones with the sound turned up to an extremely high volume exposes your ears to high intensity sounds that can immediately or eventually cause hearing loss. Engineers have crafted hearing protection equipment that protects against permanent ear damage.

FYI

Shrimp Sonic Boom

There is a mysterious creature that lives beneath the ocean surface that has a remarkable way of catching its prey. What makes this creature even more fascinating is the fact that it is no bigger than the size of a human finger. The snapping shrimp has an oversized claw that it uses to stun its prey. This shrimp does not use its claw to clamp its prey. Instead, it snaps its claw to produce a bubble that travels faster than the speed of sound. It can reach speeds of up to 100 kph. The force of the bubble strikes the prey and immobilizes it. This process occurs within 300 microseconds. The bubble not only produces a loud bang, it also transmits light, which indicates that the temperature inside the bubble is at least 4,727°C.

Pitch is how high or low a sound is, and it is related to frequency. If many sound waves pass a given point in a short period of time, they have a high frequency. A high frequency produces a high-pitched sound such as a mosquito's buzz. A low frequency produces a low-pitched sound such as a tuba note. A soprano may reach a high note with a frequency of 1,000 Hz, but a bass singer may sing a low note with a frequency of 70 Hz.

Sound frequency depends on the length of the object being vibrated. For example, men have lower voices than women because men's vocal chords are longer. Higher notes can be

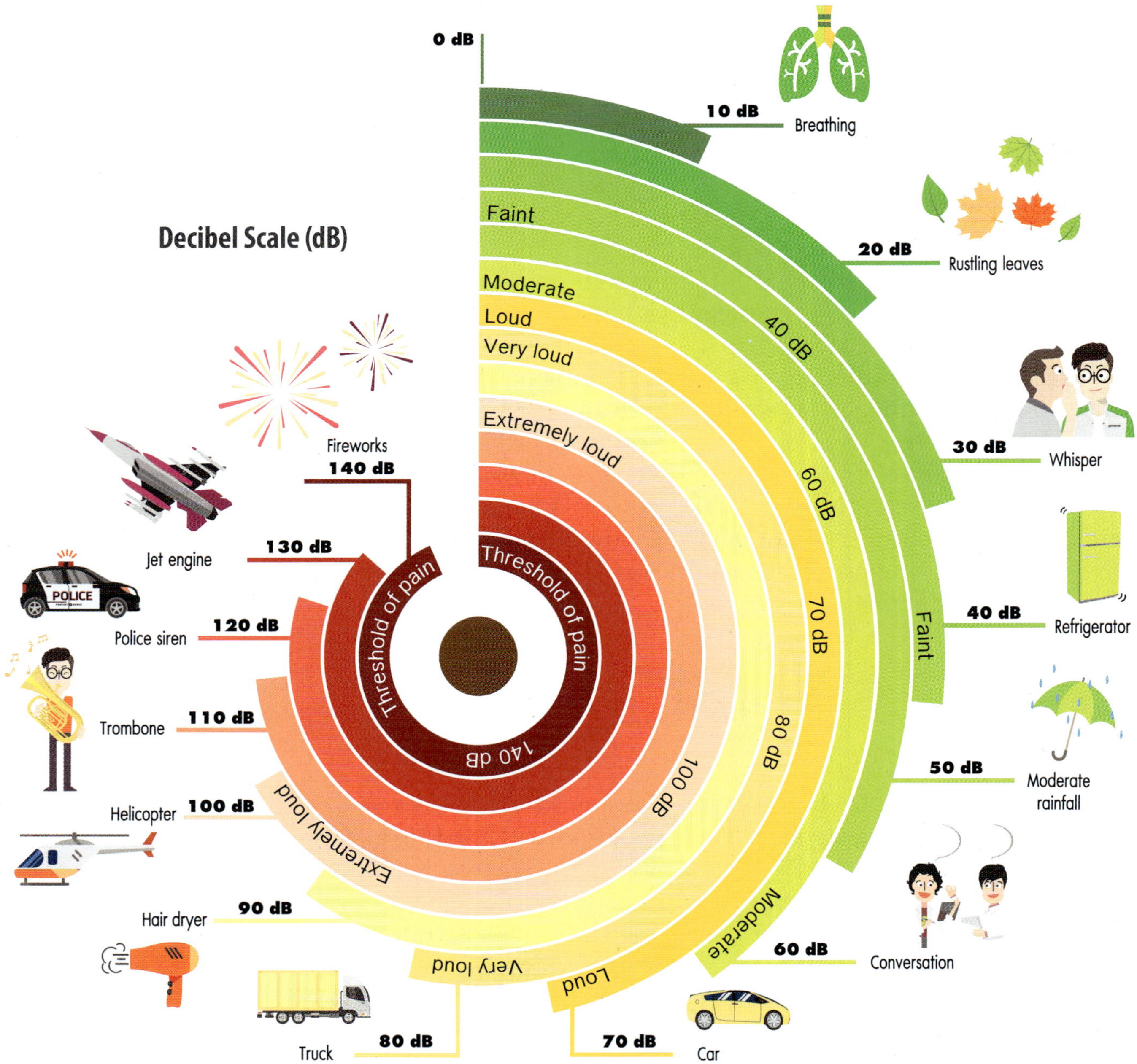

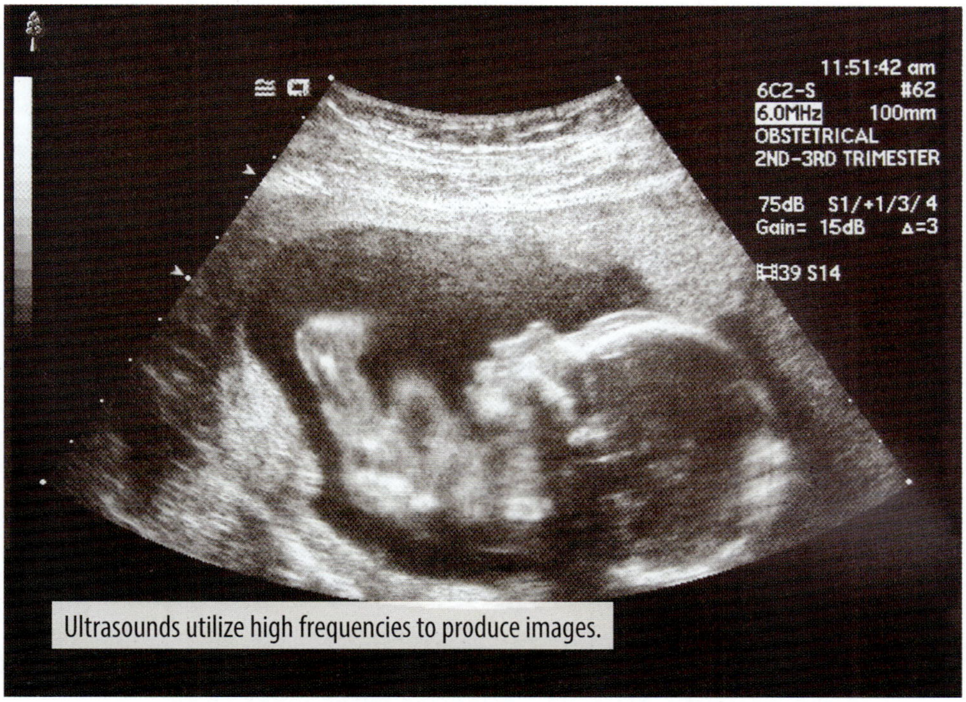

Ultrasounds utilize high frequencies to produce images.

played on a violin than on a cello because violin strings are shorter than cello strings.

Human ears are most sensitive to sounds in the 3,500–4,000 Hz range, but people can hear frequencies between 20 and 20,000 Hz. This range depends on the age of the individual and the health of his or her ears. Older people usually cannot hear very high frequencies. Other mammals hear different frequencies. Animals with a large ear separation, such as

TRY THIS

The Speed of Sound

Using a meterstick, measure and mark with a piece of colored tape 200 m from the noise production location. Have your partner stand at the noise production location, hold a timer, and practice pounding on a trash can lid or kitchen pot with a hammer at precisely one-second intervals. Help your partner adjust the strokes. When the strokes are constant, stand at the 200 m mark. You will notice that the image of the hammer strokes and the sound of the pounding do not match. Walk away from your partner until the hammer strokes and the sound of the pounding are synchronized exactly one beat apart. The distance between you and your partner is the distance it takes sound to travel in one second. Mark the place where you are standing, and measure the distance between the mark and your partner. Repeat this procedure three times and calculate an average distance. What is your calculated speed of sound? The scientifically accepted speed of sound in air is 346 m/s at 25°C. Is your calculated value close to the actual speed of sound?

Doppler Effect

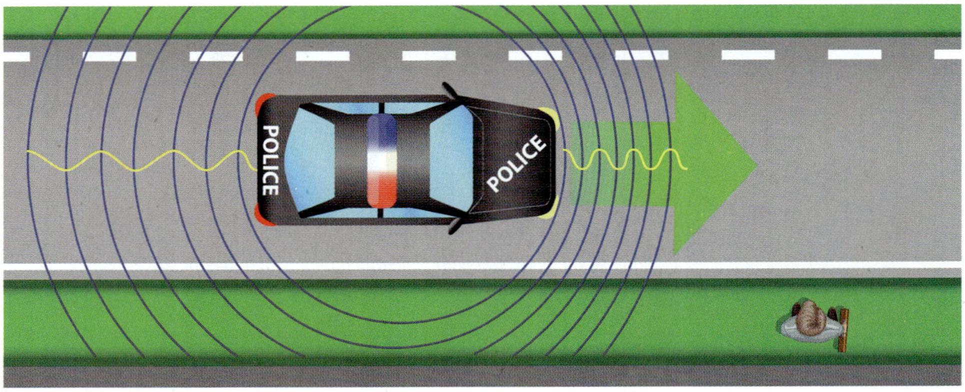

Compressions build up ahead of a moving sound source, which causes the listener to hear a high-pitched sound.

elephants, hear lower frequencies. Elephants can only hear up to 12,000 Hz. Small mammals hear very high frequencies. Cats can hear up to 64,000 Hz, and mice can hear up to 91,000 Hz.

Sounds with frequencies higher than 20,000 Hz are called **ultrasonic sounds**. Bats can produce ultrasonic sounds to locate their prey. Bats can determine if the object they are approaching is hard or soft by the frequencies they receive compared to the frequencies they emit. This method helps bats find moths and mosquitoes to eat.

Sometimes sounds change because the source or the listener moves. What happens to the sound that is produced by a blaring siren as an ambulance races by? The sound of the siren drops suddenly as the siren passes by. This is a result of the Doppler effect. The **Doppler effect** is the change in the frequency and pitch of a sound caused by the motion of the sound source or the listener. As an ambulance siren speeds toward you, more sound waves are pushed toward you every second. A buildup of compressions raises the frequency and the pitch of the siren. As the siren passes, the pitch suddenly drops. As the siren speeds away, fewer sound waves reach your ear. This lowers the frequency and the pitch of the sound.

LESSON REVIEW

1. Why do people see lightning before they hear thunder?
2. Describe the structure of a high-amplitude sound wave.
3. How are amplitude, intensity, and loudness related?
4. What intensity level can damage human ears, and what level can cause hearing loss?
5. How are pitch and frequency related?
6. Draw and describe sound waves that exhibit the Doppler effect.

The Human Use of Sound 6.2.3

Most people experience sound every day. Alarm clocks may wake you up in the morning, a beeping microwave indicates that hot food is ready, smoke alarms warn of a fire, conversation provides information, and music refreshes the soul. People use sounds to praise God, communicate, provide entertainment, and collect scientific data. Sounds may be made with hands, vocal cords, musical instruments, or machines.

The study of sound has enabled scientists to make many valuable technological contributions. For example, scientists invented sonar based on their observations of bats' use of echolocation. **Sonar** stands for **so**und **n**avigation **a**nd **r**anging—a method of using ultrasonic waves to calculate underwater distances or locate underwater objects. Sonar sends out high-frequency waves, much like a bat. These waves are reflected by solid objects under the ocean and bounce back to the ship. A detector on the ship picks up the reflected waves. The distance of the object is determined by multiplying the speed of sound in ocean water by half the time it takes for waves to make the round trip. Sonar is useful for mapping the ocean floor and finding sunken objects. Ultrasonic frequencies are frequently used because higher frequencies provide more details.

One of the most enjoyable uses of sound is music. **Music** is a sound with a pleasing quality, an identifiable pitch, and a repeated timing called *rhythm*. The quality of a sound that distinguishes it from other sounds of the same pitch and volume is called **timbre**. You can tell the difference between a cello and a trumpet, even when they are playing the same note, because they each have their own timbre. Sounds are produced by an instrument when it vibrates at a certain frequency. A musical

OBJECTIVES

- Describe the four main musical instrument families.
- Relate the length of air columns in musical instruments to frequency and pitch.
- Contrast music and noise.

VOCABULARY

- **acoustics** the scientific study of sound
- **music** a sound with a pleasing quality, an identifiable pitch, and a repeated timing called rhythm
- **noise pollution** a level of noise that causes irritation, pain, or damage to a part of the body
- **resonance** what occurs when an object vibrating at or near the natural frequency of a second object causes the second object to vibrate
- **sonar** a method of using ultrasonic waves to calculate underwater distances or locate underwater objects
- **timbre** the quality of a sound that distinguishes it from other sounds of the same pitch and volume

Sonar helps sailing vessels "see" the bottom of a body of water.

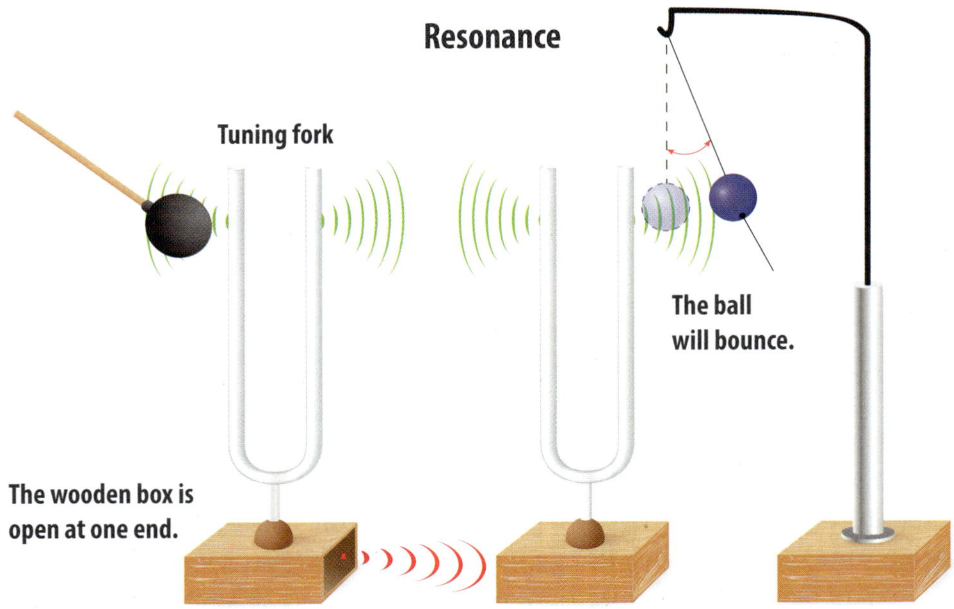

instrument produces different frequencies at the same time, and each of these frequencies produces a specific pitch. The blending of these different pitches accounts for an instrument's timbre.

FYI

Acoustic Techniques Separate Contaminants

Each year, industries in the United States extract large amounts of oil and natural gas. When these two products are extracted from the ground, water also comes to the surface. This water is called *produced water.* It is not pure. The water is usually contaminated with salt, bacteria, hydrocarbons, and radioactive materials. The extraction process also introduces chemicals into the water. In an effort to not waste the water, businesses must separate and purify it. Traditional separation and purifying processes are inefficient; they require a tremendous amount of energy input, and more chemicals are often required. A new, efficient separation technology has been introduced by a small business called *FloDesign Sonics*. FloDesign Sonics uses acoustic sound waves to separate the oil and other contaminants from water. This method is called *acoustophoresis*. Patterns of ultrasonic waves in the megahertz range bind the contaminants that can then be easily separated from the water when they sink to the bottom or rise to the surface. The current model can process 7,000 gallons each day. The company's goal is to have a system that can process 100,000 gallons per day. This process can reduce the use of chemicals and energy input by about 75%. Aside from the oil and natural gas industry, this separation process is being used in many biological aspects. One example is the ability to recover products from mammalian cells, which are very delicate and contain extremely small particles.

Musical instruments, like any other sound source, produce vibrations and have resonance. **Resonance** is what occurs when an object vibrating at or near the natural frequency of a second object causes the second object to vibrate. Resonance amplifies sounds and makes them louder. This is what makes it possible for a singer to shatter glass with a sharp, high-pitched note. If the natural frequency of the glass is the same as the singer's note, resonance causes the glass to vibrate. The glass, which is inflexible, will shatter if the vibrations are strong enough.

There are four main families of musical instruments—woodwind, brass, stringed, and percussion. Woodwind instruments have a column of air through which sound waves travel to produce musical sounds. In woodwind instruments, the column is lengthened by closing the tone holes with one's fingers and shortened by opening up successive holes, starting at the open end. The air is controlled at the other end with the mouthpiece. Most woodwind instruments have reeds at the mouthpiece that vibrate when blown. The flute has an opening through which air is blown. The vibrating air flows from the mouthpiece through the air column. Energy from these sound waves is transmitted to the outside air, creating music. The longer the air column, the longer the wave frequency and the lower the pitch. When all of the holes are closed, the music has a lower pitch than when only some holes are closed.

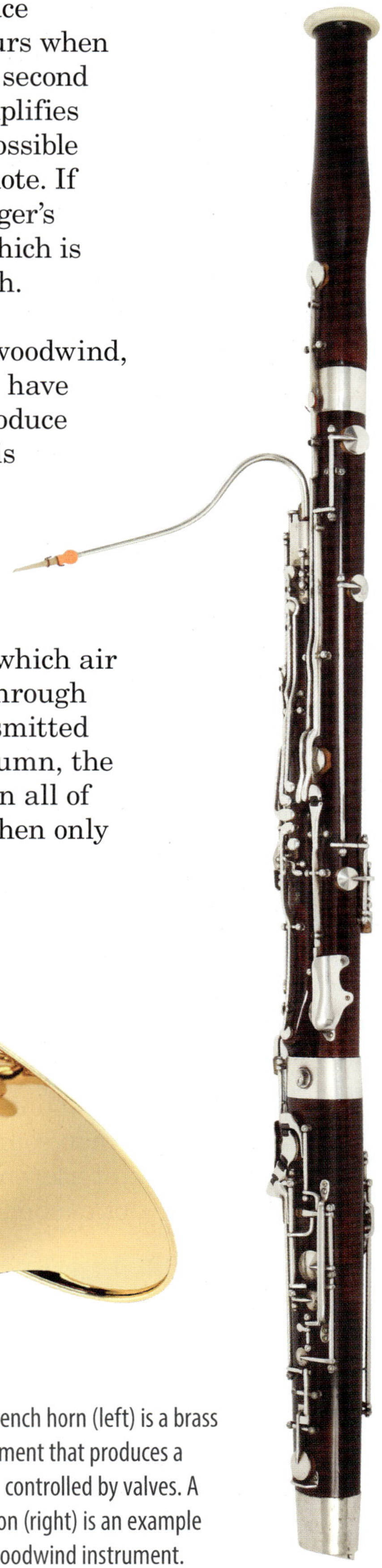

The French horn (left) is a brass instrument that produces a sound controlled by valves. A bassoon (right) is an example of a woodwind instrument.

HISTORY

Stradivarius

Many people around the world are convinced that the best violins were built by Antonio Stradivari, who lived during the 17th and 18th centuries. Some of his hand-crafted violins have sold for millions of dollars. But, what makes them so extraordinary? Perhaps the rich tone is created by the thickness of the wood top and back plates or the condition of the microscopic pores in the wood. Some researchers believe the secret is related to the mineral bath that the wood was soaked in. This process was implemented to prevent wood rot. The mineral bath gave the wood different chemical properties compared to modern violins that are made of a similar maple wood. Whatever the secret may be, people continue to enjoy the wonderful sound created by this unique musical instrument.

Brass instruments produce music because of the vibrations of the player's lips at the mouthpiece. Most brass instruments have valves that shorten or lengthen the air column. The longer the column, the lower the pitch of the note.

Stringed instruments are either plucked or rubbed to produce vibrations. The vibrations have a particular frequency and pitch. Shorter strings vibrate at higher frequencies, producing music with higher-pitched tones. Longer strings vibrate at lower frequencies, producing music with a lower pitch. Thicker strings also vibrate at lower frequencies and produce music with a lower pitch. For example, a violin has shorter strings than a cello. As a result, a violin produces higher-pitched tones than a cello. String musicians can also alter an instrument's pitch by tightening or loosening its strings. Greater tension on the instrument's strings produces higher frequencies and a higher pitch.

Musicians shake, strike, brush, or rub percussion instruments to produce sounds. Some of the most common percussion instruments are drums. They are constructed with a membrane that is stretched over a resonator—a hollow chamber that is filled with air that amplifies sound when the air vibrates. Many drums have a pitch that is fixed. However, the pitch of other drums may be adjusted by tightening or loosening the membrane.

The inside of a concert hall is usually oval shaped with high ceilings. The hall's design is meant to produce a better quality of sound. Remember, sound waves can combine in destructive or constructive interference. When sound waves are in phase, they combine constructively, producing greater intensities (louder sounds). Outdoor amphitheater structures are carefully designed to produce constructive interference to amplify music. But when sound waves are out of phase, they combine in destructive interference, and sounds are softened or deadened. Some auditoriums

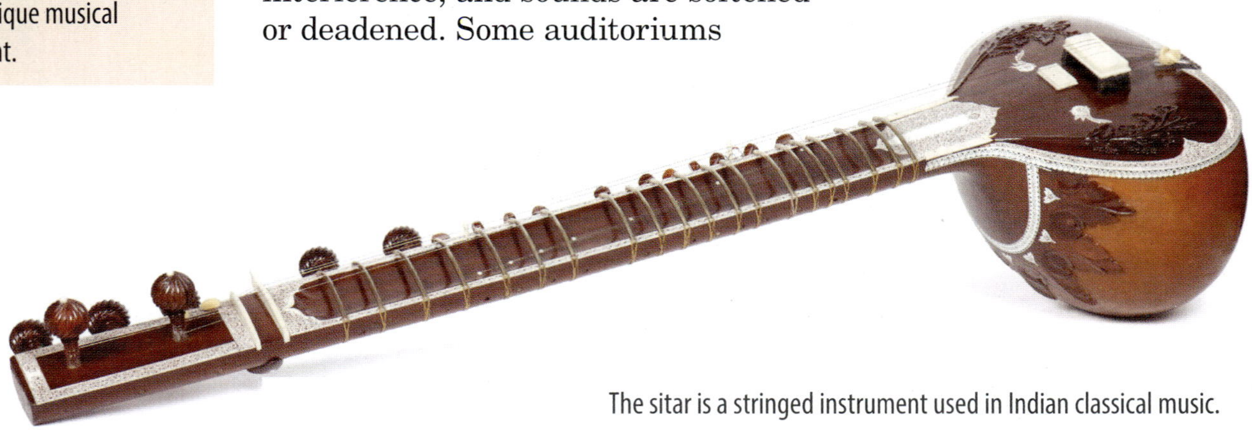

The sitar is a stringed instrument used in Indian classical music.

The tabla is a percussion instrument that is played in south Asia.

have dead spots—places where music cannot be heard because of destructive interference.

Acoustics is the scientific study of sound. Acoustic engineers design concert halls and auditoriums to be free of dead spots by using the shape, position, and materials to eliminate destructive interference.

Not all human-made sounds are satisfying. Noise is an unwanted sound. Noise has an unpleasing quality, no identifiable pitch, and no definite repeating patterns. Many machines and other devices produce bothersome noises. For example, car or jet engines create noise. A level of noise that causes irritation, pain, or damage to a part of the body is called **noise pollution**. Noise pollution can affect people's health. High levels of noise may lead to an increase in blood pressure and nervous tension. Also, loud noise can damage parts of the ear. Some communities have passed laws prohibiting noise pollution.

LESSON REVIEW

1. Describe the musical instrument families.
2. What type of air column produces a low frequency and pitch in woodwind and brass musical instruments?
3. How can a musician adjust the pitch of a percussion instrument? Include an example in your answer.
4. How is sound produced in a stringed instrument?
5. Why are there "dead spots" in some concert halls?
6. How do music and noise affect the body?

BIBLE CONNECTION

The Blast of a Trumpet

The Bible mentions many musical instruments, but none are given the same amount of attention and prominence as the trumpet. The sound of the trumpet can signify many different things. For example, it can be the alarm of war, the command to march, or the call to assemble. The musical blast that is God-ordained can reach far distances and strike the cords of people's hearts to beckon them to hear what God has to say. It is used to get people's attention. The walls of Jericho fell to the sound of a trumpet in Joshua 6:4–20. Some of the most significant mentions of a trumpet in the Bible reference the second coming of Christ. Matthew 24:31 states, "And He will send His angels with a loud trumpet call, and they will gather His elect from the four winds, from one end of the heavens to the other." Jesus will be announced with a trumpet blast. May people's ears and hearts be in tune with God's message.

6.3.1 *Electromagnetic Spectrum*

OBJECTIVES

- Distinguish between electromagnetic waves and mechanical waves.
- Relate wavelength and frequency to the regions on the electromagnetic spectrum.

VOCABULARY

- **electromagnetic spectrum** the arrangement of electromagnetic waves according to frequency or wavelength
- **radiant energy** the energy of electromagnetic waves
- **visible light** the narrow range of wavelengths in the electromagnetic spectrum that can be seen by human eyes

Modern physics has helped people better understand the light that God called into existence at the beginning of time. When God created light, He created more than the white light from the sun or the colors of light that are reflected in a sunset. These types of light are examples of a group of waves called *electromagnetic waves*. Some of these waves are visible. Others are invisible to human eyes.

Electromagnetic waves are similar to mechanical waves in many ways. Both types of waves transfer energy. However, electromagnetic waves have their own unique set of properties. As the name suggests, electromagnetic waves are comprised of electric and magnetic fields. Most matter contains electrons that are surrounded by electric fields. When the electrons vibrate, their electric fields change. Magnetic fields are created by magnets, changing electric fields, and vibrating charges. Because the electrons are moving, they are also surrounded by a magnetic field that changes in response to the movement. The electric field creates the magnetic field, and the magnetic field creates the electric field. They regenerate each other. The constantly changing fields create an electromagnetic wave.

The electric and magnetic fields that make up electromagnetic waves travel perpendicularly to each other away from the electric charge. Electromagnetic waves are transverse waves. Energy

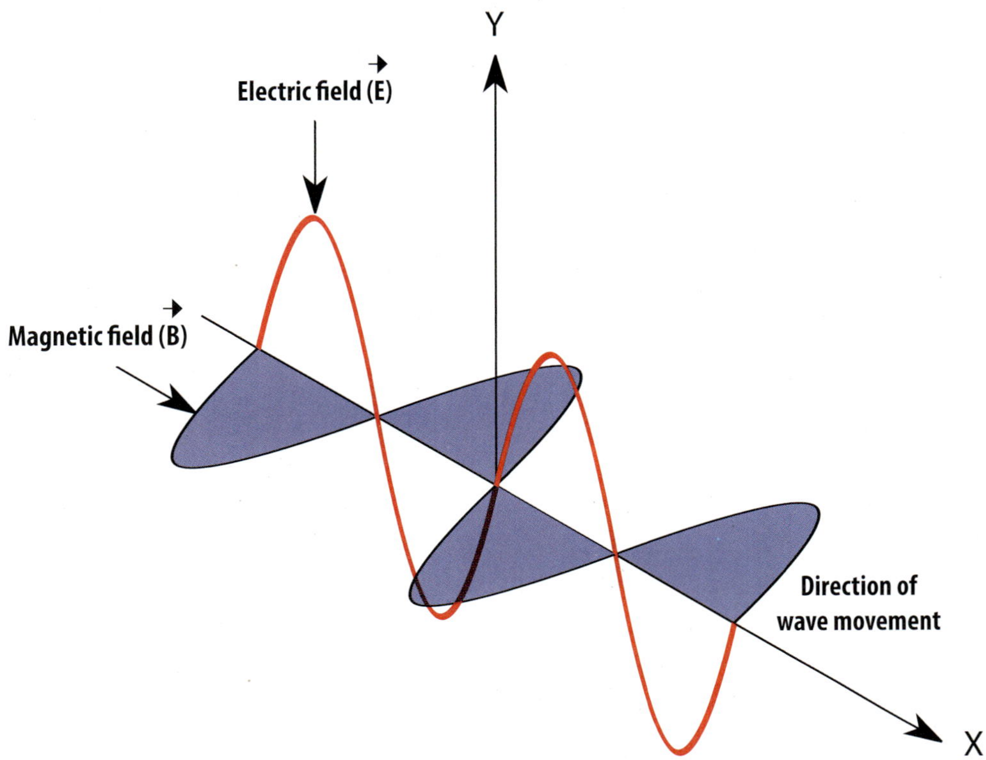

Electromagnetic waves have two parts: an electric field and a magnetic field. Both are vector quantities, indicated by the arrows.

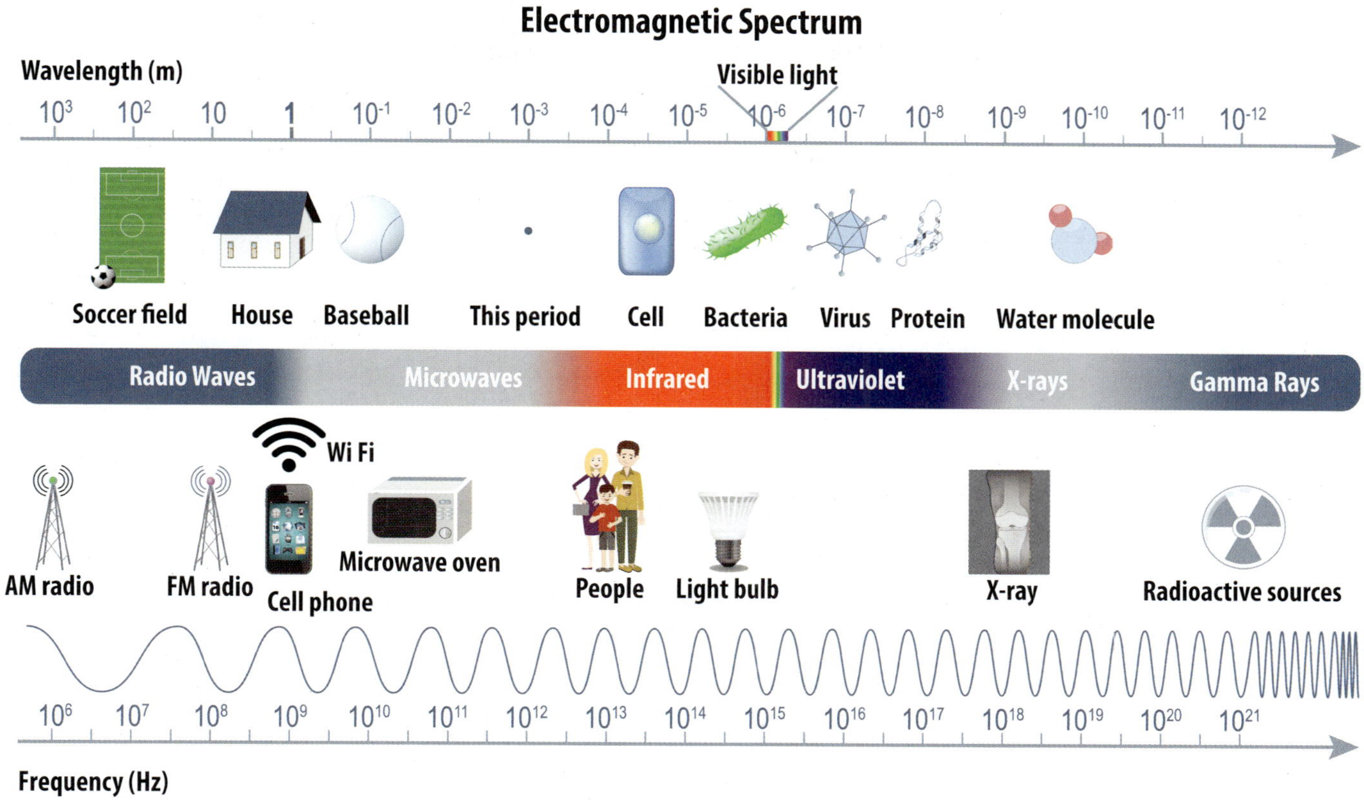

is transferred between the electric and magnetic fields. The energy of electromagnetic waves is called **radiant energy**. As the electric and mechanical fields collide with matter, energy is transferred, the particles begin to vibrate, and energy travels through the matter. Unlike mechanical waves, electromagnetic waves can transfer energy without a medium. They can travel through a vacuum.

Electromagnetic waves have a wide variety of frequencies and wavelengths. The arrangement of these waves according to frequency or wavelength is called the **electromagnetic spectrum**. The electromagnetic spectrum consists of several different types of radiation: radio waves, microwaves, infrared rays, visible light, ultraviolet rays, X-rays, and gamma rays. Electromagnetic waves with lower frequency and longer wavelengths carry less energy than electromagnetic waves with higher frequency and shorter wavelengths. An increase in temperature helps promote an increase in energy.

Visible light is the narrow range of wavelengths in the electromagnetic spectrum that can be seen by human eyes. This narrow range is also called *the visible spectrum*. Most of the light in this region is white light. White light is composed of many

TRY THIS

Emission Spectrums

Infrared **Visible** **Ultraviolet**

Each type of light source produces a specific set of wavelengths, known as *an emission spectrum.* These spectrums are unique to the source and help scientists identify the type of light that is produced. Diffraction grating has a repeating groove structure that causes each wavelength to gain a maximum intensity at a specific angle of deflection from the source. This means that the wavelengths are separated. Turn off the lights. Using diffraction grating, observe the spectrums that are produced by an incandescent light, a compact fluorescent lamp (CFL), a light-emitting diode (LED) lamp, an insect light, and an ultraviolet (UV) lamp. Fill in the emission spectrum for each light source on **WS 6.3.1A Emission Spectrums**. Use the diffraction grating to test different light sources at home. Compare the spectrums, and identify which types of sources are used.

colors: red, orange, yellow, green, blue, indigo, and violet. Red light has the longest wavelength (about 700 nm), and violet light has the shortest wavelength (about 400 nm). Because shorter wavelengths have more energy, violet light has the highest amount of energy of visible light.

The waves that are located to the left of the red portion of the visible light section on the electromagnetic spectrum include radio waves, microwaves, and infrared rays. They have long wavelengths and low frequencies. The waves that are located to the right of the violet portion of the visible light section on

Infrared image

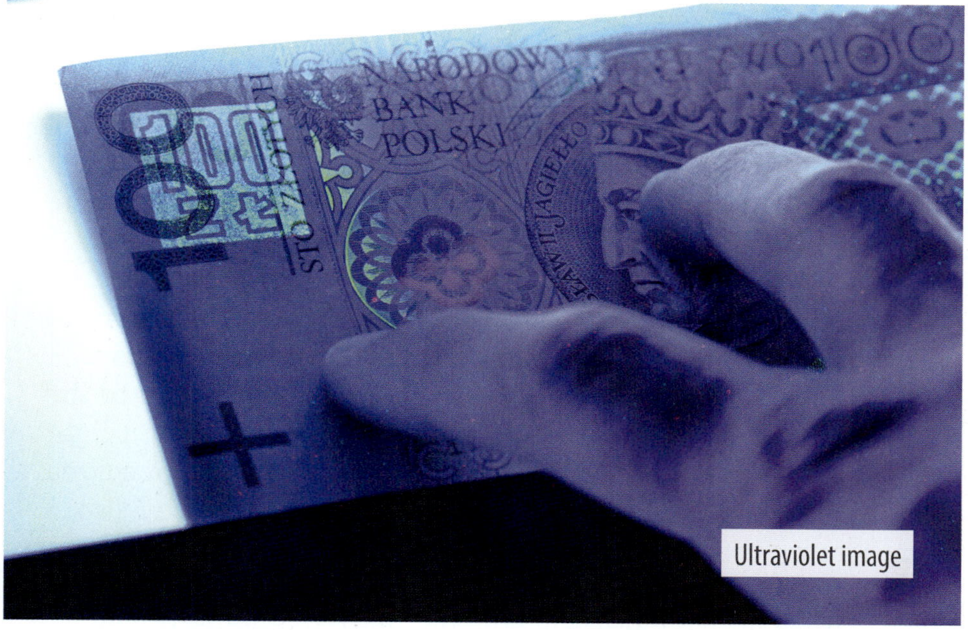

the electromagnetic spectrum include ultraviolet rays, X-rays, and gamma rays. They have relatively short wavelengths and high frequencies. The wavelengths of the electromagnetic spectrum range in length from one-trillionth of a meter to several kilometers, but the human eye can see only the band of wavelengths measuring about 400–700 nanometers.

People can also detect infrared rays, but not with their eyes. Infrared rays are too long to excite the retina in the human eye with the impression of light, but your skin detects infrared rays in the form of heat. The narrow band of white light and infrared rays are the only types of radiation that human bodies can detect.

LESSON REVIEW

1. Describe how the electric field and the magnetic field in electromagnetic waves interact.
2. How do the electric and magnetic fields travel in relation to each other? What type of wave does this produce?
3. How are mechanical waves and electromagnetic waves different?
4. What types of radiation are included in the electromagnetic spectrum?
5. Describe the frequencies and wavelengths of different regions of the electromagnetic spectrum.
6. What two types of radiation can humans detect?

6.3.2 How Light Travels

OBJECTIVES

- Differentiate between luminous and illuminated objects.
- Contrast transparent, translucent, and opaque substances.

VOCABULARY

- **illuminated object** an object that reflects light from luminous objects
- **luminous intensity** the amount of visible light that is emitted from a source
- **luminous object** an object that emits light
- **opaque substance** a substance that does not transmit light
- **translucent substance** a substance that transmits light but scatters it as it passes through
- **transparent substance** a substance through which light is easily transmitted

Have you ever wondered how light travels? Suppose that you could slow down time so much that you could actually watch light creep across the room. What would you observe? You would notice that light travels in a straight line. Because of this type of movement, the brightness of light decreases with distance. A light source seems brighter at close distances. That is why you can see well if you sit next to a lamp to read at night.

The amount of visible light that is emitted from a source is called **luminous intensity**, I, and the luminous intensity unit is a candela (cd). This measurement is inversely proportional to the square of the distance from the light source ($I = 1/d^2$). If a light source is located one meter away from a specific area, one square meter of that area is lighted. If the light source is moved two meters away, that same amount of light spreads over four square meters. Although more area is covered, the intensity of the light at a distance of two meters is only one-fourth of what it was at a distance of one meter.

When you turn on a light, the light seems to instantly reach every corner of the room. But if you could slow down time, you would see the light traveling through the air from the source across the room. The reason it seems to happen instantly is that light moves at an incredible speed—about 300,000 km/sec in a vacuum. This speed is about the same through air. At this speed, light could travel around the world in less than a second. No wonder light seems to travel instantly!

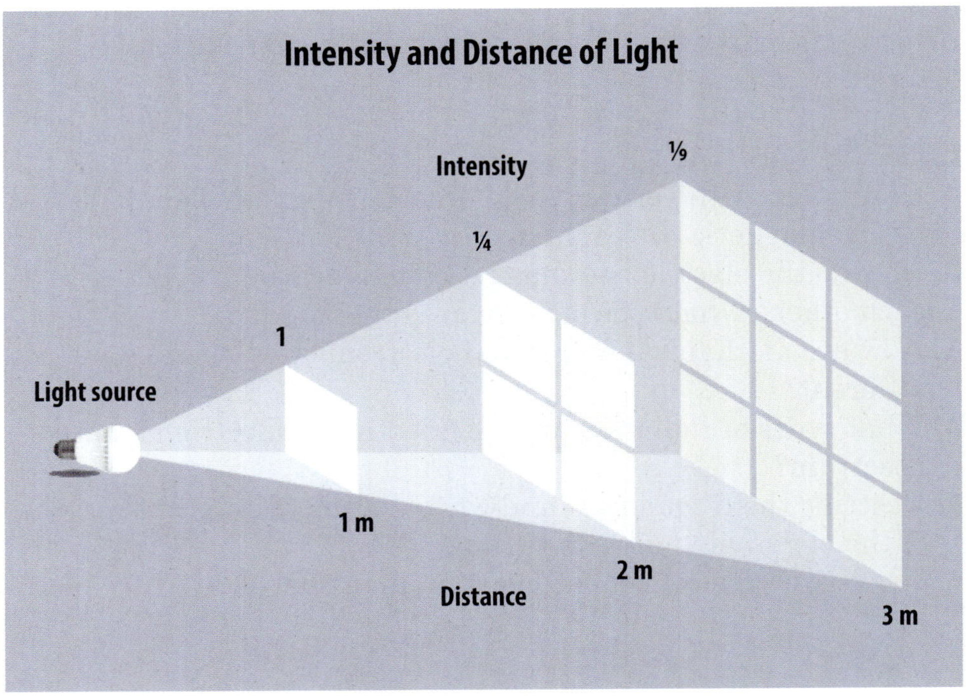

The intensity of light decreases as the distance from the light source increases.

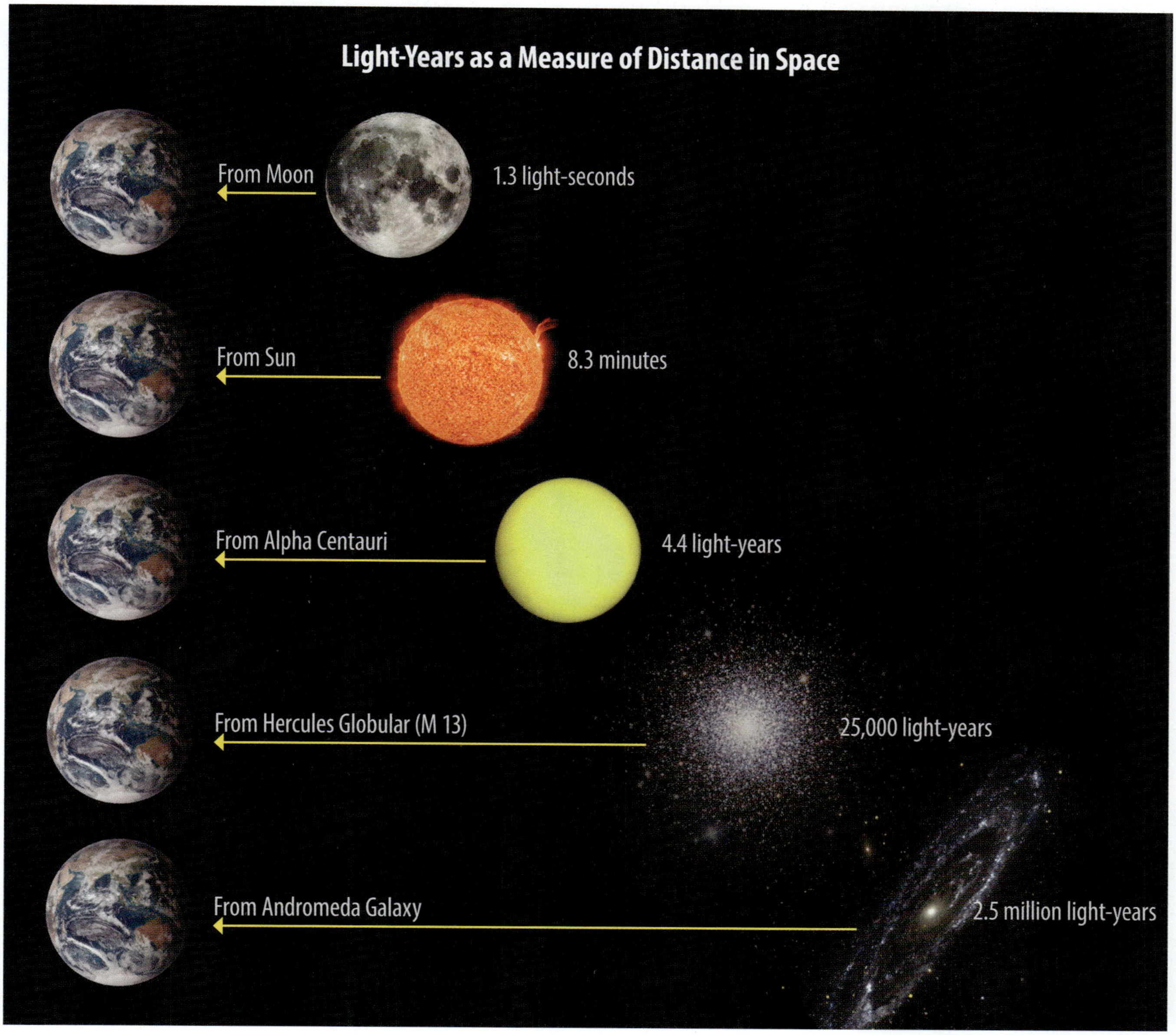

Even though light travels quickly, it takes years for the light of stars to reach Earth from space. Astronomers apply the unit of a light-year to measure these distances. A light-year is the distance that light travels in a vacuum in one year. It is equal to about 9.46×10^{12} kilometers. Astronomers utilize the light-year unit to make the great distances of stars and galaxies more manageable. The concept of light-years gives people an idea of the vastness of the universe.

Sunlight can often be seen long after the sun goes down. Imagine walking along an ocean shore on a clear night. Silvery light is gleaming on the water. Where does this light come from? The light that is emitted from the sun is reflected from the moon to Earth.

The wood is illuminated by the light bulbs, which are luminous objects.

Most of the natural light radiated to Earth comes from the sun. A small amount of light is transmitted by stars. The sun and stars are luminous objects. A **luminous object** is an object that emits light. Other luminous objects include light bulbs, fire, and lasers. Some creatures are also luminous, such as fireflies and plankton and some deep sea fish. These living things create their own light through a process known as *bioluminescence*. An object that reflects light from luminous objects is called an **illuminated object**. The moon is an illuminated object. In fact, most objects people see are illuminated objects.

When light strikes an object, it can be transmitted, scattered, absorbed, or reflected. A **transparent substance** is a substance through which light is easily transmitted—the light passes through it. Light passing through transparent substances does not scatter. This makes it possible for you to see clearly through the substances. Windowpanes, aquarium glass, clean water, and clean air are all transparent substances.

A **translucent substance** is a substance that transmits light but scatters it as it passes through, making objects on the other

TRY THIS

Transmitted and Scattered Light

Fill an aquarium with water. Shine a focused flashlight beam through the aquarium onto a white screen behind the aquarium. Slowly add powdered milk to the water. Observe the color that appears on the screen, which is transmitted light. Also, observe the color of the light beam from the side of the aquarium, which is scattered light. What accounts for these colors? How else do you experience this effect?

Light passes straight through the transparent vase without being scattered.

The stalks appear blurry in the translucent vase.

side appear unclear and blurry. Frosted glass, waxed paper, and quartz are translucent substances. An **opaque substance** is a substance that does not transmit light. It absorbs and reflects it. The way that an object absorbs and reflects light determines its color.

The stalks in the opaque vase cannot be seen.

LESSON REVIEW

1. How does the intensity and span of light change with increasing distance?
2. If a light source is located 12 m away from an area, how much of the area is illuminated? Calculate the intensity of light at this distance.
3. Use the terms *luminous* and *illuminated* to describe the visual relationship between the sun and the moon.
4. Is aquarium glass transparent, translucent, or opaque? Explain.
5. What is the difference between a translucent and an opaque substance?

6.3.3 Color

OBJECTIVES

- Explain why objects have different colors.
- Distinguish between primary colors of light and primary colors of pigment.

VOCABULARY

- **light filter** a transparent or translucent device that allows only certain wavelengths of light to pass through it
- **pigment** a material that gives a substance its color by absorbing some colors of light and reflecting others

When you were a child, did you ever ask why the grass is green or why some grapes are purple? Perhaps you were told that the grass is green because God created it that way. And God did. He created grass to interact with light in a certain way so you see green.

For a number of years, many people believed that color was the product of a mixture of black and white until the latter part of the 17th century when Sir Isaac Newton conducted prism experiments. Newton discovered that white light is a mixture of all the colors of the visible spectrum.

An object's color depends on what it does with the light that hits it. When light strikes an object, the light is transmitted through a transparent object, scattered in a translucent object, or reflected or absorbed by an opaque object. White light is composed of red, orange, yellow, green, blue, indigo, and violet.

Opaque objects do not allow light to pass through them. The light waves absorbed by an opaque object never reach your eyes, but the light waves that bounce off an opaque object are reflected into your eyes. It is the color of the wavelengths of light that the object reflects that determines the color of the object. For example, a red apple reflects the red wavelengths of white light but absorbs the orange, yellow, green, blue, and violet wavelengths. An apple is red because it reflects red and absorbs all the other colors.

Color and Reflection

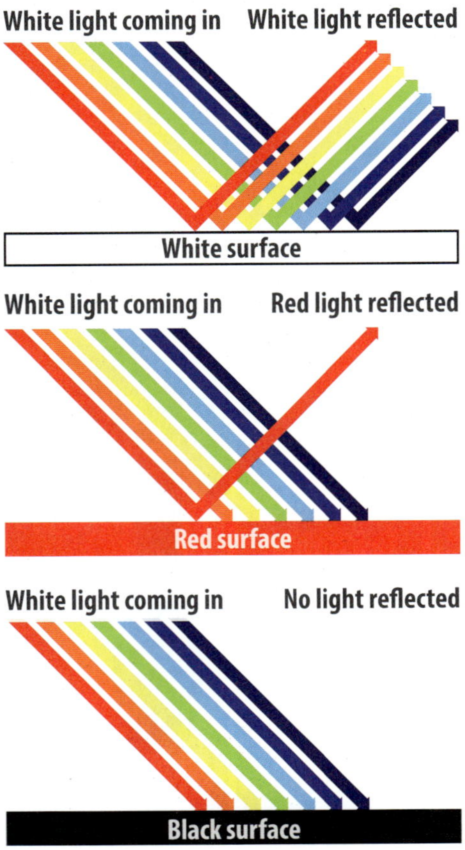

White objects reflect all the wavelengths of color back to your eyes. That is why white clothing is cooler than dark clothing; white clothing reflects all the wavelengths of light (and its heat) away from the white clothing. In contrast, black objects absorb all the wavelengths of color, reflecting no color back to your eyes. But most objects reflect more than one color. These combinations of reflected colors produce the varieties of colors that can be seen in many objects.

Each color of light has its own wavelength and frequency, but all the colors of creation are combinations of just three different colors: red, blue, and green. Red, blue, and green wavelengths of light are called *the primary colors of light*, which can be combined to produce any other color of light. When all three primary colors of light are combined, they produce white light. Yellow, magenta, and cyan (blue-green) are called *the secondary colors of light* because they are produced by equal intensities of two primary colors of light. Any two wavelengths of light that can be combined to form white light are called *complementary colors*. The three sets of complementary colors of light are yellow and cyan, cyan and red, and magenta and green.

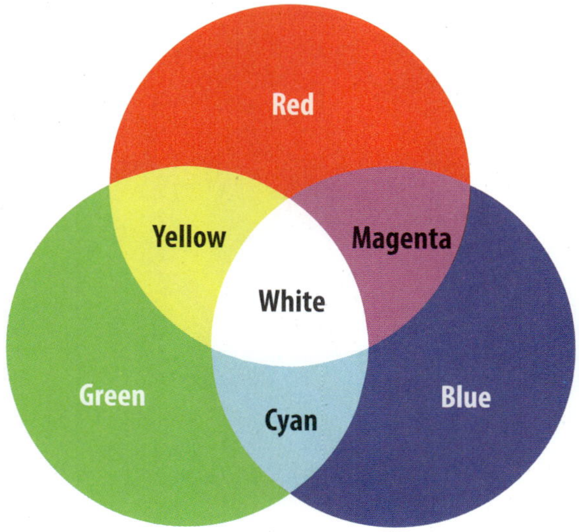

A combination of the three primary colors of light produces white light.

The primary colors of light can be separated from each other by using a filter. A **light filter** is a transparent or translucent device that allows only certain wavelengths of light to pass through. Filters absorb some light wavelengths, changing either the light's color or its intensity. They are used in photography, optical instruments, and different types of lights.

The primary colors of light are very different from the three primary colors used to produce paint or ink. The primary colors of pigment are yellow, cyan, and magenta, which can be combined to produce any other color of pigment. Cyan paint and yellow paint combine to make green paint, but cyan light and yellow light make white light. Paints are not light sources. They are pigments. A **pigment** is a material that gives a substance its color by absorbing some colors of light and reflecting others. Mixing the three primary pigments in equal amounts produces

Red and blue light filters

BIOGRAPHY

Painting with Music

Wassily Kandinsky was born in 1866 to musical parents in Russia. Kandinsky had a unique gift because he heard music when he painted. Each color was related to a different instrument and various musical notes. When the colors combined, a symphony arose in his mind from the easel. Many art enthusiasts credit Kandinsky with being the first artist to create a purely abstract work. He saw painting as a form of worship. "Color is a power which directly influences the soul."

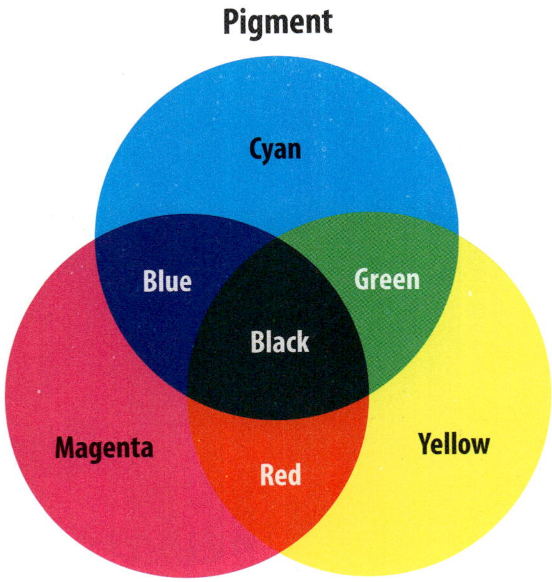

The three primary colors of pigment combine to produce black.

the color black because these three pigments together absorb all the different wavelengths of light.

People can see distinct colors because their eyes detect the difference between various wavelengths of light. Vision involves various parts of the eye and the brain. Light enters the eyes through an opening called *the pupil*. The iris, the colored area around the pupil, regulates the amount of light entering the pupil. The eye's lens refracts light onto the retina. The retina is made of light-sensitive nerves that send images to the brain. Rods are the nerve cells in the retina that are sensitive to light and dark. Cones are the nerve cells in the retina that allow people to see colors. Your amazing eyes can see the details, depth, and color of objects all around.

LESSON REVIEW

1. Why are objects certain colors?
2. All of the colors of light are created by a combination of what three colors?
3. What color is produced when equal amounts of the three primary pigments are combined? Explain.
4. Explain how people perceive objects.

BIBLE CONNECTION

The Color Purple

The color purple is an attractive part of creation. In Bible times, it was a symbol of royalty, wealth, and status. Purple had such value because it was difficult to manufacture. The most notable person to wear purple in the Bible was Jesus. Mark 15:17 states, "They put a purple robe on Him, then twisted together a crown of thorns and set it on Him." The soldiers mocked the authority of Jesus, but His purple royal garment was appropriate because He is the King of kings.

Anatomy of the Eye

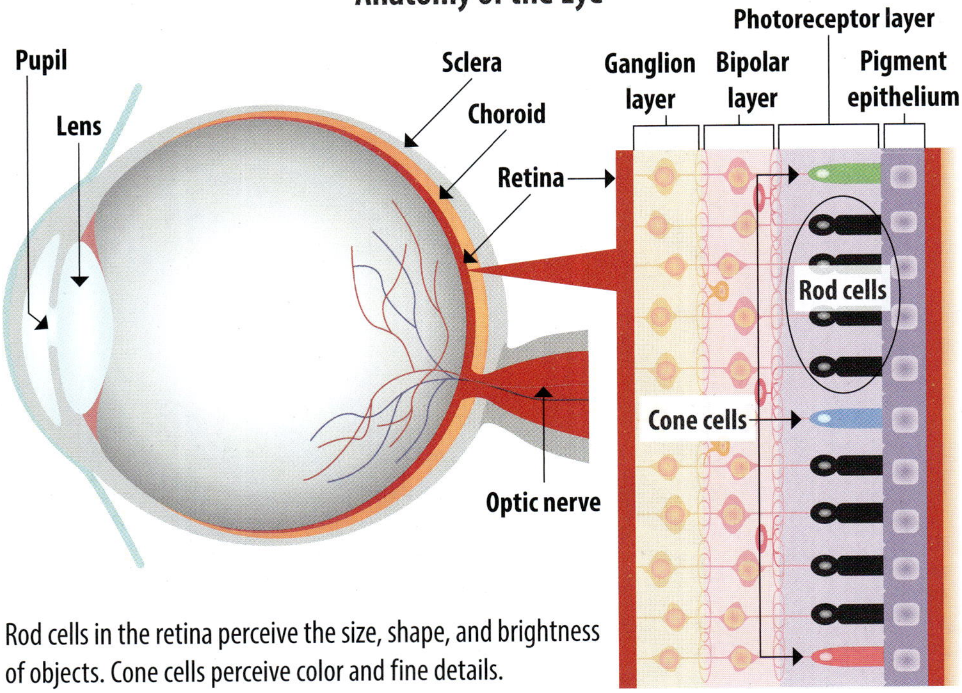

Rod cells in the retina perceive the size, shape, and brightness of objects. Cone cells perceive color and fine details.

Radio Waves 6.3.4

When tuning a radio or television to a favorite station, listeners and viewers expect to hear music or a show. But how does the music or show get from the station to the receiver? The entertainment travels on electromagnetic waves called **radio waves**, which are electromagnetic waves with long wavelengths and low frequencies. They are primarily used for radio or television broadcasts.

The sounds that are sent by radio waves pass by internal or external antennas that contain electrons. When the electrons interact with the electromagnetic wave, they begin to vibrate. Electron vibrations induce changing currents. The currents travel through an amplifier and eventually to a speaker that produces the sounds. Some radios have a switch for selecting AM or FM stations. AM (amplitude modulation) radio waves transport sound by changing the amplitude of the waves. These waves have frequencies between 535 kHz and 1,605 kHz (kilohertz). Many radio stations broadcast AM radio waves.

FM (frequency modulation) radio waves carry information by changing the frequency pattern in the waves. These waves have a frequency between 88 MHz and 108 MHz (megahertz). These numbers may sound familiar. They correspond to the call numbers of FM radio stations. Radio stations are generally located between the numbers 87 and 108 on the FM radio dial. The pictures and the sounds of television are also transported by FM waves. Once the basic signals have been separated from the

OBJECTIVES

- Explain the transmission of radio waves.
- Compare AM and FM radio waves.
- Examine uses of microwaves.

VOCABULARY

- **microwave** a type of radio wave with short wavelengths and high frequencies
- **radio wave** an electromagnetic wave with long wavelengths and low frequencies

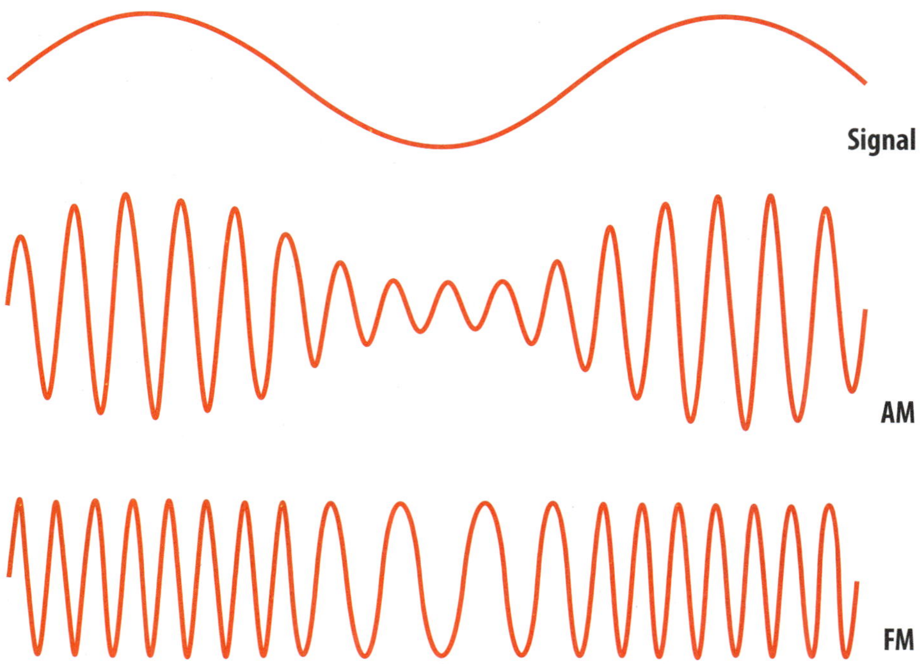

AM waves carry radio signals by changing the amplitude of the wave. FM waves carry radio signals by changing the frequency of the wave.

WiFi signals are located in the area where radio waves and microwaves are on the electromagnetic spectrum.

carrier wave, they are fed to a loudspeaker or a display device where they are converted into sound and visual images.

FM and AM radio waves have different advantages. FM radio has the advantage of a constant amplitude that keeps the strength of its signal from changing. Because AM radio works by changing its amplitude, an AM signal may be too weak to be picked up. FM radio also has less interference than AM radio. FM radio can carry more information, such as pictures and sound, than AM radio because FM radio has a higher frequency range. But, AM radio has its distinct advantages. Because AM radio operates at a lower frequency, it has longer wavelengths than FM. The longer wavelengths mean that AM radio has a much larger broadcast range. These longer wavelengths can travel through solid objects. FM radio signals may fade out in the mountains, but AM radio signals do not.

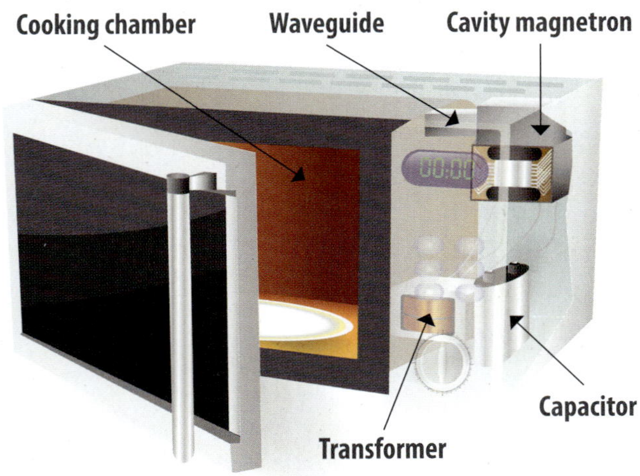

The high-frequency waves that microwave ovens generate allow food to be cooked quickly.

Microwaves are a type of radio wave with short wavelengths and high frequencies. The wavelengths of microwaves are only a few centimeters, and they have a frequency of one billion to one trillion hertz. One familiar use of microwave technology is the microwave oven. The magnetron in a microwave oven generates waves with a frequency of about 2,500 megahertz. The capacitor stores this energy. Radio waves with this frequency are absorbed by water, fats, and sugars. When the alternating electric fields inside microwaves interact with polar water molecules, they cause the water molecules to rotate back and forth at very high speeds. This movement is converted directly into thermal energy. Microwaves at this frequency are not absorbed by most plastics, glass, or glazed ceramics, so these materials will not overheat when used in the microwave oven. Metal reflects microwaves, so metal containers cannot be used in a microwave oven or they will cause a fire in or destroy the oven.

The Evolution of Radio

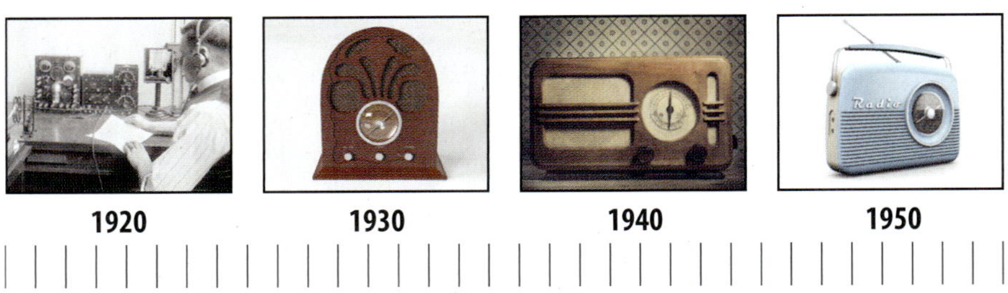

Terrestrial Cellular Phone System

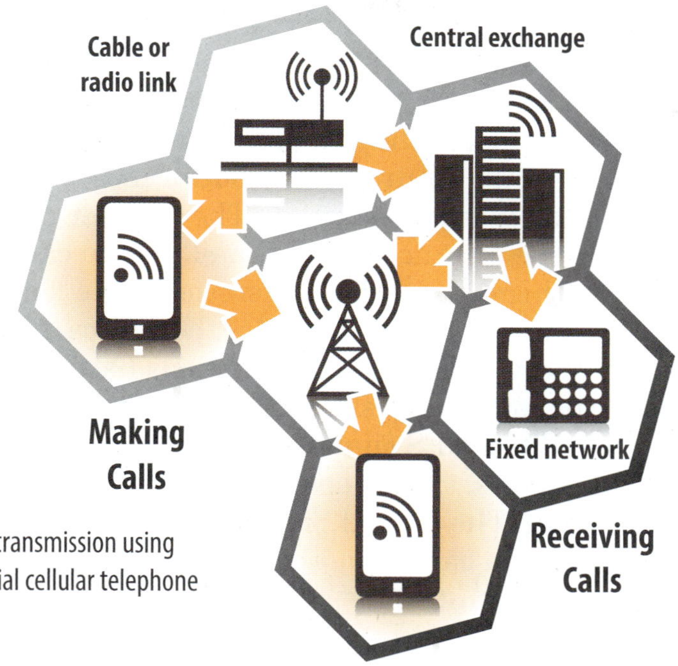

Cell phone transmission using the terrestrial cellular telephone system

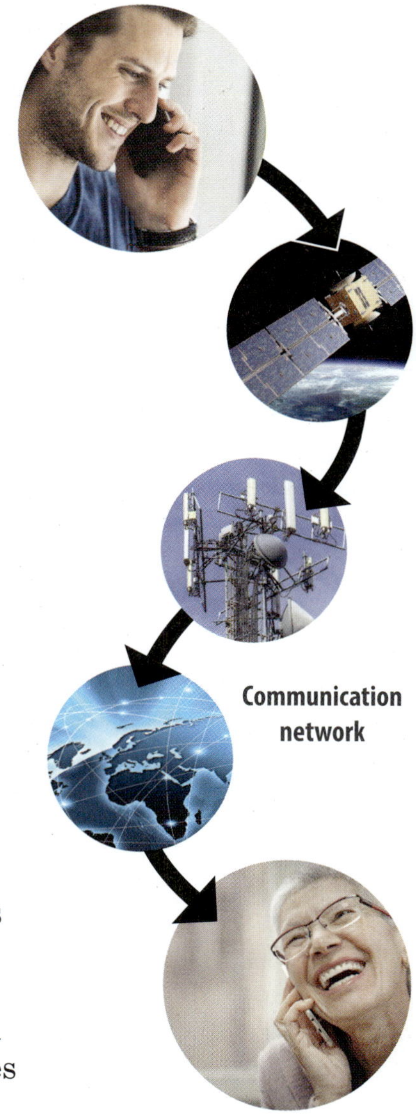

Cell phone transmission using the global satellite-based system

Many communication devices use microwaves. Cellular and cordless telephones are actually microwave radio receivers. They operate using either a cellular telephone system or a global satellite-based system. Cell phones that operate under the cellular telephone system have a microphone that converts sound waves into electrical signals. The electrical signals produce a current that creates radio waves that pass through a low-powered radio transmitter and receiver to allow communication between cell phone users. These radio waves that operate in the 800–900 MHz band are transmitted to and from microwave towers that provide service to geographic areas called *cells*.

Cell phones that operate using the global satellite-based system transmit radio waves directly to a satellite. The satellite receives the signal and sends it to a ground tower. The message travels from the tower to the communication network and the intended recipient.

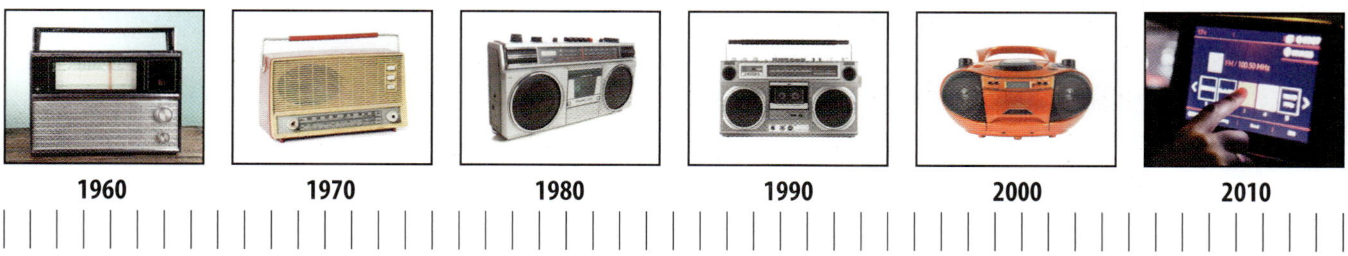

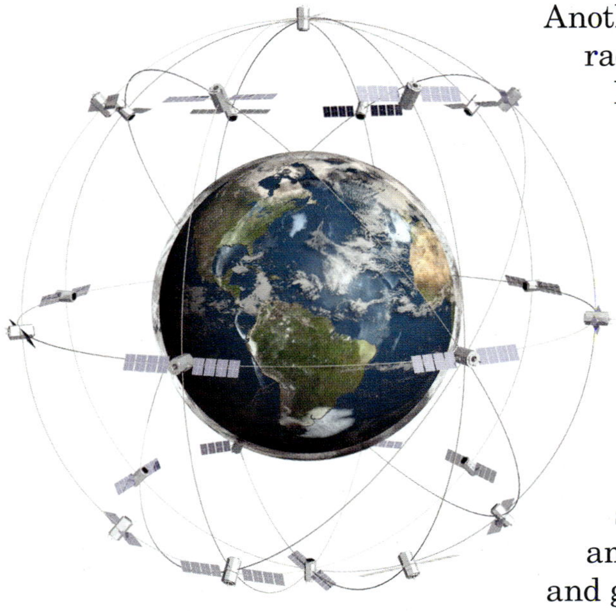

GPS uses multiple satellites to accurately determine the location of various sources.

Another system that utilizes the transmission of radio waves is the global positioning system (GPS). It is a radio-navigation system that operates primarily from space. Typically, a signal is sent from a source to a set of satellites. A receiver measures the amount of time it takes a radio signal to travel from four or more satellites to an intended location. The receiver calculates the distance to each satellite and can use this information to determine the longitude, latitude, and altitude of the source. Many individuals and organizations have access to the microwave signals that are sent out by GPS satellites. GPS signals are used in airplanes, ambulances, trains, spacecraft, cars, farm tractors, and golfing. The United States, Russia, the European Union, and China each have their own version of a global positioning system.

LESSON REVIEW

1. Explain how sound is transmitted from a radio wave to a radio speaker.
2. What are some advantages of AM radio stations?
3. How does a microwave oven heat food?
4. What is the difference between a cellular telephone system and a global satellite-based system?

TRY THIS

Speed of Sound

How fast do electromagnetic waves travel compared to the speed of sound? With a lab partner, take two walkie-talkies, a balloon, and a pin outdoors. Create a large distance between you (such as the length of a football field) and turn on the radios. One lab partner will be the observer, and the other will inflate the balloon, press the talk button, and pop the balloon. Carry out the activity. What does the observer notice?

FYI

Radar

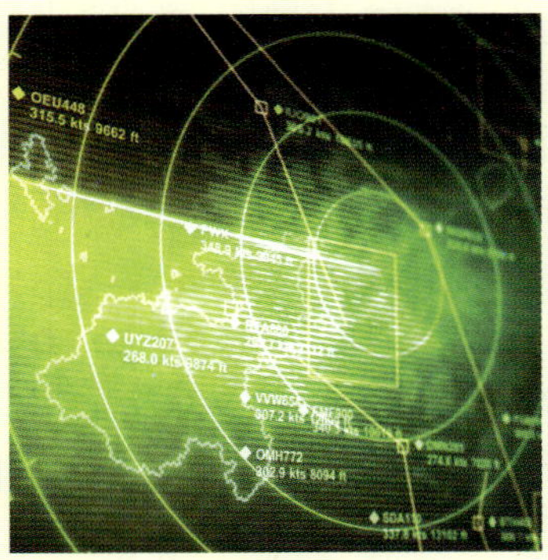

Radar, which stands for **ra**dio **d**etection **a**nd **r**anging, operates in the microwave region of the electromagnetic spectrum to detect, track, locate, and recognize various objects at close range or far distances. The sensor sends out a signal and gathers data from the echo that returns. This data can provide information about the location, velocity, size, and shape of an object. Radar technology is very precise and can detect objects at considerable distances in the rain, snow, extreme heat, or other adverse weather conditions. Police officers hold a radar device to measure the speed of drivers. Other objects that may be measured using radar include spacecraft, ships, aircraft, and birds. This technology is used in air traffic control, space surveillance, weather observation, and planetary observation.

Other Forms of Radiation 6.3.5

What comes to mind when you hear the word radiation? Many people think of X-rays or dangerous nuclear waste. Radiation, however, refers to the entire electromagnetic spectrum: radio, microwave, infrared, visible light, ultraviolet, X-ray, and gamma ray emissions. These types of radiation all travel at the speed of light but have different wavelengths and frequencies. Most people experience visible radiation. But how much do you know about invisible radiation?

OBJECTIVES

- Summarize the different forms of radiation.
- Describe uses of the different forms of radiation.

VOCABULARY

- **infrared wave** an electromagnetic wave with slightly greater wavelengths and lower frequencies than red light
- **ultraviolet (UV) wave** an electromagnetic wave with slightly shorter wavelengths, slightly higher frequencies, and more energy than violet light
- **X-ray** an electromagnetic wave with slightly shorter wavelengths and higher frequencies than ultraviolet rays

Infrared waves are electromagnetic waves with slightly greater wavelengths and lower frequencies than red light. Although people cannot see infrared waves, they can feel their energy in the form of heat. For example, when you stand in the sun or near a hot stove or fire, the infrared radiation is converted into thermal energy that warms your skin. Because all objects give off heat, all objects emit infrared waves. The amount of infrared radiation given off by an object is determined by its temperature. Warmer objects give off more infrared waves than colder objects. As the temperature increases, infrared wavelengths become shorter.

Although humans cannot see infrared radiation, some other creatures can. For example, some insects and fish see infrared radiation. Rattlesnakes have two pits on their face that help them detect infrared radiation during the day and at night. This helps them track their prey.

Scientists have developed ways for people to "see" infrared radiation. Medical professionals use a process known as *thermography* to examine areas of inflammation in the body. The infrared thermograph equipment detects infrared energy that is emitted from an object. It then converts the energy into a temperature reading and produces an image of temperature distribution.

This spring landscape would normally appear vibrant with colors of green and yellow. When viewed through the infrared portion of the spectrum, the trees and grass appear white.

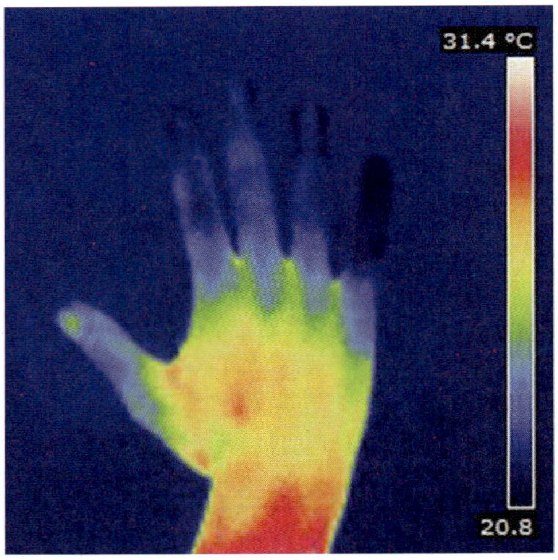

This thermogram shows a person who has neuropathy. He or she has nerve damage in the fingers, which affects the amount of heat that the fingers emit—indicated by the blue color.

Thermography can also be applied to energy conservation. Thermograms of buildings show color-coded temperature variations in one-degree increments to indicate where heat is escaping and insulation is needed. These images can also be used to save lives. With this technology, helicopters flying over an area at night or in a storm can detect missing persons. Firefighters may be equipped with thermal-imaging masks to see through smoke and find people trapped in burning buildings. The masks are helpful because the wavelengths of visible light are too short to be seen through smoke. The smoke particles scatter these light waves. Infrared rays, however, are longer and not as easily scattered.

Ultraviolet (UV) waves are electromagnetic waves with slightly shorter wavelengths and slightly higher frequencies and more energy than violet light. UV waves have frequencies high enough to be dangerous to living things. The energy of UV waves can kill cells. Hospitals use UV lamps to kill bacteria. UV waves are also used to preserve food.

UV waves come to Earth in the form of sunlight. The waves trigger the cells in a person's body to produce vitamin D, a vitamin the human body needs for healthy bones and teeth. But too much exposure to UV waves is unhealthy. These waves can burn skin and eventually may cause skin cancer.

CAREER

X-ray Technician

Radiologic technologists, also known as *X-ray technologists*, are responsible for using medical imaging equipment to produce images of bones, organs, tissues, and vessels. With additional training, these individuals may also assist with the administration of radiation therapy.

People who are interested in this field of study can specialize in areas such as magnetic resonance imaging (MRI), mammography, computed tomography (CT), sonography, radiation therapy, and nuclear medicine. X-ray technicians work closely with doctors and patients to help diagnose medical issues. They have a thorough understanding of human or animal anatomy, radiographic terminology, and pathology. The work of X-ray technicians is vital to the diagnosis, treatment, and potential cure of maladies.

TRY THIS

UV Protection

Write the word *Control* on one piece of masking tape and the names and SPF values of three different types of sunscreen on three pieces of masking tape. Place each piece of tape on the bottom of four Petri dishes. Fill each Petri dish with five UV beads. Apply a thin layer of each sunscreen on the lids of the appropriately labeled dishes. Do not apply a layer of sunscreen on the Control dish. Seal each dish with its lid. Arrange the dishes in a cardboard box and seal the box with a lid. Place the cardboard box in direct sunlight and remove the lid. With a timer, record how long it takes each set of UV beads to change color. Rank each sunscreen from most effective to least effective. Does this ranking match the indicated SPF values?

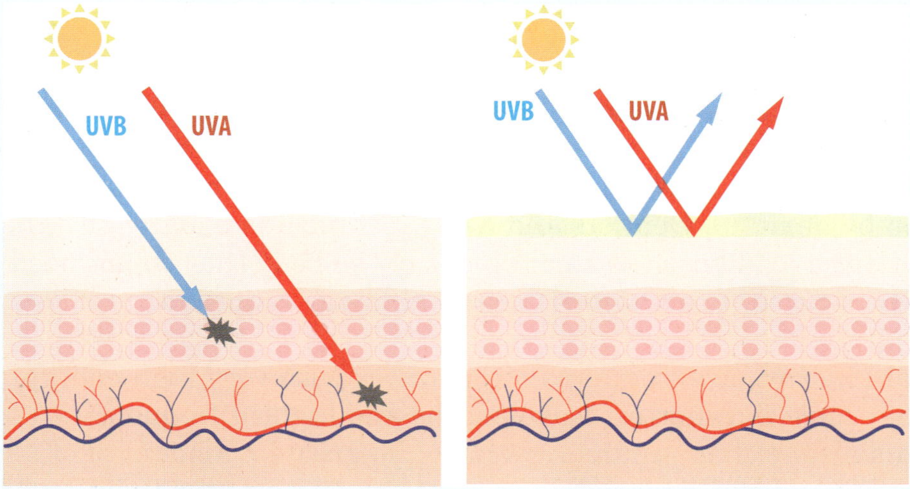

Normal skin

Skin with sunscreen

A natural shield in the atmosphere protects life on Earth by filtering out excess ultraviolet radiation. This important shield is called *the ozone layer*. Between 1928 and 1996, the ozone layer was being damaged by chemicals called *chlorofluorocarbons*, or CFCs, that were being used in refrigerators and air conditioners. In 1987, 46 countries signed the Montreal Protocol, an agreement to stop using ozone-damaging chemicals by 1996. The agreement has now been signed by nearly 200 countries.

X-rays are electromagnetic waves with slightly shorter wavelengths and higher frequencies than ultraviolet rays. How are X-ray images made? X-rays move with enough energy to pass through solid materials such as skin. Denser materials, such as bones, absorb X-rays. When an X-ray

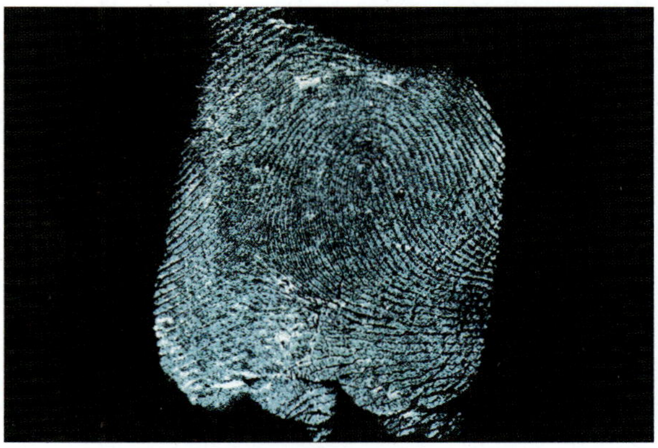

Detectives use special powder that fluoresces, or glows, under ultraviolet light, to discern fingerprints.

picture is taken, the bones absorb the radiation, but the tissues do not. Bones then appear white on the image. The soft tissue areas show up as black. Too much exposure to X-ray radiation can be harmful. Dentists and X-ray technicians usually give the patient a lead apron to shield parts of the body that do not need to be x-rayed. Lead absorbs nearly all X-ray radiation.

Gamma rays are electromagnetic waves with very high energy. They have the shortest wavelengths and highest frequencies. This means that gamma rays have the greatest amount of energy and are potentially the most destructive rays to living things. Gamma rays come from outer space and from radioactive materials used in nuclear reactors and bombs. Because they have so much energy, gamma rays can penetrate objects much easier than any other form of radiation. These rays have enough energy to pass through concrete two meters thick. Nuclear power plants build layers of protective materials around the reaction core to keep gamma radiation from leaking. Gamma radiation is used to treat cancer, but overexposure to gamma radiation can cause radiation sickness.

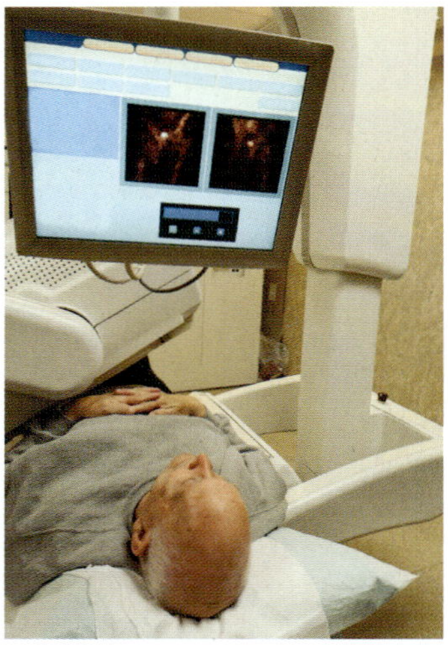

Gamma rays help doctors view internal structures, reducing the need for surgery.

LESSON REVIEW

1. What two measurements differentiate types of radiation?
2. Summarize the four different forms of radiation.
3. Describe how temperature is related to the amount of infrared radiation that an objects emits.
4. What are two uses of thermography and X-rays?
5. Identify the advantages and disadvantages of ultraviolet waves.
6. Why do gamma rays have the most energy?

Wave and Particle Properties of Light 6.3.6

Light has existed since the beginning of creation. Even though light is part of creation, scientists knew very little about it until the 19th century. Around 1865, Scottish physicist James Maxwell discovered that light was an electromagnetic wave. With this discovery, he also hypothesized that invisible forms of light existed. Maxwell's equations and observations were the foundation for the wave theory of light.

Much of what people observe about light can be explained by the wave nature of light. For example, light reflects off surfaces. When light hits a surface, some of the light is absorbed, but the remaining waves are reflected. The bouncing back of a wave after it hits a barrier that does not absorb all of its energy is called *reflection*.

How light reflects off a surface depends on the type of surface. Light waves that reflect off a smooth surface, such as a mirror, are parallel to each other. Because the light hits the mirror's smooth surface, every light wave's angle of incidence equals its angle of reflection. This type of reflection is called *regular reflection*. It explains why when you look in a mirror you see your face prominently instead of what is above, beneath, or beside you.

If a light wave strikes a rough surface, such as a rock, the reflected waves scatter in many different directions and produce a fuzzy reflection called *diffuse reflection*. The incoming light waves are parallel to each other, but the rough surface scatters the waves in different directions. The scattered waves create an image that is not clearly defined. If diffuse images did not exist, dust particles in the air would not scatter the sun's rays and objects could only be seen in direct sunlight. The sun's glare would be so strong that you could not see these objects without constantly squinting.

OBJECTIVES

- Establish the three ways that light acts as a wave.
- Describe the dual nature of light.

VOCABULARY

- **index of refraction** a property of matter that compares the speed of light in a vacuum with the speed of light in a material
- **particle theory** the theory that light is made of small bundles of energy called photons
- **penumbra** the edge of a shadow that is partially illuminated
- **photoelectric effect** the release of electrons from metals by photons of light
- **photon** a tiny bundle of light energy
- **shadow** a region of darkness formed when an opaque object comes between a space and a light source
- **umbra** the part of a shadow that is completely dark

Reflection

Regular reflection occurs when light beams are reflected at the same angle.

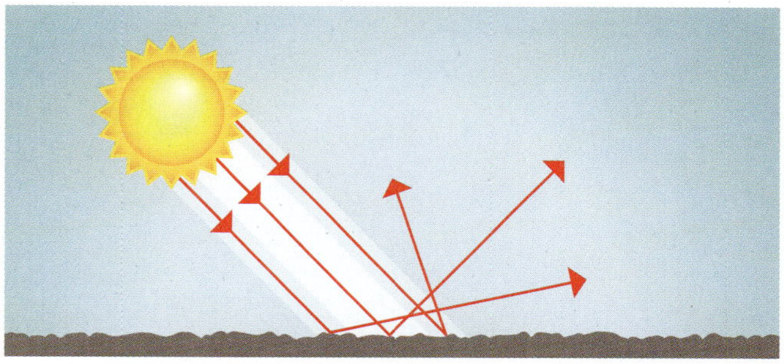

Diffuse reflection occurs when light beams reflect at many different angles.

FYI

Iridescence

Have you ever noticed how the appearance of a butterfly wing, a soap bubble, or a peacock feather seems to change color when you change your perspective? This optical illusion can be explained by iridescence. This phenomenon is caused by interaction between an object's physical structure and the incoming light waves. The light waves combine to create either constructive interference or destructive interference, depending on your view. Constructive interference causes an increased vibrancy in the reflected color, and destructive interference causes the color to appear dull.

Another wave property of light is refraction. Refraction is the bending of a wave because of a change in the wave's speed as it passes from one medium to another. All light waves travel in a straight line until they enter a different medium. Once light waves hit a different medium, the medium changes their speed and the waves bend. If you place a pencil into a glass of water, the pencil looks as though it is bending because the light waves are slowing down when they enter the water. When light passes from a less dense medium, such as air, into a denser medium like water, it slows down. When it passes from a denser medium to a less dense medium, it speeds up. This is why the pencil appears to bend in water.

Every substance has a refractive property that describes how much the speed of light changes in that substance. This property is known as the index of refraction. The **index of refraction** compares the speed of light in a vacuum with the speed of light in a material. A higher index of refraction means that the light waves bend more in a particular material. For example, ice has an index of refraction of 1.31. A diamond's index of refraction is 2.42. Light travels much slower and bends farther in a diamond compared to ice.

Diffraction is the bending of a wave around a barrier or through an opening. In 1803, English physician and physicist Thomas Young showed with his double-slit experiment that diffraction is a wave property of light. Young coated a sheet of glass with carbon and then cut two scratches into the carbon to make fine slits. He held a light source behind the glass and observed alternating dark and light bands. The dark bands showed a cancellation of light waves because of destructive interference,

Material	Index of Refraction
Hydrogen gas	1.00014
Air (0°C)	1.00029
Water (ice)	1.309
Water (liquid, 20°C)	1.33283
Ethyl alcohol	1.361
Eye lens	1.41
Mineral oil	1.470
Fiberglass	1.560
Emerald	1.576
Diamond	2.418

Double-Slit Experiment

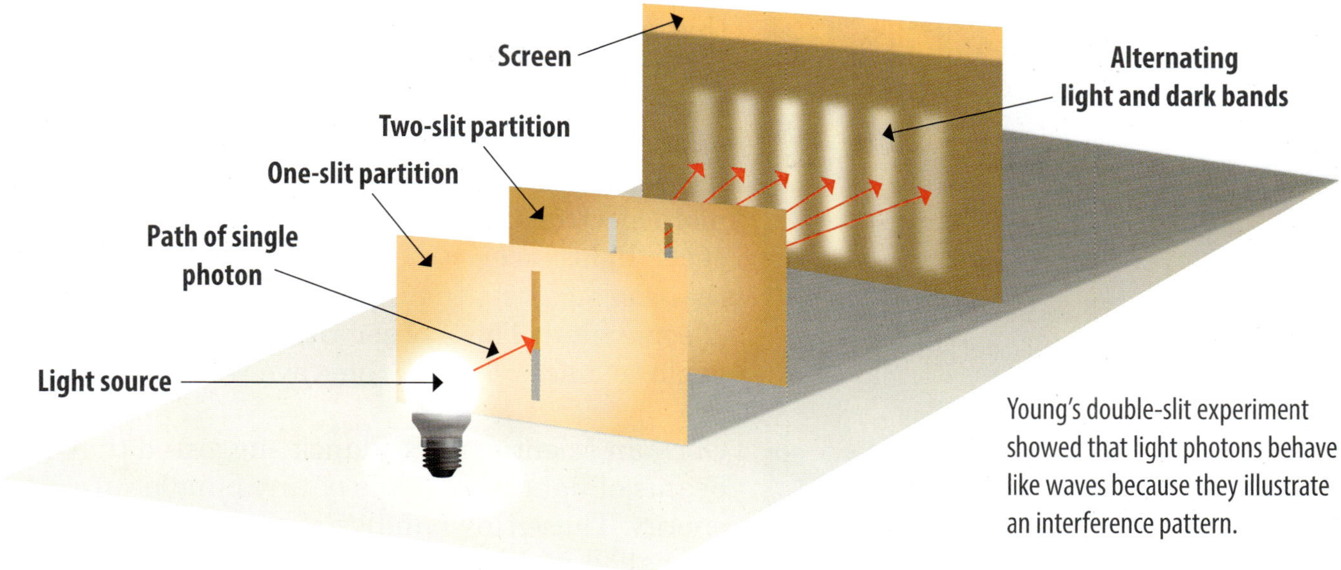

Young's double-slit experiment showed that light photons behave like waves because they illustrate an interference pattern.

a type of interference that results in a wave with smaller amplitude than the original waves. This occurs when two inverse light waves meet and the crest of one wave meets the trough of a second wave. These two waves cancel each other to form a dark spot. Bright bands appeared in regions that would have been totally dark if light did not diffract around the slits. A diffraction grating consists of many closely spaced slits that allow for a precise determination of light wavelengths.

Diffraction determines the quality of a shadow. A **shadow** is a region of darkness formed when an opaque object comes between a space and a light source. Shadows are not completely dark because light can bend around the object. For this reason, shadows have a partly illuminated fuzzy region. The part of the shadow that is completely dark is called the **umbra**. The edge of the shadow that is partially illuminated is called the **penumbra**.

It has been shown so far that light behaves like a wave. However, there are some light behaviors that cannot be explained according to the definition of a wave. This is because light has a dual nature. Sometimes light behaves like a particle instead of a wave. After James Maxwell and other scientists described the wave behaviors of light, scientists believed that light was like any other wave. They soon realized that the wave theory of light did not explain everything about light.

TRY THIS

The Disappearing Coin

Place a quarter in the bottom of an empty opaque mug. Remain seated as your lab partner moves the mug away from you until the quarter is just out of your line of vision. Have your partner slowly fill the mug with water until the quarter is visible. Using a metric ruler, measure and record the height of the water. Repeat this procedure using oil and rubbing alcohol. Which liquid allowed you to see the coin with the least amount of liquid added? Which liquid has the highest index of refraction? Explain. What does a high index of refraction imply about the speed of light in a particular substance?

The dancers' shadows are composed of two regions: the umbra and penumbra.

German physicist Heinrich Hertz discovered that light does not always act like a wave. He found that when light falls onto certain surfaces, the surface emits electrons. Scientists noticed that when they shined violet light onto the surface of a metal, the metal emitted electrons. If they increased the frequency of the light, it emitted more electrons. But if they lowered the frequency of the light, no electrons were emitted. This reaction suggests that light involves particles of some kind. These observations could not be explained according to the wave theory of light.

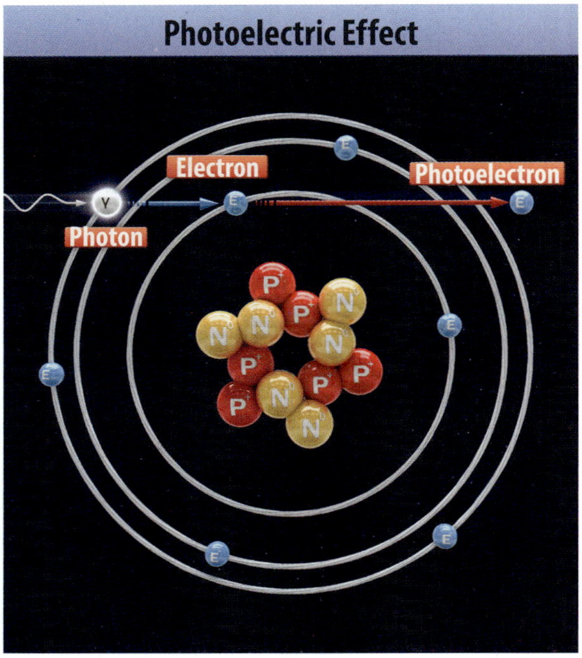

German scientist Max Planck suggested that beams of light were made of tiny bundles of energy. These tiny bundles of light energy are called **photons**. The release of electrons from metals by photons of light is called the **photoelectric effect**. Albert Einstein proposed that photon energy can knock electrons from metal. He suggested that the photoelectric effect is the production of electrons by photons of light. Red light does not produce the photoelectric effect. If light was nothing more than a wave, then the electrons of the metal should eventually soak up enough energy from the red light to be knocked off, but this does not happen. Red photons do not have enough energy to knock electrons away from the metal. The theory that light is made of small bundles of energy called photons is called the **particle theory**.

LESSON REVIEW

1. What aspect of light explains why it reflects off surfaces?
2. Contrast regular and diffuse reflection.
3. Why do light waves refract and diffract?
4. Mineral oil has an index of refraction of 1.47 and an emerald has an index of refraction of 1.576. How does light refraction compare in these two substances?
5. Describe the photoelectric effect.
6. Why do scientists believe that light has a dual nature?

Artificial Light Sources 6.3.7

Have you ever been caught in the dark without any source of light? Perhaps you were camping and forgot your flashlight. Or maybe a storm knocked out the power at your home, and you could not find matches or candles. That is when you realize how much you depend on artificial light sources.

Luminous objects are objects that can produce their own light. God created many natural luminous objects to light Earth, such as the sun, stars, fire, and bioluminescent organisms. People have also invented many luminous sources to help light up places where there is no light.

What kind of light sources are in a home? Some are incandescent lights. An **incandescent light** is a light source that generates light by producing heat. Inside an incandescent light bulb is a thin wire, called *a filament*, that is made of the metal tungsten. This metal can be heated to a temperature of 3,000°C without melting. When an incandescent light bulb is turned on, electrons flow through the tungsten wire. The resistance that these electrons encounter produces heat. The tungsten wire glows and sends out photons of visible light.

Incandescent light bulbs are not very energy efficient because about 90% of their energy is given off in the form of thermal energy. A better light alternative is a compact fluorescent light bulb, often referred to as a CFL. These bulbs use about 75% less energy than an incandescent light and still produce the same amount of illumination. Larger fluorescent lights are often found in schools, offices, medical facilities, and warehouses.

OBJECTIVES

- Compare the ways that artificial sources produce light.
- Distinguish between efficient and inefficient artificial light sources.

VOCABULARY

- **fluorescent light** a light source that produces light by flooding gas molecules with electrons under low pressure
- **incandescent light** a light source that generates light by producing heat
- **neon light** a light source that produces light by passing electrons through tubes of neon gas

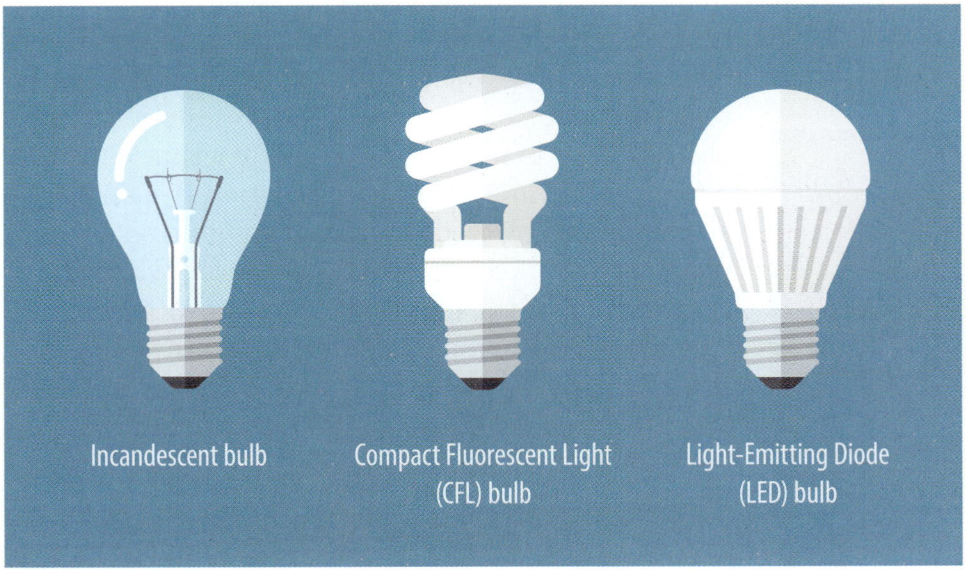

Fluorescent Light

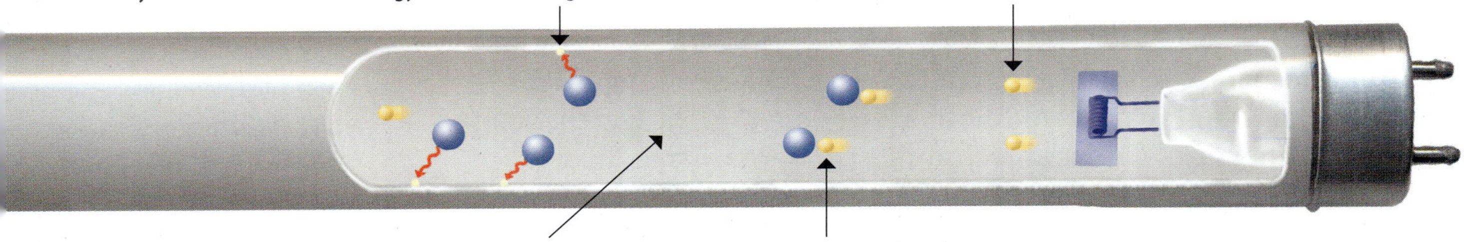

A **fluorescent light** is a light source that produces light by flooding gas molecules with electrons under low pressure. The inside of a fluorescent light tube is lined with a phosphor coating that emits visible light any time UV radiation is absorbed. At the ends of a fluorescent light tube are electrodes. When the electrodes are connected to a circuit, electrons are produced. The electrons collide with gas molecules, emitting UV radiation. This UV radiation is absorbed by the phosphor coating and visible light is produced. Fluorescent lights produce much less heat than incandescent lights, use less electricity, and have a much longer lifespan.

LED light, which stands for light-emitting diode, is a form of artificial light that uses semiconductors to produce visible or infrared light. *Diode* means that there are two terminals. Each terminal is made of a semiconductor material that has a different composition and electronic properties. The terminals are brought together to form a junction. At the junction, an electric field is induced by a battery that causes a current to flow across the junction. When the electrons are excited, photons are emitted. A wide variety of colors can be produced when the composition of the semiconductors is altered. The most common semiconductor material used in LED lights is gallium arsenide. These lights are used in automobile brake lights, billboards, television remote controls, fiber optics, and some light bulbs.

Sodium-vapor lights are used to illuminate dark streets.

There is another type of artificial light seen on signs or on the sides of buildings. These types of lights are called *neon lights* and they operate much like fluorescent lights. A **neon light** produces light by passing electrons through tubes of neon gas. When the electrons collide with the neon gas molecules, visible light is produced. If only neon gas is present inside the tube, the light will glow bright red. Other gases can be added to create a wide variety of colors.

Neon sign

Two other types of artificial lights exist that also contain gases: sodium-vapor lights and tungsten-halogen lights. Sodium-vapor lights consist of a tube that contains neon and argon gas as well as a small amount of metal sodium. When the light is turned on, the temperature of the gas mixture increases and the metal sodium sublimes into a vapor that emits a yellow-orange light. Sodium-vapor lights are commonly used as streetlights.

Tungsten-halogen lights are somewhat similar to an incandescent light. They contain a tungsten filament inside of a quartz bulb. The difference between tungsten-halogen lights and incandescent lights is tungsten-halogen light bulbs also contain a halogen gas, typically either chlorine or fluorine. The halogen gas provides an environment where the tungsten filament can become even hotter and produce an intense bright light. These lights have a much longer lifespan compared to traditional incandescent bulbs. They are used by filmmakers and underwater photographers for their brilliant quality.

LESSON REVIEW

1. Why are incandescent light bulbs not very efficient?
2. Describe how a fluorescent source produces light.
3. What types of light do LED lights produce?
4. What component of sodium-vapor lights is responsible for the emission of visible light?
5. How are tungsten-halogen lights and incandescent lights similar and different?

Underwater photographers use tungsten-halogen lights.

6.4.1 *Mirrors*

OBJECTIVES

- Distinguish between plane, concave, and convex mirrors.
- Compare virtual images and real images, and relate them to diverging and converging light rays.
- Identify practical uses for each type of mirror.

VOCABULARY

- **concave mirror** a mirror that is curved inward
- **convex mirror** a mirror that is curved outward
- **focal length** the distance between the center of the mirror and the focal point
- **focal point** the point at which a mirror or lens causes the parallel light waves to meet
- **image** a copy of an object formed by light
- **mirror** a smooth object that reflects light and forms images
- **optical axis** a straight line drawn outward from the center of the mirror
- **plane mirror** a mirror that has a perfectly flat surface
- **real image** an image through which light passes
- **virtual image** an image through which light does not pass

Imagine waking up to a beautiful, sunny day in the mountains. Looking in one direction, you see an open field stretching to the foot of a mountain. Looking in the other direction, you see the mountains not only where they stand but also reflected in a lake. The surface of the water is perfectly still, appearing as smooth as glass. Why does the water reflect the mountains but the field does not? And why do you see the reflection in the water only some of the time?

How light reflects off a surface depends on the type of surface. The ability to actually see the reflection depends on whether the line of sight is in the path of the reflection. Light waves reflecting off a smooth surface, such as a lake or mirror, are parallel to each other. When the light hits a smooth surface, every light wave's angle of incidence equals its angle of reflection.

A **mirror** is defined as a smooth object that reflects light and forms an image. An **image** is a copy of an object formed by light. Any smooth, reflective surface, such as a calm lake, can act as a mirror. A lake can reflect a clear image of a nearby mountain.

Consider the mirrors in your house. Probably all of them are flat. But the surface of a mirror can have various shapes. Each shape reflects light differently and therefore can be used for different purposes. The three main types of mirrors are plane, concave,

Law of Reflection

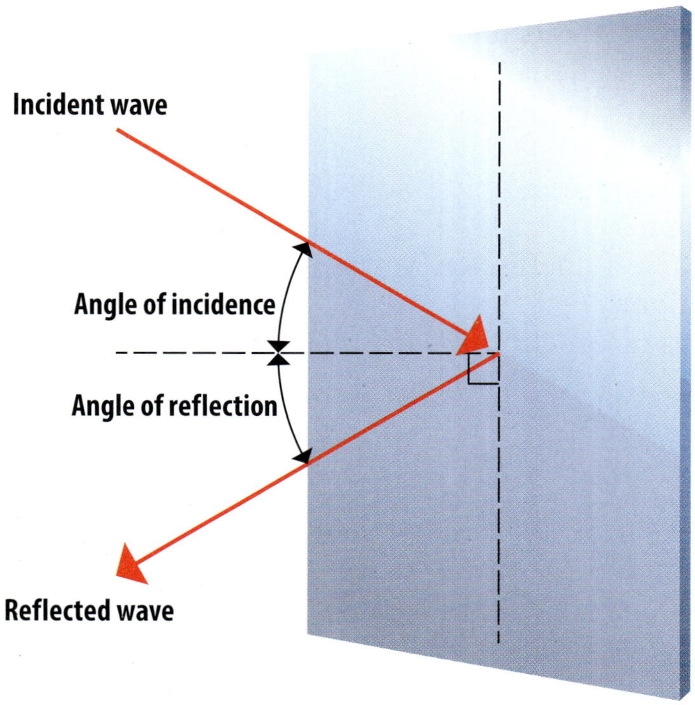

BIBLE CONNECTION

Mirror Image

"For now we see only a reflection as in a mirror; then we shall see face to face. Now I know in part; then I shall know fully, even as I am fully known" (1 Corinthians 13:12). What was the apostle Paul trying to say when he used the metaphor of a mirror reflecting an incomplete image? Paul wrote the letter to the Corinthians more than 2,000 years ago, but mirrors have changed since then. At that time, mirrors were made of polished bronze or some other metal, which reflected a distorted form. Mirrors are much improved now; however, we still see imperfectly. Soon we will see God face to face and know Him clearly and directly.

TRY THIS

Plane Mirrors

Using a lump of clay as support, mount a small plane mirror vertically on a piece of graph paper. Place a coin on the graph paper in front of the mirror. Look in the mirror and observe the image of the coin. Is it a real image or a virtual image? Count the number of squares from the object to the mirror and from the image to the mirror. How do the distances of the object and the image to the mirror compare?

and convex. A **plane mirror** has a perfectly flat surface. Plane mirrors create images that appear to be behind the mirror. Images in a plane mirror are right-side-up, but reversed, right to left. If you salute yourself with your right hand, the image seems to salute back with the left hand. The apparent distance of the image behind the mirror is the same distance as the actual object in front of the mirror, and the image is the same size as the object. The image formed by a plane mirror is a **virtual image**, an image through which light does not pass. It appears to be behind the mirror, but light never actually passes through the place where the virtual image is. The light rays do not converge, and virtual images cannot be projected onto screens.

A lake behaves like a plane mirror.

SCIENCE IN ACTION

Solar Furnace

The Odeillo solar furnace in the south of France has 9,600 mirrors. The mirrors on the front of the building form a large concave surface that converges the sun's energy on a central target only 40 cm in diameter. The furnace is used to conduct research on materials exposed to extremely high temperatures.

FYI

Fun-House Mirrors

A trip to a fun house often includes a room full of mirrors. Plane mirrors are part of everyday life and people are accustomed to looking at themselves. So what makes these mirrors so entertaining? For some reason, standing in front of an oddly shaped mirror and seeing heads and torsos stretched out of proportion generates lots of giggles. How does a fun-house mirror create distorted images? The mirrors set up in amusement parks have several different shapes in one surface. With a concave surface at the top, a plane surface in the middle, and a convex surface at the bottom, your reflection becomes huge at the top and tiny at the bottom.

Another type of mirror is a **concave mirror**—one that is curved inward. A concave surface reflects all the incoming light waves to a point in front of the surface. These reflected rays converge at the **focal point**, where the parallel light waves meet. The distance between the center of the mirror and the focal point is called the **focal length**. The image formed at or behind the focal point is smaller than the object, and it is upside-down. For example, if you look into the inside of a shiny metal spoon, you will see a small, upside-down image of yourself. This image is called a **real image**, an image through which light passes. Real images appear on the same side of the mirror as the actual object. Real images form when all the reflected light rays converge at the focal point. If a screen is set in the plane of the real image, it is visible on the screen as in a movie theater.

When an object is placed at the focal point of a concave mirror, all the light waves reflect parallel to each other and to the

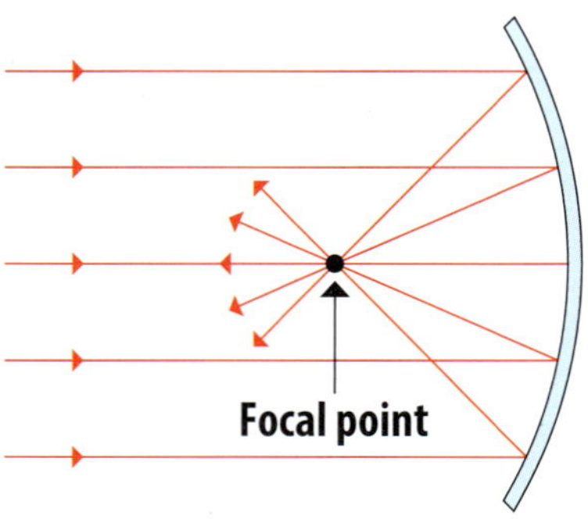

Concave Mirror

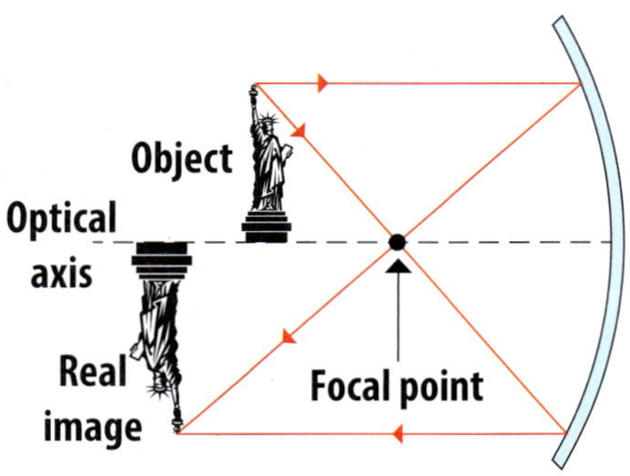

Concave Mirror

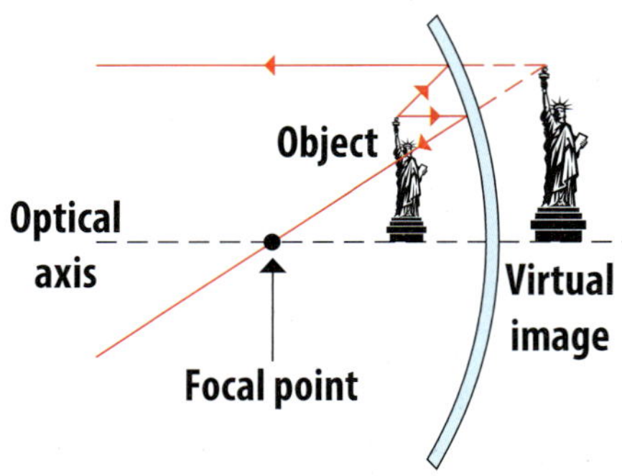

Concave Mirror

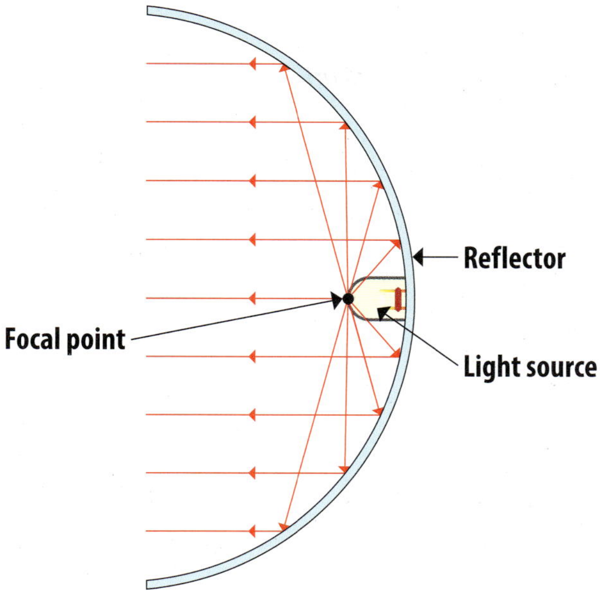

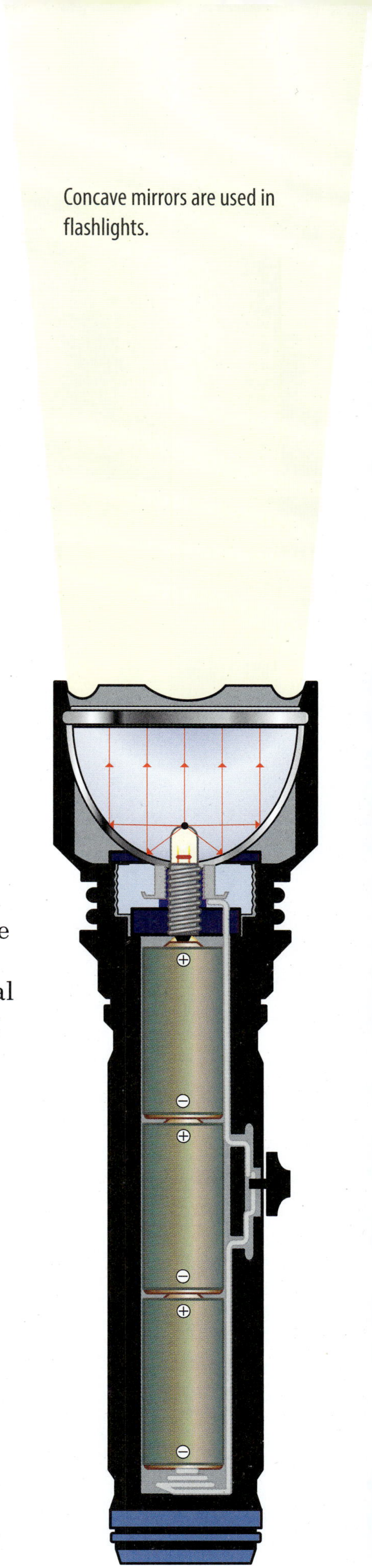

Concave mirrors are used in flashlights.

optical axis, which is a straight line drawn outward from the center of the mirror. If a light source is placed at the focal point, the reflected light waves form a concentrated beam of light. A flashlight uses a concave mirror behind a bulb. The bulb is fixed at the focal point of the mirror, producing a powerful beam. Concave mirrors are also placed behind car headlights, searchlights, and spotlights to create focused beams of light.

When an object is placed between the focal point and the concave mirror, the reflected light waves diverge, or never meet, and a **virtual image** forms. This image appears larger than the actual image. The images that you see in makeup mirrors and shaving mirrors are this type of image.

The third type of mirror is called a **convex mirror**. Its surface curves outward. Unlike concave mirrors that reflect light waves to one point, convex mirrors spread reflected rays in all

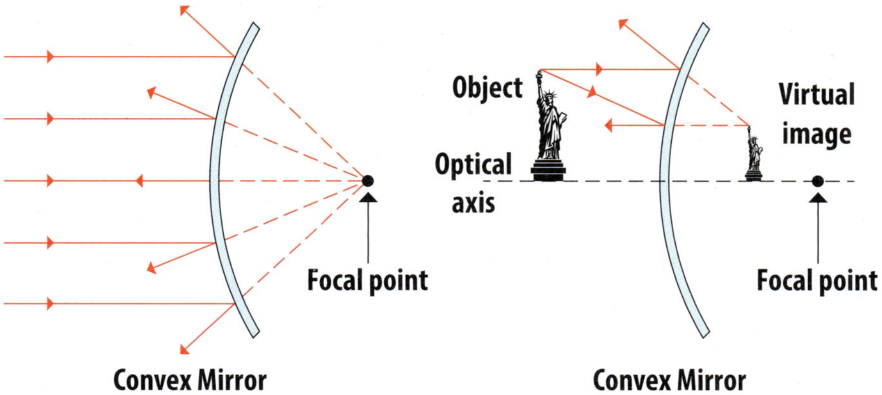

A convex mirror provides a wide view of the shop.

directions. The images that you see in convex mirrors are always right-side-up and are always smaller than the object. Convex mirrors are used for side-view and rearview mirrors in cars. Side-view mirrors often include the warning "Objects are closer than they appear." Because a convex mirror makes objects appear smaller, drivers think objects are farther away than they actually are. Sometimes convex mirrors are mounted on ceilings and walls of stores to provide a complete view of the shopping area for security reasons. Also, they are used in large warehouses to help workers see whether anything is coming around a blind corner.

LESSON REVIEW

1. What is a mirror?
2. Describe plane mirrors, concave mirrors, and convex mirrors and the images each kind produces.
3. What is the difference between virtual images and real images?
4. How is each type of image formed?
5. Give an example of a practical use for each type of mirror.

BIBLE CONNECTION

Bible Reflections

Have you ever heard the superstition that breaking a mirror will bring you seven years of bad luck? Where did that superstition come from? Long before mirrors were invented, people believed that a shiny surface was a tool of the gods. They did not understand the scientific principles of light reflection. When they saw a person's reflection in the water of a pond or lake, they thought that the reflection was that person's soul or other self they saw in dreams. To avoid injuring this other self, they never dropped rocks on their reflection in the water or swished the water.

When metal and glass mirrors came into use, people thought that breaking a mirror would harm their other self. Around the first century AD, the Romans decided that breaking a mirror meant not only bad luck, but seven years of it! The Romans believed that life renewed itself every seven years, so it would take seven years to recover health "broken" by a broken mirror.

There is no need to think superstitious thoughts, but it is important to consider what the heart of a person reflects. Christians are to reflect the Lord's glory (2 Corinthians 3:18). Also, Proverbs 27:19 says, "As water reflects the face, so one's life reflects the heart." Now that is worth thinking about.

Lenses 6.4.2

Have you ever used a camera, binoculars, a microscope, or a telescope? Do you wear contacts or glasses? If so, you have used a lens. A **lens** is a transparent object that refracts light to produce an image.

Lenses have one or two curved edges to refract light in a certain way. Most lenses are made of glass or plastic. The lens of the human eye is made of a transparent membrane filled with a transparent fluid.

The two basic types of lenses are convex lenses and concave lenses. A convex lens is thicker in the center than at the edges and brings light waves together to converge at one point. Just like mirrors, lenses have a focal point at which the reflected light rays meet. The distance from a lens to its focal point is called *the focal length*. A convex lens refracts light twice—once as it enters the lens and again as it leaves the lens. Convex lenses with more of a curve refract light more than lenses with a slight curve.

A convex lens produces three different types of images depending on where the object is placed. If an object is placed more than two focal lengths away from the lens, the image is real, upside-down, and smaller than the object itself. Camera lenses create this type of image. If an object is placed between one and two focal lengths away from the lens, the image is real, upside-down, and larger than the object. Microscopes create this type of image. If an object is placed between the lens and the focal point, the image is virtual, right-side-up, and larger than the object. Hand lenses use convex lenses in this way.

OBJECTIVES

- Identify the two types of lenses.
- Describe the three types of images produced by convex lenses.
- Explain how lenses compensate for variations in eye shape.

VOCABULARY

- **lens** a transparent object that refracts light to produce an image

Camera

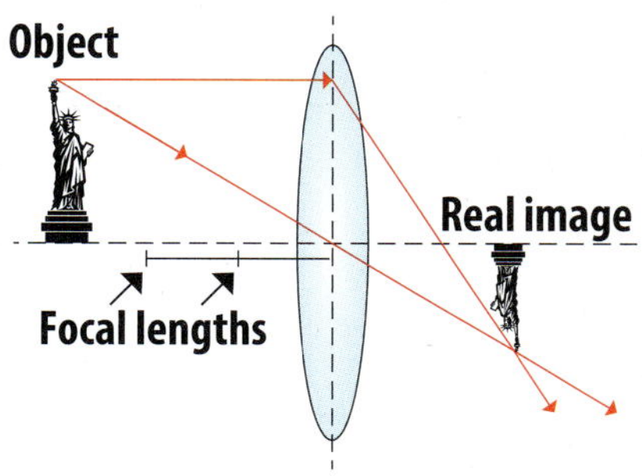

If the object is more than two focal lengths away from the lens, a reduced image is formed.

Hand Lens

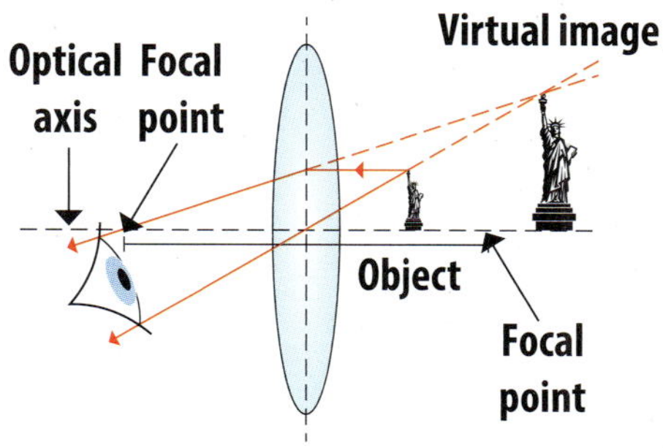

The image is larger than the object when you look into a hand lens.

Projector

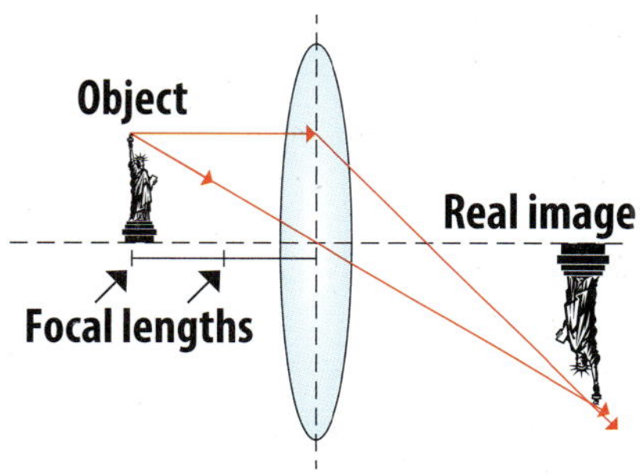

An enlarged image is formed when the object is two focal lengths away from the lens.

Human Eye Anatomy

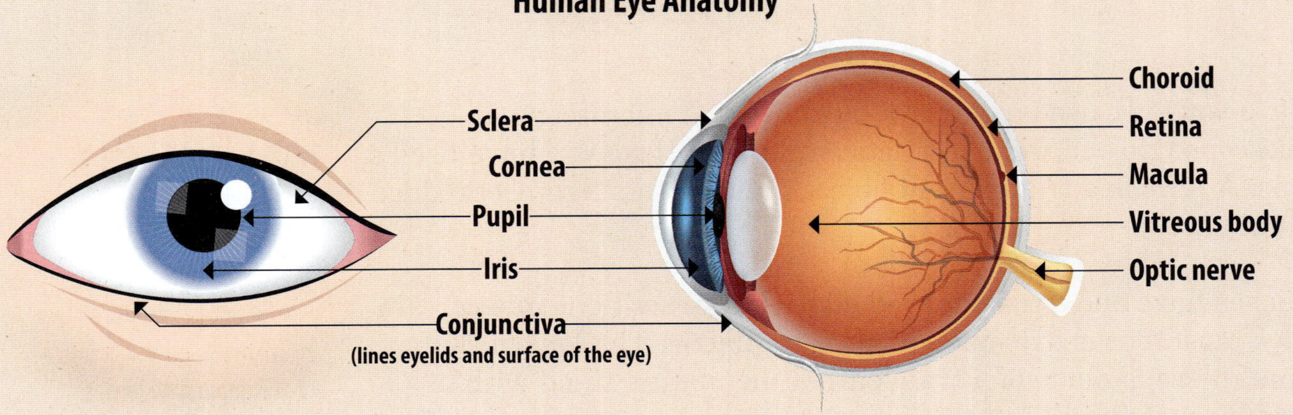

Inverted Image on the Retina

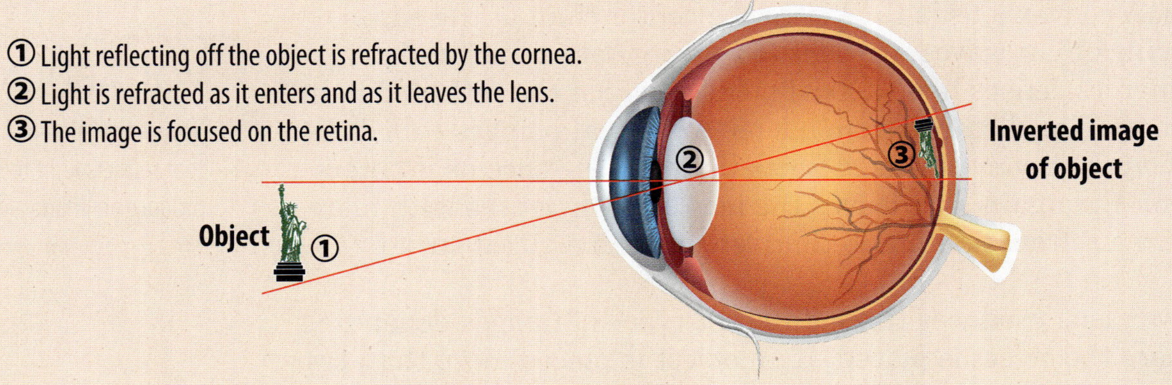

Vision Disorders

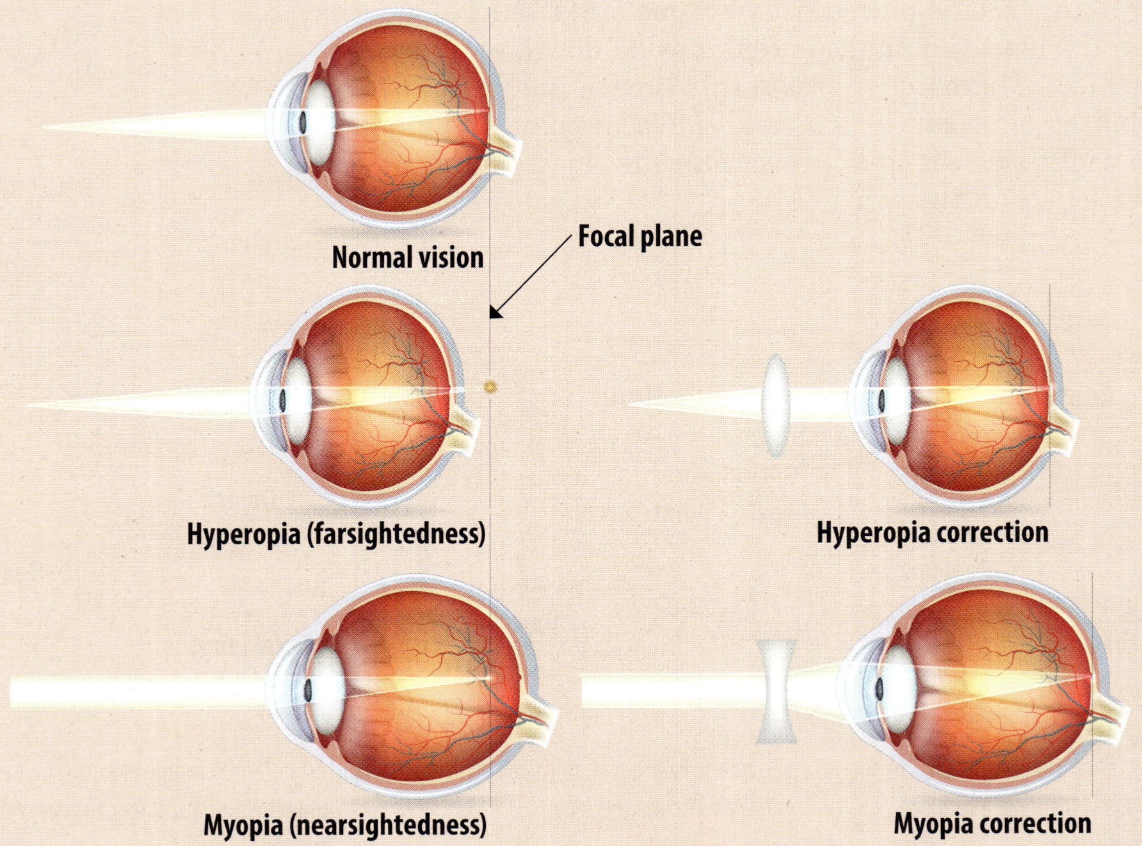

A scientist uses a light microscope. Pairs of convex lenses provide magnification.

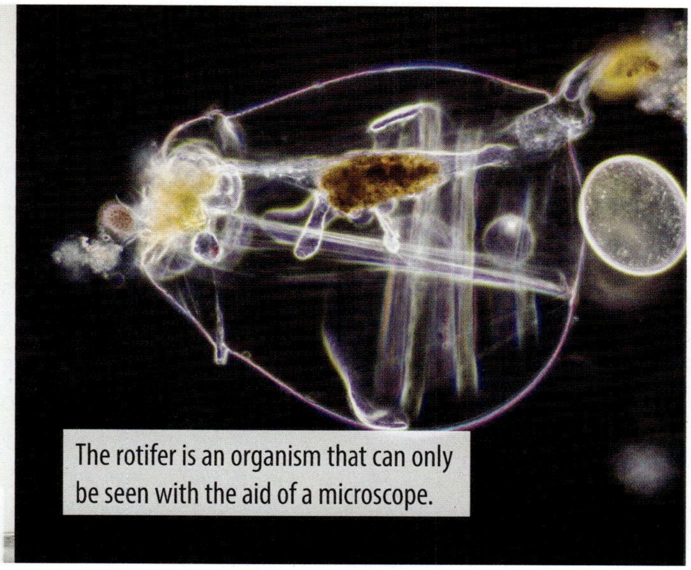

The rotifer is an organism that can only be seen with the aid of a microscope.

A concave lens is thicker at the edges than in the center and spreads light waves outward. It produces virtual, right-side-up images that are smaller than the object.

The lens of the human eye is a convex lens. Eye lenses are flexible membranes that automatically change shape to focus objects that are close and far away. For someone to have normal vision, the image formed by the eye's convex lens must fall directly on the retina. If the eyeball is too long, the image falls in front of the retina, causing nearsightedness, or myopia. Nearsighted people cannot see far-away objects clearly, but they can see objects nearby. Nearsighted people wear glasses or contacts with concave lenses to correct this problem. These concave lenses refract light waves outward before they enter the eye.

Farsighted people can see far-away objects clearly but cannot see nearby objects clearly. Nearby objects appear blurry to someone who is farsighted because the eyeball is too short, and the light waves come together behind the retina. Convex lenses correct farsightedness, or hyperopia, by refracting light waves closer together before they enter the eye.

Image Types Created with a Convex Lens

Distance	Image Type	Orientation	Magnification	Usages
<1 focal length	virtual	right-side-up	larger	magnifying glass
1–2 focal lengths	real	upside-down	larger	microscope, projector
>2 focal lengths	real	upside-down	smaller	camera lens

TRY THIS

Light Refraction

Use a laser ray box and several differently shaped prisms. Choose one prism and use the laws of reflection and refraction to predict the path of the light from the laser ray box, through the lens, and out the other side. Then test the prediction. Repeat the procedure with other prisms. How are the laws demonstrated?

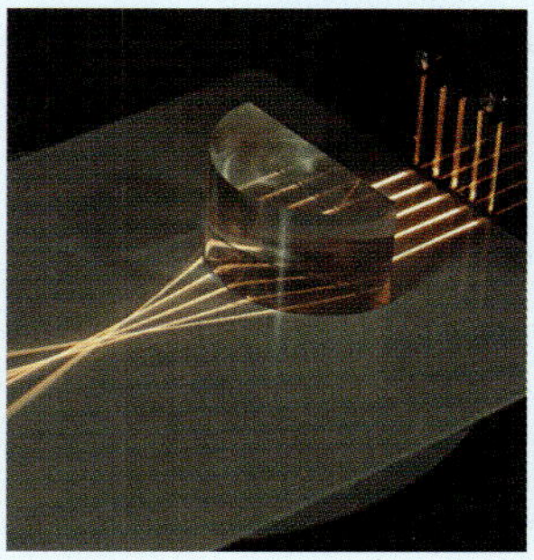

FYI

Bionic Lens

Imagine looking across a river at a clock and being able to read it as though it were right in front of you. Picture looking up in a tree and identifying the species of a beetle walking across a branch. Think about watching a video on your phone without looking at your phone. These thoughts may seem like something out of a science fiction novel, but the technology could be a reality in the near future. Scientists are developing a bionic lens that could replace the human lens. It has the potential to provide three times the standard 20/20 visual clarity. It is being designed to monitor certain conditions, such as glaucoma, and detect levels of enzymes, proteins, and lipids, which are biomarkers for many eye diseases. Scientists are also expecting to tie artificial intelligence technology to the bionic lens, which would project images directly onto the retina.

LESSON REVIEW

1. What is the difference between convex and concave lenses?
2. What is the focal point of a convex lens?
3. List the three types of images produced by a convex lens and how they are produced.
4. What image is produced by a concave lens?
5. Explain how nearsightedness is corrected.
6. Explain how farsightedness is corrected.

Looking through a convex lens

Optical Instruments 6.4.3

Why do people enjoy looking at photographs? Photographs from places never visited, like the Amazon, the Arctic, or the moon, can help people visualize those places in a way that words never could. Photographs of last summer's vacation or last month's birthday party or last week's field trip can bring back fond memories. Maybe you like to look at family photographs taken when you were a small child. Photographs from long ago can help you understand the past. Photographs are among some people's most cherished possessions.

OBJECTIVES

- Explain how optical instruments use lenses and mirrors.
- Compare the uses of different optical instruments.

Without an understanding of light, there would be no photographs. To produce cameras and many other optical instruments, it is necessary for inventors to know how light behaves. Optical instruments use lenses and mirrors to refract and reflect light, producing images of a desired size and quality. Common optical instruments include the camera, telescope, hand lens (magnifying glass), and microscope.

A camera uses one or more convex lenses to produce images on photographic film or on a sensor. Camera lenses work like the human eye, except that the eye works with only one lens and focuses even more quickly than an autofocus camera. Camera lenses are convex lenses that produce a real, upside-down image that is smaller than the object. Its size on the film or sensor depends on the focal length of the lens. Lenses with short focal lengths produce small images. Lenses with large focal lengths, such as telephoto lenses, produce larger images. The opening through which the light enters a camera is called *the aperture*. The size of the aperture determines how much light reaches the film.

Vintage camera that used photographic plates

Vintage camera that used black and white film

Analog camera that uses 35 mm color film

Digital camera that projects image onto a sensor

Yerkes Refracting Telescope

A telescope uses mirrors and lenses to produce enlarged images of distant objects. Astronomers study the night sky using two basic types of telescopes: refracting and reflecting. A refracting telescope employs a series of lenses to gather and focus light. The lens at one end of the tube is a convex lens that focuses the light as it enters the tube, forming a real, upside-down image. This lens is called *the objective lens*. The eyepiece lens at the other end of the tube is the lens that the viewer looks through. This convex lens acts as a hand lens, enlarging the real image formed by the objective lens. The eyepiece lens forms an enlarged, virtual image.

When a telescope has a larger lens, it can gather more light and magnify more distant objects. The lens of the refracting telescope at the Yerkes Observatory in Wisconsin is 1 m in diameter. This telescope has a light-gathering power 40,000 times greater than the human eye.

A reflecting telescope uses a series of mirrors to gather light. It focuses light with a concave mirror instead of a lens. The concave mirror is located at one end of the tube. Light enters at the other end, travels the length of the tube, and strikes the concave mirror, which forms a real image. A small plane mirror reflects this image to the eyepiece on the other side of the tube. The eyepiece enlarges the image. The largest reflecting telescope

Refracting Telescope

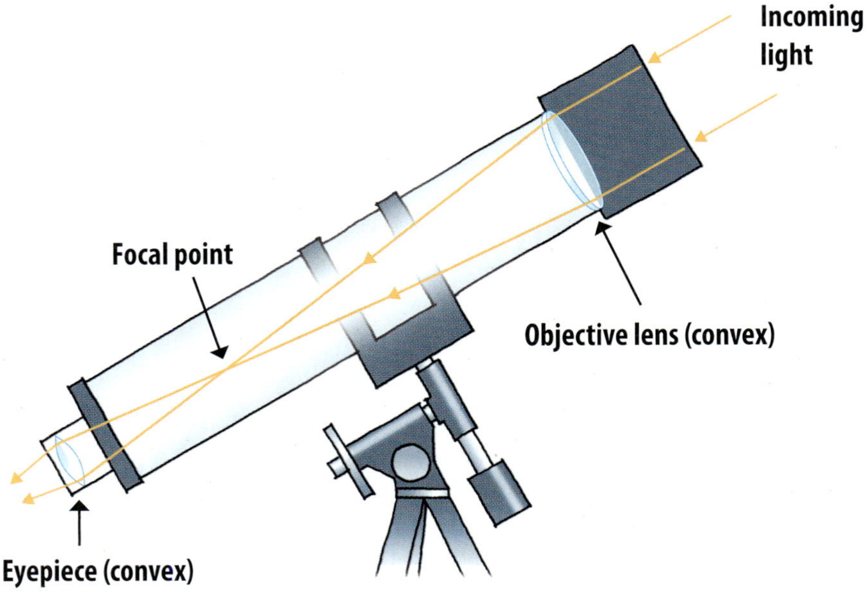

Reflecting Telescope

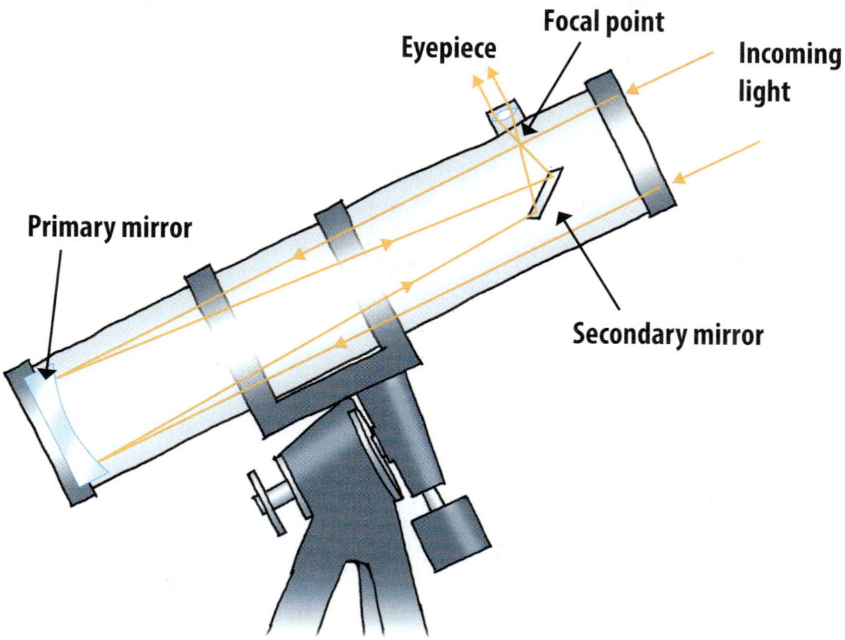

is the Gran Telescopio Canarias in La Palma, Spain. Its concave mirror is 10.4 m in diameter.

Some optical instruments create images of distant objects; others magnify small details or objects close up. The smallest detail or object human eyes can discern without an optical instrument is about 0.1 mm. A hand lens has a simple convex lens and can enlarge an image by two or three times. It is sometimes referred to as a simple microscope.

Electron microscope

A microscope combines two or more lenses to magnify tiny, nearby objects. Most microscopes have two convex lenses. One lens is placed at the end of the microscope tube; the other lens serves as the eyepiece. The power of magnification is determined by multiplying the magnification of the eyepiece lens by the magnification of the other lens (the objective lens). The maximum useful magnification of an optical or light microscope is about 2,000 times.

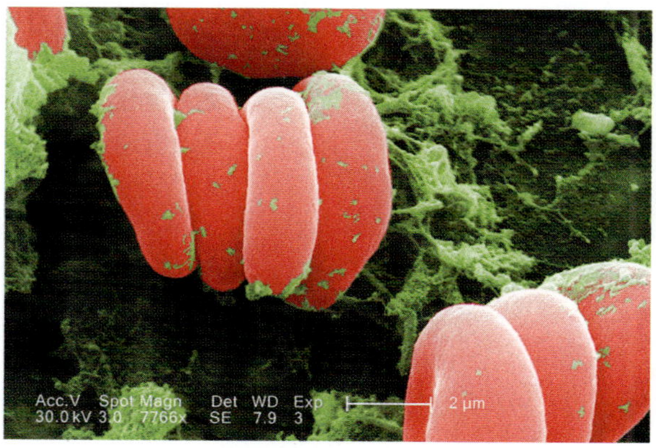

Electron microscope image of red blood cells

Another type of microscope is the electron microscope. It does not use lenses to focus light. Instead, it uses electromagnets to focus

FYI

Giant Magellan Telescope

Telescopes provide a view of space that could never be seen with the naked eye. Bigger and bigger telescopes provide clearer windows deeper and deeper into the vast expanse of the universe. What does it take to build mirrors big enough and precise enough to see so many hundreds of thousands of kilometers away? It takes years of precision work.

The University of Arizona is the only location in the world that has the facility to create giant mirrors. First, individual ceramic cores are arranged inside an 8-meter circle to create a mold for the glass mirrors. The Giant Magellan Telescope will use seven of these mirrors.

Second, glass chunks manufactured in Japan are placed on the ceramic cores. Twenty tons of glass are needed for each mirror. Once all the spaces in the mold are filled, the furnace is set to a temperature near 1,100 degrees Celsius. The glass remains in the furnace for three months, melting and cooling. Finally, the glass must be polished to specific measurements with a precision measured in nanometers. This can take years to accomplish.

The Giant Magellan Telescope will be built on a mountaintop at the Las Campanas Observatory in northern Chile. The work is the result of an international consortium that includes Australia, the United States, Brazil, South Korea, and Chile.

electrons. Therefore, in the strictest sense of the term, it is not an optical instrument. But it does produce an image. The electron microscope can magnify an object up to 100,000 times its original size, and it can provide images of molecules. The light microscope cannot. Why use a light microscope if the electron microscope can do so much more? The light microscope is less expensive, much smaller, and it can be used to view live organisms. The electron microscope requires that the specimen be placed in a vacuum, so it is only useful for viewing nonliving samples.

LESSON REVIEW

1. What is an optical instrument?
2. Compare a camera to the human eye.
3. How does a camera produce a photograph?
4. What is the difference between a reflecting telescope and a refracting telescope?
5. How does a light microscope produce an image?
6. Order the following optical instruments from greatest viewing distance to smallest viewing distance: microscope, camera, telescope, hand lens.

Applications of Light 6.4.4

Beams of light can bore holes in metal. Beams of light can also help doctors perform delicate eye surgery. Beams of light detect changes in the surface of a DVD and transmit the information to be translated into an electric signal. How are these things possible? They are possible because of the way that light behaves. Ordinary light consists of waves that are out of phase. The crests and troughs are out of step, much like members of a marching band who are out of step. The energy of the waves is dispersed over a wide area. Light that contains waves of different frequencies and amplitudes moving out of phase is called **incoherent light**.

Scientists have developed a way to make the waves in a beam of light travel in step with waves that have identical frequency and amplitude. Such light waves are said to travel in phase. Light that contains waves of identical amplitude and frequency that are in phase is called **coherent light**.

A **laser** (**l**ight **a**mplification by **s**timulated **e**mission of **r**adiation) produces and transmits a narrow, intense beam of coherent light. White light is made of many different frequencies of light, but laser light is composed of only one frequency, or color. Because all the aspects of the waves are the same, the energy is also focused into a narrow area. Lasers are used in medical research, DVD players, surveying, computers, printers, and military guidance systems. Lasers are also used for some types of surgery and for accurately measuring long distances. For example, a laser beam can be used to determine the distance from Earth to the moon. *Apollo* astronauts placed mirrors on the moon. Scientists directed

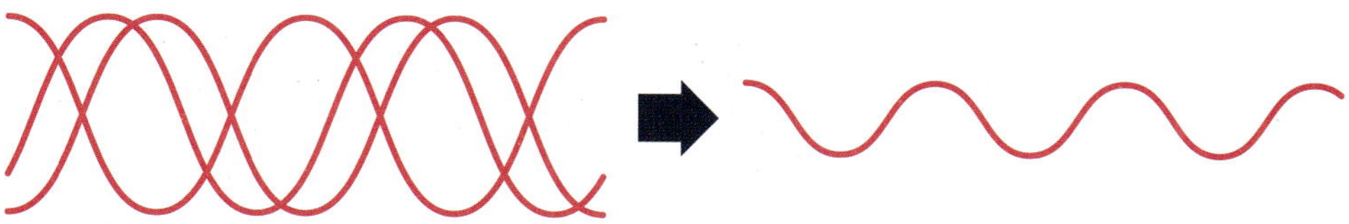

Coherent waves have the same wavelengths and travel in one direction, and their amplitude and frequency are in phase.

Incoherent waves have several wavelengths and travel in many directions, moving out of phase.

OBJECTIVES

- Compare incoherent and coherent light.
- Describe different applications of light.
- Evaluate the benefits of each application.

VOCABULARY

- **coherent light** light that contains waves of identical amplitude and frequency that are in phase
- **fiber optics** the science or technology of light transfer through very thin, flexible glass or plastic fibers
- **hologram** a three-dimensional photograph or image produced by lasers and a series of mirrors
- **incoherent light** light that contains waves of different frequencies and amplitudes moving out of phase
- **laser** a device that produces and transmits a narrow, intense beam of coherent light
- **optical fiber** a fiber made of glass or plastic along which light can be transmitted to carry information
- **polarized light** light with waves vibrating in only one direction

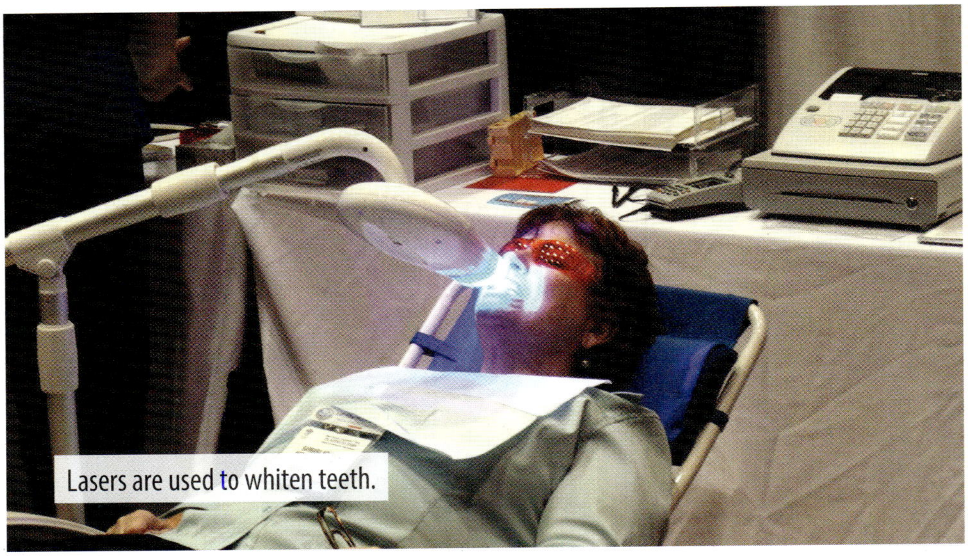
Lasers are used to whiten teeth.

a laser beam at the mirrors and measured the amount of time it took for the laser beam to reflect back to Earth. The reflected mirror calculations resulted in a very precise measurement. The use of laser beams to measure shifting of rock helped predict that Mount Saint Helens would erupt. Vulcanologists placed mirrors on the side of the volcano to reflect laser beams. As the volcano grew, the mirrors moved, causing the angles of reflected laser beams to change.

Laser lights are also used in the formation of holograms. A **hologram** is a three-dimensional photograph or image produced by lasers and a series of mirrors. The hologram is created from the interference of two laser beams that are formed when an initial laser beam passes through a beam splitter. These two beams are reflected off mirrors. One laser beam shines onto a photographic plate coated with light-sensitive material. The other beam shines onto an object and then onto the plate. When the two beams meet, they are out of phase with each other because one has had to travel farther. So, they interfere with each other. The interference creates patterns that are similar to the ripples formed when two stones are dropped into a pond. The parts that are in phase expose the photographic plate. The parts that are out of phase cancel each other out in destructive interference and leave that part of the plate unexposed. The patterns are recreated when a laser is later shone onto the plate. The result is a three-dimensional hologram. Holograms are used on posters, greeting cards, and other merchandise. They are used for security purposes

In 1969, a mirror was placed on the moon by the *Apollo 11* crew for lunar laser ranging.

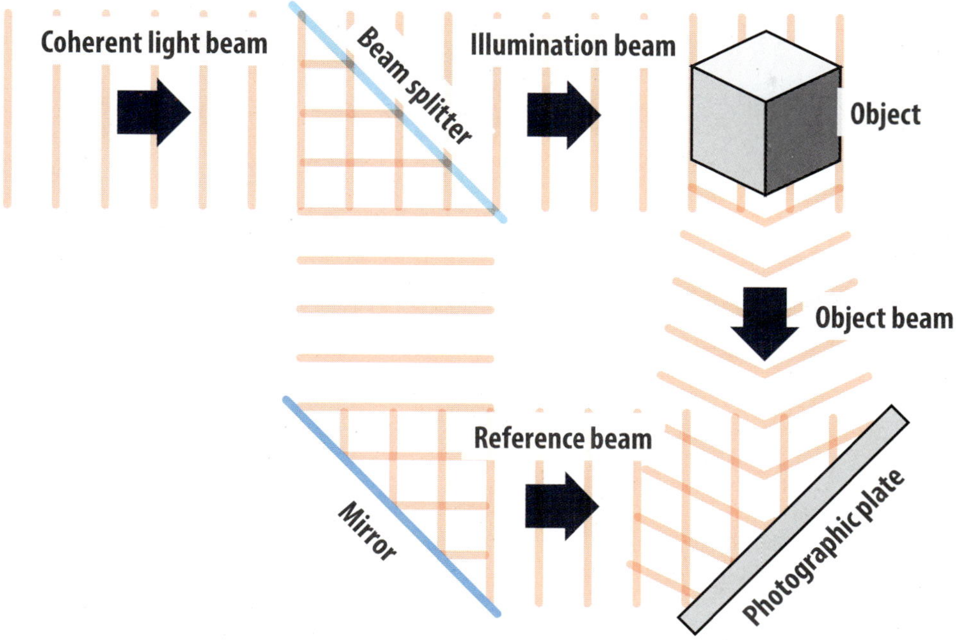

Credit card hologram

on some credit cards and tamper-evident stickers. Bar code scanning also uses holograms. Lasers can recognize the patterns of these holograms and translate them into the appropriate information, such as whether a credit card is genuine or how much a clothing item costs.

Have you ever stood squinting in bright sunlight? Perhaps you quickly put on a pair of polarizing sunglasses to cut the glare. Why do these sunglasses work so well? Recall that light waves

FYI

A Laser, a Parrot, and Safety Goggles

Who would use a 3-D printer to make safety goggles for a parrot? Anyone who needs their parrot to fly through a laser beam, that's who! A group of researchers tested standard calculations for lift, speed, wake flow, and path of a parrot named Obi-Wan. In order to test the accuracy of traditional mathematical calculations, the team needed accurate observations of a parrot's flight. Lasers were the right tool for the job. And to protect Obi-Wan's eyes, a special pair of safety goggles was made. The team spent months training Obi-Wan to fly through a flattened vertical laser projected through a cylindrical lens. The laser light was needed to give an accurate 3-D view of the bird's motion. The experiments demonstrated that the equations did not give accurate analysis of the bird's flight. Two of the equations consistently underestimated the lift values. If the research proves conclusive and better equations are created, the results could improve the development of flying robots.

Polarizing film with vertical slits allows light waves with vertical crests to pass. Polarizing film with horizontal slits allows light waves with horizontal crests to pass.

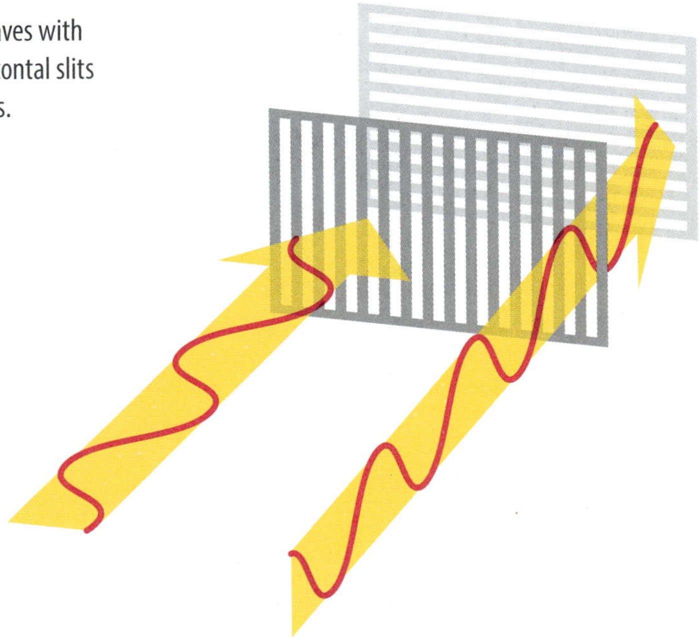

are transverse waves—waves traveling at right angles to the direction of the wave. Usually the different waves in light vibrate in many different directions. But passing light through certain substances can produce waves that vibrate in only one direction. Light with waves vibrating in only one direction is called **polarized light**. A substance that causes polarization is called *a polarizing filter*. Polarized sunglasses act as a polarizing filter.

A polarizing filter can be compared to a picket fence. The filter is made of long chains of molecules that act like narrow slits. The slits are vertical, so light waves that are traveling horizontally cannot pass through; only light waves traveling vertically can get through. Because glare is reflected light from objects and reflected light is polarized, polarizing filters can reduce glare. Polarizing sunglasses make it easier to see by reducing glare from water, snow, the road, or vehicles. Photographers also use polarizing filters to reduce glare in their pictures.

Without a polarizing filter, the flash produces glare. With a polarizing filter, only light waves traveling in one direction pass through and the glare is reduced.

Another use of polarizing filters has less to do with practicality and more to do with beauty. Polarizing filters create a colorful effect when used with substances called *optically active materials*. If polarized light passes through an optically active material and then through another polarizing filter, the orientation of the light is rotated. Because different frequencies of light are uniquely turned, it is possible to see

TRY THIS

Polarizing Sunglasses

Hold two pairs of similar sunglasses perpendicularly to each other with a horizontal lens directly in front of a vertical lens. Look through the overlapping lenses and record your observation. Adjust the angles of the two pairs of glasses, and observe the amount of light able to pass through them. Record your observations. Holding the first two pairs of glasses in a perpendicular position, insert a third pair between them, and observe the amount of light passing through. Record your observations. What happened to the light when two pairs were held perpendicular to one another? Why? What happened when the two pairs were not perpendicular? How much light was transmitted using three pairs of polarized sunglasses?

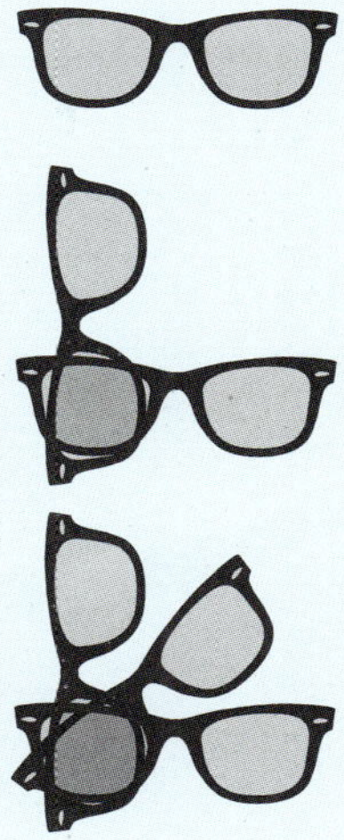

some of the separate wavelengths, or colors, of the light. Corn syrup, sugar solutions, turpentine, and amino acids are all examples of optically active materials.

In addition to measuring distances, creating holograms, and polarization, light can transmit data. Light makes it possible to talk on the phone or explore the Internet. Transmitting voice, text, and video used to require metal wires and radio waves to carry electric signals. Newer technology uses light. **Fiber optics** is the science or technology of light transfer through very thin, flexible glass or plastic fibers. Think of strands of glass 10 times thinner than a human hair. Imagine that the strands of glass are so pure that you could pile them several kilometers thick and still see through the pile. Because of their purity, these fibers can carry light signals for many kilometers.

Glass or plastic fibers that transmit light and carry information are called **optical fibers**. Light pulses carry messages through these fibers by reflecting in a continuous, zigzagging path. The core of the fiber and the surrounding glass have different indexes of refraction. The core has a high index to perpetuate the reflection, and the layer around the core has a low index to prevent the light from leaving the fiber.

The applications of fiber optics include communication and medicine. Optical fibers are replacing electric wires in telephone and computer systems because they have three advantages over copper wires—they have less attenuation or signal loss, meaning information can travel over optical wires nearly 10 times farther

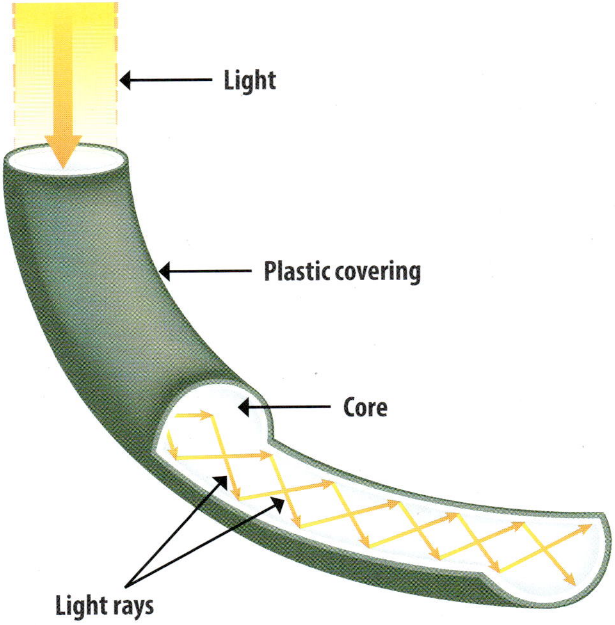

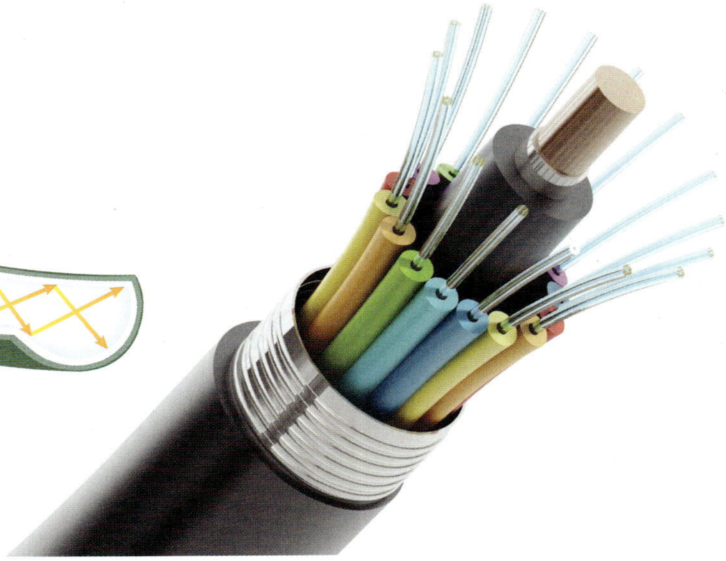

Cross section of optical fiber

than it can through copper; optical wires are not subject to interference from electromagnetic signals created by other wires; and they have a higher bandwidth, which means a single fiber optic cable can carry much more data than a copper cable of the same diameter.

Doctors can use optical fibers to look inside a human body without performing surgery. For example, optical fibers can be used to detect tumors in the throat. A set of fibers carries the light to the throat and another bundle brings the reflected images back to the doctor.

LESSON REVIEW

1. What is the difference between incoherent light and coherent light?
2. What is a laser and where is it used?
3. How are laser beams different from polarized light?
4. Why is polarizing film used in sunglasses?
5. In what ways are optical fibers better than copper cables?
6. What is one benefit of using optical fiber in the medical field?

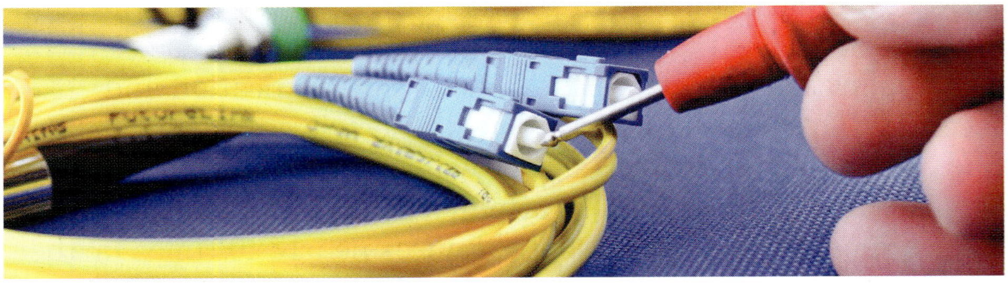

An engineer is testing fiber optic cables.

UNIT 7

Electricity and Magnetism

Chapter 1: *Electricity*
Chapter 2: *Magnetism*

Vocabulary

alternating current
amplifier
conduction
conductor
direct current
electric charge
electric current
electric discharge
electric motor
electric power
electrode
electrolyte
electromagnet
electromagnetic induction
electromagnetism
electronics
generator
insulator
magnetic domain
magnetic field
magnetism
resistance
semiconductor
static electricity
superconductivity
superconductor
transformer
voltage

Key Ideas

- Systems, order, and organization
- Evidence, models, and explanation
- Change, constancy, and measurement
- Form and function
- Abilities necessary to do scientific inquiry
- Understandings about scientific inquiry
- Motion and forces
- Transfer of energy
- Structure of atoms
- Chemical reactions
- Conservation of energy and increase in disorder
- Abilities of technological design
- Understandings about science and technology
- Natural and human-induced hazards
- Risks and benefits
- Science and technology in society
- Science as a human endeavor
- History of science

SCRIPTURE

For as lightning that comes from the east is visible even in the west, so will be the coming of the Son of Man.

Matthew 24:27

7.1.1 *Static Electricity*

OBJECTIVES

- Connect the behavior of charged particles with their movement across electric fields.
- Explain how uncharged objects can become electrically charged.
- Summarize charging by contact and charging by induction.

VOCABULARY

- **conduction** the movement of electrons through a material
- **electric charge** a property of matter that lets an object exert a force of repulsion or attraction on other objects that are near but not necessarily touching
- **electric discharge** the removal of a charge from a charged object because of a buildup of static electricity
- **static electricity** a stationary electric charge that is caused by the imbalance of positive and negative charges

Have you ever heard the expression "opposites attract"? Lightning strikes because opposites attract. You may get a shock from a doorknob because opposites attract. This expression is observed in nature between positive and negative charges. Atoms are composed of protons, neutrons, and electrons. Protons have a positive charge and electrons have a negative charge. Neutrons do not have a charge and contribute nothing to the overall charge, or net charge.

An **electric charge** is a property of matter that lets an object exert a force of repulsion or attraction on other objects that are near but not necessarily touching. As a rule, like charges repel and opposite charges attract. These charges can influence other objects across an area known as *an electric field*. The greatest force is located in the region of the field that is closest to the charged particle. An electric force between two objects is directly proportional to the net charge on each object and inversely proportional to the square of the distance between them. If the net charge on an object doubles, the strength of the electric force doubles. If the distance between two objects doubles, the strength of the electric force between the charged particles is one-fourth the original strength.

Simply picking up a pen or taking milk out of the refrigerator does not cause a shock because most objects are electrically neutral. The atoms that make up these objects have equal numbers of electrons and protons. They have no net charge. However, not all objects remain neutral because electrons do not remain stationary. They are free to move about and can be transferred from one object to another. This transfer creates a charge imbalance.

A stationary electric charge that is caused by the imbalance of positive and negative charges is called **static electricity**.

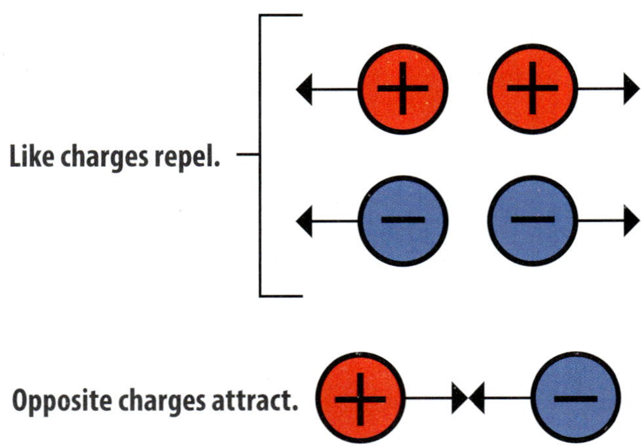

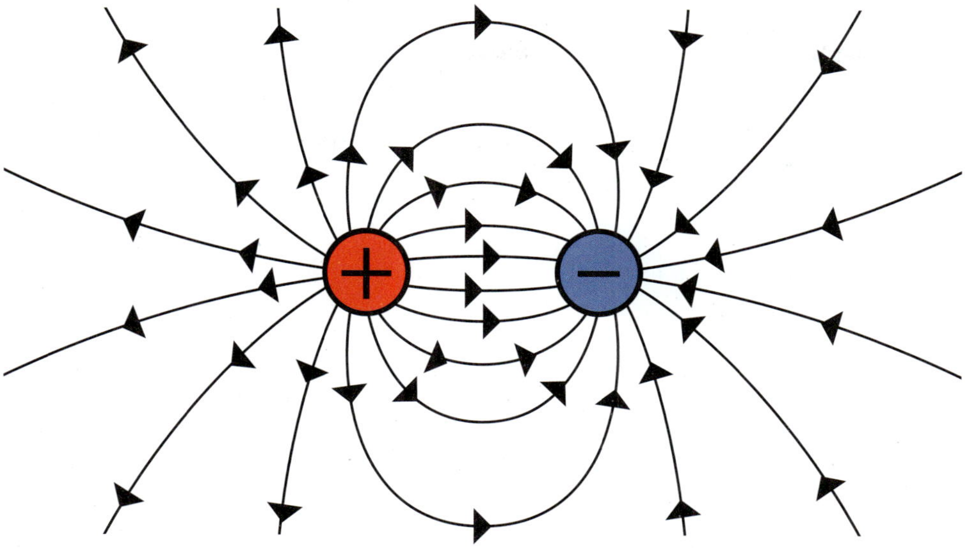

Charged particles exert the strongest force in the region of the electric field that is closest to the particle.

Uncharged objects can become electrically charged in one of three ways: they can come in contact with charged objects, they can be near a charged object, or they can come in contact with other uncharged objects that are made of certain materials. When an uncharged object touches a charged object, some of the electrons in the charged object may be more attracted to the atoms in the uncharged object. As a result, the electrons move from the charged object to the uncharged object. The movement of electrons through a material is called **conduction**. This process of transferring charge through touch is called *charging by contact*.

Two neutral objects can also participate in charging by contact. When you scuff across carpet, the rubber soles of your shoes pick up electrons from the carpet by conduction. Your shoes now have an overall negative charge and the carpet has an overall

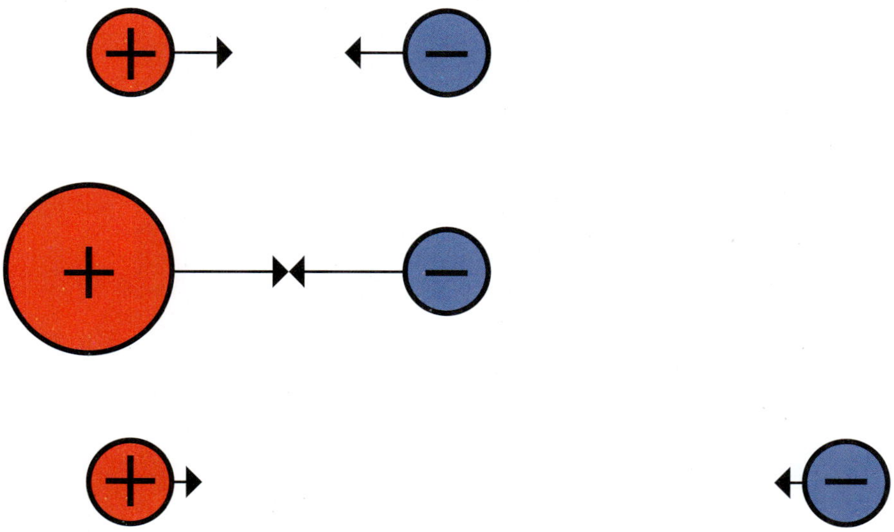

An increased charge causes the strength of the electric force to increase. An increase in distance between charged particles causes the strength of the electric force to decrease.

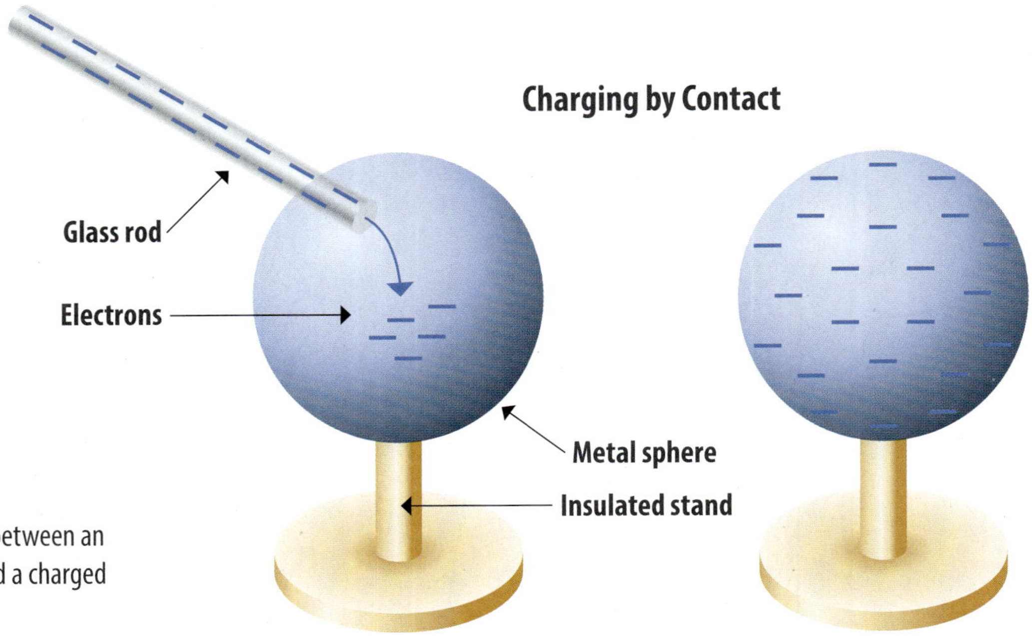

Charging by contact between an uncharged sphere and a charged rod

positive charge. The electrons spread throughout your body. If you do not come in contact with an object that greatly attracts electrons, such as one made of metal, the excess charge will flow through your body into the air. The air contains water molecules, which attract negative charges. If you touch a metal object, the electrons flow from your body through your hand to the object. This transfer of electrons produces a shock and possibly even a spark. The same process occurs when you rub a balloon on your dry hair or when you remove clothes from the dryer.

Uncharged objects can also become charged when they are near, but not touching, a charged object. *Charging by induction* is the rearrangement of electric charges of a neutral object caused by the influence of a nearby charged object. When an electrically charged glass rod comes near a neutral balloon, the electrons in the balloon travel away from the rod, leaving a positively

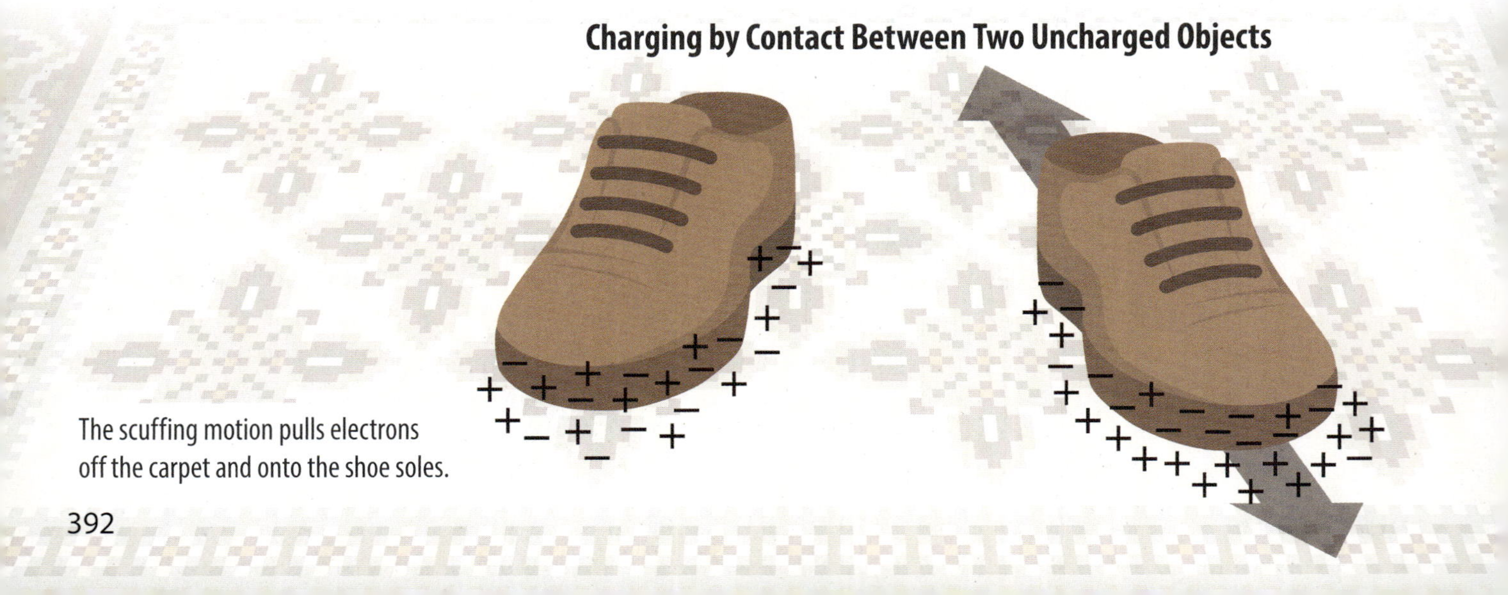

The scuffing motion pulls electrons off the carpet and onto the shoe soles.

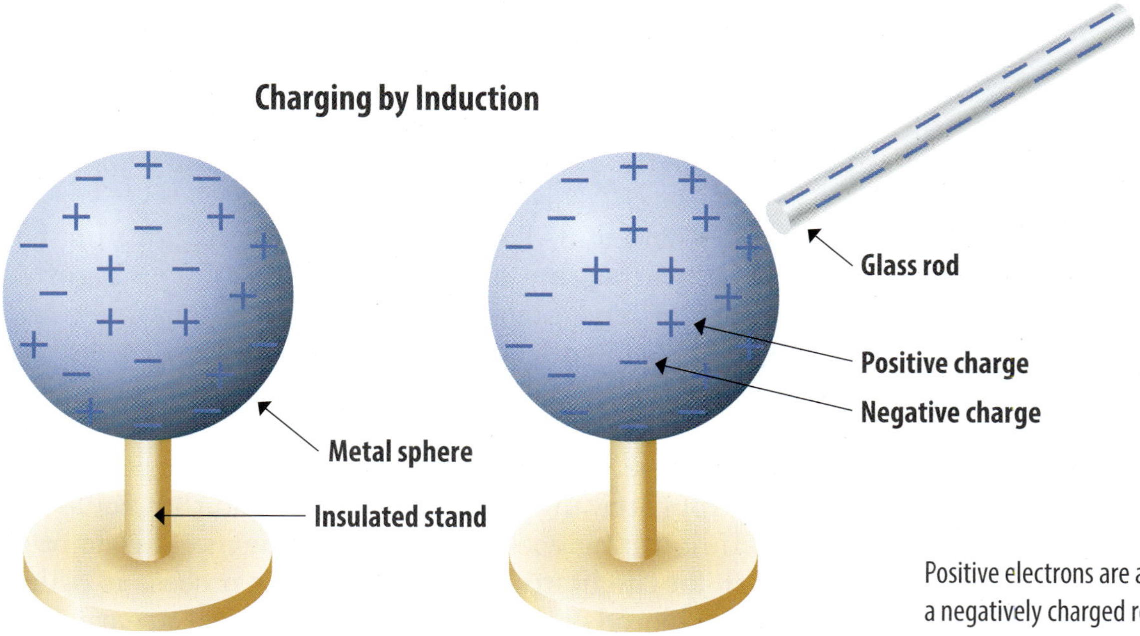

Positive electrons are attracted to a negatively charged rod.

charged area on the side of the balloon near the rod. As soon as the glass rod is moved away, the electrons return to their original positions. This is an example of a temporary charge induction.

Sometimes, the electrons are not able to return to their original positions. If so, they maintain a stable static charge. Consider a charged glass rod that approaches two neutral balloons that are in contact with each other. The electrons from the balloon nearer the glass rod move as far away from the rod as possible. This means that they travel to the opposite balloon. If the balloons are separated before the glass rod is moved away, the electrons remain in the balloon farther away and are not able to return to the balloon near the glass rod.

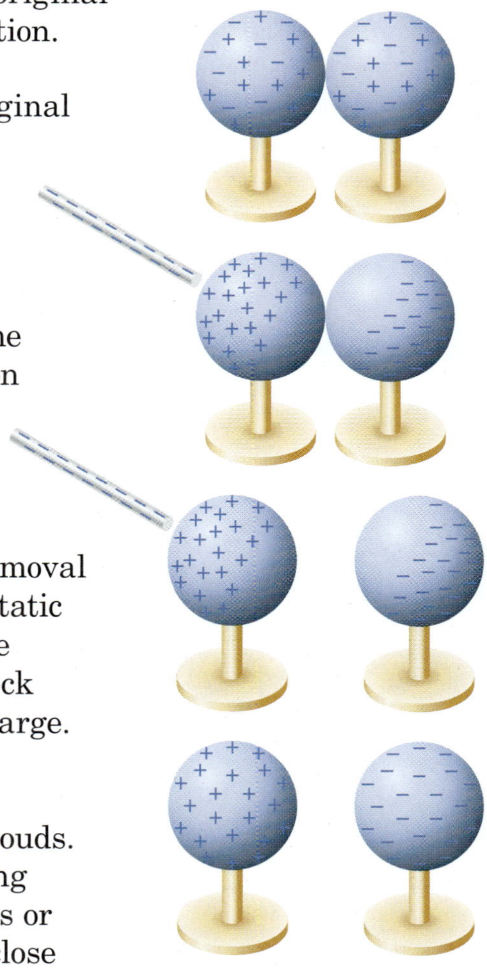

A stable static charge is created if the rearranged electrons cannot return to their original position.

The electrons that moved from one object to another and caused static electricity eventually leave the object. The removal of a charge from a charged object because of a buildup of static electricity is called **electric discharge**. It is caused by the repulsion or attraction between charged particles. The shock that you feel from touching a doorknob is an electric discharge. Lightning is another example of electric discharge.

Lightning is the electric discharge of energy from storm clouds. Electric charges within a cloud become unbalanced, causing the positive and negative charges to separate within clouds or in separate clouds. When negatively charged clouds form close to Earth's surface, objects on the ground become electrically charged by induction. A stream of electrons then leaps from the cloud to the object. As the electrons leap through the air, they

Lightning rods help to neutralize electric charges by directing the charges to the ground where they can dissipate over a large area.

produce large quantities of heat and light. Lightning can also occur within clouds. The heat from the lightning causes the air to expand rapidly. This creates the sound that is called *thunder*.

Lightning is powerful. The air that surrounds lightning bolts can reach temperatures that are hotter than 22,000°C. Lightning bolts also carry huge amounts of electricity. Benjamin Franklin performed experiments that provided evidence that lightning is a result of static electricity. Franklin proposed the idea of lightning rods—metal rods attached to building roofs. Lightning rods are still in use today. The major purpose of the rods is to prevent lightning from striking objects on the ground that could cause damage. If lightning does strike Earth, the metal rods provide a safe path for the lightning to travel. This process is called *grounding*. The rods are attached to cables that are connected to the ground. The energy of the electric current is directed harmlessly from the rods through the cables and down into the ground. This prevents damage to buildings.

One way that scientists detect electric charges is through the use of an instrument called *an electroscope*. A simple electroscope consists of a metal rod with a pair of thin metal strips at one end. The rod fits into a rubber stopper, and the stopper is inserted

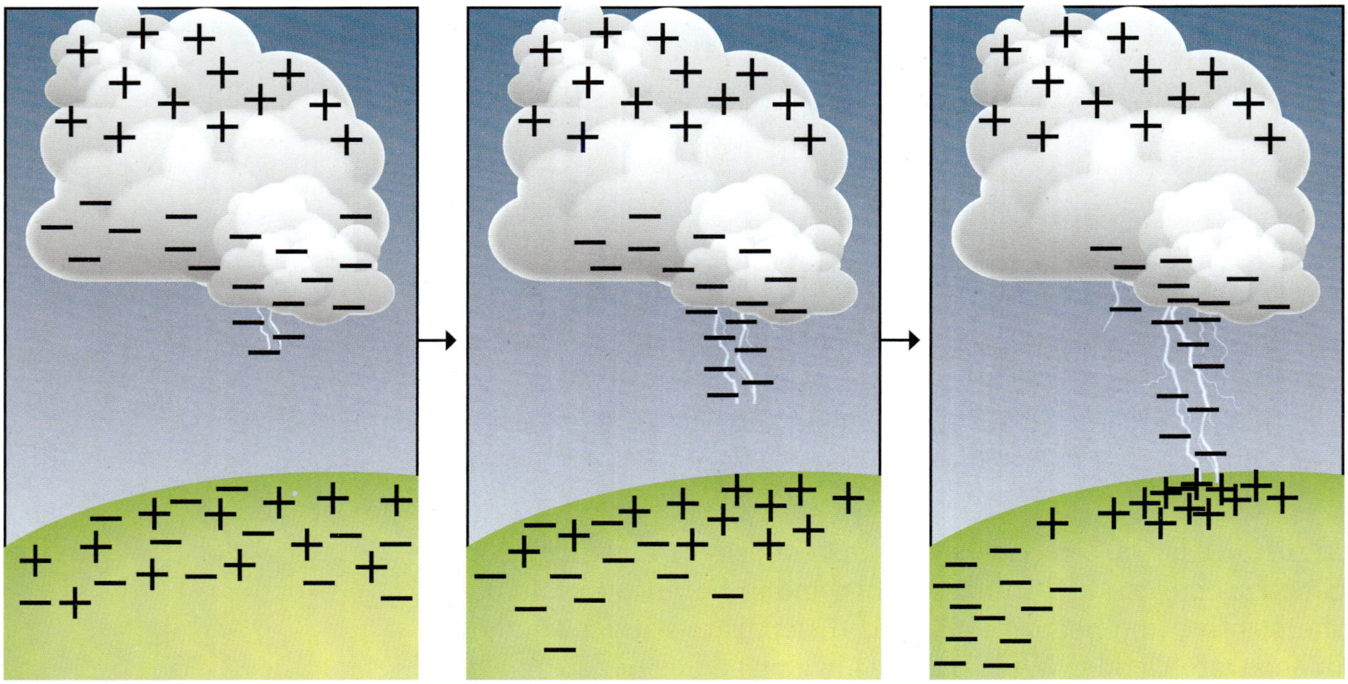

Convection currents in the cloud cause a charge separation.

The negative charges in the lower portion of the cloud attract positive charges in the earth.

A buildup of enough negative and positive charges creates a strong attraction that results in a lightning strike.

Electroscope

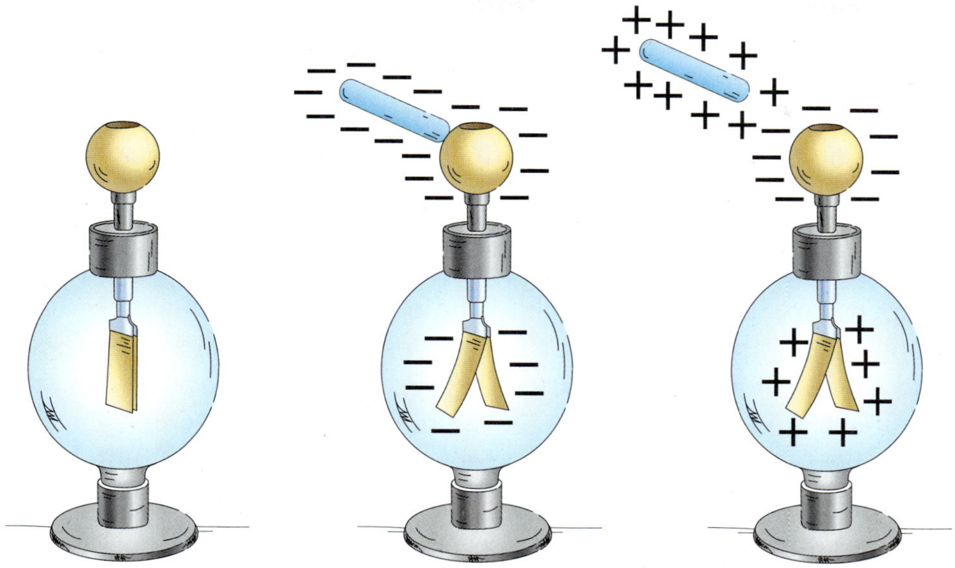

into a flask. The lower part of the rod with the metal strips is lowered into the flask. When the electroscope has no charge, the metal strips hang down. When the instrument detects an electric charge, the metal leaves separate. The two strips repel each other because they both have the same charge.

LESSON REVIEW

1. How does an object become charged?
2. How does the electric force between two objects change if the charge on one of the objects increases?
3. Why does an uncharged object become charged when it touches a charged object? What is this process called?
4. Explain charging by induction.
5. How is a stable static charge produced?
6. Compare the appearance of a charged electroscope to that of an electroscope that is uncharged.

TRY THIS

Static Electricity

Sprinkle about one teaspoon of salt onto the surface of a desk. Charge a comb by running it through your clean, dry hair for about 10 seconds. Bring the comb near the salt. What do you observe? Sprinkle about one teaspoon of pepper onto the surface of the desk. Thoroughly mix the salt and pepper. Charge the comb again. Bring the comb near the salt and pepper mixture. What do you observe? Form a hypothesis to explain why the salt or pepper was picked up more easily in the presence of an electric charge.

7.1.2 Conductors and Insulators

OBJECTIVES

- Explain why a substance would be classified as a conductor or an insulator.
- Appraise conductors, insulators, semiconductors, and superconductors.

VOCABULARY

- **conductor** a substance that allows electricity to flow through it easily
- **insulator** a substance that does not allow electricity to flow through it easily
- **semiconductor** a substance that conducts electricity better than an insulator but not as well as a conductor
- **superconductor** a substance that conducts electricity without losing energy when it is cooled to extremely low temperatures

The flow of electrons creates an electric current in substances. Electric current flows through certain substances better than others. Why? The flow of current is dependent on a few factors that include the amount, shape, and type of substance used. Scientists classify these substances into four basic categories: conductors, insulators, semiconductors, and superconductors.

A **conductor** is a substance that allows electricity to flow through it easily. The best conductors are metals. At room temperature, silver is the most conductive metal, followed by copper, gold, and aluminum. The atoms of metals do not tightly hold on to their valence electrons. These outer electrons form a sea of electrons that surrounds the metal atoms. Because the outer electrons move easily around in the sea, the electrons in an electric current also flow easily through the sea.

When current is flowing through a conductor, such as a copper wire, the electrons are moving through the atoms of the conductor. These atoms block some of the electrons, and some electric energy is converted into heat energy, which causes some current-carrying wires to feel hot. Metal wires conduct electric currents to power many devices such as computers and blenders. The metal conductors that are connected to automobile batteries allow electric current to flow and generate vehicle movement.

An **insulator** is a substance that does not allow electricity to flow through it easily. Insulators' atoms hold tightly to valence electrons. As a result, there is no sea of electrons for an electrical current to flow through. Insulators include wood, glass, rubber, plastic, and air. Some of these insulators, such as plastic or rubber, cover electrical wires to prevent heat and electricity from escaping and causing fires or electric shock.

A **semiconductor** is a substance that conducts electricity better than an insulator but not as well as a conductor. Semiconductors do not have as many free electrons as conductors, so they carry much less current. The two most common semiconductors are silicon and germanium.

The computer age began with the development of semiconductors. Before semiconductors were available, vacuum tubes were used in computers. Semiconductors are smaller and more dependable than vacuum tubes, and they last longer. They also consume less power and

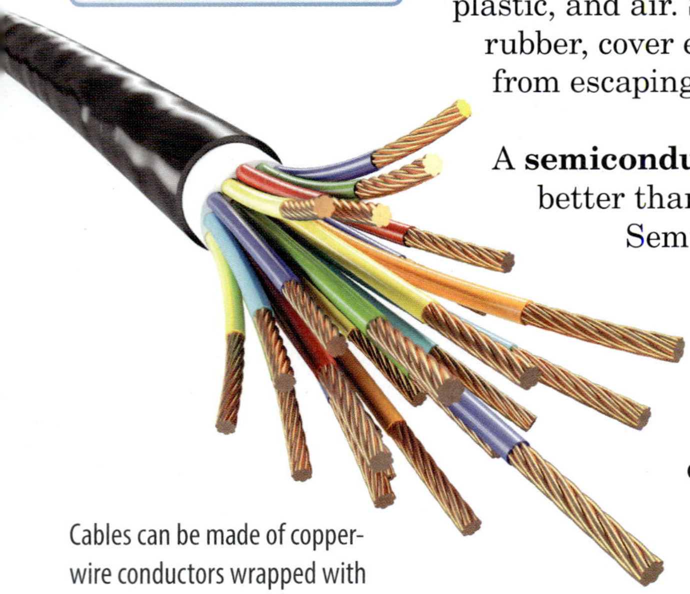

Cables can be made of copper-wire conductors wrapped with plastic or rubber insulators.

give off much less heat. Without semiconductors, the scope of computers would be very limited. The Internet, word processing, and other valuable computer services simply would not exist. Semiconductors are also used in vehicles, telephones, televisions, and many other modern electric devices that have computer chips.

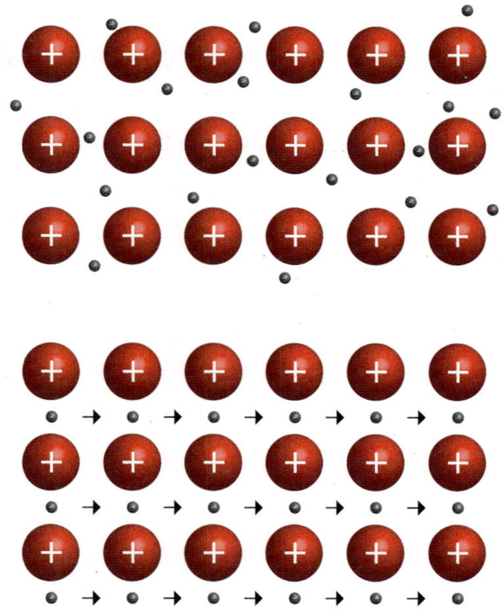
Metal atoms form a sea of electrons. When a current is applied to the conductor, the electrons start to travel in the same direction.

Imagine a substance that can conduct electricity without losing any of its energy in the form of heat, which is what happens with copper. A **superconductor** is a substance that conducts electricity without losing energy when it is cooled to extremely low temperatures. Superconductors do not waste any power.

Super conducting magnets are used in MRI (magnetic resonance imaging) machines, which make it possible for doctors to see inside the body, reducing the need for many exploratory surgeries. The MRI machines use a large amount of liquid helium to keep the magnets near absolute zero. Superconductors have also been used in power lines, computers, and high-speed trains.

One limiting factor to the use of superconductors is that they must be kept at extremely low temperatures. Prior to 1985, scientists could only produce superconductors at a temperature of around –250°C. Expensive liquid helium was required to keep them cold. In 1987, a superconductor was developed to work at –175°C. Liquid nitrogen, which is much less expensive, can be used to store this type of superconductor. Physicists are currently looking for ways to produce superconductors that will work at room temperature.

Insulators are positioned where power lines connect to utility poles or transmission towers.

Microchips are made of semiconductor materials, which allow electrical currents to travel much faster through the chips at high temperatures.

TRY THIS

Lightning Strikes

Obtain two metal pie pans and a foam plate. Turn one pan over. Secure a piece of modeling clay in the center of each pan, which will act as a handle. Without touching the foam plate, rub the bottom of the plate with a piece of wool for at least 30 seconds. Grasp the modeling clay on the right-side-up pie pan and place the bottom of the pan on the bottom surface of the plate. Darken the room. Grasp the modeling clay on the pie pan that is upside-down and gently lower it onto the first pan. Record your observations. Repeat this procedure, except do not place the upside-down pan on top of the right-side-up pan. Darken the room. Straighten a paper clip and bring it close to the edge of the pie pan. Record your observations. What created the sparks? Compare your observations in both procedures. Describe how the charge on the right-side-up pan changed throughout the experiment.

LESSON REVIEW

1. Compare conductors and insulators. Include examples of each.
2. Rank semiconductors, insulators, and conductors according to their ability to conduct electricity.
3. What are the two most commonly used semiconductors? Why are they beneficial? What type of products use semiconductors?
4. What type of substance can conduct electricity without losing any of its energy in the form of heat? What types of products utilize this substance?

BIOGRAPHY

Shirley Ann Jackson

Shirley Ann Jackson is a woman of many firsts. She is the first African-American woman to receive a doctorate from the Massachusetts Institute of Technology. Jackson earned her degree in theoretical physics. Her specialty was in condensed matter physics, specifically layered systems and the physics of opto-electronic materials. She is also the first African-American woman to be president of Rensselaer Polytechnic Institute—the oldest technological research university in the United States. Jackson is the first African-American woman to be elected president and chairman of the board of the American Association for the Advancement of Science (AAAS). She has assisted members of the White House in formulating scientific and technological policies. Her work helped to strengthen the economy and provide more employment and innovation opportunities for many people. She also served as the Chairman of the US Nuclear Regulatory Commission. In this capacity, Jackson helped to license, regulate, and safeguard the use of nuclear reactor by-product materials. She was awarded the National Medal of Science in 2016. Jackson's drive and passion for science prompted her to encourage other students, specifically women and ethnic minorities, to pursue scientific study. Some consider her the ultimate role model for women in science.

A generator in a power plant is the source of the electricity conducted through the wires of a home. This generator causes the electrons to flow back and forth through the wires so electric appliances can be used. However, there are times when electricity is needed but a wire from a power plant as a pathway for electrons cannot be used or is not accessible. This is the case for most cars. A vehicle's motor requires electricity to start. Other car parts need electricity to function. But, it is not practical to run a wire to the car every time to start it. So, a car uses a battery as a source of electrons.

A Canadian company, GNB, manufactures very large batteries that are used to operate forklift trucks and railway locomotives. These vehicles are used in factories, in warehouses, and deep inside mines—enclosed areas where regular gas engines and their exhaust fumes are too dangerous. In Europe, battery-operated (electric) delivery trucks have been used to deliver food and merchandise for years. These battery-operated vehicles run silently.

Electric vehicles have one big drawback—they cannot travel long distances before the batteries need to be recharged. The average electric vehicle has a range of between 64 and 322 kilometers before needing to be recharged. Another drawback is that the electricity needs to be made from other fuel sources that are often nonrenewable. Making longer-lasting batteries is a new and exciting field of research.

OBJECTIVES

- Examine the structure of batteries.
- Compare dry cell and wet cell batteries.

VOCABULARY

- **electrode** an electric conductor that conducts current into and out of an electrolyte
- **electrolyte** a mixture of chemicals that conducts an electric current

Most electric cars have to recharge every 64 to 320 kilometers.

SCIENCE IN ACTION

How to Jump-Start a Car Battery

Imagine turning the ignition switch and nothing happens. You hear a few clicks, but the car will not start. It is probably a dead battery. How will you get to school to take final exams? Then, you remember the jumper cables in the trunk of the car. But, what is the right way to connect the cables to the car's battery? It is important to remember that jumper cables with a low gauge value indicate greater strength. The standard jumper cable size is a gauge of six. Also, some cars require extra steps in order to have a successful jump. Consult the car manual first to see if the car will permit a jump. It is always important to utilize safe practices when handling jumper cables. Once the cables are connected to a battery, never touch the metal clamps to anything other than the intended vehicle target. Here are the steps to follow to get the car running again:

1. Position two cars so they are facing each other but not touching. Ideally, they will be separated by a distance of about 45 centimeters.
2. Place automatic transmission cars in park and manual transmission cars in neutral. Set the parking brakes.
3. Both cars should be turned off with the keys removed from the ignition.
4. Open the hood of both cars and locate the batteries' terminals. For each battery, there will usually be both a positive and a negative symbol located near the terminals.
5. Attach the red, positive cable clamp to the positive battery terminal of the dead battery.
6. Attach the other red, positive cable clamp to the positive battery terminal of the car that is functioning.
7. Connect the black, negative cable clamp to the negative battery terminal of the car that is functioning.
8. Attach the other black, negative cable clamp to an unpainted, metal part of the car with the dead battery, such as a nut on the engine block. Do not connect the negative cable clamp to the dead battery.
9. Start the functioning car. It may be helpful to wait a few minutes before attempting to start the car with the dead battery.
10. After a few minutes, try starting the car with the dead battery. If the car does not start right away, wait a few more minutes and try again. Revving the functioning car's engine may also help revive the dead battery.
11. Once the car that had a dead battery is running, disconnect the jumper cables. Remove the black, negative cable clamps first. Do not let the clamps touch each other.
12. Drive the car that had a dead battery for at least 15 minutes to build up a charge in the battery. This will help prevent the battery from dying again once the car is shut off.

Parts of a Dry Cell

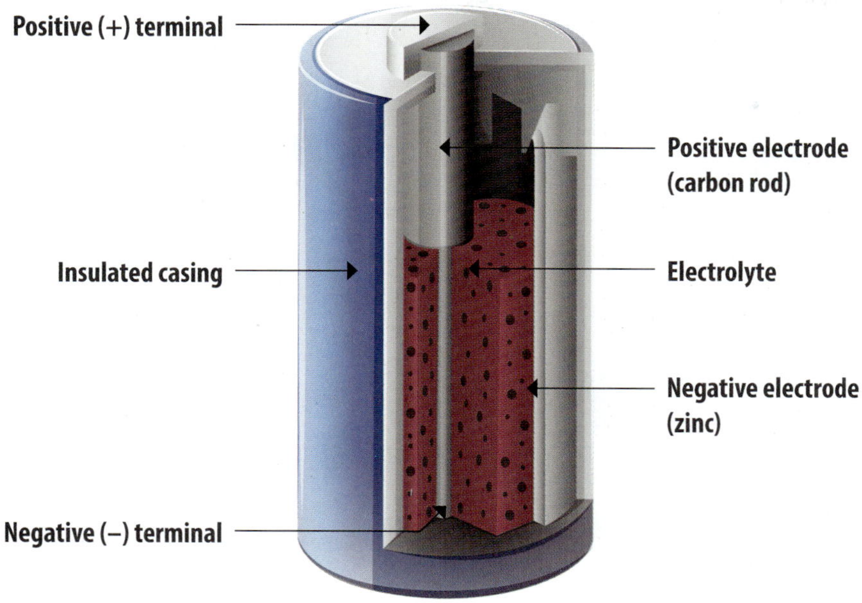

A battery is a combination of electric cells strung together. An electric cell is a device that provides a steady supply of electrons to create an electric current. Sometimes this device is called *an electrochemical cell.* Chemical reactions in electric cells convert chemical energy into electrical energy. The two basic types of electric cells are dry cells and wet cells.

Dry cell batteries are made of two metal plates, or electrodes, that are surrounded by an **electrolyte**—a mixture of chemicals that conducts an electric current. An **electrode** is an electric conductor that conducts current into and out of the electrolyte. The electrolyte allows charges to pass from one plate to another. Dry cells are not actually dry. They were given this name because the electrolyte is a moist paste and not a liquid solution.

Think about the battery that is used to power a TV's remote control. The center of the battery is made of a carbon rod and the outer layer of the battery is made of zinc. When the power switch is pressed, the battery is connected in a circuit and the chemicals that are in the electrolyte that separates these two electrodes react with the zinc layer. These reactions cause a separation of charges. The positive charges attach to the carbon rod and the negative charges attach to the zinc layer. This charge separation encourages the constant current flow. Dry cells are often the power sources for radios and other electronic devices, including laptop computers.

Parts of a Wet Cell

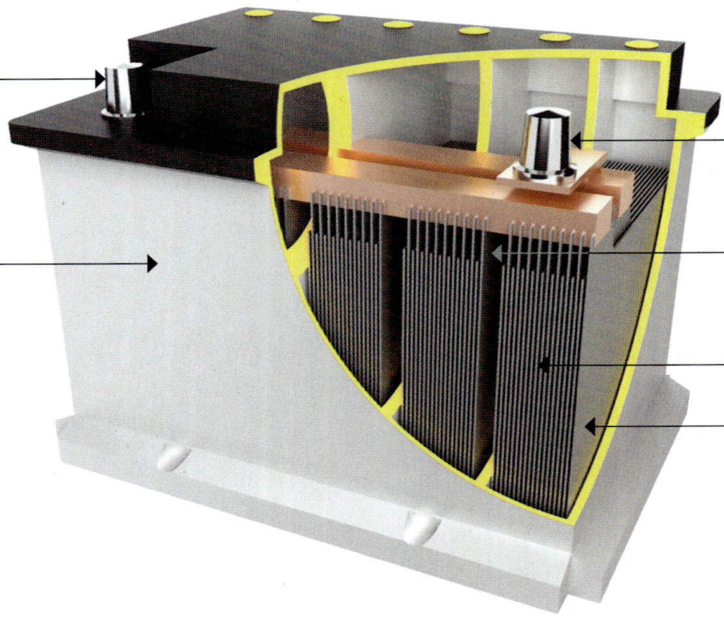

A wet cell consists of two metal plates or metallic compounds that are submerged in a liquid electrolyte. Hydrochloric acid is an example of a wet cell electrolyte. In some wet cells, the electrodes are made of copper and zinc. When the hydrochloric acid and the zinc react inside the cell, each zinc atom releases two electrons. Extra electrons build up on the zinc electrode, and electric current flows through the wire and the load and back into the battery. The cycle repeats itself as more chemical reactions produce more electrons. Car batteries are examples of wet cells. They use lead and lead dioxide plates for electrodes. Typical car batteries contain a set of six wet cells. A sulfuric acid solution serves as the electrolyte.

LESSON REVIEW

1. What is the main difference between a dry cell and a wet cell battery?
2. What is the purpose of an electrolyte?
3. How is a constant current produced in a dry cell battery that powers a remote control?
4. Describe how a wet cell battery produces an electric current.
5. What structures do dry cell batteries and wet cell batteries share?

Electric Currents and Circuits 7.1.4

> **OBJECTIVES**
> - Relate electrical currents and circuits.
> - Examine each component of a simple circuit.
> - Contrast series and parallel circuits.

> **VOCABULARY**
> - **alternating current** an electric current in which the electrons alternate from one direction to another at regular time intervals
> - **direct current** an electric current in which the electrons always flow in the same direction
> - **electric current** the net flow of electrons

The electrical energy of all the lightning that strikes Earth in a year is enough to supply the electrical needs of a significant fraction of households in the United States. Why do people not take advantage of that energy instead of putting so much effort into generating electricity by burning fossil fuels? One reason is that lightning is unpredictable. It can strike anywhere at any time. Another reason is lightning must be controlled to be considered useful. At present, there is no known way to control lightning's moving charge.

Moving charges produce an **electric current**—the net flow of electrons. The current direction moves in opposition to the flow of electrons, from the positive terminal of a battery to the negative terminal. Once an electric force has been applied to electrons to initiate movement, they can be made to flow continuously as long as there is an electron source. Electrons take the path of least resistance, so they travel from a high voltage location where there is a high concentration of negative charges to a low voltage location. This idea is similar to water pressure. Water molecules under a large amount of pressure relieve the pressure by moving to a location of less pressure.

Electric current is measured in amperes or amps, A, and is represented by the symbol *I*. An ampere indicates the amount of charge that passes a given point in a certain amount of time. The unit for an electric charge is a Coulomb, C, and time is usually measured in seconds, s. An ampere can be broken down into C/s.

Effects of Electric Current on People

Current	Effect
0–0.5 mA	None
0.5–2 mA	Threshold of feeling
2–10 mA	Threshold of pain, muscle contractions
10–20 mA	Muscle Injury
20–100 mA	Respiratory paralysis
100 mA–3 A	Ventricular fibrillation
>3 A	Cardiac arrest

There are two basic types of electric current: direct current (DC) and alternating current (AC). A **direct current** is an electric current in which the electrons always flow in the same direction. A direct current is created when a path of electrons exists between the terminals of a battery or a DC generator. The battery or generator continually pumps electrons through

TRY THIS

The Right Answer

Electric quiz games can be a lot of fun. Someone asks a question and if it is answered correctly, the light bulb illuminates or the buzzer sounds. Using your knowledge of open and closed circuits, build an electric quiz game board using a 40 cm × 30 cm piece of cardboard, a piezoelectric buzzer or miniature light bulb, a D cell battery, nuts and bolts, various lengths of insulated wire that have been stripped at the ends, and two lengths of insulated wire with alligator clips. Formulate questions and answers about electricity and see who can score the most points. Remember to arrange the answers in a random order.

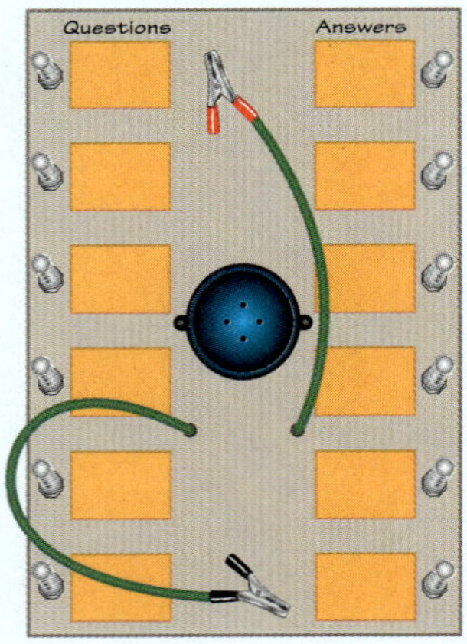

Electrical Circuit Symbols

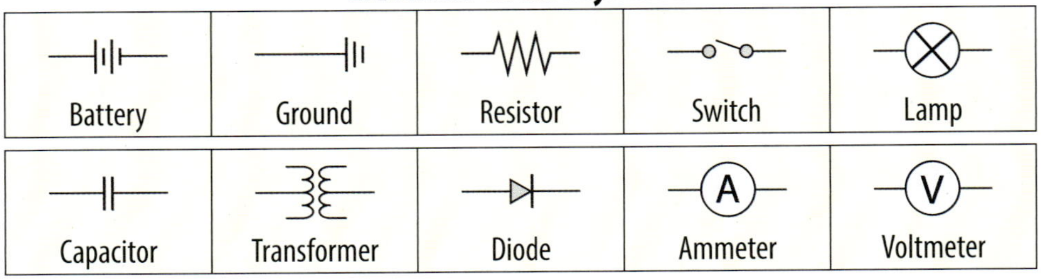

the pathway and maintains an electric current by converting chemical energy into electrical energy.

An **alternating current** is an electric current in which the electrons alternate from one direction to another at regular time intervals. Alternating current is the most commonly used electrical current in the United States. This type of electric current runs through the wires of buildings.

For electrons to flow in an electric current, there must be a closed path through which they can travel. An **electric circuit** is a closed path through which electrons can flow. Electric circuits typically have three parts: a voltage source, an electrical device, and a conductor. Circuits may have one or more of each of these parts. A switch is often included as well. The voltage source for direct currents may be a battery or an electrical outlet. The voltage source for alternating currents is a generator or power plant. An electrical device converts the electrical energy into another form of energy, such as light or heat. Light bulbs, toasters, car motors, and computers are examples of electrical devices. A conductor creates the connection between the voltage

Simple Circuit

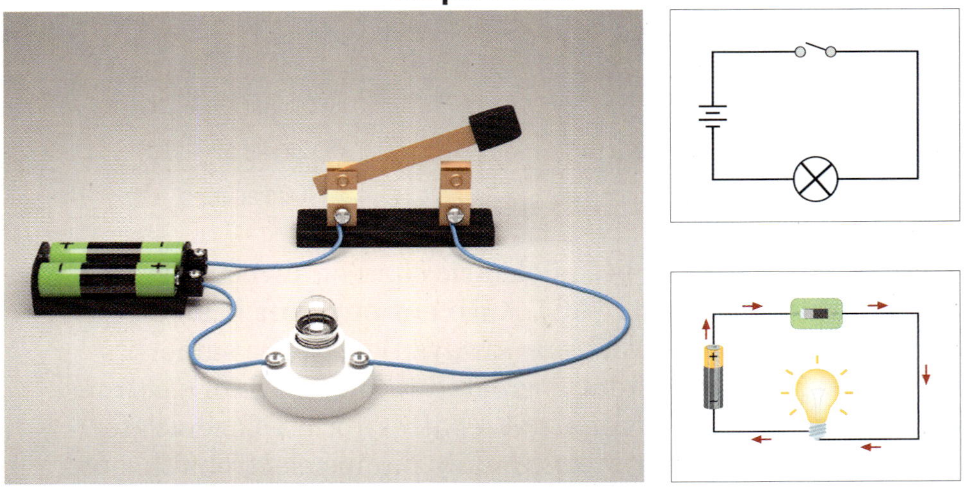

source and electrical device. Metal wires are the most common conductors in an electric circuit.

A simple circuit might consist of a set of three wires that connect a battery, switch, and light bulb. In this example, the battery is the voltage source and the light bulb is the electrical device. If the switch is closed, an electric current flows through the closed circuit and the light bulb illuminates. If the switch is open, the circuit is broken, and the bulb remains unlit.

Not all circuits are set up the same way. A circuit can be either a series circuit or a parallel circuit. In a series circuit, all of the components are connected one after the other. This type of circuit creates only one path for the flow of electricity. A flashlight is an example of a series circuit. If a break occurs in any part of a series circuit, the flow of electricity is disrupted. Older strings of Christmas lights are connected in a series circuit. If one bulb burns out or is loose, the entire string of bulbs goes dark. Each bulb has to be tested until the defective bulb is located.

In a parallel circuit, the components are on separate branches. The branches create several paths for the flow of electricity. If one branch of a parallel circuit breaks, the flow of electricity continues through the other branches. Christmas tree lights operating on parallel circuits have a definite advantage over lights with a series circuit. If one bulb on the parallel circuit goes out, the flow of electricity takes another path, and the rest of the bulbs stay lit. Houses and automobiles utilize parallel circuits. In houses, wires branch out from a breaker or fuse box to connect to appliances, wall sockets, and lights. If home electrical systems

Series Circuit

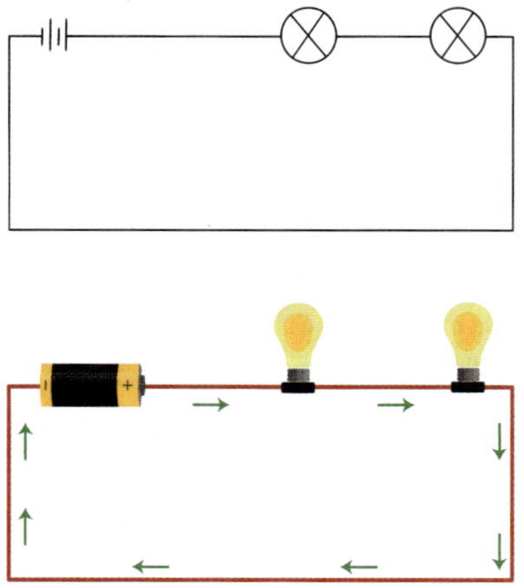

Parallel Circuit

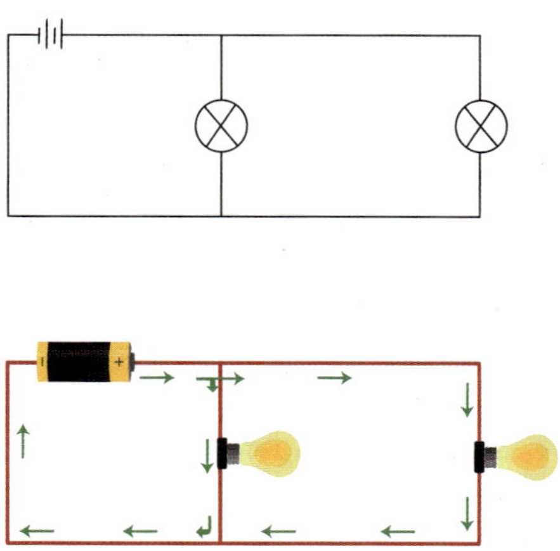

FYI

Vampire Appliances

Even when some appliances are turned off, they still use energy. These devices are called *vampire appliances*. Common vampire appliances include televisions, game consoles, computers, and kitchen appliances. The energy wasted accounts for about 10% of the energy used in an average American home. Energy used by televisions that are turned off can cost up to $160 a year. Game consoles can cost about $75 a year; computers, $40 a year. A savings solution would be to unplug the devices, but many people would not consider this option. An alternative would be to use an advanced power strip. Experts suggest that the sleep mode on a computer should be adjusted to 30 minutes or less and a monitor should be set to shut down after 10 to 15 minutes. By reducing the amount of wasted energy used by these vampire appliances, people could save a lot of money.

HISTORY

Tesla, Edison, and Westinghouse

George Westinghouse

Throughout history, there have been many rivalries in the science sector. One of the most notable rivalries occurred between Thomas Edison and George Westinghouse. In the United States, there was a race to provide electricity across the country. Edison developed a system that used direct current (DC). Westinghouse created a system that offered electricity through alternating currents (AC). This battle was known as *The War of Currents*. There were some limitations to DC systems that Edison recognized. The main limitation was the difficulty to transmit direct currents over large areas without significant energy loss. He enlisted the help of Nikola Tesla to design a better power transmission system. Tesla indicated that the best way would be to use alternating currents because they contain high-voltage energy that could be delivered over large areas without using an excessive amount of current. Edison discounted Tesla's ideas. Tesla stopped working for Edison in 1885 and sold some of his patents to Westinghouse. This shift in production led to the expansion of Westinghouse Electric AC generators. In an effort to ruin Westinghouse, Edison strongly promoted the idea that AC generators were extremely unsafe and could cause incidents of electrocution. By 1893, despite all of Edison's attempts, Westinghouse became the industry leader in electricity.

worked on a series circuit, everything would turn off with just the flip of a light switch. Parallel circuits enable components of home electrical systems to operate independently of one another.

Circuits can convert electrical energy into a variety of different energy types: light, sound, and heat. Electric circuits transfer electrical power through high-voltage power lines and transformers—electrical devices used to raise or lower voltage from one electric circuit to another—or through distribution systems in houses and businesses. Electric circuits relay electrical energy to communication devices, processing and storage devices, motors, generators, lamps, and loudspeakers.

LESSON REVIEW

1. What type of path is required for electrons to flow in an electric current?
2. What are the three main parts of an electric circuit?
3. What role do electrical devices play in an electric circuit? Name an example of an electrical device.
4. What role do conductors play in an electric circuit? Name an example of a conductor.
5. Compare series and parallel circuits.

Energy is required to move any object. Just as a bowling ball does not roll down an alley and knock over pins without being thrown, electrons do not flow through a wire without energy. Electrons are much smaller than a bowling ball, but they still need energy to move.

An electric current is measured by counting the number of electrons that pass a given point in a certain amount of time. Higher electric currents send a greater number of electrons past a given point each second. Lower currents send fewer electrons each second. An electric current is determined by two properties: voltage and resistance.

Voltage is the difference in energy per unit charge as a charge moves between two points in the path of a current. It is called *the potential difference*. Picture a waterwheel being turned by a stream of water. The water is similar to the flowing electric charge, the waterwheel is like an electric motor, and the voltage represents the height of the stream above the wheel. The water current would not exist if there was not a height difference and no work would be done by the waterwheel. In the same way, without a voltage difference across an electric motor, there would be no electric current and no work performed by the motor. The

OBJECTIVES

- Describe factors that affect electrical resistance.
- Determine how current, voltage, and resistance are related using Ohm's law.

VOCABULARY

- **resistance** the opposition to the flow of an electric charge
- **voltage** the difference in energy per unit charge as a charge moves between two points in the path of a current

TRY THIS

Incandescent Lamp

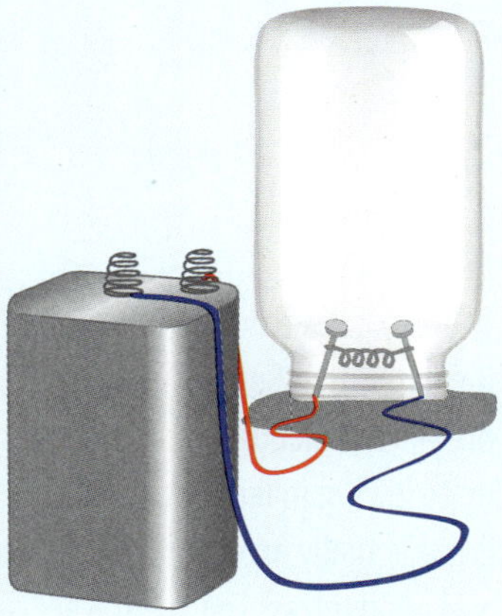

Wrap 10 cm of nichrome wire tightly around an iron nail. Spread the coils so they are not touching. Leave about 2 cm of straight wire on either side of the coil. Flatten a sample of modeling clay to form a circle that is 3 cm high with a 15 cm diameter. Carefully remove the nail from the wire coil. Insert two nails into the clay, several centimeters apart. Tilt the heads of the nails toward each other. Attach one of the straight ends of the nichrome wire to the upper portion of one of the nails. Attach the other straight end of the nichrome wire to the other nail. Using a pair of scissors or wire strippers, remove about 2 cm of insulation from both ends of two pieces of wire. Attach the copper portion of one of the wires to the base of one of the nails. Attach the other wire to the base of the second nail. Secure the free ends of each wire to the terminals of the battery. Caution: Do not touch the nichrome wire when the circuit is closed. Invert the jar over the nichrome wire and nails. Secure the mouth of the jar in the clay so there are no gaps. Wait several minutes and record your observations. Does the nichrome wire have high or low resistance?

Multimeters are used to measure voltage, resistance, and current.

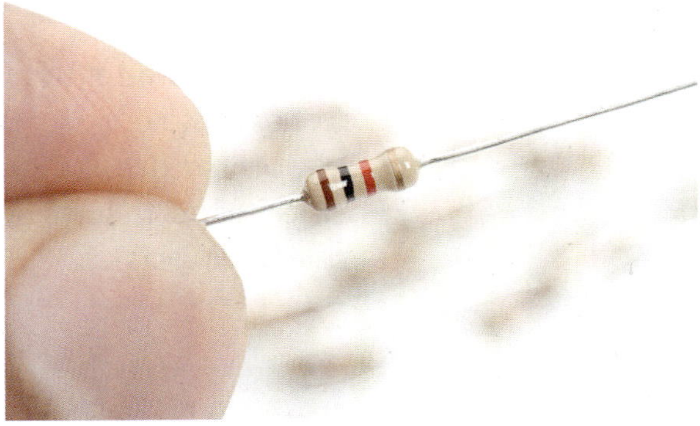

Different colored bands indicate the degree of resistance. The value of this resistor is 1 kΩ with ±5% tolerance.

higher the voltage, the more energy is released per charge—just as water from a higher hill above the waterwheel supplies greater energy. The unit used to express voltage is the volt, V. An instrument called *a voltmeter* is used to measure voltage.

How does an electric current illuminate a light bulb or run an appliance? How is electrical energy converted into other forms of energy, such as heat or light? Many electronic applications work because of resistance. For example, there is a thin strip of wire called *a filament* inside incandescent light bulbs. When an electric current flows through the filament, the metal resists the flow of electrons. This resistance converts some of the electrical energy to heat and light. These secondary forms of energy cause the filament to glow, and light is produced as a result. The opposition to the flow of an electric charge is called **resistance**, R. The ohm is the unit used to express resistance and is represented by the symbol Ω.

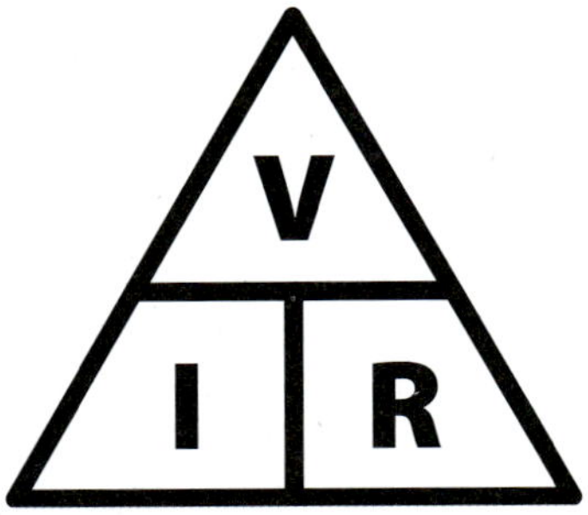

The mathematical operation is revealed when you cover the unknown variable. For example, to calculate voltage, cover the *V* and the illustration shows to multiply the current (*I*) and resistance (*R*). Covering the *I* indicates that voltage (*V*) needs to be divided by resistance (*R*).

	Unit	Symbol
voltage	volt	V
current	ampere or amp or A	I
resistance	ohm or Ω	R

A wire's resistance depends on several factors: composition, thickness, length, and temperature. Conductors have much less resistance and are more efficient than insulators. Thin wires have little room for electrons to flow and therefore offer greater resistance. A thin garden hose has little room for water to flow

FYI

Electric Eels

There exists a fish that inhabits the Amazon and Orinoco River basins in South America that uses electricity to stun its prey—the electric eel. It is long and scaleless, and it can grow up to 2.75 m in length and weigh up to 22 kg. The electric eel is not actually an eel. It belongs to the same family as piranhas and neon tetras. The electric eel is slow-moving and prefers to exist in muddy or stagnant water. Scientists believe that this fish uses electric shock when capturing prey to protect its sensitive mouth, to keep prey from escaping, or to cause the hiding prey to reveal its location by twitching. When the electric current moves through the prey's body, it is shocked long enough for the eel to swallow the prey whole. Three organs in the electric eel generate high- and low-voltage pulses to create the electric field between its head, which acts as the positive pole, and its tail, which acts as the negative pole. It can discharge 300–650 V, which is powerful enough to jolt a human.

LAWS & PRINCIPLES

Ohm's Law

A current of 5 amps flows through a 40 W incandescent light bulb and the voltage difference is 30 V. What is the resistance of the filament?

Known data: current (I) = 5 amps
voltage (V) = 30 V

Unknown variable: resistance (R) = ?

Formula: $\mathbf{I = V \div R}$

Rearrange formula to solve for unknown variable: $R = V \div I$

Solve for unknown variable: $R = 30\text{ V} \div 5\text{ A} = 6$ ohms

and therefore offers greater resistance to the flow of water. Likewise, longer wires force electrons to travel greater distances, so they have a greater resistance to the flow. An increase in temperature usually causes the resistance of most materials to increase.

Current, voltage, and resistance are related. The electric current, I, flowing through a wire equals the voltage, V, of the current divided by the resistance, R. This is known as *Ohm's law* and is represented by the following formula: $I = V \div R$. The electric current is directly proportional to the voltage and inversely proportional to the resistance. If the voltage stays the same, a decrease in resistance causes an increase in the current. If the resistance remains constant, an increase in voltage causes an increase in the current.

LESSON REVIEW

1. What two properties determine an electric current?
2. How does a high voltage affect the energy in a current?
3. Describe how resistance is affected by the thickness and length of a wire.
4. If resistance remained the same, how would the voltage be affected if the amount of current decreased?
5. A current of 5.3 amps runs through a circuit with a resistance of 40 ohms. What is the voltage?

CAREER

Electrician

Electricians put their knowledge of electricity to practical use so people can benefit from artificial light, power, refrigeration, and air conditioning. They install and maintain electrical systems in houses, factories, offices, and other buildings. Electricians read blueprints, which are technical diagrams of electrical systems, that indicate the location of circuits, outlets, load centers, panel boards, and other equipment. They may wire a building's electrical system or they may install telephone and computer wiring, security systems, or other electrical equipment. Electricians also connect motors to electrical power and install electronic controls. They regularly inspect equipment to correct hazards and prevent problems or repair electrical equipment. Electricians often work with complex electronic devices such as circuit breakers, fuses, switches, and wiring.

7.1.6 Electric Power and Safety

OBJECTIVES

- Calculate electric power and electrical energy.
- Describe how fuses and circuit breakers work.

VOCABULARY

- **electric power** the rate at which electrical energy performs work

Power is defined as the rate at which work is done or energy is used. This definition can be applied to electric power as well. **Electric power** is the rate at which electrical energy performs work. The electrical energy used to power household appliances is converted into some other form of energy, such as light, heat, sound, or the movement of parts (kinetic energy).

The power in a direct current is calculated by multiplying voltage by current. The formula is: $P = V \times I$. A watt, W, is the unit of measurement used to express electric power. One watt of power is delivered to a light bulb when one ampere of current flows through a circuit that has a voltage of one volt. A watt is a joule of energy per second. As an example, consider a television that uses 2 amps when connected to a voltage of 120 V. The amount of electric power used can be determined by multiplying the voltage by the current. $P = 120\text{ V} \times 2\text{ A}$; $P = 240\text{ W}$.

One watt is a relatively small unit of electric power. A light bulb, for example, may use 100 watts of power. Kilowatts, kW, are used to describe larger quantities of power. Different appliances have different power ratings. Appliances with higher ratings require more electric power to operate. Running a 300 watt television for two hours uses 600 watt-hours of energy.

The electric meter in homes measures the total amount of electrical energy used by all the lights, computers, and other appliances. The total electrical energy used depends on the total amount of power needed by all these things and the total amount of time used. Electrical energy usage is calculated by using the formula $E = P \times t$ (electrical energy = power × time).

SCIENCE IN ACTION

Time to Unplug

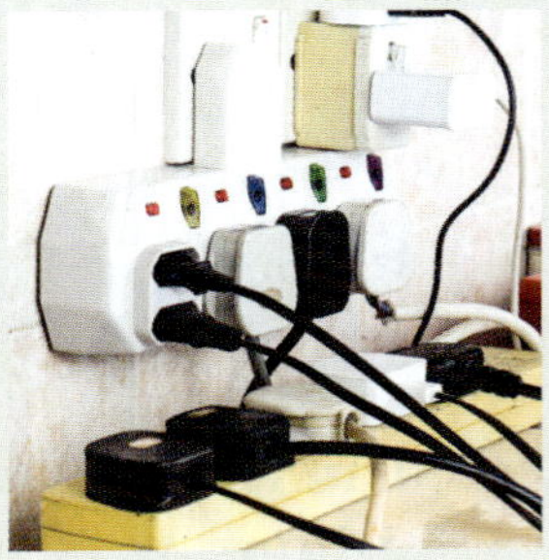

Electricity is a good thing. Sometimes there can be too much of a good thing. This applies to overloaded power outlets. First, the phone charger is plugged into the outlet. Then, the television is added followed by 10 strings of Christmas lights and a space heater. This is a fire danger. It is estimated that an average of 5,300 American homes experience fires from overloaded outlets annually. Outlets can handle a certain amount of current. If that limit is exceeded, fuses or circuit breakers are used to prevent dangerous situations. But, they are not always 100% reliable. To prevent electrical fires, unplug some of those devices, check for faulty wiring, and invest in quality electric products.

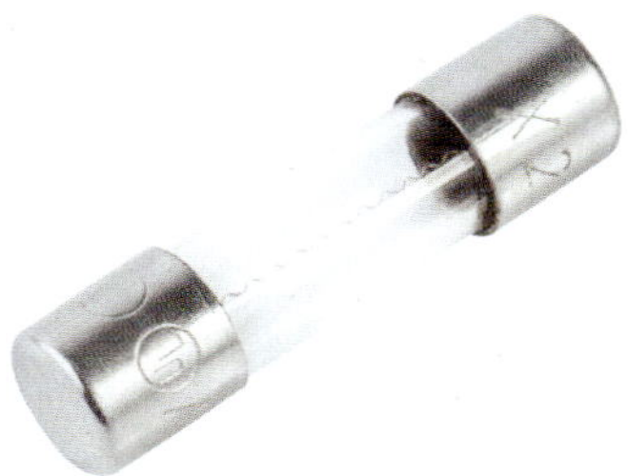

Electrical fuse

LAWS & PRINCIPLES

Electric Power

An electric stove is plugged into an outlet where the voltage difference is 240 V. How much power does the electric stove use if the current is 20 A?

Known data:
voltage (V) = 240 V
current (I) = 20 A

Unknown variable:
power (P) = ?

Formula:

$$P = V \times I$$

Solve for unknown variable:
$P = 240 \text{ V} \times 20 \text{ A} = 4{,}800 \text{ W}$

Electrical Energy

If a refrigerator uses 10 kWh of energy over a period of 14 hours, what is the power rating of the refrigerator?

Known data:
electrical energy (E) = 10 kWh
time (t) = 14 hours

Unknown variable:
power (P) = ?

Formula:

$$E = P \times t$$

Rearrange to solve for unknown variable:
$P = E \div t$

Solve for unknown variable:
$P = 10 \text{ kWh} \div 14 \text{ hours} = 0.71 \text{ kW}$

For small amounts of energy, the unit watt-second, or joule, is used. Kilowatt hours, kWh, refer to larger amounts of electric energy, such as the amount of electricity used by a household in one month. A kilowatt hour is equal to 1,000 watts of electric power used in one hour. If ten 100-watt light bulbs are lit for one hour, they consume one kilowatt hour of electric energy. Utility companies charge for electricity per kilowatt hour.

Fuses and circuit breakers protect against overloaded circuits and possible fires. A fuse is a strip of metal on an electric circuit that melts when the flow of current is too high. When the metal melts, it opens the circuit and stops the flow of electric current. A fuse acts as an emergency switch. A circuit breaker is similar to a fuse. A circuit breaker is an automatic switch that flips open when the flow of electric current becomes too high, creating an open circuit. Once the flow of energy has been reduced, a circuit breaker can be easily reset and used again. Because they are reusable, circuit breakers are easier to use than fuses.

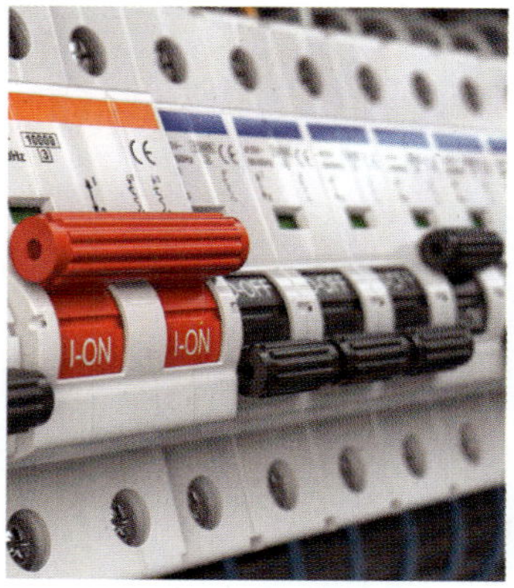

Circuit breakers

LESSON REVIEW

1. What types of energy can electrical energy be converted into?
2. How much electrical power does a television use if it uses 1.5 A when connected to a voltage of 120 V?
3. How much electrical energy does a hair dryer with a power rating of 1,000 W use in 2 minutes?
4. Describe how fuses and circuit breakers work.

7.2.1 *Magnets and Magnetic Fields*

OBJECTIVES

- Describe magnetism.
- Distinguish between a temporary magnet and a permanent magnet.
- Appraise the benefits of magnetism.

VOCABULARY

- **magnetic domain** a region in which the magnetic fields of individual atoms are lined up
- **magnetic field** the region around a magnet in which magnetic forces can act
- **magnetism** a force of attraction or repulsion of substances, especially those that contain iron, that results from the arrangement of their electrons

Imagine the year is 600 BC. You are walking through a field with your iron hoe over your shoulder, looking for a place to plant a crop. You begin to hoe the ground to break up the dirt. When you try to lift the hoe, the ground "pulls" back. Is the ground literally grabbing the hoe? With an extra-hefty yank, you manage to move the hoe. What just happened? History tells a similar tale about the discovery of a strange rock that attracted metal objects. The Greeks named this mysterious rock *magnetite*. They soon discovered that the rock had another mysterious quality—when it was cut into the shape of a bar and suspended from a string, it always pointed toward the northern star, Polaris. Although they did not know it then, they had discovered the first magnetic material.

Magnetite also became known as *lodestone*. Lodestones acted as the first compasses to help ancient people navigate at sea. Eventually, people found other valuable uses for magnets. The properties of magnets produce the spin in electric motors for powering transportation and in generators for creating electricity. Machines lift tons of scrap metal without actually grabbing the metal. In fact, the world welcomed a whole new industry that produced many things that are now taken for granted—vehicles, electric lights, curling irons, ovens, televisions, and computers. Without magnets, none of these things would be possible. The potential that God created in magnets is amazing!

What is a magnet and what causes magnetism? A magnet is a substance that exhibits a property called **magnetism**. Magnetism is a force of attraction or repulsion of substances, especially those that contain iron, that results from the arrangement of their electrons. Because magnetism depends on the arrangement of electrons, magnetism is related to electricity.

Magnetite

All magnets have certain properties—they have two magnetic poles, exert magnetic forces that are either attractive or repulsive, and are surrounded by a magnetic field.

The ability of a magnet to attract magnetic materials is strongest at the ends of the magnet, the magnetic poles, but the magnetic force extends along the whole length of the magnet. The region around a magnet in which magnetic forces can act is called the **magnetic field**.

Attractive and Repulsive Fields of Bar Magnets

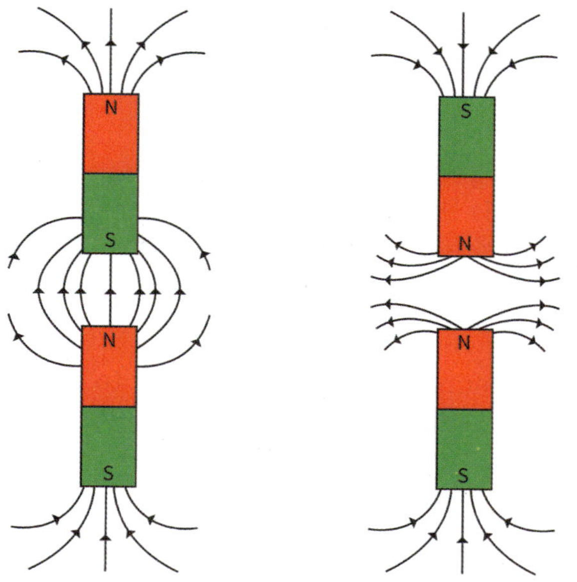

Attraction between opposite poles

Repulsion between like poles

> **FYI**
>
> **Electricity and Magnetism**
>
> *Similarities*
>
> - Electric charges give rise to electric forces; magnetic poles give rise to magnetic forces.
> - Both electric and magnetic forces can be attractive and repulsive.
> - The strength of the force depends on the distance between the objects. The closer together the objects, the stronger the force; the farther apart, the weaker the force.
>
> *Differences*
>
> - Positive and negative electric charges can exist independently.
> - North and south magnetic poles always occur together.

If a magnet is allowed to turn freely, one end will point north and the other south. The end that points north is called *the north pole* of the magnet; the end that points south is called *the south pole*. When the north pole of one magnet approaches the north pole of another magnet, the poles repel each other. When the north pole of one magnet approaches the south pole of another magnet, however, the magnets attract each other. The reason for this goes back to the "opposites attract" rule. Similar to electric charges, like poles of two magnets repel each other and the opposite poles attract each other.

What gives a magnet two poles and a magnetic field? The atoms of magnetic materials act like tiny, individual magnets. These

Iron filings collect along the magnetic field lines of a bar magnet (below left) and a horseshoe magnet (below right).

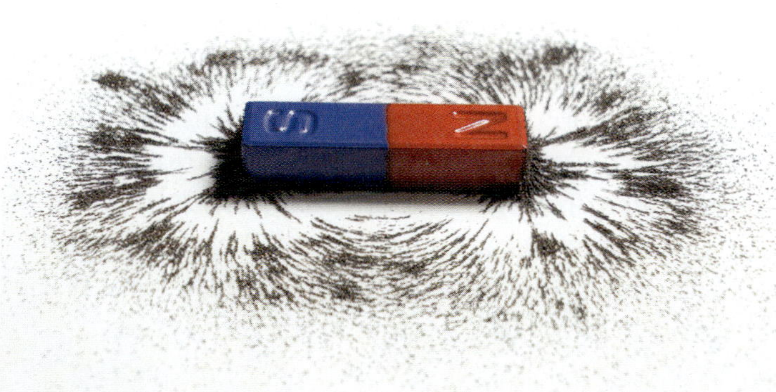

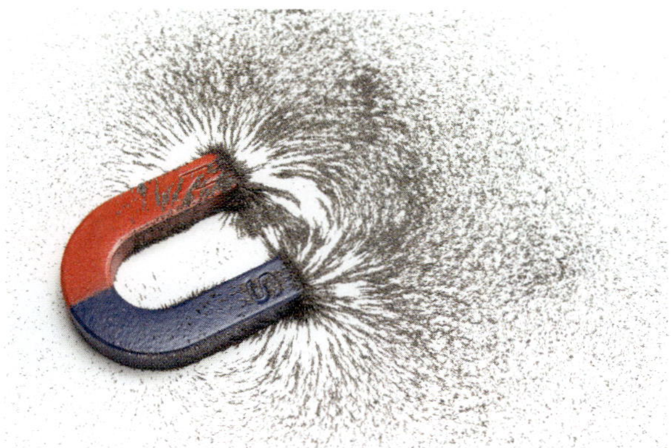

TRY THIS

Magnetic Fields

Sprinkle iron filings on a piece of thick paper or cardboard. Lower the paper over a bar magnet. Gently shake the paper a couple times. The filings align themselves into the pattern of the magnet's magnetic field. More filings collect where the field is the strongest. Illustrate the magnetic field and label the areas of strongest and weakest strength.

tiny magnets, or atoms, attract other atoms to line up so their north poles face the same direction. Within a magnet, a region where the magnetic fields of individual atoms are lined up is called a **magnetic domain**. When the atoms in a material line up, the combined magnetic field makes the material a magnet. In most materials, the magnetic domains face many directions and cancel each other out. But in magnets the domains face the same direction. The combined forces of the lined-up atoms give a magnet its magnetic force.

Magnetic domains do not line up in an unmagnetized piece of iron, but the iron can be magnetized by lining up these domains. Stroking the iron with a magnet or just leaving the iron close to it for a time brings the domains into alignment. The forces of the magnet rearrange the domains, and the iron becomes a magnet, but only temporarily. Once the magnetic domains return to their random positions, the iron is no longer a magnet. Soft iron is easy to magnetize, but it also loses its magnetism quickly. Magnets made of materials unable to retain their magnetic properties are called *temporary magnets*.

An iron nail is a temporary magnet. The magnetic domains are not aligned. When it is placed near a permanent magnet, the magnetic domains align, and it becomes a magnet.

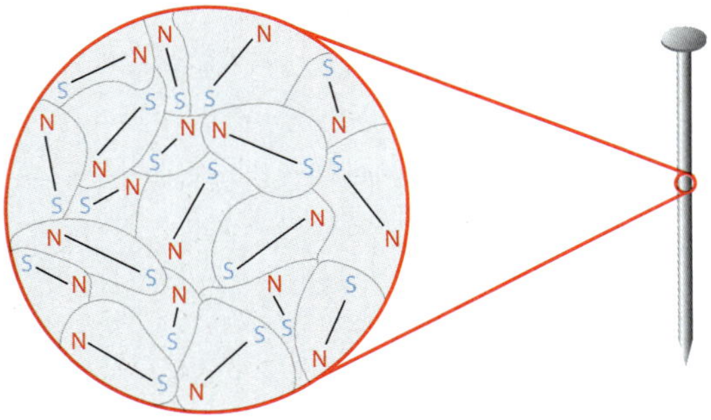

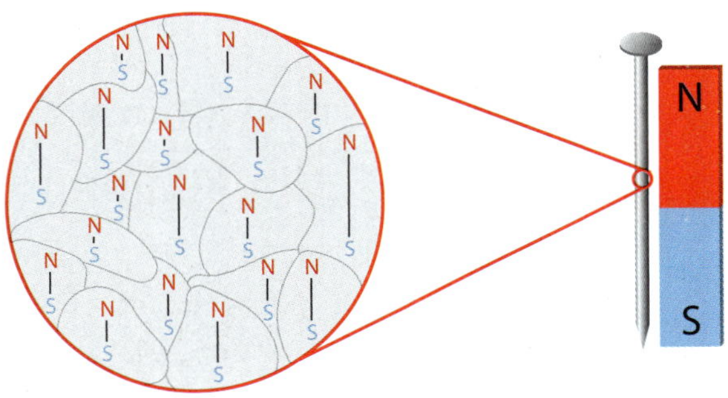

Other materials form permanent magnets; these materials stay magnetized. Strong permanent magnets have strong attractions between their poles, so the domains stay lined up. Permanent magnets are made from the elements iron, nickel, and cobalt or from alloys of aluminum, nickel, iron, and cobalt. But even the domains of strong, permanent magnets can be pushed out of line, demagnetizing the magnet. A magnet can be demagnetized by heating it, hammering it, or bringing it close to another strong magnet.

Magnetosphere

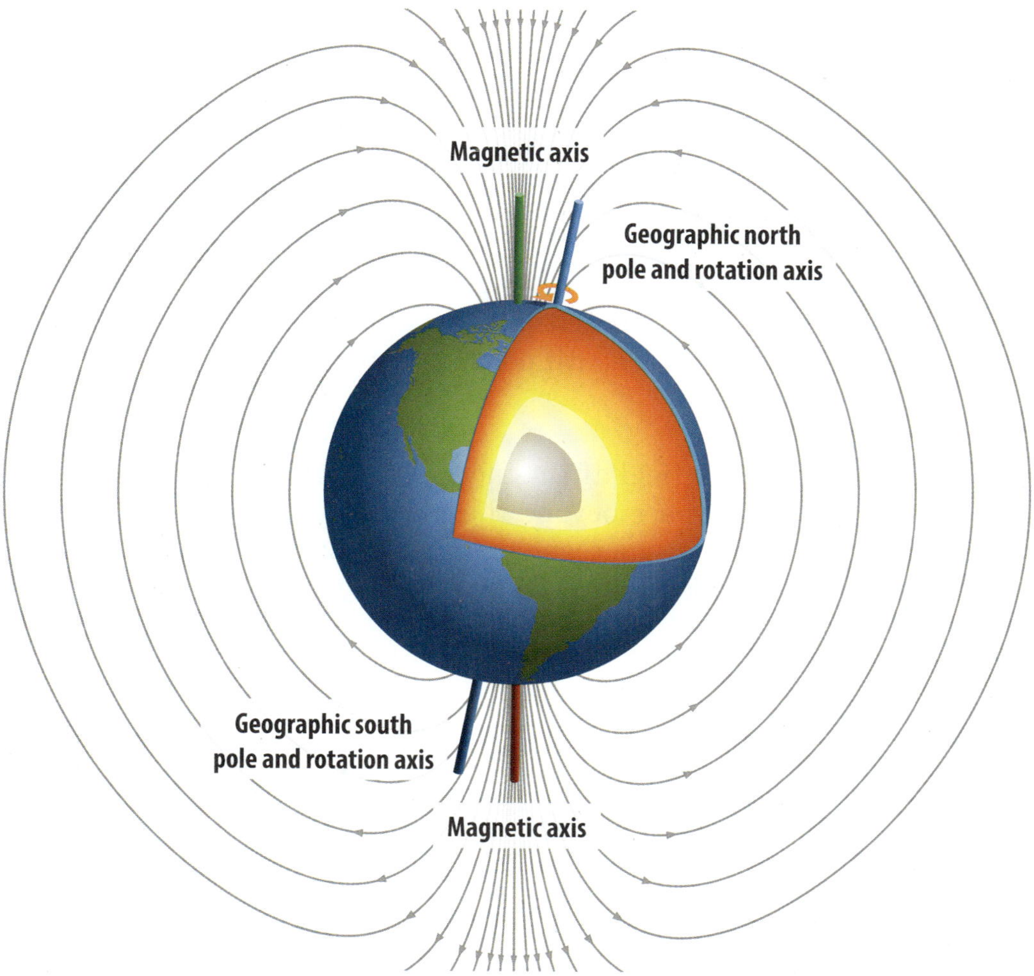

FYI

Ferrofluid

Ferrofluid is a combination of particles of iron oxide, oil, and a surfactant, which prevents pieces of iron oxide from clumping together. When ferrofluid is exposed to a strong magnet, beautiful, symmetrical structures are formed. What prompted the creation of ferrofluid? In the 1960s, NASA researchers were looking for a way to control liquid fuel in a low-gravity environment. They experimented with adding nanoparticles of magnetic substances to oil. Today, ferrofluids form a tight seal in the hydraulic machinery used to manufacture semiconductors. The ferrofluid increases the power of the machinery and the sealing power prevents contamination. Ferrofluids are also used in speakers and hard drives to prevent excess vibrations.

TRY THIS

Magnetic Domain

Hold an iron nail above some small paper clips. Are the paper clips attracted to the nail? Move the nail away from the paper clips and then stroke the nail 50 times in the same direction with a permanent bar magnet. Place the nail above the paper clips. What happens? Remove the nail from the paper clips. Stroke the nail 50 times in the opposite direction with the magnet. Hold it above the paper clips again. What happens? What happens to the nail when the magnet strokes it? What does changing the direction of the motion of the magnet do?

FYI

Built-in Compasses

A wide variety of living creatures use Earth's magnetic field to navigate.

- Honeybees depend on magnetic fields to navigate to their hives and to lead each other to food.
- Magnetic bacteria align themselves along the earth's magnetic field. These bacteria normally swim downward, seeking the soft mud they use as a habitat; but when magnets are used to cancel out the earth's magnetic field, they swim upward. Magnetic bacteria in the southern hemisphere, where the vertical alignment of the earth's magnetic field is opposite to the northern hemisphere, swim upward when brought to the northern hemisphere.
- European robins caged during migration season align themselves to the magnetic field of the cage.

Imagine hiking in the woods with your friends. Nobody knows how to get back to the main road, and you suspect you may be walking in circles. What can you do? You can take out your compass and align it to the north. This will help you walk in a straight line—east, for example, until you are out of the woods. Why does this work? Because the earth is a giant magnet. The earth's magnetic field exerts a force on the magnet's needle, just as it did on the lodestones of early sailors. Compass needles do not point exactly to the earth's geographic poles (the earth's rotation axis); they point to the magnetic poles. The magnetic north pole is in Canada, more than 500 km from the geographic north pole. The magnetic south pole is located near the coastline of Antarctica.

The earth's magnetic field affects not only the atmosphere but also a region in space beyond the atmosphere called *the magnetosphere*. The magnetosphere plays an important role in protecting life on Earth by deflecting dangerous charged particles from the sun.

LESSON REVIEW

1. What is a magnet? What is magnetism?
2. List the properties of magnets.
3. What is a magnetic field? Where is the magnetic field strongest?
4. Why will stroking an iron object with a magnet in one direction magnetize the object? What happens if the direction of motion is random?
5. What is the difference between a temporary magnet and a permanent magnet?
6. What benefits does magnetism provide?

Magnetism to Electricity 7.2.2

In 1820, Danish teacher Hans Christian Oersted was giving a public demonstration lecture when something surprised him. He noticed that whenever he brought his compass close to a live electrical wire, the needle of the compass jumped. The needle moved 90° when the electric current was turned on, but the needle did not move at all when the current was turned off. With his curiosity piqued, Oersted reversed the direction of electric current to see if he could repeat the event. To his surprise, the needle turned 90° in the opposite direction. What did this mean?

It meant that electric currents produce magnetic fields. A looped or coiled wire carrying electric current acts like a magnet. The strength of the magnetic field increases with the amount of current that is passing through the wire and with the number of coils that the wire has. The relationship between electricity and magnetism is known as **electromagnetism**.

If a bar of iron is wrapped with a coil of wire carrying an electric current, the coil's magnetic field magnetizes the iron bar. Such a magnet is called an **electromagnet**. The magnetic fields of the coil and iron combine to produce a strong magnetic field. When the current is turned off, the coil and iron lose their magnetism. This makes electromagnets useful for picking up large metal objects in one area and releasing them in another. The strength of an electromagnet can be increased simply by adding more coils of wire around the iron core or increasing the current.

Electromagnets have many uses besides picking up and dropping scrap metal. They can, for example, be used with permanent magnets to produce movement. Electromagnets were used in telegraphs, and today they are used in doorbells, washing machines, motors, and telephones.

OBJECTIVES

- Explain how to create an electromagnet.
- List uses of electromagnets.
- Identify principles of electromagnetic induction.
- Describe how a generator produces electricity.

VOCABULARY

- **electromagnet** a bar of iron wrapped with a coil of wire carrying an electric current
- **electromagnetic induction** the process by which electric current is produced by a changing magnetic field
- **electromagnetism** the relationship between electricity and magnetism
- **generator** a machine that converts mechanical energy into electrical energy

Electromagnet used in a junk yard

Hans Christian Oersted

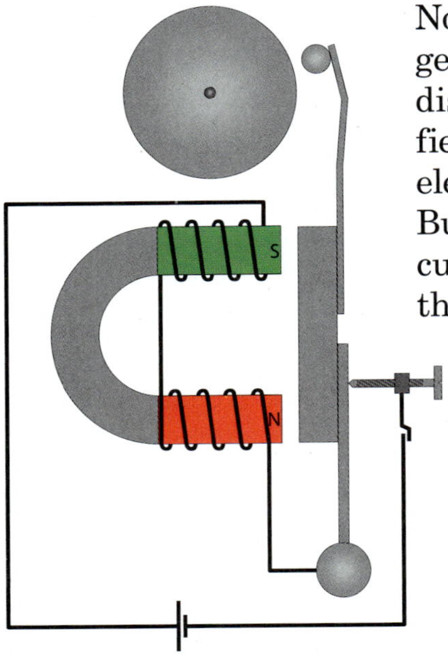

The bell rings when the electric circuit is made and the electromagnet is magnetized.

Not only can electricity create magnetism, magnetism can generate electricity. In 1831, English scientist Michael Faraday discovered that the motion of an electric wire through a magnetic field could generate electricity. Faraday tried to create an electric current in a wire by connecting it to an electromagnet. But no matter how strong the electromagnet was, no electric current was produced. As Faraday experimented, he noticed that the instant he connected the wires of the electromagnet to the battery, the galvanometer (an instrument that measures small electric currents) indicated that electric current was present. The same thing happened the instant he disconnected the wires. As long as the wires were connected, no electric current was present. Faraday concluded that electric current in the second wire was produced only when the magnetic field was changing, in this case being turned on and off.

The process by which electric current is produced by a changing magnetic field is called **electromagnetic induction**. Faraday did more experiments and discovered the following principles of electromagnetic induction:

1. An electric current is produced when a magnet is moved through a coil of wire in a closed circuit or when a wire is moved through a magnetic field.

BIOGRAPHY

Michael Faraday

Michael Faraday (1791–1867) opened the door to our modern, technological world. Born to a London blacksmith, Faraday had only an elementary school education when he became an apprentice to a bookbinder at the age of 14. His employer allowed him to read the books in his shop. Through his reading, Michael Faraday educated himself and became particularly interested in chemistry and electricity. He eventually obtained a position in the Royal Institute, an organization designed to give the general public a knowledge of valuable scientific inventions. In 1825, Faraday became the director of the institute.

As Faraday continued his career in science and research, he was offered the presidency of the Royal Institute and a position as professor of chemistry at the University in London. He preferred a modest life of research and involvement in his church where he was an elder for more than 20 years. When he retired from the Royal Institute, he thanked those he had worked with, and he thanked God for the all the gifts he had been given.

Michael Faraday was one of the greatest experimentalists. One of his discoveries was that a moving magnet could create an electric current in a coil of wire. This was the first generator; Faraday also invented the electric meter. This and other experiments provided the basics for the theory of electromagnetism.

2. The faster the magnet or wire is moved, the greater the electric current.
3. More electric current is produced if more loops of wire are added to the coil.

The most important application of Faraday's discovery is the **generator**, a machine that converts mechanical energy into electrical energy. Generators perform the opposite task of electric motors, which convert electrical energy into mechanical energy. A simple generator can be made with a horseshoe magnet and an insulated loop of wire. The wire is rotated between the two poles of the magnet. As the wire rotates, it first moves parallel to the magnetic lines of force. As the wire rotates further, it interrupts the lines of force, creating a current. Still further rotation cuts the lines of magnetic force again, but from the opposite direction. During one complete rotation of the wire, an alternating current (AC) is produced because the direction of the current reverses with each rotation of the wire. The generator can then act as a source of electrical energy.

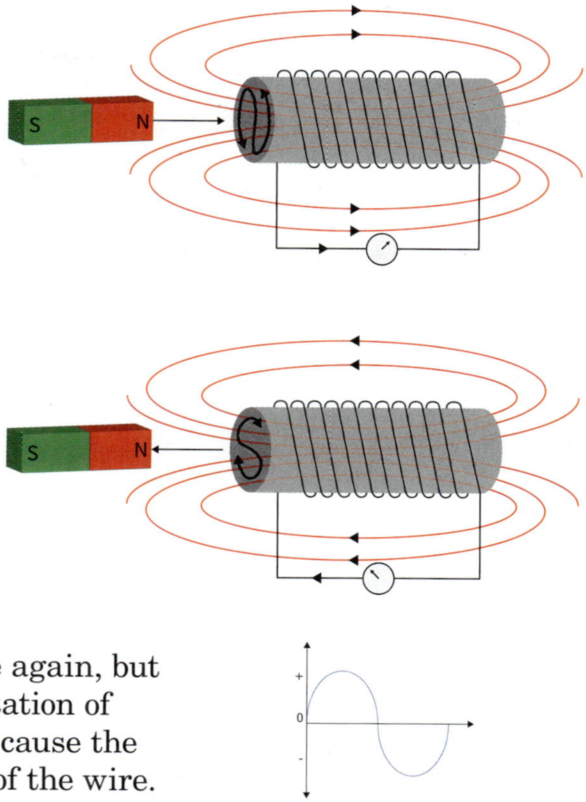

Electromagnetic Induction

Large generators located at power plants supply electricity for thousands of people. Turbines act as the mechanical energy

Simple Generator

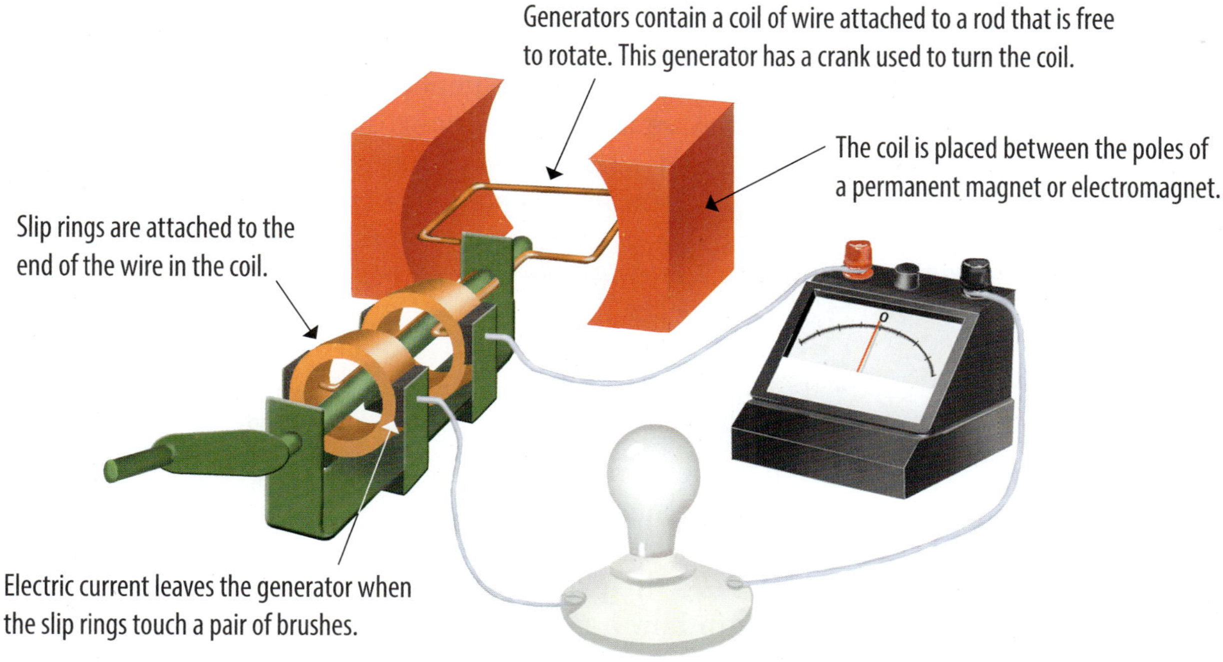

HISTORY

Wireless Communications

In 1888, Heinrich Hertz discovered electromagnetic waves (also called radio waves). In the 1890s, Nikola Tesla developed a device that could send signals through the air without wires; he had invented radio—the "wireless." In 1901, Guglielmo Marconi sent the first long-distance radio message. The message was sent from England across the Atlantic Ocean to Newfoundland. (The message was the letter S in Morse Code.) With this invention, ships could send messages to shore. The "wireless" made many successful sea rescues possible.

Heinrich Hertz

The first radio broadcast was sent by Canadian Reginald Aubrey Fessenden. Improving on Marconi's ideas, he devised a way to transmit voice and music. On Christmas Eve of 1906, Fessenden startled wireless operators on ships in the Atlantic Ocean. Instead of hearing the usual series of dots and dashes, they heard Christmas music and Bible verses. The world's first radio program ended with the words *Merry Christmas*.

source for power plant generators. The turbines are turned by running water released from a dam at a hydroelectric power plant, by wind power, or by steam generated from water that is heated by burning coal, oil, or nuclear fission. As the turbine rotates, it moves a coiled wire inside the generator through the magnetic lines of force of a powerful magnet. This movement of a conductor through a magnetic field generates electrical energy.

Large generators in power plants have many loops of wire rotating inside giant electromagnets. In the United States and Canada, the speed of the generators is carefully controlled to create a current that reverses direction exactly 120 times a second. A hertz (Hz) is a unit equal to one cycle per second. Two direction reversals complete one cycle; therefore, alternating current has a frequency of 60 Hz. Appliances are developed to be compatible with 60 Hz frequency.

LESSON REVIEW

1. Describe the relationship between magnetism and electricity. How can an electromagnet be created?
2. What are some uses of electromagnets?
3. How can the strength of an electromagnet be increased?
4. What is electromagnetic induction?
5. What are the principles of electromagnetic induction?
6. Explain how a generator works.

Electric Motors and Electronics 7.2.3

Life would not be the same without electromagnetism. Its applications range from generators to electric guitars. One very important application of electromagnets is the electric motor. An **electric motor** is a device that uses an electromagnet to convert electrical energy into mechanical energy, or the energy of movement. For instance, you might use an electric motor on a boat to move the boat across the water. Or, instead of fanning yourself with a handheld fan, you might use an electric fan to cool off.

The traditional electric motor consists of a loop of wire that is mounted on an armature, which rotates. Carbon blocks, which are referred to as *brushes*, feed the current to metal rings. The metal rings transfer the energy to the loop. A magnet supplies the magnetic field around the loop. This magnetic field turns the loop when current flows through it. If the motor is an alternating current (AC) motor, the current is set to reverse direction when the loop's plane is perpendicular to the magnetic field and the magnetic force exerted on the loop is the greatest. The loop's momentum carries it around until the current is again supplied, resulting in continuous motion.

If a device requires direct current (DC), the metal rings are replaced with specialized parts called *split-ring commutators*. The split in the rings switches the direction of the current each half rotation so the shaft keeps moving in the same direction.

Both AC and DC motors convert electrical energy into mechanical energy, but each is built differently and uses a different power source. AC motors use alternating current, and

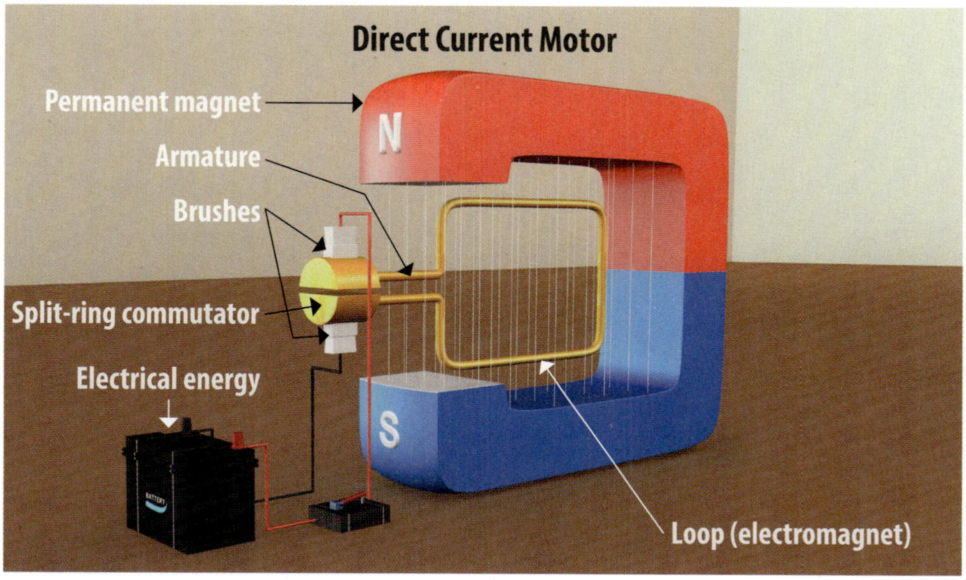

OBJECTIVES

- Explain how an electric motor works.
- Compare DC motors to AC motors.
- Describe the purpose of a transformer and how it works.
- Identify the uses of various electronic devices.

VOCABULARY

- **amplifier** an electronic device that increases the strength of an electric signal
- **electric motor** a device that uses an electromagnet to convert electrical energy into mechanical energy
- **electronics** the study and technology devoted to developing and improving devices that use a flow of electrons
- **transformer** a device that increases or decreases the voltage of alternating current

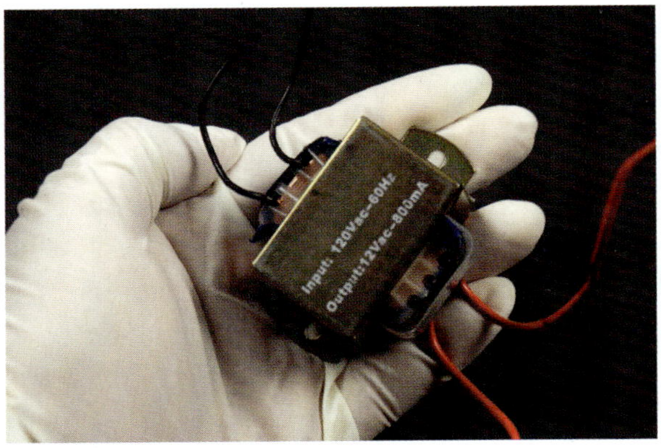

Transformer used in an electrical component

Transformer at a power substation

DC motors use direct current. AC motors can be powered by batteries whereas DC motors need a converter. AC motors have rings; DC motors have split-ring commutators. Varying the frequency changes the speed of an AC motor, and changing the current of the armature in a DC motor controls the motor's speed.

An AC motor is designed specifically for one application, and an AC motor is less expensive to make. Vehicles have AC devices, called *alternators*, which charge the vehicles' batteries. Vehicle manufacturers use very specific alternators for each model of vehicle. When an alternator needs to be replaced, it is imperative that the mechanic knows the car's year, make, model, and engine. Putting the wrong alternator in the car could be disastrous. Vehicles have batteries that produce DC current. The current from the alternator has to be converted to DC before being sent to the battery.

Another devastating situation would be to send hundreds of thousands of volts of electricity from a power station directly to a business or home. Power stations generate enormous amounts of electricity in order to provide power to entire communities. The problem is that the voltage they create is too high for a copier, a single light bulb, or a phone charger. One of the most important inventions for the widespread distribution of electricity is the transformer. A **transformer** increases or decreases the voltage of alternating

Step-Up Transformer

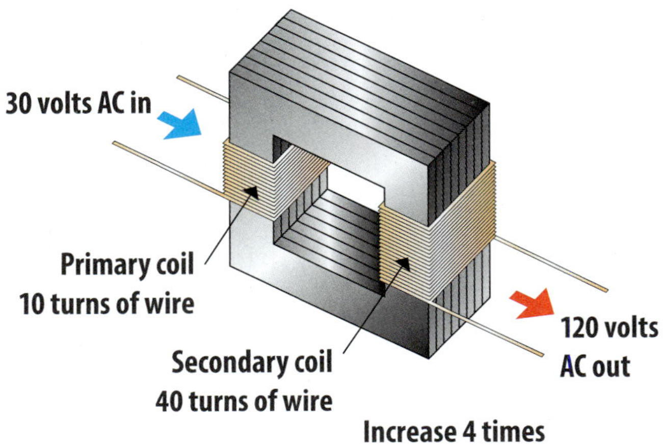

Step-Down Transformer

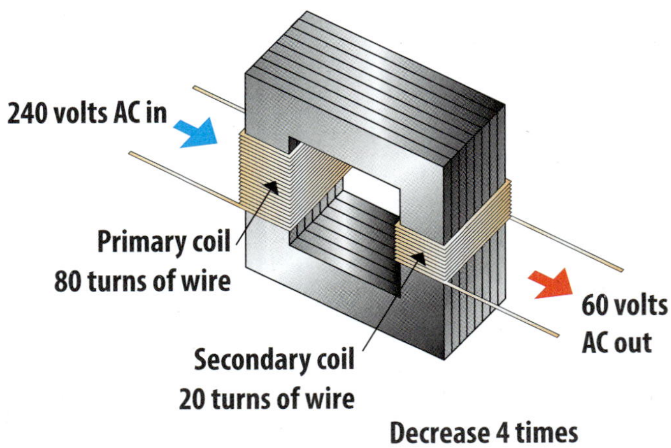

LAWS & PRINCIPLES

Transformers

In a transformer, the ratio of the output voltage to the input voltage equals the ratio of the turns in the secondary coil to the turns in the primary coil.

$$\frac{\text{output voltage}}{\text{input voltage}} = \frac{\text{turns in secondary coil}}{\text{turns in primary coil}}$$

For example: A transformer has 180 turns in its primary coil and 60 turns in its secondary coil. If the input voltage equals 120 V, what is the output voltage? Is this a step-up or a step-down transformer?

$$\frac{\text{output voltage}}{120\text{ V}} = \frac{60}{180} \qquad \frac{120\text{ V} \times 60}{180} = \frac{7{,}200\text{ V}}{180} = 40\text{ V}$$

This is a step-down transformer.

current. A transformer functions on the basis of the relationship between electricity and magnetism. The transformer has a primary side, or initial side, that creates a magnetic field; one wire is wound around this side of the iron core. The other side, or secondary side, of the transformer has another wire around it a different number of turns. The electromagnetic current from the primary wire induces a current in the second wire. Both currents alternate at the same frequency. The strength of the current in the secondary wire is either less than or greater than the current in the primary wire, depending on the relative number of turns in the coil.

Various transformers function differently. A step-down transformer has fewer turns in the secondary wire than in the primary wire. The ratio of turns from secondary to primary is directly proportional to the ratio of the secondary voltage to the primary voltage. Step-down transformers reduce the voltage created by a power plant so it is safe for home use. Step-down transformers are also used in doorbells, laptops, tablets, and cell phones. Step-up transformers, on the other hand, increase the voltage of the current. Suppose a television requires 24,000 V and the standard voltage supplied to the house is 120 V. A step-up transformer would increase the voltage from 120 V to 24,000 V in order to produce enough current to power the television. The turns in the coils of the transformer would have the same ratio as the change in voltage. In this example, the secondary coil would have 200 times as many turns as the first coil. Step-up transformers are also used in X-ray machines and fluorescent lights.

FYI

Current Trends

Most of today's electricity is supplied through AC, but computers, cell phones, solar cells, and electric cars require DC. Companies are now taking the long-distance benefits of AC and combining them with the stability of DC in a hybrid current called *high voltage direct current* (HVDC).

SCIENCE IN ACTION

Magnetic Nanostructures

Researchers at Uppsala University are studying the collective dynamics of a structure composed of interacting magnetic nano-islands. The scientists have manipulated the magnetic nano-islands to imitate nature and create 2-dimensional patterns. The goal is to use these structures instead of magnetic atoms because they can be tuned with more precision. These nanostructures can potentially improve measurement analysis, and they could be used to create magnetic logical circuits that would maintain a desired logical state without power.

The study and technology devoted to developing and improving devices that use a flow of electrons is called **electronics**. Electronics started more than 100 years ago when electricity was applied to lamps, heaters, and welding arcs. Today, scientists manipulate electrons to control electrical signals to carry messages, create colored pictures, and perform calculations.

Vacuum tubes can amplify electric current and convert AC to DC. But vacuum tubes are large and give off heat. Early radios, which used vacuum tubes, were bulky because the vacuum tubes not only took up a lot of room but also needed space to give off thermal energy.

Vacuum tubes are not used as much as they once were, but they are still used in some amplifiers. An **amplifier** is an electronic

BIOGRAPHY

Nikola Tesla

Croatian-American inventor Nikola Tesla (1856–1943) was the genius responsible for discovering the rotating magnetic field and making it useful with alternating current. He was at the forefront in research and design of dynamos, induction motors, radar technology, and X-ray technology.

After emigrating to the United States from Croatia, he worked with Thomas Edison. His time with Edison did not last long, and he soon left and opened his own laboratory. In 1887, he filed several patents for AC-based inventions.

Thomas Edison was working with DC power, which could be transmitted only about a mile. Tesla's alternating-current system could transmit electricity hundreds of miles. His system was used to light the Chicago World's Fair in 1891. Soon after, Tesla's designs were used in building the first hydroelectric power plant at Niagara Falls. In 1896, that power plant was used to power the city of Buffalo, New York.

This is a promotional photo stunt showing Tesla sitting in his laboratory in Colorado Springs, Colorado, while his transmitter was producing huge bolts of electricity. Even before Photoshop, photographers could take two separate pictures and put them together, giving a false representation of events. The first photo was taken when the electricity was moving through the air. The second photo was taken when Tesla was sitting in the room and the machine was off.

Tesla held 112 patents. Among his inventions were the induction motor, new types of generators and transformers, fluorescent lights, and a new kind of steam turbine. His Tesla coil is still used today and is the foundation of wireless technology.

In 2003, Tesla Motors was formed for the purpose of building a fully electric-powered car. In 2008, Tesla Motors introduced its first electric car, the Roadster. The more budget-friendly Model S rolled out in 2014. It set a world record in 2017, accelerating from 0–60 mph in 2.28 seconds. In that same year, the company's name was changed to Tesla, Inc. as the company broadened its scope to become an energy company.

device that increases the strength of an electric signal. Small electric currents entering the amplifier are converted into large currents leaving the amplifier. The larger output current makes a stronger signal. Amplifiers are used to strengthen weak sound or picture signals in wireless communications, broadcasting, and audio equipment.

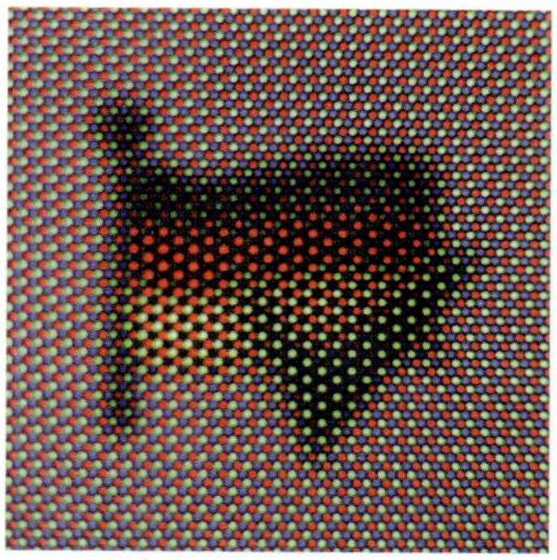
Close-up view of a CRT screen displaying a German flag

Another type of vacuum tube, called *a cathode ray tube* (CRT), creates picture images for older televisions, computer displays, video games, and radar devices. A sweeping beam of electrons on a fluorescent screen gives off visible light. The light's brightness depends on the number of electrons that strike a particular part of the screen. The electron beam moves rapidly both vertically and horizontally, creating patterns of light that come through as pictures. Television CRT beams create 30 new pictures per second. The human eye translates this as movement.

CRT monitors

Technology has changed drastically for televisions and computer screens. Gone are the days of CRTs. Now screens are designed to operate using LCDs (liquid crystal displays), plasma, or LEDs (light-emitting diodes). Some TVs are even 3-D. In 2009, all television signals went from analog to digital, and many people switched from their older CRT TVs to newer HD (high definition) flat screen TVs. Personal electronic devices are sure to keep changing rapidly.

LCD monitor

Electronic devices can send not only light waves but also sound waves. A device called *a transmitter* converts sound waves into patterns of electric waves. Alexander Graham Bell, the inventor of the telephone, used grains of carbon as a transmitter in his telephone's microphone. Today's telephones use semiconductor crystals as a transmitter. The receiver in a landline phone converts the electronic waves back into sound using an electromagnet, although semiconductor crystals are becoming the preferred material, rapidly replacing the electromagnets in receivers.

LESSON REVIEW

1. How do electric motors differ from generators?
2. How does an electric motor work?
3. What is the difference between an AC motor and a DC motor?
4. What is the purpose of a transformer and how does one work?
5. What are some functions of vacuum tubes?
6. Identify uses of electronic devices.

LED TV

7.2.4 *Superconductivity*

OBJECTIVES

- Compare conductors to superconductors.
- Identify advantages and disadvantages of superconductors.
- Explain how a maglev train works.

VOCABULARY

- **superconductivity** the perfect conductivity shown by certain substances at extremely low temperatures

Picture a train moving noiselessly at a high speed; think about traveling 300 km in less than an hour without having to board a plane. Scientists and engineers have imagined it and have succeeded in creating many of these magnificent trains. Currently, three countries (Japan, China, and South Korea) have noiseless, frictionless trains, thanks to the way God designed magnets. The key to this modern mode of transportation is superconducting magnets. But before superconducting trains could be built, the problem of cold had to be overcome.

What does cold have to do with it? Everything. Imagine the coldest possible temperature—a temperature so cold that it is not possible to get any colder. The coldest possible temperature in the universe is 0 Kelvin, or –273°C. This temperature is known as *absolute zero.* In 1911 as Heike Kamerlingh Onnes was researching how mercury conducted electricity at temperatures very near absolute zero, he noticed something strange. He noticed that mercury had no resistance to current flow at a temperature of 4.2 Kelvin. This observation led to the discovery of superconductivity. **Superconductivity** is the perfect conductivity shown by certain substances at extremely low temperatures.

A superconductor is a substance that conducts electricity without losing energy when it is cooled to a low enough temperature. Copper, the most common conductor, loses electric power as heat. Even though copper is a conductor, its atoms still have some resistance to the flow of electrons. In contrast, superconductors conduct electric current without wasting any energy. This makes the production and transmission of electrical energy cheaper and more efficient than when common conductors are used. Scientists have identified many metals that can act as superconductors. Several more elements can become superconductors in certain situations.

TRY THIS

Repulsion

Hold one strong bar magnet above another, aligning the two south poles. Let go. Try to set the top magnet on the bottom one. What happens? Use rulers to set up barriers along the sides of the magnet. How is this setup like a maglev train?

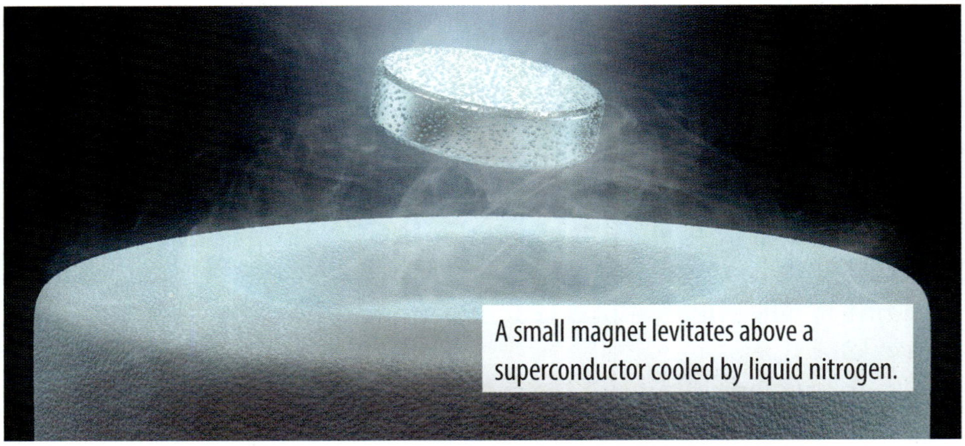

A small magnet levitates above a superconductor cooled by liquid nitrogen.

In 1986, Karl Alexander Müller and J. Georg Bednorz created a superconductor at a temperature of –243°C. Before that time, superconductors had been produced only at temperatures below –250°C. These early superconductors had to be kept cold with liquid helium, a very expensive substance.

Although helium is still sometimes used in superconductors, less expensive coolants are more often chosen. Currently, superconductors are being used in geological sensors that find new oil deposits and in magnetic resonance imaging (MRI) machines, which generate images of the body that doctors use to diagnose medical issues.

Semiconductors changed life through the use of the computer chip; superconductors may also change life dramatically in the near future. Physicists are experimenting with ways to produce superconductors at room temperature. If and when they succeed, a burst of new technologies is sure to follow.

The dream of creating superconducting (maglev) trains has become a reality. China began commercial service of the Shanghai Transrapid maglev train in 2004. Japan began operation of its Tobu Kyuryo Line in 2005, and South Korea opened a maglev train line in 2016. Maglev is short for magnetic levitation. A maglev train levitates, or lifts, off the tracks and uses superconductors as a power source. It lifts off the track, which is called *a guideway*, because the electromagnets on the guideway repel the magnets on the bottom of the train. Maglev trains run by electricity from nearby power plants. An alternating electric current is fed through power lines to wire coils along the guideway's walls. The current causes each coil to change its polarity (S to N, N to S) as each train magnet passes. When the magnets on the guideway and train line up S-N or N-S, they attract. This pulls the train forward. When the magnets line up S-S or N-N, they repel. This pushes the train forward. This pushing-and-pulling action causes the maglev train to speed along the guideway. To make the train accelerate or decelerate, the guideway engineers increase or decrease the amount of electric power to the coil in the guideway.

Because maglev trains speed along on a cushion of air, they are not slowed down by friction, a contact force that resists movement. The maglev trains are built of lightweight materials, and are bullet-shaped, which cuts down on air resistance. For these reasons, maglev trains have the potential to travel at 550 kph and beyond.

FYI

MRI Machine

An MRI machine uses extremely powerful magnets to create images of the inside of a person's body. The images may show tumors or other medical problems that might otherwise require surgery to find. Most of the magnets are superconducting magnets. The wire in the machine is immersed in liquid helium to keep it very cold. The wire is well-insulated in a vacuum, so the temperature of the wire does not affect the patient. The cold causes the resistance in the wire to drop to zero. As a result, less electricity is required, and the machine is much less costly to operate. But how does an MRI machine work? The machine's main magnet covers the patient with a stable and very intense magnetic field. Another part of the machine transmits radio waves into the patient's body. These and other parts work together to create images of the inside of the patient's body.

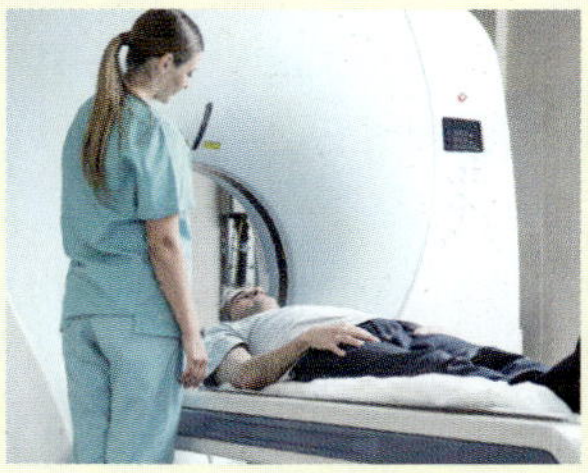

Magnetic Levitation

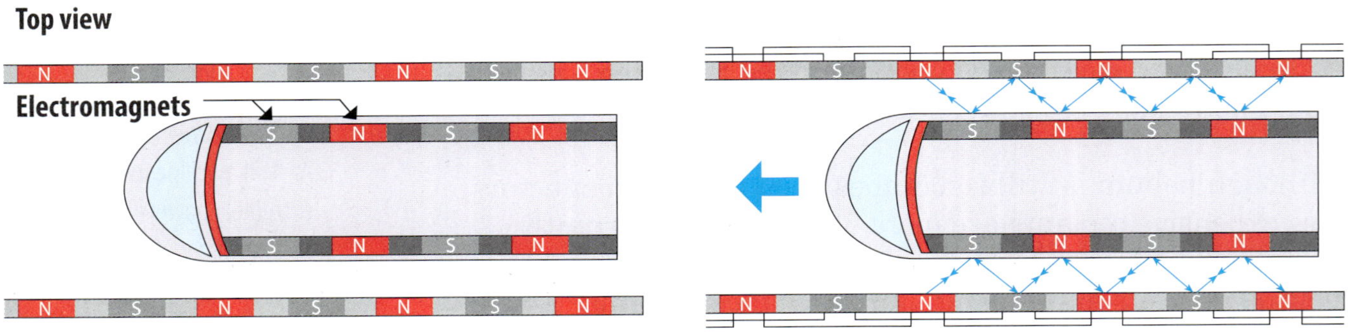

Electromagnets are under the train and on the guideway. The maglev train is pushed forward as the like poles attract and the unlike poles repel between the guideway and the train. The levitation occurs from the repulsion of like magnetic poles under the train.

In addition to maglev trains, some other benefits that superconductors provide include the following:

1. *Superconducting electric cables that carry electricity without losing energy.* Some of the energy transferred by today's electric cables is converted (or lost) to heat energy. Using superconducting cables would perhaps mean that energy costs would drop.
2. *Smaller and more efficient motors as a result of the use of superconducting wire.* These motors could be used in automobiles and ships.
3. *Generators that are lighter and produce more power with less fuel.* Replacing the iron cores of generators with superconducting wire would increase efficiency.
4. *Faster and smaller computers that use superconducting parts.*
5. *Lighter, faster, less energy-consuming parts for space vehicles.*

LESSON REVIEW

1. How are superconductors different from conductors?
2. What are the advantages and disadvantages of superconductors?
3. How are electromagnets used to propel a maglev train?
4. What causes the maglev train to change speed?

China's maglev train

Japan's test track for a future maglev train, scheduled to begin operation in 2027

Acceleration

$$a = \frac{v_f - v_i}{t_f - t_i}$$

(a = acceleration, v = velocity, t = time)

Actual Mechanical Advantage (AMA)

$$AMA = efficiency \times IMA$$

(IMA = ideal mechanical advantage)

Boyle's Law

$$P_1V_1 = P_2V_2$$

(P = pressure, V = volume)

Charles's Law

$$\frac{V_1}{T_1} = \frac{V_2}{T_2}$$

(V = volume, T = temperature)

Concentration

$$concentration = \frac{volume\ of\ solute}{volume\ of\ solution} \times 100\%$$

Density

$$D = m/V$$

(D = density, m = mass, V = volume)

Efficiency

$$efficiency = (work\ output/work\ input) \times 100\%$$

Electrical Energy

$$E = Pt$$

(E = electrical energy, P = electrical power, t = time)

Electrical Power

$$P = VI$$

(P = electrical power, V = voltage, I = current)

Force of Gravity

$$F = mg$$

(F = force, m = mass, g = acceleration of gravity = 9.8 m/sec^2)

Free Fall

$$d = \tfrac{1}{2}at^2$$

(d = distance, a = acceleration, t = time)

Gravitational Potential Energy

$$PE = mgh$$

(PE = gravitational potential energy, m = mass, g = acceleration of gravity = 9.8 m/sec^2, h = height)

Kinetic Energy

$$KE = \tfrac{1}{2}mv^2$$

(KE = kinetic energy, m = mass, v = velocity)

Mechanical Advantage (MA)

$$MA = resistance\ force/effort\ force$$

MA of an inclined plane

$$MA = length\ of\ the\ incline/height\ of\ the\ plane$$

MA of a lever

$$MA = effort\ arm\ length/resistance\ arm\ length$$

MA of a screw $MA = circumference/pitch$	**Pressure** $P = F/A$ (P = pressure, F = force, A = area)
MA of a wedge $MA = length\ of\ slope/height$	**Specific Heat** $Q = mc\Delta T$ (Q = heat added, m = mass, c = specific heat, ΔT = change in temperature)
MA of a wheel and axle $MA = wheel's\ diameter/axle's\ diameter$	**Speed** $s = d/t$ (s = speed, d = distance, t = time)
Momentum $p = mv$ (p = momentum, m = mass, v = velocity)	**Wave Period** $T = 1/f$ (T = wave period, f = frequency)
Newton's Second Law of Motion $F = ma$ (F = force, m = mass, a = acceleration)	**Wave Speed** $v = f\lambda$ (v = velocity, f = frequency, λ = wavelength)
Ohm's Law $I = V/R$ (I = current, V = voltage, R = resistance)	**Work** $W = Fd$ (W = work, F = force, d = distance)
Power $P = W/t$ (P = power, W = work, t = time)	**Work Output and Input** $work\ output = resistance\ force \times distance$ $work\ input = effort\ force \times distance$

Metric Unit Conversion Factors

Length

1 meter (m) = 1,000 millimeters (mm)
1 meter (m) = 100 centimeters (cm)
1,000 meters (m) = 1 kilometer (km)

Volume

1 liter (L) = 1,000 milliliters (mL) or
1,000 cubic centimeters (cm^3)

Mass

1 gram (g) = 1,000 milligrams (mg)
1,000 grams (g) = 1 kilogram (kg)

SI Base Units

Quantity	Metric Unit	Unit Symbol
electric current	ampere	A
electric resistance	ohm	Ω
length	meter	m
mass	kilogram	kg
thermodynamic temperature	kelvin	K
voltage	volt	V

Relationship Between Metric and English Units

1 inch = 2.54 cm	1 centimeter = 0.393 in.
1 foot = 0.3048 m	1 meter = 3.28 ft
1 yard = 0.9144 m	1 meter = 1.09361 yd
1 mile = 1.609 km	1 kilometer = 0.621371 mi
1 gallon = 3.78 L	1 liter = 0.264172 US gal

Density of Common Substances

Substance	Density (g/cm^3)
gasoline	0.7
ethanol	0.785
wood	0.85
water (solid)	0.92
water (liquid)	1.0
salt	2.2
aluminum	2.7
steel	7.8
iron	7.9
copper	9.0
silver	10.5
lead	11.3
mercury (liquid)	13.5
gold	19.3

Melting/Freezing and Boiling Points of Common Substances

Substance	Melting Point/ Freezing Point (°C)	Boiling Point (°C)
hydrogen	–259	–253
oxygen	–218	–183
mercury	–78	–33
water	0	100
salt	801	357
copper	1,085	2,562
iron	1,538	2,861

Common Elements with Multiple Oxidation Numbers

Element Name	Oxidation Numbers
Chromium, Cr	2+, 3+
Copper, Cu	1+, 2+
Iron, Fe	2+, 3+
Lead, Pb	2+, 4+
Tin, Sn	2+, 4+

Covalent Greek Prefixes

Number	Prefix
1	mono-
2	di-
3	tri-
4	tetra-
5	penta-
6	hexa-
7	hepta-
8	octa-
9	nona-
10	deca-

General Group Oxidation Numbers

Group Number	Oxidation Number
IA	1+
IIA	2+
IIIA	3+
IVA	4+
VA	3–
VIA	2–
VIIA	1–
VIIIA	0

Common Polyatomic Ions

Ion Name	Ion Formula	Ion Charge
Ammonium ion	NH_4^+	1+
Acetate ion	$C_2H_3O_2^-$	1–
Carbonate ion	CO_3^{2-}	2–
Chlorate ion	ClO_3^-	1–
Chlorite ion	ClO_2^-	1–
Cyanide ion	CN^-	1–
Hydroxide ion	OH^-	1–
Hypochlorite ion	ClO^-	1–
Nitrate ion	NO_3^-	1–
Nitrite ion	NO_2^-	1–
Perchlorate ion	ClO_4^-	1–
Permanganate ion	MnO_4^-	1–
Phosphate ion	PO_4^{3-}	3–
Sulfate ion	SO_4^{2-}	2–
Sulfite ion	SO_3^{2-}	2–

Solubility of Solutes at Different Temperatures

Solute	Formula	0°C	60°C	100°C
potassium nitrate	KNO_3	13.9	106	245
cerium sulfate	$Ce_2(SO_4)_3$	20.8	3.87	n/a
table salt	NaCl	35.7	37.1	39.2
potassium bromide	KBr	53.6	86	104
sodium chlorate	$NaClO_3$	79.6	135	204
table sugar	$C_{12}H_{22}O_{11}$	179	287	487

Metal Activity Series

Most reactive
Li
Rb
K
Ca
Na
Mg
Al
Mn
Zn
Fe
Ni
Sn
Pb
Cu
Ag
Pt
Au
Least reactive

Specific Heat of Common Substances

Substance	Specific Heat (cal/g°C)	Specific Heat (J/g°C)
gold	0.0301	0.126
lead	0.0305	0.128
mercury	0.033	0.14
silver	0.06	0.23
copper	0.09	0.39
zinc	0.09	0.39
aluminum	0.22	0.90
ice (−10°C)	0.49	2.05
alcohol (ethyl)	0.58	2.4
water	1.00	4.18

(F) *refers to the pages in the very front of the book.*

A

absorption the conversion of a wave's energy into another form of energy 6.1.3

acceleration the rate of change in velocity over time 4.1.2

acid a compound that releases hydrogen ions when it is dissolved in water 3.3.1

acoustics the scientific study of sound 6.2.3

activation energy the minimum amount of energy that is required to start a chemical reaction 3.2.4

adhesion the attractive force between unlike substances that holds them together 1.2.4

alkali metal the elements in group IA on the periodic table, except for hydrogen 2.1.3

alkaline-earth metal the elements in group IIA on the periodic table 2.1.4

alkane a saturated hydrocarbon that has single bonds between the carbon atoms 2.3.1

alkene an unsaturated hydrocarbon with at least one pair of carbons joined by a double covalent bond 2.3.1

alkyne an unsaturated hydrocarbon with at least one pair of carbons joined by a triple covalent bond 2.3.1

alloy a solid metallic solution or compound made of a metal and one or more elements 2.4.1

alpha particle a particle with two protons and two neutrons released during alpha decay 5.2.2

alternating current an electric current in which the electrons alternate from one direction to another at regular time intervals 7.1.4

amalgam an alloy that contains mercury 2.4.1

amorphous solid a solid with a random particle arrangement 1.2.3

amplifier an electronic device that increases the strength of an electric signal 7.2.3

amplitude the maximum distance a wave vibrates from its rest position 6.1.2

antinode the place where constructive interference causes the largest wave 6.1.3

aqueous solution a solution in which water is the solvent 3.1.2

atom the smallest unit into which an element can be divided and still maintain its properties 1.1.2, 1.3.1

atomic number the number of protons in one atom of an element 1.3.2

atomic weight the weighted average mass of the isotopes of an element 1.3.2

B

base a compound that releases hydroxide ions when it is dissolved in water 3.3.2

beta particle an electron or a positron released from a nucleus during beta decay 5.2.2

biogas a gas fuel derived from the decay of organic matter 2.3.2

biomass an energy source that is obtained from organic matter, such as crops or agricultural waste products, used for the production of fuel 2.3.2

boiling point elevation the raising of a solvent's boiling point by the addition of a solute 3.1.5

buoyant force the upward force exerted on objects in a fluid 4.1.6

C

calorie the amount of heat needed to raise the temperature of 1 g of water by 1°C 5.1.2

carbon dating a method used to determine the age of things of organic origin by measuring the radioactivity of their carbon-14 content 5.2.4

catalyst a substance that increases the reaction rate without being permanently changed 3.2.2

Celsius scale a temperature scale in which 0° represents the freezing point of water and 100° represents the boiling point of water (F)

centripetal acceleration the change in velocity of an object traveling in a circular path 4.1.4

centripetal force the force that continuously pulls objects traveling in a circular path toward the center of the circle 4.1.4

ceramic a nonmetal, inorganic substance that contains high proportions of silicon and oxygen 2.4.2

chemical bond a strong force of attraction that holds the atoms of a compound together 1.4.2

chemical change a change that occurs when a substance changes into an entirely different substance with unique properties 1.2.7

chemical equation a description using chemical symbols and formulas to show the reactants and products in a chemical reaction 3.2.3

chemical formula an expression that identifies the types and numbers of elements present in a compound 1.4.1

chemical reaction the process in which one or more substances are transformed into different substances 3.2.1

chemistry the study of how atoms, molecules, and chemicals interact with each other 1.1.1

chromatography a method of separating mixtures according to how the different components are absorbed 1.1.3

coefficient a whole number written in front of a chemical formula in a chemical equation 3.2.3

coherent light light that contains waves of identical amplitude and frequency that are in phase 6.4.4

cohesion the force of attraction between particles of the same kind 1.2.3

colloid a mixture in which the particles of a substance are dispersed throughout and are not large enough to settle 1.1.4

combustion the burning of substances 1.2.7

combustion reaction a reaction that occurs when a substance reacts with oxygen 3.2.5

composite a material made of two or more substances that are combined to produce materials with useful properties 2.4.4

compound a pure substance consisting of atoms of two or more elements that are chemically combined in a fixed proportion 1.1.2

compression a section in a longitudinal wave where the particles are tightly packed together 6.1.1

compressional force a force that results from pushing or squeezing an object 4.1.1

concave mirror a mirror that is curved inward 6.4.1

concentration a measure of the amount of solute dissolved in a certain amount of solvent 3.1.4

condensation the changing of a substance from a gas to a liquid 1.2.6

conduction the transfer of heat from one substance to another through direct contact of the substances' molecules 5.1.3; the movement of electrons through a material 7.1.1

conductor a substance that allows electricity to flow through it easily 7.1.2

control the sample in an experiment in which the variables are kept at a base level (F)

convection the heat transfer that occurs in moving fluids (liquids or gases) caused by the circulation of currents from one region to another 5.1.3

convex mirror a mirror that is curved outward 6.4.1

corrosion the gradual wearing away of a metal as a result of a chemical reaction in which the metal is changed into a metallic compound 2.1.1

covalent bond the type of bond in which valence electrons are shared between atoms 1.4.3

crest the highest point of a transverse wave 6.1.2

critical mass the minimum amount of isotope needed to sustain a chain reaction 5.2.3

crystalline solid a solid whose particles are arranged in a regular, repeating three-dimensional pattern 1.2.3

D

decomposition reaction a reaction where a substance is broken down into two or more products 3.2.5

density the mass per unit of volume of a substance 1.2.1

deposition the changing of a substance from a gas directly to a solid 1.2.6

desalination the process of removing salt from ocean water to obtain freshwater for drinking, irrigation, or industrial use 3.1.3

diatomic molecule a molecule made of two atoms of the same element 1.4.3

diffraction the bending of a wave around a barrier 6.1.3

direct current an electric current in which the electrons always flow in the same direction 7.1.4

distillation a method of separating a liquid mixture by heating it so the liquid evaporates and condenses 1.1.3

Doppler effect the change in the frequency and pitch of a sound caused by the motion of the sound source or the listener 6.2.2

drag the frictional force of air molecules exerted on a flying aircraft 4.1.7

ductility a substance's ability to be drawn or pulled into the form of a wire 2.1.1

E

efficiency a measure of the amount of useful work done by the machine compared with the amount of work done to make the machine operate 4.2.5

effort force the force needed to do the same work with a machine 4.2.4

electric charge a property of matter that lets an object exert a force of repulsion or attraction on other objects that are near but not necessarily touching 7.1.1

electric current the net flow of electrons 7.1.4

electric discharge the removal of a charge from a charged object because of a buildup of static electricity 7.1.1

electric motor a device that uses an electromagnet to convert electrical energy into mechanical energy 7.2.3

electric power the rate at which electrical energy performs work 7.1.6

electrode an electric conductor that conducts current into and out of an electrolyte 7.1.3

electrolyte a mixture of chemicals that conducts an electric current 7.1.3

electromagnet a bar of iron wrapped with a coil of wire carrying an electric current 7.2.2

electromagnetic induction the process by which electric current is produced by a changing magnetic field 7.2.2

electromagnetic spectrum the arrangement of electromagnetic waves according to frequency or wavelength 6.3.1

electromagnetism the relationship between electricity and magnetism 7.2.2

electron a negatively charged subatomic particle outside the nucleus of an atom 1.3.2

electronics the study and technology devoted to developing and improving devices that use a flow of electrons 7.2.3

element a pure substance that cannot be broken down into simpler substances by physical or chemical means 1.1.2

endergonic reaction a reaction that absorbs more energy at the beginning of a reaction than is released 3.2.4

endothermic reaction a chemical reaction that absorbs heat 3.2.4

end point the time when an indicator changes color during a titration 3.3.4

energy the ability to do work 4.2.1

energy farming the growing of organic substances to produce fuel 2.3.2

equivalence point the moment when the concentration of hydrogen ions equals the concentration of hydroxide ions 3.3.4

evaporation the slow vaporization at the surface of a liquid 1.2.6

exergonic reaction a reaction that releases more energy during bond formation than the activation energy used to begin the reaction 3.2.4

exothermic reaction a chemical reaction that releases heat 3.2.4

F

fiber optics the science or technology of light transfer through very thin, flexible glass or plastic fibers 6.4.4

filtration the use of paper or other materials to remove solid particles from a liquid mixture 1.1.3

fluid any substance that flows and takes the shape of its container 4.1.6

fluorescent light a light source that produces light by flooding gas molecules with electrons under low pressure 6.3.7

focal length the distance between the center of the mirror and the focal point 6.4.1

focal point the point at which a mirror or lens causes the parallel light waves to meet 6.4.1

force a push or pull exerted on an object 4.1.1

fossil fuel a source of energy formed from the buried remains of dead plants and animals 5.3.1

freezing point depression the lowering of a solvent's freezing point by the addition of a solute 3.1.5

frequency the number of complete waves that pass a particular point in a unit of time 6.1.2

friction a force that resists motion in two objects that are moving against each other 4.1.1

fulcrum the point at which a lever pivots 4.3.2

G

gamma ray an electromagnetic wave with the shortest wavelength and highest frequency released during gamma decay 5.2.2

gas the state in which matter does not have a definite shape or a definite volume 1.2.5

gaseous solution a solution in which two or more gases are mixed together evenly 3.1.2

generator a machine that converts mechanical energy into electrical energy 7.2.2

glass a type of ceramic that is composed mainly of silicates that are bound to alkali metals under high temperature conditions 2.4.2

gram the standard unit for mass in the metric system (F)

gravity an attractive force between two objects with mass 1.2.1

group a vertical column of elements in the periodic table 1.3.3

H

half-life the time it takes for one-half of the atoms of an isotope in a given sample to decay into atoms of another element 5.2.2

halogen an element that belongs to group VIIA on the periodic table 2.2.3

heat the energy that is transferred between substances that are at different temperatures 5.1.1

heat engine a machine that converts heat energy into mechanical energy in order to do work 5.1.4

heterogeneous mixture a combination of substances in which the particles are not uniformly distributed 1.1.3

hologram a three-dimensional photograph or image produced by lasers and a series of mirrors 6.4.4

homogeneous mixture a combination of substances in which the properties and appearance are the same throughout 1.1.3

hydraulic device a device that uses liquids to transfer pressure from one point to another 4.1.6

hydrocarbon an organic compound containing only hydrogen and carbon atoms 2.3.1

hypothesis a prediction of what you think will happen and which can be tested to see if it is true (F)

I

illuminated object an object that reflects light from luminous objects 6.3.2

image a copy of an object formed by light 6.4.1

incandescent light a light source that generates light by producing heat 6.3.7

inclined plane a simple machine that has a slanted surface 4.3.1

incoherent light light that contains waves of different frequencies and amplitudes moving out of phase 6.4.4

index of refraction a property of matter that compares the speed of light in a vacuum with the speed of light in a material 6.3.6

indicator a substance that changes color in the presence of an acid or a base 3.3.3

inertia the tendency of an object to resist a change in its movement 4.1.3

inference an educated guess based on observation (F)

infrared wave an electromagnetic wave with slightly greater wavelengths and lower frequencies than red light 6.3.5

inhibitor a substance used to slow down or stop a chemical reaction 3.2.2

insulator a substance that does not allow electricity to flow through it easily 7.1.2

intensity the average rate of energy that flows through a defined area 6.2.2

interference the effect produced by the combination of two or more waves 6.1.3

ion an atom or group of atoms that has a charge 1.4.4

ionic bond the force of attraction between oppositely charged ions 1.4.4

irradiation the exposure of foods to radiation in order to kill harmful bacteria 5.2.4

isomer a compound that has the same chemical formula as another compound but has a different molecular structure 2.3.1

isotope an atom of an element that has the same atomic number but a different number of neutrons and a different mass number 1.3.2

K

kinetic energy the energy contained in moving objects 4.2.1

L

laser a device that produces and transmits a narrow, intense beam of coherent light 6.4.4

lens a transparent object that refracts light to produce an image 6.4.2

lever a simple machine made of a rigid bar that pivots at a fixed point 4.3.2

lift the force that pushes an aircraft up 4.1.7

light filter a transparent or translucent device that allows only certain wavelengths of light to pass through it 6.3.3

liquid the state in which matter has a definite volume but not a definite shape 1.2.4

liquid solution a solution in which a solid, gas, or other liquid is dissolved into a liquid solvent 3.1.2

liter the standard unit for the volume of liquid in the metric system (F)

longitudinal wave a wave in which the movement of the medium's particles is parallel to the direction of the wave's movement 6.1.1

luminous intensity the amount of visible light that is emitted from a source6.3.2

luminous object an object that emits light 6.3.2

M

machine a device that makes work easier by changing the size or direction (or both) of a force 4.2.4

magnetic domain a region in which the magnetic fields of individual atoms are lined up 7.2.1

magnetic field the region around a magnet in which magnetic forces can act 7.2.1

magnetism a force of attraction or repulsion of substances, especially those that contain iron, that results from the arrangement of their electrons 7.2.1

malleability a substance's ability to be flattened into thin sheets by hammering or by the application of intense pressure 2.1.1

mass the amount of matter in an object 1.2.1

mass number the sum of the number of protons and neutrons in an atom's nucleus 1.3.2

matter anything that has mass and takes up space 1.1.2

mechanical advantage a measure of how much a machine increases a force applied to it 4.2.4

medium any substance through which a wave can travel 6.1.1

meltdown what occurs when the fuel in a reactor melts and radiation leaks out of the reactor 5.2.3

metal an element that is usually hard, shiny, malleable, ductile, and a good conductor of heat and electricity 1.3.3

metallic bond the force of attraction between the metal atoms and the surrounding sea of electrons in the metal 2.1.1

metalloid an element that has similar physical and chemical properties to both metals and nonmetals 2.2.1

meter the standard unit for length in the metric system (F)

metric system a universal system of measurement based on the number 10 used by scientists around the world (F)

microwave a type of radio wave with short wavelengths and high frequencies 6.3.4

mirror a smooth object that reflects light and forms images 6.4.1

mixture a combination of two or more substances that are not chemically combined 1.1.3

molecule a particle consisting of two or more atoms chemically tied together 1.1.2

momentum the property of a moving object that depends on its mass and velocity 4.1.5

music a sound with a pleasing quality, an identifiable pitch, and a repeated timing called rhythm 6.2.3

N

neon light a light source that produces light by passing electrons through tubes of neon gas 6.3.7

net force the total of all the forces acting on an object 4.1.2

neutralization reaction the reaction between an acid and a base in an aqueous solution that yields water and a salt 3.3.4

neutron a subatomic particle that has no charge in the nucleus of an atom 1.3.2

newton the unit of force that equals 1 kg × m/sec^2 4.1.1

noble gas a chemically stable element that belongs to group VIIIA on the periodic table and does not readily react with other elements under ordinary conditions 2.2.4

node the point where destructive interference causes no displacement 6.1.3

noise pollution a level of noise that causes irritation, pain, or damage to a part of the body 6.2.3

nonmetal an element that is usually soft, not malleable or ductile, and a poor conductor of heat and electricity 1.3.3

nonpolar molecule a molecule with an equal charge distribution 1.4.3

nonpolar substance a substance that has molecules whose electric charges are equally distributed 3.1.1

nonrenewable resource a resource that cannot be replaced once it is used, or can be replaced only over long periods of time 4.2.3

nuclear chain reaction a continuous series of nuclear fission reactions 5.2.3

nuclear radiation the particles released during radioactive decay 5.2.2

nucleus the center of an atom that contains protons and neutrons and is surrounded by an electron cloud 1.3.2

O

observation something noticed through the senses (F)

opaque substance a substance that does not transmit light 6.3.2

optical axis a straight line drawn outward from the center of the mirror 6.4.1

optical fiber a fiber made of glass or plastic along which light can be transmitted to carry information 6.4.4

ore a naturally occurring rock from which a useful metal or mineral is recovered 2.1.2

organic compound a compound that contains carbon 2.3.1

oxidation a chemical change in which a substance combines with oxygen 1.2.7

P

particle theory the theory that light is made of small bundles of energy called photons 6.3.6

penumbra the edge of a shadow that is partially illuminated 6.3.6

period a horizontal row of elements in the periodic table 1.3.3; the time it takes two successive wave crests or troughs to pass a given point 6.1.2

periodic table an organized arrangement of elements in order of increasing atomic number with repeating physical and chemical property patterns 1.3.3

phase change the changing of a substance from one phase to another 1.2.6

photoelectric effect the release of electrons from metals by photons of light 6.3.6

photon a tiny bundle of light energy 6.3.6

pH scale a scale that measures the concentration of hydrogen ions in solution 3.3.3

physical change a change in a substance's physical properties that does not change the type of matter in the substance 1.2.6

physical science the study of the nonliving things that make up the world 1.1.1

physics the study of matter, energy, motion, and forces 1.1.1

pigment a material that gives a substance its color by absorbing some colors of light and reflecting others 6.3.3

pitch how high or low a sound is 6.2.2

plane mirror a mirror that has a perfectly flat surface 6.4.1

pneumatic device a device that uses compressed gas to transfer pressure from one point to another 4.1.7

polarized light light with waves vibrating in only one direction 6.4.4

polar molecule a molecule with a balanced charge that is not equally distributed 1.4.3

polar substance a substance that has molecules whose electric charges are unequally distributed 3.1.1

polyatomic ion a group of atoms that are covalently bound and have a charge 1.4.4

polymer a very large compound that is made of small, repeating units called monomers 2.4.4

potential energy the energy that is stored by way of an object's position, chemical makeup, or elasticity 4.2.1

power the amount of work done or energy provided in a period of time 4.2.4

pressure the amount of force exerted on an object per unit of area 1.2.5

product a new substance that is formed following a chemical reaction 3.2.1

projectile motion the curved path an object follows when it is thrown or propelled near Earth's surface 4.1.3

proton a positively charged subatomic particle in the nucleus of every atom 1.3.2

pulley a simple machine consisting of a rope or cable that turns around a wheel 4.3.3

pure substance a substance of constant composition that possesses definite qualities 1.1.2

R

radiant energy the energy of electromagnetic waves 6.3.1

radiation the transfer of energy through space or matter by waves or rays 5.1.3

radioactive decay natural transmutation 5.2.2

radioactivity the ability of some elements to give off nuclear radiation as a result of a change in the atom's nucleus 5.2.1

radio wave an electromagnetic wave with long wavelengths and low frequencies 6.3.4

rarefaction a section in a longitudinal wave where the particles are loosely packed together 6.1.1

reactant a substance that is present before a chemical reaction that then undergoes changes during a chemical reaction 3.2.1

reaction rate a measure of the change in concentration of the reactants over a period of time 3.2.2

real image an image through which light passes 6.4.1

reduction the process of converting an ore into a free, uncombined metal 2.1.2

reference point a point against which the position of an object is measured 4.1.2

reflection the bouncing back of a wave after it hits a barrier that does not absorb all of its energy 6.1.3

refraction the bending of a wave because of a change in the wave's speed as it passes from one medium to another 6.1.3

renewable resource a resource that is constantly available, or that can be replaced in a relatively short period of time 4.2.3

resistance the opposition to the flow of an electric charge 7.1.5

resistance force the force needed to do work without a machine 4.2.4

resonance what occurs when an object vibrating at or near the natural frequency of a second object causes the second object to vibrate 6.2.3

S

salinity the amount of salt in water 3.1.3

salt a compound formed from the positive ions of a base and the negative ions of an acid 3.3.4

scientific law a generalization based on observations that describe the ways an object behaves under specific conditions (F)

scientific method the series of steps that scientists follow when they investigate problems or try to answer questions (F)

screw a simple machine that consists of an inclined plane wrapped around a cylinder 4.3.4

semiconductor a substance that conducts electricity better than an insulator but not as well as a conductor 7.1.2

shadow a region of darkness formed when an opaque object comes between a space and a light source 6.3.6

solid the state in which matter has a definite shape and a definite volume 1.2.3

solubility a measure of the amount of solute needed to make a saturated solution using a given amount of solvent 3.1.4

solute the substance in a solution that is dissolved 3.1.1

solution a homogeneous mixture of two or more substances 3.1.1

solvent the substance in a solution that is present in the largest amount and acts as the dissolving agent 3.1.1

sonar a method of using ultrasonic waves to calculate underwater distances or locate underwater objects 6.2.3

sound a mechanical disturbance produced by an object's vibrations 6.2.1

specific heat the amount of heat needed to raise the temperature of 1 g of that substance 1°C 5.1.2

speed the rate at which an object moves 4.1.2

stable nucleus a nucleus that does not decay and fall apart 5.2.1

static electricity a stationary electric charge that is caused by the imbalance of positive and negative charges 7.1.1

strong acid an acid that produces many hydrogen ions when it is dissolved in water 3.3.1

strong base a base that produces many hydroxide ions when it is dissolved in water 3.3.2

sublimation the changing of a substance from a solid directly to a gas 1.2.6

superconductivity the perfect conductivity shown by certain substances at extremely low temperatures 7.2.4

superconductor a substance that conducts electricity without losing energy when it is cooled to extremely low temperatures 7.1.2

surface tension the greater force acting on particles at the surface of a liquid 1.2.4

surface wave a wave in which the medium's particles move in a circular motion 6.1.1

suspension a mixture in which the particles of a substance are mixed throughout but are large enough to settle 1.1.4

synthesis reaction a reaction where two or more substances combine to form a product 3.2.5

T

temperature the measure of the average kinetic energy of the particles in an object 5.1.1

tensile force a force that results from a pull exerted on an object 4.1.1

terminal velocity the constant velocity after which an object stops accelerating 4.1.3

theory an explanation of the scientific laws (F)

thermal conductor a substance that conducts heat well 5.1.3

thermal expansion the increase in the volume of a substance resulting from an increase in temperature 1.2.2

thermal insulator a substance that does not conduct heat well 5.1.3

thrust the force that pushes an aircraft forward 4.1.7

timbre the quality of a sound that distinguishes it from other sounds of the same pitch and volume 6.2.3

titration a method by which a solution of known concentration is used to determine the unknown concentration of a second solution 3.3.4

tracer a radioactive isotope whose pathway can be traced through the body 5.2.4

transformer a device that increases or decreases the voltage of alternating current 7.2.3

transition metal the elements in groups IIIB through IIB on the periodic table 2.1.5

translucent substance a substance that transmits light but scatters it as it passes through 6.3.2

transmutation the process of change within an atom's nucleus that produces a different element 5.2.2

transparent substance a substance through which light is easily transmitted 6.3.2

transverse wave a wave in which the medium's particles move in a direction that is perpendicular to the direction of the wave's movement 6.1.1

trough the lowest point of a transverse wave 6.1.2

Tyndall effect the scattering of light by colloids 1.1.4

U

ultrasonic sound a sound with a frequency higher than 20,000 Hz 6.2.2

ultraviolet (UV) wave an electromagnetic wave with slightly shorter wavelengths, slighter higher frequencies, and more energy than violet light 6.3.5

umbra the part of a shadow that is completely dark 6.3.6

unstable nucleus a nucleus that naturally decays and falls apart 5.2.1

V

vaporization the changing of a substance from a liquid to a gas 1.2.6

variable a changeable factor that could influence an experiment's outcome (F)

velocity the speed of an object in a particular direction 4.1.2

virtual image an image through which light does not pass 6.4.1

viscosity a liquid's resistance to flow 1.2.4

visible light the narrow range of wavelengths in the electromagnetic spectrum that can be seen by human eyes 6.3.1

voltage the difference in energy per unit charge as a charge moves between two points in the path of a current 7.1.5

volume the amount of space an object or substance takes up 1.2.1

W

water-soluble solid a solid that can be dissolved in water to form an aqueous solution 3.1.2

wave a disturbance that transfers energy through matter or space 6.1.1

wave height the distance between a trough and a crest 6.1.2

wavelength the distance between corresponding points on a wave 6.1.2

weak acid an acid that produces a small number of hydrogen ions when it is dissolved in water 3.3.1

weak base a base that produces few hydroxide ions when it is dissolved in water 3.3.2

wedge a simple machine that consists of two back-to-back inclined planes 4.3.4

weight the measure of the amount of gravity exerted on an object 1.2.1

wheel and axle a simple machine that consists of a wheel attached to an axle that has a smaller diameter 4.3.4

work the measure of the energy transferred when a force is applied to an object that causes it to move, and at least part of that force is in the direction of the movement 4.2.1

X

X-ray an electromagnetic wave with slightly shorter wavelengths and higher frequencies than ultraviolet rays 6.3.5

Index

G

H

I

J

K

L

M

N

O

P

U

V

W

X

Y

Z

Credits

Unit 1

Chapter 1

SLS wind tunnel testing, **page 3**, Dr. Patrick Shea, NASA/Ames/Dominic Hart

James Webb Space Telescope, **page 8**, NASA/Chris Gunn

exploded red dye pack, **page 12**, Colin Brown/CC-BY-2.0

Chapter 2

solar eruption, **page 21**, NASA/GSFC/Solar Dynamics Observatory

portrait of the Honourable Robert Boyle, **page 22**, Wellcome Library, London/CC-BY-4.0

Shaefer's anglerfish, **page 33**, Expedition to the Deep Slope 2007, NOAA-OE/Department of Commerce

Chapter 3

Prof. Ernest Rutherford, **page 46**, Library of Congress, Prints & Photographs Division, LC-DIG-ggbain-03392

Niels Bohr, **page 47**, Library of Congress, Prints & Photographs Division, LC-DIG-ggbain-35303

nanotube, **page 47**, CSIRO, CC-BY-3.0

sodium metal, **page 55**, Dnn87/CC-BY-SA-3.0

reaction of sodium and water, **page 55**, Tavoromann/CC-BY-SA-3.0

Unit 2

Chapter 1

HAWC+ instrument on SOFIA aircraft, **page 92**, NASA/AFRC

Chapter 2

Hindenburg disaster, **page 103**, Naval History and Heritage Command

Chapter 4

kiln, **page 129**, ©Deborah Smith, used with permission

Dr. Percy Lavon Julian, **page 139**, Science Source/Science Photo Library

Unit 3

Chapter 1

oiled bird, **page 144**, Brocken Inaglory/CC-BY-SA-3.0

reverse osmosis plant in Barcelona, Spain, **page 153**, James Grellier/CC-BY-SA-3.0

Trematomus bernacchii, **page 161**, Hectonichus/CC-BY-SA-3.0

Chapter 2

lead(II) iodine precipitating out of solution, **page 164**, PRHaney/CC-BY-SA-3.0

magnesium ribbon burning, **page 176**, Captain John Yossarian/CC-BY-SA-3.0

Unit 4

Chapter 1

NASA Aqua satellite, **page 214**, NASA/JPL-Caltech

Wallops sounding rocket, **page 215**, NASA

Pascal's calculator, **page 222**, Tieum512/CC-BY-3.0

Amelia Earhart, **page 224**, Library of Congress, Prints & Photographs Division, LC-H2-B-10065

Chapter 2

hovercraft, **page 246**, MesserWoland/CC-BY-SA-3.0

Chapter 3

exact replica of a Gutenberg printing press, **page 248**, Museum of the Bible

Jan Ernst Matzeliger, **page 249**, NYPL/Science Photo Library

Bandaloop--vertical dancing, **page 256**, Paul VanDerWerf/CC-BY-2.0

penny-farthing bicycle, **page 264**, Fjmustak/CC-BY-SA-4.0

Unit 5

Chapter 1

mosaic of images from NASA's WISE observations, **page 266**, NASA/JPL-Caltech/UCLA

Irish white pudding nutrition label, **page 273**, O'Dea/CC-BY-SA-3.0

Chapter 2

Pierre and Marie Curie, **page 285**, Wellcome Library, London/CC-BY-4.0

Lise Meitner, **page 286**, Acc. 90-105 - Science Service, Records, 1920s–1970s, Smithsonian Institution Archives

IAEA Experts at Fukushima, **page 295**, IAEA Imagebank/CC-BY-SA-2.0

Chapter 3

fossilized lycopsid at Joggins, **page 305**, Michael C. Rygel via Wikimedia Commons/CC-BY-SA-3.0

peat, **page 306**, Donna Pizzarelli/USGS

lignite coal, **page 306**, Donna Pizzarelli/USGS

bituminous coal, **page 306**, Donna Pizzarelli/USGS

anthracite coal, **page 306**, Donna Pizzarelli/USGS

analysis of oil spill samples from Deepwater Horizon Drilling Platform, **page 309**, Leslie C. Gordon/USGS

Unit 6

Chapter 2

Captain Charles E. Yeager, **page 331**, US Air Force/Air Force Historical Support Division

Chapter 3

US currency in UV, visible, and IR light, **page 346**, Steve Pomeroy/CC-BY-2.0

1920s radio, **page 356**, Library of Congress, Prints & Photographs Division, LC-H25-3029

Chapter 4

solar furnace of Odeillo, France, **page 372**, H. Zell/CC-BY-SA-3.0

funhouse mirror, **page 372**, Janine and Jim Eden/CC-BY-2.0

security mirror, **page 374**, Nichole.s/CC-BY-SA-4.0

plano convex lens focus and aberration, **page 377**, Lebazele/iStock by Getty Images

image seen through a convex lens, **page 378**, AntanO/CC-BY-SA-4.0

red blood cells, **page 381**, Janice Haney Carr, USCDCP

Giant Magellan Telescope, **page 382**, GMTO Corporation

teeth whitening with lasers, **page 384**, Cory Doctorow/CC-BY-SA-2.0

mirror on the moon, **page 384**, NASA

polarizing filter, **page 386**, Esquilo/Johan Fredriksson/CC-BY-SA-3.0

Unit 7

Chapter 1

George Westinghouse, **page 406**, Joseph G. Gessford/Library of Congress, Prints & Photographs Division, LC-USZ62-93492

Chapter 2

Nikola Tesla with his equipment, **page 424**, Wellcome Collection/CC-BY-4.0

CRT monitor, **page 425**, Politikaner/CC-BY-SA-3.0

China's maglev train, **page 428**, Andreas Krebs/CC-BY-SA-2.0

Japan's maglev train test track, **page 428**, Maryland GovPics/CC-BY-2.0